Extremophiles
General and Plant Biomass Based Biorefinery

Editors

Pratibha Dheeran
Department of Botany
Maharaj Singh College
Saharanpur, Uttar Pradesh, India

Sachin Kumar
Biochemical Conversion Division
Sardar Swaran Singh National Institute of Bio-Energy
Kapurthala, Punjab, India

CRC Press
Taylor & Francis Group
Boca Raton   London   New York

CRC Press is an imprint of the
Taylor & Francis Group, an **informa** business

A SCIENCE PUBLISHERS BOOK

Cover credit: The picture taken by the editor.

First edition published 2022
by CRC Press
6000 Broken Sound Parkway NW, Suite 300, Boca Raton, FL 33487-2742

and by CRC Press
4 Park Square, Milton Park, Abingdon, Oxon, OX14 4RN

Library of Congress Cataloging-in-Publication Data (applied for)

ISBN: 978-0-367-85652-6 (hbk)
ISBN: 978-1-003-01418-8 (ebk)

DOI: 10.1201/9781003014188

Typeset in Times New Roman
by Radiant Productions

Preface

Extremophiles are microorganisms which can thrive in extreme conditions such as either high or low temperature, pressure, salt concentration, alkaline/acidic, etc. These extremophiles may be harnessed for their applications in extreme conditions where normal microorganisms fall short to survive. Moreover, they are capable of degrading the different wastes efficiently under the extreme conditions desired by many industrial and biorefining processes. Extremophiles may play an important role in biorefining of waste materials such as lignocellulosic wastes, industrial wastes, biomedical wastes, etc. Most of these wastes need extreme conditions for their processing. Biorefinery is an integration of processes for generating more than one product to make the process economically and environmentally sustainable. The book 'Extremophiles: General and Plant Biomass Based Biorefinery' covers the extremophiles in general and applications of extremophiles for generating value-added products from plant biomass in a biorefinery manner, and has thirteen chapters as described below:

Chapter 1 describes the diversity and applications of extremophiles along with current trends. The authors have focused on different types of extremophiles with examples, their habitats, and proteins and metabolic adaptations to their environments. The chapter encapsulates the understanding about different extremophiles and their pertinent role in biotechnological and pharmaceutical industries.

Chapter 2 describes the diversity, enzymology and applications of extremophiles for sustainable bioenergy production. The authors have compiled the information about adaptation capabilities adopted by different extremophiles for their prospective utilization and applications in bioenergy production. The diversity, enzymology, cell wall properties, cell-transport mechanism, DNA/protein synthesis in extremophiles are also discussed in much detail to improve the present understanding and advances in this area of research.

Chapter 3 describes life of microorganisms in extreme environments. The authors have discussed the investigations of extremophiles, their habitats, biomolecules and products of extraordinary potential in various biotechnological operations and utilization of extremophiles in their creation or production.

Chapter 4 understands the insights into the enzymology and the capability of a thermophilic genus *Geobacillus* to degrade plant biomass through genomics and metagenomic analyses. The authors have summarized the current state of knowledge about the thermophilic genus *Geobacillus*, including genomic, metagenomic, and other omics analysis to highlight the extraordinary functional diversity of members of *Geobacillus* as a rich source of biomass-degrading enzymes of potential biotechnological significance.

Chapter 5 discusses the bioprospecting of extremophiles/extremozymes by employing various techniques such as omics analysis for the identification of countless novel enzymes and their application for improving sustainability of existing biotechnological processes towards a greener biobased industry. Furthermore, existing technologies for the production of value-added products and their limitations and strategies to overcome those limitations are also discussed.

Chapter 6 explores the potential applications of extremophiles in biorefinery process for the production of different bioactive value-added products like enzymes, biofuels, pharma products, etc. This chapter also sheds light into the immerging bioremediation technologies using extremophiles.

Chapter 7 elaborates the advancements in extremozymes and their potential applications in biorefinery. The authors have discussed the potential applications of extremophiles and their extremozymes with major focus on the conversion of lignocellulosic and starchy wastes into valuable products.

Chapter 8 explores the adaptations, importance and industrial relevance of psychophilic enzymes. This chapter also dictates the various modifications in the psychrophilic enzymes to sustain their lifestyle in hash conditions and summarizes their contribution in the development of various sustainable energy and environmental efficient industrial processes.

Chapter 9 discusses the basic concept of hemicellulase and α-L-arabinofuranosidase enzymes as extremozymes including their structure, biochemical properties (pH, temperature, etc.), microorganisms involved in their production along with their potential applications in the different industries as extremozymes.

Chapter 10 discusses the various aspects of aviation biofuel or jet biofuel productions based on ASTM standard using extremozymes. The author has emphasized on understanding of metabolic dynamics of extremophiles for the production of aviation biofuel or jet biofuel at an economical rate.

Chapter 11 explains an overview of different biofuels and extremophiles, current development on application of extremophiles for different type of biofuel production.

Chapter 12 discusses the recent advances in using microbial catalysts under extreme conditions and possible new routes that extremophilic microorganisms open for biofuels production.

Chapter 13 deals with the latest advancements and current scenario mainly focusing on the major challenges for biochemical conversion of agricultural waste to biofuels. Besides this, it also deals with the commercial facets and valorization of by-products of hemicellulose, lignin and silica. The authors have also discussed the various strategies for the development of a bioprocess using solvent-tolerant and inhibitor-tolerant saccharifying and (co)-fermenting strains, and productivity, and situate the duct for novel cutting edge prospects for biofuel production in biorefinery mode.

India

Pratibha Dheeran

Sachin Kumar

Acknowledgements

We thank all the authors, who made painstaking contributions in the making of this book. Their contributions brought the desired lusture to the quality of this book. Their patience and diligence in revising the initial draft of the chapters after incorporating the comments/suggestions of the reviewers are highly appreciated.

We would also like to acknowledge the contributions of all the reviewers for their constructive and valuable comments and suggestions to improve the quality of the contributions of various authors.

Contents

1

Extremophiles
Diversity, Biotechnological Applications and Current Trends

Yogita Lugani[1,2] and *Venkata Ramana Vemuluri*[1,*]

1. Introduction

Extremophiles are adapted to thrive under myriad extreme environmental conditions such as high (55°C to 121°C) or low (–2°C to –20°C) temperature, acidic (pH < 4) or alkaline (pH > 8) pH, high pressure (> 500 atm), nutrient shortage niche (like ice, rock, soil, air and water), radiations (ultraviolet resistance > 600 J/m^2), chemical extremes of heavy metals (like zinc, copper, cadmium, arsenic, etc.) and high salt concentrations (5% to 30%) (Navarro-Gonzalez et al. 2009). Some extremophiles may be found in combination of extreme environments such as high temperature and high alkalinity, low temperature and high pressure, and high temperature and high acidity. Extremophiles represent a key of research for multitude disciplines due to their ability to thrive under harsh environmental conditions. For the first time, the term extremophile was used in 1974 by MacElroy (MacElroy 1974). The first International Congress on Extremophiles was held in Portugal in 1996 and the scientific journal on "extremophiles" was launched in 1997. In 2002, an International Society for Extremophiles (ISE) was formed to share ideas, information and development in the field of extremophiles all over the world (Parihar and Bagaria 2020). Most of the extremophiles identified till date belongs to archaea domain; however, many extremophiles from eubacterial and eukaryotic kingdoms have also been identified and characterized (Rothschild and Manicinelli 2001). On the basis of their niche, the extremophiles can be classified as psychrophiles, thermophiles, acidophiles, alkaliphiles, halophiles, radiophiles, metallophiles, xerophiles, barophiles, and basophiles (Dumorne et al. 2017, Parihar and Bagaria 2020). In such organisms, specific proteins and enzymes mediate the metabolic processes under extreme environmental conditions. Quorum sensing (QS) is the major mechanism which helps in the regulation of various physiological and virulence process in most bacteria including extremophiles for biofilm formation, lowering the freezing point, persistent cell formation, cold adaptation, and oxidative stress resistance (Kaur et al. 2019). Thermophilic proteins, acidophilic proteins, halophilic proteins, and barophilic proteins have been gaining more attention

[1] Microbial Type Culture Collection and Gene Bank (MTCC), CSIR-Institute of Microbial Technology, Chandigarh-160036, India.
[2] Department of Agriculture and Environmental Sciences, National Institute of Food Technology Entrepreneurship and Management (NIFTEM), Sonepat, Haryana-131028, India.
  Email: yogitalugani9@gmail.com
* Corresponding author: venkat@imtech.res.in

for industrial and biotechnological applications (Yildiz et al. 2015). Presently available mesophilic enzymes are not able to withstand under harsh industrial reaction conditions; therefore, isolation and characterization of extremophiles has received much attention to provide extremophilic biocatalysts for industrial applications. Extremophilic enzymes or extremozymes are highly resistant to extreme conditions which provides a new line of research in biocatalysis and biotransformation for developing nation's economy. In addition to extremozymes, extremophiles have great potential in producing several other metabolites such as extremolytes, biosurfactants, biopolymers, and peptides (Raddadi et al. 2015, Ijaz et al. 2017, Donati et al. 2019, Parihar and Bagaria 2020). With the development of biotechnology and biobased economy, extremozymes are gaining intense importance for research by scientific and industrial communities due to their stability, high temperature tolerance, wide pH ranges, and high reproducibility (Haki and Rakshit 2003, Raddadi et al. 2015). Each group of extremozyme possesses unique features which allow the possibility of their use for diverse industrial applications (Table 1, Fig. 1).

Extremozymes have been proven to show multilateral applications in synthesis of various thermostable enzymes (like Taq DNA polymerases, protease S, Pfu DNA ligase, and alkaline phosphatase used in molecular biology), biofuel production, biomining, soap and detergent industry, leather processing industry, production of natural pigments, starch processing, paper-pulp bleaching, production of animal feedstock, chemical synthesis, waste transformation and degradation, biosensor development, drug delivery, cosmetic packaging, bioremediation, biomineralization, and pharmaceuticals (Coker 2016, Fakruddin 2017, Parihar and Bagaria 2020). The exploration of extremophiles is growing in search of natural products, particularly those showing anti-microbial activities against multi-drug resistant pathogens (Pettit 2011, Charlesworth and Burns 2016). The dramatic advancement in molecular and computational biology, biochemical engineering of enzymes, and new sources of enzymes has stimulated the interest of researchers towards enzyme technology (Gomes and Steiner 2004). Some advanced techniques such as protein engineering, directed evolution, recombinant DNA techniques are used continuously for developing novel and efficient enzymes with high-yield heterologous protein expression for developing industrially viable technology (Rigoldi et al. 2018).

According to Markets and Markets (2020), the industrial enzymes market is estimated at United States Dollars (USD) 5.9 billion in 2020, which is projected to reach USD 8.7 billion by 2026 at a

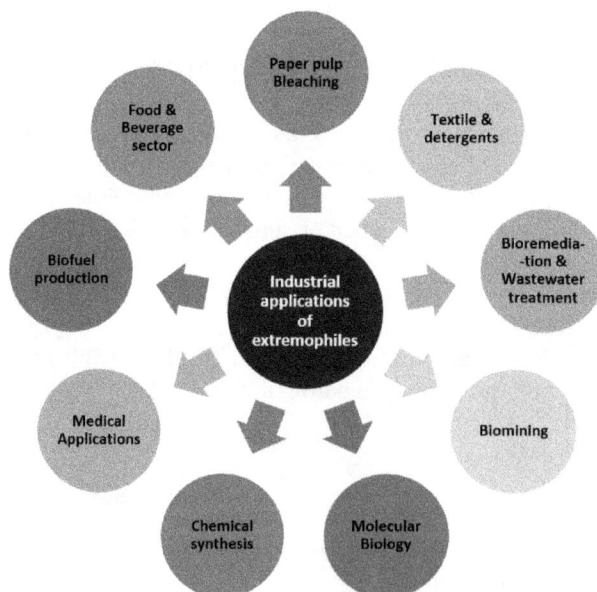

Fig. 1. Industrial applications of extremophiles.

Table 1. Industrial applications of different products from extremophiles.

S. No.	Type of extremophile	Microorganism	Product	Application	Reference
1.	Psychrophile	*Cryptococcus cylindricus*	Pectinase	Clarification of fruit juices	Nakagawa et al. 2004
2.	Thermophile	*Thermococcus fumicolans*	DNA ligase	Ligase chain reaction (LCR), Diagnostics	Rolland et al. 2004
3.	Halophile	*Halomonas boliviensis* LC1	Poly-β-hydroxybutyrate	Synthesis of bioplastics	Quillaguaman et al. 2005
4.	Haloalkaliphile	*Natrialba magadii*	Protease	Aqueous organic solvent biocatalysis	Ruiz and Castro 2007
5.	Alkaliphile	*Bacillus* sp. PN5	Amylase	Starch saccharification, Detergent formulation	Saxena et al. 2007
6.	Psychrophile	*Pseudoalteromonas haloplanktis* TAH3A, *Flavobacterium* sp. MSY-2	Xylanase	Bread Making	Dornez et al. 2011
7.	Thermophile	*Bacillus licheniformis* T4SS10	Biosurfactants	Bioremediation of hydrocarbon contaminated sites	Mnif et al. 2011
8.	Thermo-alkaliphile	*Bacillus subtilis* DR8806	Glucoamylopullulanase	Leather industry	Asoodeh and Lagzian 2012
9.	Psychrophile	*Thalassospira* sp. 3SC-21	β-galactosidase	Manufacturing of lactose free dairy products	Ghosh et al. 2012
10.	Radiophile	*Rhodanobacter* sp.	Whole microorganism	Bioremediation of radio-nucleotide contaminated sites	Green et al. 2012
11.	Halophile	*Salimicrobium halophilum* strain LY20	Serine protease, β-amylase	Food, and detergent industries	Li and Yu 2012
12.	Thermophile	*Bacillus smithii* ET138	Transformed strain of *Bacillus smithii* ET138	Conversion of renewable resources to lactic acid and green chemicals	Bosma et al. 2014
13.	Psychrophile	*Cladosporium* sp.	Xylanase	Bioremediation, Food and textile industries	Del-Cid et al. 2014
14.	Halophile	*Halomonas elongata*	Compatible solute, Ectoine	Cosmetics, health care products	Kunte et al. 2014
15.	Thermophile	Oil degrading bacterium AR80	Surfactants and Polymers	Enhanced oil recovery	Purwasena et al. 2014
16.	Thermotolerant	*Malbrancheacinnamomea* NFCCI 3724	Xylanase	Paper and Pulp industry, Animal feedstock, beverage making	Ahirwar et al. 2017
17.	Metalophile	*Pseudomonas* sp. W6	Exopolysaccharide	Bioremediation of lead contaminated wastewater	Kalita and Joshi 2017
18.	Acidophile	*Acidothiobacillus* sp., *Pseudomonas* sp., *Leptospirillum* sp.	Microbial consortium	Bioleaching	Emmanuel et al. 2017

Table 1 contd. ...

...Table 1 contd.

S. No.	Type of extremophile	Microorganism	Product	Application	Reference
19.	Halophile	*Haloarcula marismortui*	Bacteriorhodopsin	Optical switches, Photocurrent generators in bioelectronic	Alsafadi et al. 2018
20.	Thermophile	*Geobacillus stearothermophilus PV72/p2*	Surface (S)-layer proteins, Lipids	Component for fabrication of bioinspired S-layer protein based biosensors	Schuster 2018
21.	Thermophile	*Streptomyces mutabilis*	Amylase	Starch processing	Temsaah et al. 2018
22.	Piezophile	*Piezobacterrhemophilus*	Whole microorganism	Antibiotics production	Tortorella et al. 2018
23.	Alkaliphile	*Bacillus subtilis* subsp. *Subtilis S13*	Collagenase	Meat tenderization	Khamson et al. 2019
24.	Thermophile	*Geobacillus thermoleovorans KNG 112*	Amylase	Alcohol fermentation, Maltooligosaccharide production, Sugar syrup formulation	Kotresh et al. 2020
25.	Thermophile	*Anoxybacillusrupiensis T2*	Amylase, Protease	Detergent industries	Tuysuz et al. 2020

Psychrophiles	Thermophiles		Acidophiles	Halophiles
Eg: *Psychrobacter okhotskensis* *Flavobacterium psychrophilum*	Eg: *Methanococcus jannaschii* *Thermus aquaticus*	**Extreme habitats: Loving niches for extremophilic microbes**	Eg: *Acidophilium angustum* *Ferroplasma acidiphilum*	Eg: *Haloferax mediterranei* *Halomonas meridian*
Mountain with snow	Hot springs		Acid drain	Salt pan
Radiophiles	**Metallophiles**	**Alkaliphiles**	**Xerophiles**	**Barophiles**
Eg: *Deinococcus radiophilus* *Deinococcus radioduran*	Eg: *Ralstonia metallidurans* *Pseudomonas syringae*	Eg: *Alkaliphilus metalliredigens* *Bacillus halodurans*	Eg: *Arthrospira platensis* *Methanococcus doii*	Eg: *Shewanella benthica* *Thermococcus piezophilus*
Waste dumped deep underground	Mining site	Alkaline spring	Desert	Deep sea

Fig. 2. Extreme environmental niches harbouring extremophiles.

compound annual growth rate (CAGR) of 6.5%. The key players in industrial enzyme market are DuPont (USA), BASF (Germany), Novozymes (Denmark), Chr. Hansen (Denmark), Kerry Group (Ireland), and Advanced Enzymes (India) (https://www.marketsandmarkets.com/Market-Reports/industrial-enzymesmarket%20237327836.html?gclid=Cj0KCQjw17n1BRDEARIsAFDHFewuSD4J_xNPzbdtadpvFALuFI7XCjVc10U46Ic1D0iPDIgDzqgugoUaAu8XEALw_wcB). According to a report, "India Industrial Enzymes Market Forecast & Opportunities, 2020" published by Pharmaion, the Indian industrial enzymes market is projected to reach USD 361 million by 2020 because of increasing tanneries and textile manufacturing facilities, and food processing facilities in the country. The leading industrial enzymes manufacturers in India are E.I. DuPont India Pvt. Limited, Advanced Enzymes, and Novozymes South Asia Pvt. Limited. The other regional market players are Lumis Biotech Pvt. Ltd., Maps Enzymes Pvt. Ltd., Anthem Cellutions India Ltd., and DSM Nutritional Products Pvt. Ltd. (https://www.prnewswire.com/news-releases/india-industrial-enzymes-market-to-surpass-us-361-million-by-2020-says-pharmaion-532434072.htm).

The current chapter focuses on different types of extremophiles with examples, their habitats (Fig. 2), proteins and metabolic adaptation to their environments. Thereafter, the applications of extremozymes in diverse sectors have been discussed.

2. Types of Extremophiles

Extremophiles show stability at low and high temperatures, range of pH, salinities, ionic strength, and hence provide various biotechnological opportunities. They show higher reaction rates, reproducibility, high performance and economic viability, and capability of eliminating xenobiotics, and hence prevent the accumulation of pollutants, radionucleotides, and heavy metals. Due to their diversity and exceptional properties, extremozymes have found various applications in different biotechnological and commercial processes (Dumorne et al. 2017). Examples of different types of extremophiles, their growth characteristics, and endogenous compounds are given in Table 2.

Table 2. Different types of extremophiles, their growth characteristics, and endogenous compounds (Source: Sharma et al. 2012, Preiss et al. 2015, Dumorne et al. 2017, Parihar and Bagaria 2020).

S. No.	Type of Extremophile	Examples	Growth Characteristics	Source/Geographic Location	Enzymes/Endogenous Compounds
1.	Psychrophile	*Cystofilobasidiumlarimarini, C. capitatum, C. capitatum, Cryptococcus adeliae, C. cylindricus, C. macerans, C. aquaticus, Flavobacterium psychrophilum, Pseudoalteromonas haloplanktis, Psychrobacter okhotskensis, Penicillium chrysogenum*	Organisms which can grow at optimum temperature of <15°C, and a maximum temperature of 20°C	Arctic ocean, Antarctic ice, Glaciers, Mountains, Ocean deeps, Snow, Upper atmosphere	Alkaline phosphatase, Amylases, Cellulases, Hemi-cellulases, Lipases, Proteases, Renin, Dehydrogenase, β-galactosidase, Polyunsaturated fatty acids, Ice proteins, Subtilisin
2.	Thermophile	*Desulfurococcus mucosus, Methanococcus jannaschii, Picrophilus oshimae, Pyrococcus furiosus, P. woesei, Pyrodictium abyssi, Rhodothermus marinus, Staphylothermus marinus, Sulfolobus solfataricus, Thermococcus litoralis, T. celer, T. profundus, Thermoplasma acidophylum, Thermotoga maritima, Thermus aquaticus*	Organisms which can thrive at an optimum temperature of 60–80°C (for thermophiles), and 80–110°C (for hyperthermophiles)	Hot Spring, Grand Prismatic Spring, Hydrothermal vents, Volcanic island, Yellowstone National Park (USA)	Amylases, Glucoamylases, Pullulanase, Amylopullulanase, Glucose/Xylose Isomerase, α-Glucosidases, Alcohol dehydrogenases, DNA ligases, DNA polymerases, Cellulases, Xylanases, Lipases, Proteases, Antibiotics, S-layer proteins and lipids
3.	Acidophile	*Acidophilium angustum, A. symbioticum, A. multivorum, A. rubrum, Alicyclobacillus acidocaldarius, A. cyclohepatanicus, Desulfurolobusambivalens, Ferroplasma acidiphilum, Leptospirillum ferrooxidans, Sulfolobus solfataricus, Sulfobacillus thermosulfidooxidans, Thermoplasma acidophilum, Thiobacillus caldus,*	Organisms which grow at or below pH range of 3–4	Acid mine drainage, Volcanic springs (USA)	Amylases, Glucoamylases, Proteases, Cellulases, Chalcopyrate concentrate
4.	Alkaliphile	*Alkaliphilus metalliredigens, Bacillus pseudofirmus, B. halodurans, Caldalkali bacillus thermarum, Ilyobactertartaricus, Natranaerobiusthermophilus, Pseudomonass aeruginosa, Propionigeniummodestum, Rhizobium meliloti, Serpentinomonas raichei, Staphylococcus aureus, Vibrio cholerae*	Organisms with optimal growth at pH above 10.0	African Rift Valley, Alkaline hydrothermal vents, Groundwater springs, Hind gut of insects, Soda lakes, Utah (USA), Deserts, Waldi El Natrun (Egypt)	Cellulases, Proteases, Pectinases, Xylanases, Lipases, Pullulanases, Keratinases, Elastases, Collagenase, Cyclodextrins

5.	Halophile	Organisms which can grow at a salt concentration of more than 1 M	Salt lakes, Utah (USA), El-Djerid salt lake (Tunisia), soda lakes, deep sea, alkaline soils, carbonate springs	Proteases, Dehydrogenases, Nucleases, Amylases, Polyhydroxyalkanoates, Rheological polymers, Myco-oncogene products, Bacteriorhodopsin, Compatible solutes (Ectoine, Betaine, Trehalose, Sucrose, Mannitol, Proline), Linoleic acid, β-carotene, cell extracts (*Spirulina, Dunaliella*), Glycerol
	Aspergillispenicillicides, Aquisalibacillus elongates, Bacillus dipsosauri, B. methylotrophicus, Dunaliellasalina, Flammeovirgapacifica, Haloarcula hispanica, Haloferax mediterranei, Halorhabdus utahensis, Halorubrum xinjiangense, Halomonas meridian, Halothermothrix orenii, Salimicrobium halophilum, Tetraselmis elliptica, Zunongwangia profunda			
6.	Radiophile	Organisms which can grow in the presence of high ionizing (X-rays, Gamma rays) and non-ionizing (Ultraviolet) radiations	Antarctic dry valley	Whole microorganism
	Chroococcidiopsis sp., Deinococcus radiophilus, D. radiodurans, D. hohokamensis, Nostoc sp., Thermococcus radiotolerans, T. gammatolerans, T. marinus			
7.	Metallophile	Organisms that are able to tolerate high levels of heavy metals such as arsenic, cadmium, copper, lead, mercury, zinc	Heavy metals, Latin America and Europe, Deep hydrothermal vents, volcanic areas, Ultramafic soil, Nickel rich (New Caledonia)	Whole microorganism
	Gymnostomadeplancheanum, Pseudomonas syringae, Pyrococcusfuriosus, Ralstonia metallidurans, Thermococcus littoralis, Xanthomonascampestris, Yersinia enterocolitica			
8.	Piezophile	Organisms showing optimal growth at hydrostatic pressure of 40 MPa or above	Antarctic ice, Deep ocean, Mariana trench, Sea beds	Whole microorganism
	Psychromonasprofunda, Pyrococcusabyssi, P. horikoshii, P. yayanosii, Shewanella benthica, Sulfolobus solfataricus, Thermococcus piezophilus			
9.	Xerophile	Organisms able to grow at low water activity, and resistant to high desiccation	Atacama Desert in Chile, Brackish sea water, Great Salt Lake, Rock surfaces, Soda lakes (Egypt, Kenya, Tanzania), Salterns (Spain, Mexico), Subterranean halite	Whole microorganism, Compatible solutes/Osmolytes
	Arthrospira platensis, Bacillus halophilus, Chromohalobacter canadiensis, C. marismortui, C. israeliensis, Dunaliella spp., Halospirulinatapeticola, Halomonas campisalis, H. elongata, H. salina, H. halophila, Methanococcusdoii, Natrococcuscocultus, Oreniasalinaria, Salinibacterruber, Zygogaccharomycesrouxii			

2.1 Psychrophiles

Psychrophiles are the microorganisms which are able to grow at < 15ºC, and are found mainly in glaciers, mountainous zones, polar zones, ocean deeps, upper atmosphere, subterranean systems, Arctic and Antarctica ice beds. They are found amongst Archaea, Bacteria and Eukarya domains of life, and possess different metabolic pathways such as chemoautotrophic, photosynthesis, and heterotrophic pathways (Sarmiento et al. 2015). These organisms are widespread among genera of *Arthrobacter, Alteromonas, Bacillus, Candida, Colwellia, Crystococcus, Cladosporium, Gelidibacter, Halobacterium, Halorubrum, Methanogenium, Methanolobus, Methanococcoides, Moraxella, Pseudomonas, Pseudoalteromonas, Penicillium, Psychrobacter, Shwanella,* and *Vibrio* (Anupama and Jayaraman 2011, De Maayer et al. 2014, Raddadi et al. 2015). There are many psychrophiles which live under more than one stress factor like high salt concentration and low temperature in sea ice (halo-psychrophile), and low temperature and high pressure in deep seas (piezo-psychrophile) (Kumar et al. 2011). Such organisms are able to grow in cold temperatures (ranging from –20ºC to +10ºC), and they maintain an active metabolism at freezing point. *Planococcus halocryophilus* Or1, an isolated bacterium from high Artic permafrost, showed optimum growth and reproduction at –15ºC (Mykytczuk et al. 2013). Psychrophiles face many challenges to thrive under the conditions of low temperature like decreased membrane fluidity, altered transport systems, reduced enzymatic activities, and protein cold-denaturation (DÁmico et al. 2006). Therefore, psychrophiles have adapted diverse adaptations such as high amounts of unsaturated and methyl-branched fatty acids with shorter acyl-chain fatty acids to enhance membrane fluidity, presence of cold-shock proteins and antifreeze proteins to maintain protein folding and inhibit the formation of ice crystals, and maintain high enzyme activity at low temperature (Sarmiento et al. 2015). Psychrophilic proteins possess less hydrogen bonds, disulphide linkages and salt bridges, and large hydrophobic interface for enhancing their kinetic energy to finish the reaction in low energy state. Their proteins have negative charge, decreased hydrogen bonds and ionic interactions, some surface loops, and some proline residues to enhance the flexibility, and stability of proteins (Maayer et al. 2014, Michetti et al. 2017). The presence of ice nucleation proteins in psychrophiles help them in ice nucleation, and snow making which results in their potential roles in food industry (Charlesworth and Burns 2016). Novozymes and Genencor are already providing cold-adapted proteases, and amylases for removal of starch stains (Kumar et al. 2011). The cold adapted psychrophilic enzymes have many genetic adaptations to keep high flexibility, and high enzyme activity at low temperature such as change in amino acid composition (more glycine residues for better conformation motility, more proline residues in α-helices but lesser in loops, lower arginine/lysine ratio, and more non-polar residues on protein surface), increased surface hydrophobicity but decreased core hydrophobicity, increased conformational entropy of unfolded protein, weak protein interactions (less metal binding sites, less disulphide bridges, fewer hydrogen bonds, less electrostatic interactions, few inter-domain and inter-subunit interactions), and decreased oligomerization and secondary structures but increased size and number of loops (Cavicchioli et al. 2011a). Enzymes from psychrophilic microbes offer great industrial potential due to their ability to catalyse reactions under low or moderate temperatures (Table 3).

Psychrophilic enzymes such as alkaline phosphatase, amylase, cellulase, hemicellulose, lipase, protease, renin, and dehydrogenases are replacing mesophilic enzymes in various industrial processes to decrease energy consumption, and thus proved their importance in food industry, detergents, cosmetic industry, pharmaceutical industry, waste management, and production of sugars, lipids, and enantiomeric peptides synthesis (Marasco et al. 2012, Rolli et al. 2015, Dumorne 2018).

2.2 Thermophiles

Thermophiles are microorganisms which grow optimally at habitats having higher temperatures such as hot springs, deep sea hydrothermal vents, cavern systems, decaying plant matter and compost, and volcanic islands. According to the optimum growth temperature, thermophiles are classified into

Table 3. List of commercially available Psychrozymes and Thermozymes (Sarmiento et al. 2015).

S. No.	Extremophile	Extremozyme	Application	Commercial product
1.	Psychrophile	Cellulases	Wash of cotton fabrics, Bio-finishing combined with dyeing of cellulosic fabrics	Celluclean®, Celluzyme® (Novozymes), UTA-88, UTA-90 (Hunan Youtel Biochemical), Rocksoft TM Antartic, Antarctic LTC (Dyadic), Retrocell Recopand, Retrocell ZircoN (EpyGen Biotech) PrimaGreen®EcoFade, PrimaGreenEcoLight 1, Primafast® GOLD, HSL IndiaAge®NeutraFlex (Genencor/DuPont)
		Amylases	Breakdown of starch based stains, De-sizing of woven fabrics	Purafect®OxAm (Genencor), Stainzyme® Plus (Novozymes), Preferenz™ S100 (DuPont), Optisize NEXT, Optisize® COOL (Genencor/DuPont)
		Alkaline phosphatases	Dephosphorylation of 5' end of a linearized fragment of DNA	Antarctic phosphatase (New England Biolabs Inc.)
		Pectinases, Pectate lyases	Fermentation of wine and beer, Bread making, Fruit juice processing, Pectin stain removal activity	Pectinasse 62L (Biocatalysts), Novoshape® (Novozymes), Lallzyme® (Lallemand), XPect® (Novozymes)
		Catalase	Textiles, and cosmetics	Catalase (CAT) (Swissaustral)
		Mannanases	Degradation of mannan gum	Effectenz™ (DuPont), Mannaway® (Novozyme)
2.	Thermophile	Glucoamylases, Amylases	Used on liquefied starch-containing substrates, Applied in baking, brewing and production of digestive aids, Enzymatic starch hydrolysis to make syrups, Production of fruit juices and cakes	AlphaStar Plus (Dyadic), Liquozyme®, Avantec®, Fungamyl®, Novamyl®, Liquozyme®, Termamyl® SC, Spirizyme® (Novozymes), Fuelzyme® (Veremium)
		Proteases	Applied in brewing to hydrolyse most proteins	Protease PLUS
		Xylanases	Bio-leaching, Hydrolyse hemicellulose in low molecular weight polymers in brewing	Xylacid® (VarunaBiocell), Luminase® PB-100, PB-200 (Verenium), Xyn 10A (Megazyme), Dyadic xylanasse PLUS, Xylanase 2XP CONC (Dyadic)
		Laccases	Bio-leaching	Laccase (Novozyme)
		Lipases and Esterases	Pitch control	Optimyze® (Buckman Laboratories), Lipase A NovoCor® AD L, Resinase A2X, Resinase™ HT, Lipase B Lipozyme® CALB (Novozymes)
		Glucose oxidases	Used to obtain stronger gluten in bakery	Gluzym® (Novozymes)

various classes like moderate thermophiles (optimum growth at 50°C–60°C), extreme thermophiles (optimum growth at 60°C–80°C), and hyperthermophiles (optimum growth at 80°C–110°C) (Irwin and Baird 2004). The multicellular plants and animals can't tolerate temperature above 50°C; similarly, most of the eukaryotic microbes are also not able to tolerate temperature more than 60°C (Kumar et al. 2011). Extreme thermophiles are distributed among species belonging to the genera of *Bacillus, Clostridium, Thermoanaerobacter, Thermus, Fervidobacterium, Rhodothermus, Thermotoga* and *Aquifex*, and hyperthermophiles are distributed among four phyla of archea, i.e., Crenarchaeota (genera of *Acidianus, Pyrobaculum, Desulfurococcus, Pyrolobus, Pyrodictium, Staphylothermus,*

Sulfolobus, Thermoproteus, and *Thermofilum*), Euryarchaeota (extreme halophiles such as *Halobaculum, Halobacterium, Halococcus, Haloferax,* and *Halorubrum*, extremeacidophiles such as *Picrophilus,* and *Thermoplasma,* extreme thermophiles such as *Archaeoglobus, Ferroglobus, Methanopyrus, Pyrococcus,* and *Thermococcus,* and methanogens such as *Methanobacterium, Methanobrevibacter, Methanococcus, Methanosphaera,* and *Methanothermus*), Korarchaeota, and Nanoarchaeota (*Nanoarchaeum equitans*) (Leveque et al. 2000, Vieille and Zeikus 2001, Fujiwara 2002, Huber et al. 2003). The 16S rRNA gene of the members of the phylum Korarchaeota have been isolated from environmental samples but isolation of pure strains of the same phylum has not been reported (Barns et al. 1996). Hyperthermophilic microbes like *Methanopyrus kandleri* strain 116, and *Geogemma barossii* strain 121 maintains cell integrity by contrasting effects of high temperature and pressure (Merino et al. 2019). The genome size of hyperthermophiles is comparatively smaller than those of mesophiles and their chromosomes are densely packed with genes of essential functions (Fujiwara et al. 1996). The cellular components like enzymes of thermophiles are thermostable, which can withstand high temperature, extremely acidic and alkaline conditions, and are highly resistant to proteolysis and denaturation (Vieille and Zeikus 2001, Kumar and Nussinov 2001). Thermophiles possess adaptive proteins with high number of ionic bonds, disulphide bonds, large hydrophobic cores, prosthetic groups, and co-factors, which broaden flexibility, stability, and functioning of such proteins (Reed et al. 2013). Special type of proteins called chaperonins are produced by thermophiles which are thermostable and resistant to proteolysis and denaturation. The cell membrane of these microbes possesses saturated fatty acids which results in enhancing protein core hydrophobicity which keeps the cells rigid enough to survive under high temperatures (Paiardini et al. 2002, Liszka et al. 2012). Thermophilic enzymes have large number of industrial applications due to their extreme stability at high temperature and high biodegradability (Table 3). Thermophiles are also energetic at high temperature and possess the capacity to keep their configuration and function under extreme temperature due to high quantity of hydrophobic deposits, and are capable of accepting extreme conditions like organic solvents, denaturing agents, and high salinity. The use of thermophilic enzymes helps in keeping a low adhesiveness, great substrate solubility and low adhesiveness which promises their use for industrial processes (Raddadi et al. 2015, Dumorne et al. 2017). Thermozymes like lipase, protease, amylase, pullulanase, cellulose, xylanase, chitinase, esterase, DNA polymerase, dehydrogenase, and mannanase have shown high efficiency in processing of milk, food, beverage, paper, pharmaceuticals, dyed products, biodiesel, and leather industry (Haki and Rakshit 2003, Schmid et al. 2009, Sarmiento et al. 2015). Thermophilic microbial consortia have also proved its potential for wastewater treatment and methane production (David et al. 2018).

2.3 Acidophiles

Acidophiles are organisms which can grow at an optimum pH below 4.0 usually found in hot springs and mine drainage systems such as Iron Mountain (Shasta Country, CA, Unites States), and their examples are *Sulfolobus solfataricus,* and *Ferroplasma acidiphilum* (Golyshina and Timmis 2005, Sharma et al. 2012). Acidophiles are clustered with thermophiles because most of the acidophilic environments are also considered as thermophilic in nature (Elleuche et al. 2014). The two hyperacidophilic Archaea *Picrophilus oshimae* and *P. torridus*, isolated from a sulfataric hot spring in Noboribetsu, showed lowered pH_{min} of –0.06 with pH optimum of 0.7 (Schleper et al. 1996). These organisms are adapted to survive under different environmental conditions such as hyper saline plains, high temperatures, and heavy toxic metal concentrations. Their geographic location is acid mine drainage and volcanic springs (Dopson and Holmes 2014). Recently, Colman et al. (2018) have reported the co-evolution of thermoacidophilic archaea and their habitats after the evolution of oxygenic photosynthesis, which is a significant example of niche engineering and biosphere-geosphere co-evolution. The acidophiles have adapted different mechanisms to withstand under low pH such as unique transporter systems, high internal buffer capacity, positively charged membrane surface, and over expression of H⁺ exporting enzymes (Gomes and Steiner 2004). Acidothermophiles grow at low pH and high temperature such as *Sulfolobus solfataricus* which grows at pH of 3.0 and temperature of 80°C. The true acidophilic archea such as *Picrophilus oshimae* and *P. torridus* show optimal growth and produce starch hydrolysing enzymes like amylases, glycoamylases, glycosidases, and pullulanases at pH and temperature of 0.7 and 60°C, respectively (Serour and Antranikian 2002). Proteins from acidophiles have negative surface charge, ether linkages, membrane channel internal buffers, and large isoprenoid core (Baker-Austin and Dopson 2007). The enzymes from acidophiles show more stability at low pH value (< 5.0) and their activity decreases significantly above it. Acidophilic enzymes like amylase, glucoamylase, protease, cellulases, and xylanases possess wide biotechnological and industrial applications in biomining, biofuel and ethanol production (Golyshina and Timmis 2005, Jackson et al. 2007, Shafiei et al. 2011, David et al. 2018, Parihar and Bagaria 2020).

2.4 Alkaliphiles

Alkaliphiles are microorganisms with optimal growth at pH value above 10.0, and they thrive in alkaline environments like soda lakes (Gorka Lake, Chrzanow region, Poland) and gypsum-based soils through passive and active mechanisms (Van den Burg 2003, Huang et al. 2009). The bacteria belonging to genera *Bacillus, Holomonas* and *Pseudomonas*, and archaea of different genera such as *Halorubrum, Haloalkalicoccus, Halobiforma, Natrialba, Natronorubrum* and *Natronococcus* are some of the members of alkaliphiles (Bowers and Wiegel 2011, Sarethy et al. 2011). Alkalithermophilic microbes show optimal growth at high pH (pH of 8.0 or above) and high temperature (50°C to 85°C), and some of them require additional salt concentration for their growth (Wiegel and Kevbrin 2004). A mesophilic, aerobic, and alkaliphilic bacterium *Serpentinomonas* sp. B1, isolated from a terrestrial serpentinizing system of The Cedars (CA, Unites States), showed optimum and maximum pH of 11.0 and 12.5, respectively (Suzuki et al. 2014). These kinds of microbes have special adaptive feature of maintaining internal neutral pH via adapted cell membranes and transporter proteins for maintaining their proton motive force. The hydrogen bonds of their catalytic residues provide shielding to resist change in total charge on protein (Ma et al. 2004). Alkaliphiles have negatively-charged polymers in their cell wall, and the predominant fatty acids are saturated and mono-unsaturated straight chain fatty acids (Gomes and Steiner 2004). Alkaliphiles and thermoalkaliphiles are good source of alkaliphilc enzymes (such as amylases, proteases, pectinase, catalase, peroxidase, oxidoreductase, collagenase, chitinases, esterase, keratinase, lipases, pullulanase, xylanases, and cellulases) which have applications in many sectors like detergents, food and feed, bakery, pulp bleaching, oil recovery, pharmaceuticals, and food and beverage (Verma et al. 2017, Tortorella et al. 2018).

2.5 Halophiles

Halophiles are microorganisms which are able to live in high salt concentrations (minimum 1 M NaCl/KCl concentration), and they are classified into three categories based on optimal salt concentration, i.e., slight halophiles (grow at 200 to 500 mM NaCl), moderate halophiles (grow at 500 to 2500 mM NaCl), and extreme halophiles (grow at 2500 to 5200 mM NaCl) (Pikuta et al. 2007). The highest salt-tolerant bacterium *Halarsenatibacter silvermanii* SLAS-IT, isolated from alkaline hypersaline Searles Lake (CA, United States), showed the utilization of 35% (w/v) NaCl for optimum growth (Blum et al. 2009). They possess several adaptive traits like compatible solute production, biofilm formation, and enzyme adaptation to work under ionic stress conditions to prevent cell desiccation under saline conditions (Frols 2013, Charlesworth and Burns 2016). Halophiles are found in all three domains, i.e., archaea, bacteria and eukarya, and belong to genera of *Acinetobacter, Bacillus, Halobacillus, Halobacterium, Halogeometricum, Haloferax, Halorhabdus, Halothermothrix, Marinococcus, Micrococcus, Natrialba, Natronococcus, Nesterenkonia, Salimicrobium,* and *Salinivibrio* (Taylor et al. 2004, Dumorne et al. 2017). These organisms are adapted to assimilate sugars, amino acids, and other biomolecules until their cellular osmolarity become equal to outer ionic concentration. They require high salt concentration for proper folding of proteins, and large number of acidic amino acids are displayed on protein surface to increase their flexibility (Amoozegar et al. 2017). Halophilic microorganisms have established different physiological, structural, and chemical modifications which allow stability of their enzymes at low water activity, in the presence of salt and organic solvents (Datta et al. 2010, Raddadi et al. 2013). Proteins from such microorganisms have biased amino acid composition (large number of negative charged amino acids on surface) which help them to remain stable at high ionic strength, and acidic amino acids are exposed on the surface of halophilic proteins (Madern et al. 2000). The acidic amino acids like aspartate and glutamate are absent in amylases from thermophilic halophile *Halothermothrix orenii* (Mijts and Patel 2002). Some amino acids like serine, and threonine are specific in halophilic proteins which include polysaccharide-hydrolyzing enzymes such as lipase, protease, xylanase and amylase (Raddadi et al. 2013, Elleuche et al. 2014). Proteins from halophilic microbes stabilize their active conformation due to low affinity of salt for specific sites on surface of proteins (Mevarech et al. 2000). The haloalkaliphilic microorganisms are able to grow under high alkaline areas, and discovered from soda lakes in India, China, Egypt, Kenya and western Unites States. These organisms possess complex cellular envelope with large number of glycosylated proteins which help them to maintain a neutral alkaline pH, and high proportion of acidic residues in proteins (Reed et al. 2013). Organic solvents and xenobiotic compounds have been released in effluents by companies which produce dyes, herbicides, pesticides, and explosives. Reverse micelles encapsulating halophilic extremozymes are efficient in reducing the accumulation of such hazardous compounds in the environment (Marhuenda-Egea et al. 2002). Halophilic enzymes like proteases, nuclease, and dehydrogenases show opportunities in bioremediation, biosynthetic processes, cosmetics, textile, and food industries (Dumorne et al. 2017, Czech et al. 2018).

2.6 Radiophiles

Radiophilic microorganisms can tolerate high radiation dosage due to their ability to survive in high radiations (cosmic rays, X-rays, gamma rays, and ultraviolet radiations) and oxidative stress. These organisms such as *Deinococcus radiophilus, D. radiodurans, D. hohokamensis, Thermococcus radiotelerans, T. gammatolerans*, and *T. marinus* are highly resistant to ionizing and ultraviolet radiations (Sandigursky et al. 2004, Jolivet et al. 2004, Krisko and Radman 2013, Merino et al. 2019). They are distributed among various genera of bacteria (*Bacillus, Deinococcus, Rubrobacter, Kineococcus*), and cyanobacteria (*Chroococcidiopsis, Nostoc*) (Gabani and Singh 2013). The radiophiles have adapted to high radiations by altering their DNA repair mechanisms, gaining more genome copies for genome redundancy, production of pigments, accumulation of Mn (II), condensed nucleoid, and utilization of smaller amino acids (Krisko and Radman 2013, Byrne et al. 2014). Radiophiles exhibit their potential application in nuclear waste remediation. *Deinococcus radiodurans* is the toughest radiophile which can survive under high level of radiations (5 Mrad, 3000 times higher than what would kill a human), high pressure, high temperature, and many other extreme conditions. There are multiple genes which have been isolated from this radiophile that encode for multiple activities of DNA repair. The purified enzyme superoxide dismutase from *D. radiophilus* is stable at temperature up to 40°C and pH ranging from 5.0–11.0. These extremophiles are used for detoxification of halogenated organics, toxic heavy metals, and radionuclides from nuclear waste (Yun and Lee 2004, Gabani and Singh 2013, Raddadi et al. 2015, Merino et al. 2019).

2.7 Metallophiles

Metallophiles are the organisms which can grow in the environment of high metal concentrations. Heavy metals like copper (Cu), cadmium (Cd), cobalt (Co), lead (Pb), mercury (Hg), and zinc (Zn) are used highly in different industries which results in environmental pollution and serious health hazardous effects (Carapito et al. 2006). Most of the metallophiles are also acidophiles and belong to bacterial genera *Acidithiobacillus, Alicyclobacillus, Acidiphilium, Acidimicrobium, Ferrimicrobium, Leptospirillum, Sulfobacillus*, and archaea from genera *Acidiplasma, Acidianus, Ferroplasma, Metallosphaera*, and *Sulfolobus* (Johnson 2014). *Ralstonia metallidurans* is a Gram-negative, non-spore forming *Bacillus* which can survive in millimolar concentration of heavy metals. This organism contains two large megaplasmids which contain genes for resistance against Cd, Co, Cu, Cr, Ni, Hg, Pb, and Zn (Gomes and Steiner 2004). They are able to accumulate and remove such toxic heavy metals outside the cells by specific efflux pumps ormetabolize these heavy metals resulting in production of less toxic or new harmless compounds (Parihar and Bagaria 2020). Heavy metal pollution is one of the major threats to aquatic and terrestrial life forms, and metallophiles are promising in their use to overcome this public threat through alkylation, methylation, oxidation, and reduction mechanisms. These organisms play a significant role in bioaccumulation, biobleaching, bioremediation, and biomining for reducing the concentration of toxic metals (Navarro et al. 2013, Raddadi et al. 2015, Merino et al. 2019).

2.8 Barophiles

Barophiles are organisms that adapt optimally at high hydrostatic pressure of deep-sea environments such as volcanic areas and deep-sea vents by reducing their cellular structure, and molecular dynamics. They are distributed among the genera of *Colwellia, Methanococcus, Moritella, Pyrococcus, Shewanella*, and *Thermus* (Yano and Poulos 2003). Some barophilic microbes are *Pyrococcus abyssi, P. horikoshii*, and *Sulfolobus solfataricus* (Fusi et al. 1995, Takai et al. 2009). A thermophilic archaeon *Thermococcus piezophilus* is isolated from deep sea, and able to survive up to 125 MPa (Dalmasso et al. 2016). Barophilic microorganisms adapt themselves to thrive under low temperature and high pressure, and their proteins also show stability at high pressure (Gomes et al. 2000). The culturing of these microbes requires complex systems to maintain the pressure to 38 MPa using complex gas systems, and hydraulic pumps (Pettit 2011, Zhang et al. 2015). Such microorganisms have several adaptations such as more unsaturated fatty acids in cell membrane to enhance membrane fluidity at high pressures, modification of respiratory chain, upregulation of chaperone-encoding genes, production of osmolytes, and expression of different porins (Jebbar et al. 2015). The proteins from such microorganisms possess compact, dense hydrophobic cores, low molecular weight and highly branched amino acids such as glycine, and proline for inhibiting the protein flexibility by minimizing the conformational space. Barophilic enzymes have been isolated from many psychrophilic and thermophilic microbes, and they are highly stable at high pressure. There is a great advantage of enzymes which are catalytically active at high temperature and high pressure in biotechnological applications. The thermal stability of barophilic enzymes is due to increased charged dipoles on helices, surface electrostatics, shorter loops, and more interactions between cations and aromatic rings (Yano and Poulos 2003). These enzymes show great potential in food industry, chemical synthesis industry, and detergent industry (Cavicchioli et al. 2011b, Martinez et al. 2016). The properties of barophilic enzymes, and their cellular components are not yet fully understood due to difficulty in their cultivation under high-pressure conditions using current technology (Merino et al. 2019).

2.9 Xerophiles

Xerophiles are the organisms which can grow in dry environmental conditions including deserts, and dry stones by developing anhydrobiosis, i.e., ability to survive under very less intracellular water content, and these organisms are generally metabolically inactive. Some specialized microbes belonging to algae, fungi, and lichens possess the ability to grow under dry environmental conditions (Rothschild and Manicinelli 2001). The enzymes from xerophilic organisms are highly efficient in agriculture sector for improving water management in desert plants (Rolli et al. 2015).

3. Biotechnological Applications of Extremophiles

The catalytic properties of several enzymes have already been exploited by various industrial sectors, for their cost-effective and efficient bioconversions. Most of the enzymes present today are not observed to be available for commercial applications due to their limited stability under harsh industrial environment. Therefore, various academic groups, researchers, and industries are continuously trying to explore novel enzymes having biotechnological applications using complementary approaches such as mimicking evolution at laboratories for developing stable enzyme variants, mining enzyme variants from extremophiles, and rational and computer-assisted enzyme engineering strategies (Rigoldi et al. 2018). Extremozymes are a new class of enzymes which have emerged in the recent past, and they are highly stable under extreme conditions as compared to other enzymes. Thermozymes are one such kind of extremozymes which are continuously being used extensively for food and beverages, paper and pulp, detergents and soaps, leather, textiles, pharmaceuticals, biofuels, bioremediation and wastewater treatment.

3.1 Application in Food and Beverage

The growing awareness of customers about nutritional and healthy food, improved legislations and safety regulations, and increasing demand for processed and improved quality food products result in increasing demand for environment-friendly manufacturing processes for food and bakery. The size of global food enzymes was valued at 1.944 billion USD and it is expected to reach 3.056 billion USD by 2026 with a CAGR of 5.6% from 2019 to 2026 (https://www.alliedmarketresearch.com/food-enzyme-market). Food enzymes are used in processed foods, dairy products, bakery products, confectionary products, beverages, and other products like meat and poultry for food preservation, enhancing texture, flavour and fragrance, tenderization, and coagulation. The largest share in the market of food enzymes is provided by North America due to increased customer consciousness for low-calorie and low-fat foods. The Asia Pacific market is growing at a highest CAGR due to enhanced demand from India and China. The European Food Safety Authority (EFSA) has approved the use of several enzymes like amylase, cellulase, lactase, pectinase, xylanase, polygalacturonase, pullulanase, gluco-oxidase, protease, lipase, transferase, and isomerase in food and beverage industry, and these enzymes have already been evaluated before marketing in European Union (EU) (https://www.marketsandmarkets.com/Market-Reports/food-enzymes-market-800.html?gclid=Cj0KCQjwzN71BRCOARIsAF8pjfjTlyOwxoknblMUva02qejGcWa28iQRfj_RBowxPjCZc3BGlswuaZ0aAqs8EALw_wcB). The world major players for providing food enzymes are Advanced Enzyme Technologies, Amano Enzyme Co. Ltd., Associated British Foods Plc. (ABF), BASF, Chr. Hansen Holding A/S, DowDuPont, Kerry Group PLC, Novozymes, Royal DSM N.V., and Aum Enzymes (https://www.alliedmarketresearch.com/food-enzyme-market).

Enzymes are used in dairy industry for preparation of cheese and curd, in beverage industry to reduce sulfur content and maintain wine colour and clarity, and in baking industry to improve the quality of bread (Dewan 2014). The two pectinolytic enzymes, i.e., methylesterases and depolymerases (lyases and hydrolases), are widely used in food processing for fruit juice extraction, and wine making to improve colour extraction from fruit skin, enhance juice yield, and reduce viscosity. Lactase from *Kluyveromyces lactis* is used in the production of lactose-free products by hydrolysing lactose (Mateo et al. 2004). A starch saccharification process using amylolytic enzymes, i.e., α-amylase, glucoamylase, amylopullulanase from thermophiles was reported by Satyanarayana et al. (2004). Xylanases are used in bread making to produce fluffy and strong dough by converting insoluble hemicellulose of dough into soluble sugars. Some other enzymes such as catalase, amylase, lipase, and peptide hydrolase are used to confer specific characteristics to food during its processing. The first archaeal amylase was characterized from *Pyrococcus furiosus* in 1990, which has an optimum temperature of 110°C and residual activity of 130°C (Koch et al. 1990). Thermophilic pectinolytic enzymes are very promising for commercial applications in food and beverage processing (Antranikian et al. 2005). The commercially available thermostable enzymes in the market are amylases (Avantec®, Termamyl® SC, Liquozyme®, Novamyl®, Fungamyl® (Novozymes), Fuelzymes® (Verenium), AlphaStar Plus (Dyadic)), glucose/xylose isomerases (Sweetzyme® (Novozyme)), amyloglucosidases (GlucoStar Plus (Dyadic)), proteases (Protease PLUS), glucoamylases (Spirizyme® (Novozymes)), glucose oxidases (Gluzym® (Novozymes)), and lipases and xylanases (Lipopan® and Pentopan® (Novozymes)) (Sarmiento et al. 2015).

The current trend of food and beverage industry is to replace high-temperature processes with low-temperature processes to minimize undesirable side reactions which occur at high temperatures, prevention of spoilage and contamination, energy saving processing, retention of volatile and labile flavour compounds, and high degree of control over psychrophilic enzymes (Pulicherla et al. 2011). A patent has been granted to Novozymes in 2004 for the variant of α-amylase from *Bacillus licheniformis* with enhanced specific activity from 10°C to 60°C (Borchert et al. 2004). There are several companies like Enzyme Development Corporation (New York, NY, USA), DSM/Valley Research (South Bend, IN, USA), DuPont/Danisco (Wilmington, DE, USA), and Biocatalysts (Wales, UK) which are providing lactases or β-galactosidases that show certain activities at low temperatures (Horner et al. 2011). In a previous study, psychrophilic xylanases from *Pseudoalteromonas haloplanktis* TAH3A, and *Flavobacterium* sp. MSY-2 improved the dough properties, and final bread volume by 28% when compared with the mesophilic xylanases from *Aspergillus aculeatus*, and *Bacillus subtilis*, respectively (Dornez et al. 2011). Several novel xylanases from psychrophilic microbes have also been reported by various authors (Chen et al. 2013, Del-Cid et al. 2014, Asenjo et al. 2014). Pectinases from psychrophiles are used inbreadmaking, fruit juice processing and fermentation of beer and wine, and is marketed as Novoshape® (Naovozyme), Pectinase 62L (Biocatalysts), and Lallzyme® (Lallemand) (Sarmiento et al. 2015). A thermostable amylase, produced from mutant *Pyrococcus furiosus*, showed increase production of maltoheptose from β-cyclodextrin (Park et al. 2013), thereby enhancing the value of final food product. Extremolytes are similar molecules which are produced by extremophiles and are used for the production of various food products. Cheese is treated with *Brevibacterium linens* for the surface ripening which results in accumulation of ectoine (a functional food ingredients) up to 89 mg/100 g of product (Klein et al. 2007, Cencic and Chingwaru 2010, Raddadi et al. 2015).

3.2 Application in Paper Pulp Bleaching

The extensive application of enzymes in paper and pulp processing industry like bio-bleaching, fibre-modification, de-inking, and pitch removal leads to enhanced global pulp and paper enzymes market which was around 160 million USD in 2019, and is expected to rise with a CAGR of 6.5% by 2026 (https://www.gminsights.com/industry-analysis/pulp-and-paper-enzymes-market). The enzymes which have been used extensively in pulp and paper industry are cellulases, hemicellulases, amylases, xylanases, lipases, esterases, and laccases. Pulping is a crucial step in paper production process which is carried out by hot-alkali treatment for hydrolysing woodenfibres and most of the lignin removal. Different chemical, and mechanical methods have been employed for pulp formation;these methods involve the use of strong chemicals like chlorine, sodium sulphide and sodium hydroxide, and are conducted at high temperatures (over 80°C), and alkaline pH. Multistep bleaching process by using chlorine or chlorine dioxide at high temperature is used for removing rest of the lignin content; however, this process generates large volume of waste. Therefore, enzymatic pulping is gaining more attention by offering a safer, eco-friendly, and profitable solution for the paper and pulp industry. Both pulping and bleaching steps are carried out at high temperatures; hence, paper industries need thermophilic hemicellulases for pre-treatment of paper pulp to minimize the pollution caused by the chemical methods. Hyperthermophilic hemicellulases from *Thermotogals* were initially characterized and found that the endoxylanase from *Bacillus* 3D is approximately 100 times more active than that of the *Thermotoga thermarum* (Schiraldi and Giuliano 2002). Xylanases from actinomycetes, and fungi have shown high thermal and pH stabilities, which play a major commercial rolein paper and pulp industries (Haltrich et al. 1996). Xylanases from *Actinomadura* sp. FC7 and *Nonomuraea flexuosa* have shown high thermal stability (Brzezinski et al. 2014), and high levels of xylanases are produced by *Streptomyces* sp. which provide efficient bio-bleaching (Rifaat et al. 2006). Hyperthermophilic xylanases isolated from *Streptomyces* sp., *Thermoascus aurantiacus*, *Thermotoga thermarum*, and *Thermotoga maritima* have already been proven to be effective for bio-bleaching processes (Milagres et al. 2004, Jiang et al. 2006, Shi et al. 2013). Novel hyperthermophilic lipases and esterase are used in pitch control by degrading sticky compounds including coatings and adhesives for improving the quality of final product. Commercially, they are available in the form of Lipase B Lipozyme®, CALB L, Lipase A, NovoCor® AD l, Resinase™ HT and Resinase A2X (Novozymes), and Optimyze (Buckman Laboratories). Hyperthermophilic cellulases are used to enhance strength, and brightness of paper sheets to improve overall efficiency of refining process. Cellulases and hemicellulases are used in recycling of pulps, and these enzymes are marketed as FiberZyme® G5000, FiberZyme® LBL CONC, FiberZyme™ LDI, and FiberZyme™ G4 (Dyadic). The bio-bleaching is done by using xylanases and laccases, and their products are marketed under the trade name Luminase®PB-100 and PB-200 (Verenium), Xylacid® (Verula Biocell), Xyn 10A (Megazyme) and Laccase (Novozyme) (Sarmiento et al. 2015).

3.3 Application in Textile and Detergent

Biologically derived enzymes are common components of detergent formulations in developed and developing countries. According to the report titled 'Global Detergent GradeEnzymes Market, Size, Share, Report and Forecast 2019–2024' published by Expert Market Research, the global detergents' grade enzymes market reached to 908 kilo metric tons in 2018, and is projected to grow with annual CAGR of 6.6% during the period 2019–2024. The two popular enzymes in biological laundry detergents are lipases and proteases, which account for a major portion of global enzymes used especially in Asia-Pacific countries like India and China. The enzymes market of Asia-Pacific region is expected to grow with a CAGR of 10% with countries like India, Indonesia, China and Malaysia (http://www.researz.com/global-enzymes-market-overview-industry-analysis-type-application-region-forecast-till-2024). The efficacy of cold-water household and industrial laundry has been improved by many cold-adapted enzymes like amylases, cellulases, lipases, mannases, pectinases, and proteases (Sarmiento et al. 2015). Proteases catalyse the hydrolysis of peptide bonds, thereby digesting the protein stains like human sweat, blood, cocoa, egg and grass into shorter fragments (Joshi and Satyanarayana 2013). Since serine alkaline proteases are highly resistant to denaturation against detergents and are alkaline in nature, they are widely used in detergent industries for laundering (Alqueres et al. 2007). Bacillus protease is the first commercial product of Novozyme (Bagsvaerd, Denmark) for use in detergents, and the mass production and marketing of this enzyme began in 1960s (Sarmiento et al. 2015). The first serine protease was released by Kannase® or Liquanase® in 1998 for washing laundry at cold temperature (10°C and 20°C). The isolation of a novel serine protease from *Glaciozyma antartics* strain P112 was done by Alias et al. (2014), which showed high activity at 20°C after cloning and expression of protease gene into *Pichia pastoris*. Several cellulases also find application in textile industry for polishing of fabric (Hill et al. 2006). This enzyme is also used in detergent industry, and a thermophilic cellulase from *Streptomyces thermoviolaceus* is proved to be more active than commercial cellulases in the presence of detergents (Jones et al. 2000). Amylases are widely used in detergent formulations to hydrolyze starch-based stains from various foods like fruits, pasta, gravy, cereals, and potatoes into water oligosaccharides and water soluble dextrins. A novel cold-active, detergent stable α-amylase was characterized from *Bacillus cereus* GA6, and this extracellular enzyme was observed to be stable at low temperature (4°C–37°C, with optimal activity at 22°C), and alkaline pH (pH 7.0–11.0) which demonstrates its compatibility with laundry detergents (Roohi et al. 2013). Similarly, industrially important amylase producing marine bacterium *Zunongwangia profunda* was isolated by Qin et al. (2014). A high-performance lipase has been identified from *Pseudomonas stutzeri* PS59 for use in detergent formulations, and it showed optimal activity at 20°C and pH of 8.5 in the presence of different oxidizing agents and surfactants (Li et al. 2014). Mannanase and pectinase are other enzymes used in laundry and dish washing industry to degrade mannan, and pectin, respectively (Sarmiento et al. 2015).

There are many psychrophilic enzymes like lipase, protease, amylase, pectate lyase, cellulase, catalase, and mannase which have shown applications in textile and detergent industries, and their commercial products are already available in the market. The commercial products of lipases are Lipoclean®, Lipex®, Lipolase® Ultra, Kannase, Liquanase®, Polarzyme® (Novozymes), and this enzyme is used in breakdown of lipid stains. The protein stains are broken down by proteases, and it is commercially available as Purafect® Prime, Properase® and Excellase (Genencor). Similarly, amylases, mainly responsible for breakdown of starch-based stains, are produced commercially by Stainzyme® Plus (Novozymes), Preferenz™ S100 (DuPont), and Purafect® OxAm (Genencor). Amylases are used in textiles for de-sizing of woven fabrics, and their commercial products are Optisize® COOL and Optisize NEXT (Genencor/DuPont). The commercial product of pectate lyase available in the market is Xpect® (Novozymes), and it is used for the removal of pectin stains. Cellulases are used for the washing of cotton fabrics, and the products available in the market are Celluclean® (Novozymes), Cellulozyme®, Rocksoft™ Prima, UTA-88 and UTA-90 (Huan Youtell Biochemical), Antarctic, Antarctic LTC (Dyadic), Retrocell Recop and Retrocell ZircoN (EpyGen Biotech). The bio-finishing and dyeing of cellulosic fabrics is also done by using cellulases, and the products available in the market for this application are Primafast® GOLD HSL, IndiAge® NeutraFlex, PrimaGreen® EcoLight 1, and PrimaGreen® EcoFade LT100 (Genencor/DuPont). Catalase is used in textiles for removing hydrogen peroxide from fabrics, and its commercially available product is catalase (CAT) (Swissaustral). The degradation of mannan or gum is catalysed by mannanase and the commercially available products are Mannaway® (Novozymes), Effectenz™ (DuPont) (Sarmiento et al. 2015).

3.4 Application in Biofuels Production

The current global energy crisis due to rising population, continuous depletion of fossil fuel reserves, increasing price of crude oil, global warming, and to fulfil Kyoto protocols, necessitates an urgent need for alternative fuel sources. Biofuels are renewable and sustainable sources of fuels which are obtained from agricultural, industrial and municipal wastes. Different types of biofuel includes hydrogen, methane/biogas, biodiesel, bioethanol, and biobutanol (Barnard et al. 2010). In 2014, the global revenue for biofuel enzymes was 623.0 million USD, which increased to 652.1 million in 2015, and it is further expected to rise to 1.0 billion USD by 2020 with a CAGR of 10.4% (https://www.bccresearch.com/market-research/energy-and-resources/biofuels-enzymes-global-markets-technologies-report.html). The global competitive players in biofuels enzyme market are Novozymes A/S, E.I. du Pontde Nemours and Company, SunsonIndustry Group Co. Ltd., SinoBios, Specialty Enzymes & Biotechnologies, Transbiodiesel Ltd., Enzyme Supplies Limites, Koinklijke DSM N.V., and Advanced Enzyme Technologies Ltd. The major enzymes involved in biofuel production are cellulases, amylases, lipases, and xylanases (https://www.zionmarketresearch.com/report/biofuels-enzyme-market). The current market trend focuses on the production of second-generation biofuels which hydrolyzes complex lignocellulosic materials by microorganisms and/or their purified enzymes. There are several advantages of using extremophiles, and extremozymes in biofuel production due to their stability towards extreme environmental conditions such as temperature and pH, simultaneous fermentation of pentoses and hexoses from biomass, requirement of less energy inputs, along withless chances of microbial contamination and product inhibition. Most of the steps involved in biofuel production are carried out at high temperature and extremes of pH; therefore, thermophiles are suitable candidates to replace mesophilic microorganisms for biofuels production. Methanogens (typically thermophilic anaerobic archaea), and psychrophiles (producing cold-adapted lipases) have also been exploited for the production of biogas, and biodiesel, respectively (Barnard et al. 2010). Both thermophilic (*Methanobacterium* sp., *Methanosarcina thermophila*, and *Methanothermococcus okinawensis*), and psychrophilic (*Methanosarcina lacustri, Methanolobus psychrophilus*, and *Methanosaeta*) microbial strains are involved in the production of biomethane (McKeown et al. 2009). Psychrophilic lipases from *Rhizopus delemar, R. miehei, Candida rugose, C. lipolytica, Klebsiella oxytoca, Pseudomonas fluorescens, P. cepacia*, and *Penicillium camembertii*show their potential biotechnological application in biodiesel production (Al-Zuhair 2007). *Caldicellulosiruptor saccharolytics* is an extreme thermophile which contains genes involved in degradation of lignocellulosic biomasss, and production of hydrogenase and biohydrogen (Van de Werken et al. 2008). Thermophiles which have been exploited for commercial bioethanol production are *Thermophilic clostridia, Thermoanaerobacterium saccharolyticum, Thermoanaerobacter ethanolicus, Geobacillus stearothermophilus, G. thermoglucosidasius, Saccharomyces cerevisiae*, and *Zymomonas mobilis*. There are several thermophilicmicrobial strains like *Clostridium thermocellum, C. stercorarium, C. thermopapyrolyticum, Coniochaeta ligniaria, Chaetomium thermophile, Humicola insolens, H. grisea, Melanocarpus albomyces Talaromyces emersonii, Thermotoga* sp., and *Thermoascus aurantiacus,*and halophiles including *Alkalilimnicola* sp. NM-DCM1, *Bacillus* sp. BG-CS10, *Gracibacillus* sp., *Saccharomyces cerevisiae, Salinivibrio* sp. NTU-05, *Thalassobacillus* sp. LY183, which are involved in production of lignocellulose-degrading enzymes such as cellulase, lignase, lignin peroxidase, manganese peroxidase, and xylanase (Rastogi et al. 2009, Barnard et al. 2010, Yu and Li 2015, Mesbah and Wiegel 2017, Amoozegar et al. 2019). Similarly, amylases producing extremophilic organisms are *Alteromonas* sp., *Bacillus acidoaldarius, B. stereothermophilus, Pyrococcus furiosus* and *Nesterenkonia* sp. (Boyer et al. 2006, Barnard et al. 2010, Amiri et al. 2016). The commercial synthesis of bioethanol, and biobutanol was done by using chemical processes supplemented with use of mesophilic microorganisms including *Clostridium* sp., and *Saccharomyces cerevisiae*. Recently, large scale microorganisms-based systems using thermophilic microbes *Thermotoga elfii* and *Caldicellulosiruptor saccharolyticus* have been developed. There are several companies like Green Biologies, Gevo, Joule Unlimited, Solazyme, and Sapphire Energy which are using thermozymes to produce large volumes of bioethanol, and biodiesel (Coker et al. 2016). The extremophilic microalgae are becoming more economical for the production of biodiesel due to their ability to accumulate large amounts of lipids and oils, rapid growth rate, ability to grow under saline conditions, throughout the year cultivation, and production of valuable by-products such as pigments, biopolymers, and polysaccharides. The extremophilic algae like *Botryococcus braunii, Cyanidium caldarium, Galdieria sulphuraria, Nannocloropsis salina*, and *Ochromonas danica* are mainly involved in biodiesel production. Solazyme (San Francisco) is already using commercial fermentation technologies for commercial production of Soladiesel® (Chisti 2008, Barnard et al. 2010). A halophilic green microalga *Dunaliella salina*, found in saline lakes, marine waters, and salt ponds, plays an important role in biodiesel production due to its ability to synthesize high lipid contents, mainly palmitic, and linoleic acids (Rasoul-Amini et al. 2014). Similarly, among the 21 halophilic microalgae, isolated from hypersaline Bardawil lagoon, a green microalga *Tetraselmis elliptica* showed the capacity to produce large lipid content, and predominant fatty acids like oleic and palmitic acids, which prove its potential for biodiesel production (Abomohra et al. 2017). There are several previous studies which have been carried out to prove the potential of extremophilic microorganisms and their enzymes in the production of biofuels (Ameri et al. 2017, Ai et al. 2019, Amoozegar et al. 2019).

3.5 *Applications in Medicines*

Bacteriorhodopsin, produced by halophilic archaeon, is a light-sensitive protein of 26.5 kDa, and has a transmembrane domain having several helical protein segments. This protein has excellent photochemical and thermodynamic stability, hence can be used in artificial retina, neural network optical computing, associative and volumetric optical memories (Koyama et al. 1994, Alqueres et al. 2007). Archaeosome is a liposome which is unique in Archaea domain with high degree of stability, and is used as a delivery vehicle for many drugs, antibiotics, vaccines and enzymes. The safety of these vesicles for medical applications has already been reported by conducting several mouse models' studies by their oral, intravenous, and subcutaneous administration (Patel et al. 2004, Alqueres et al. 2007). Bacterioruberin, produced by *Halobacterium* and *Rubrobacter*, has proved its potential in combating human cancers by repairing mutated DNA strands (Singh and Gabani 2011). A synthetic nucleoside analogue has been synthesized using thermostable nucleoside phosphorylase purified from hyperthermophilic aerobic crenarchaeon *Aeropyrum pernix* K1, which showed its applications in antiviral therapies (Zhu et al. 2013). Another study conducted by Choi et al. (2014) proved the chemotherapeutic action of deinoxanthin, isolated from radioresistant bacterium *Deinobacillus radiodurans* for apoptosis of cancer cells. Chitinase helps in hydrolysis of chitin (a linear homopolymer of N-acetylalucosamine), and this enzyme is used for synthesis of various biologically active compounds. Some chitooligosaccharides can be used for inhibition of some tumours, and activation of phagocytes. Chitinases have been characterized from some thermophilic microorganisms; however, their potential for economical chitin hydrolysis needs to be tested for industrial applications (Pandey et al. 2016). Some novel compounds are produced by thermophilic fungi *Talaromyces thermophilus* which have shown novel anti-nematode activity (Guo et al. 2012). The presence of diverse range of non-ribosomal peptide synthesis (NRPS) and polyketide synthesis (PKS) genes in *Actinobacter*, isolated from thermal springs in Tengchong, China, provided the possibility of discovery of novel compounds from thermophiles (Liu et al. 2016). Extremolytes possess the ability to stabilize proteins and nucleic acids, and are hence used for storage and stabilization of proteins, administering therapeutic protein-based medicines, and development of drugs for many diseases (Faria et al. 2013, Avanti et al. 2014, Raddadi et al. 2015). QS mechanism of extremophiles is used for developing whole cell microbial biosensors for detecting pathogenic microbes, anti-cancer therapeutics, and preparation of synthetic homologs of autoinducers (Choudhary and Schmidt-Dannert 2010, Kaur et al. 2019).

3.6 Applications in Molecular Biology

Enzymes from extremophiles have shown immense importance in molecular biology such as for editing and altering nucleic acid sequence, genotypic identification, DNA fingerprinting, recombinants production for cloning experiments, site-directed mutagenesis, single nucleotide polymorphism (SNP) genotyping, and also to control carry-over contamination in polymerase chain reaction (PCR). A recombinant version of cold adapted UNG enzyme, isolated from Atlantic cod (*Gadus morhua*), has been produced commercially in *Escherichia coli* by ArcticZymess; it is completely and irreversibly inactivated at high temperature, i.e., 55°C (Lanes et al. 2002). Double stranded DNase catalyses the digestion of double-stranded DNA, making it useful to remove genomic DNA from RNA preparations and decontaminate PCR mixes. ArcticZymes and Affymetric (Santa Clara, CA, USA) offer recombinant double stranded DNase for molecular applications. Cryonase is another novel enzyme from cold-tolerant microbes, which is derived from *Shewanella* sp. and can digest all types of DNA and RNA (single-stranded, double-stranded, linear or circularized). This enzyme is produced commercially by Takara-Clontech, and becomes completely inactive after incubating 30 min at 70°C (Awazu et al. 2011). The cold active recombinases from psychrophiles are used by TwistDx (Cambridge, UK) in recombinase polymerase amplification (RPA) PCR kits. RPA is a method of isothermal amplification by inserting the oligonucleotide primers onto complementary strand of DNA in a manner similar to PCR, and this method has been successfully applied for detection of malaria causing protozoan parasite *Plasmodium falciparum* and AIDS causing human immunodeficiency virus (HIV) (Sarmiento et al. 2015). DNA polymerases from thermophilic organisms such as *Thermus aquaticus* (Taq), *Pyrococcus furiosus* (Pfu), and *Thermococcus litoralis* (Vent) are highly stable, and survive beyond the denaturation temperature of long DNA fragments. Therefore, proofreading forms of DNA polymerases are used around the world in all molecular biology laboratories (Rigoldi et al. 2018). Taq polymerase shows optimal activity at 75°C and pH 9.0 with high processivity, which makes this enzyme a suitable choice for sequencing and polymerase chain reaction (PCR) technologies (Pandey et al. 2016). DNA polymerase from *Thermus thermophiles* with reverse transcriptase activity is used in various molecular studies (Fakruddin 2017). DNA ligases are utilized in molecular biology for catalysing the synthesis of phosphodiester bonds in nicked molecules of double stranded DNA. Thermostable DNA ligases show optimal activity in the range from 45°C–80°C, and their ligating property can be used in PCR technology, ligase chain reaction, mutational analysis, and gene synthesis (Pandey et al. 2016). The current market of DNA ligases is ruled by *Escherichia coli* DNA ligases, bacteriophage derived (T4 and T7 phage) DNA ligases. Psychrophilic DNA ligases are preferable in maintaining high specific activity and minimum interference with ligation process at low temperature. A novel DNA ligase has been cloned, overexpressed, and characterized from psychrophilc *Pseudoalteromonas haloplanktis*, which showed its optimum activity below 4°C (Georlette et al. 2000). The carboxypeptidase from *Sulfolobus salfatricus* shows broad specificity and stability which is used for C-terminal protein sequencing. The protease S (from *Pyrococcus furiosus*) shows optimal activity between 85°C–95°C, and is used for peptide sequencing. Serine protease from *Thermus* strain Rt41A shows optimal activity at 90°C, have applications in DNA, RNA purifications and cellular structure degradation prior to PCR (Fakruddin 2017). Alkaline phosphatase (AP) is the most demanding enzyme in molecular biology, which is used in cloning techniques for dephosphorylation of 5' end of linear fragment of DNA to avoid its recircularization. The first heat labile AP was isolated and purified from Antarctica in 1984 (Kobori et al. 1984), and the first cold adjusted AP was produced from arctic shrimp *Pandalus borealis* commercially by ArcticZymes (Tromso, Norway) in 1993. Recently, a novel psychrophilic AP was isolated from ocean-tidal flat sediments from the west coast of Korea after constructing its metagenomic library, whose purified enzyme showed comparable efficiency and characteristics similar to other commercial APs (Lee et al. 2015, Sarmiento et al. 2015).

3.7 Application in Bioremediation and Wastewater Treatment

The widespread human and industrial inputs lead to contamination of soil, water, and sediments which poses threat to terrestrial as well as aquatic habitats. Bioremediation is a beneficial and green approach for efficient removal of contaminants (Krzmarzick et al. 2018). Extremophilic microorganisms provide stable enzymes, and biosurfactants which can be used for wastewater treatment, degradation of heavy metals, xenobiotics, and pollutants from terrestrial and aquatic environments. Halophilic microorganisms are reported to degrade vast range of lignocellulosic materials, hydrocarbons, formaldehyde, chlorophenols, and nitroaromatic compounds (Garcia et al. 2004). The degradation of a variety of hydrocarbon compounds by *Halomonas organivorans* has been reported by Garcia et al. (2004). Some halophilic microbes such as *Halobacterium salinarum*, *Halomonas* spp., *Haloferax denitrificans*, *H. mediterranei*, and *Haloanaerobium lacusrosei* are successfully being used for treatment of saline and hypersaline wastewaters (Venkata Mohan et al. 2007, Oren 2010). Laccases are blue multicopper oxidases which help in degradation of aromatic compounds and polymers, and hence they are used in wastewater treatment and detoxification. They are also used for bioremediation of pesticide and herbicide contaminated soils and in water purification systems (Shraddha et al. 2011). Several psychrophilic microorganisms and their enzymes (catalase, peroxidase, and oxidase) have been exploited for wastewater (polluted with oils, lipids, and hydrocarbons) treatment, and bioremediation of solid wastes. The management of nuclear waste pollutants is carried out by using radiophiles, and metallophiles/acidophiles, which is currently being used in bioremediation due to the microbial adaptation to high concentration of heavy metals (Johnson 2014, Raddadi et al. 2015). The two bacterial strains *Pseudomonas fluorescens* and *Serratia odorifera*, isolated from arsenic polluted river from Atacama Desert, were reported to tolerate arsenic ranging from 800–1000 mM (Campos et al. 2009). A thermophilic *Geobacillus* sp. strain ID17 was isolated from Deception Island, Antartica, which is able to produce intracellular gold nanoparticles when exposed to Au (III), and this strain is useful in bioremediation of gold-bearing wastes (Correa-Llaten et al. 2013). The members of *Chroococcidiopsis* (hypolithic Cyanobacteria) are useful in bioremediation of hydrocarbon of aquatic environments (Al-Bader et al. 2013). *Deinococcus* sp. UDEC-P1 was isolated from Tempanos Lake, Chile, and it was proved as a potential candidate to degrade radioactive wastes due to its natural resistance to high levels of radiations (Guerra et al. 2015). Similarly, a thermostable cyanide degrading nitrilase, isolated from hyperthermophilic archaea *Pyrococcus* sp. M2D13, is useful in cyanide remediation from contaminated water streams (Dennett and Blamey 2016). *Pseudomonas putida* strain ATH-43 is able to tolerate different heavy metals such as copper (Cu), chromium (Cr), cadmium (Cd) and selenium (Se), and *Psychrobacter* sp. ATH-62 is resistant to mercury (Hg) and tellurite (TeO_3^{2-}) (Rodriguez-Rojas et al. 2016). Different extremophiles like acidophiles (*Acidithiobacillus ferroxidans*, and *Leptospirillum ferriphilum* from Atacama Desert and Central Chile ores), alkaliphile (*Exiguobacterium* sp. SH31 from Altiplano), halophile (*Shewanella* sp. Asc-3 from Altiplano, *Streptomyces* sp. HKF-8 from Patagonia), psychrophile (*Pseudomonas putida* ATH-43 from Antartica), thermophile (*Methanofollis tationis* from Tatio Geyser), UV- and Gamma-resistant bacteria (*Deinococcus peraridiltoris* from Atacama Desert), and xerotolerant bacteria (*Streptomyces atacamensis* from Atacama Desert) isolated from diverse regions of Chile have shown molecular and physiological properties of diverse extremophiles, and reported as an alternative source of catalyst for bioremediation processes (Orellana et al. 2018). The degradation of diverse hydrocarbon compounds and production of extracellular rhamnolipids (which are biosurfactants used for bioremediation and improved oil recovery) from *Halomonas* sp. KHS3 was reported by Corti Monzon et al. (2018). Recently, Donati et al. (2019) have reported the recent advancement of extremophiles in bioremediation.

3.8 *Application in Biomining*

Another major application of thermophiles is in mining sector, also called bioleaching, which involves the removal of insoluble metal oxides and sulphides by microorganisms or their purified enzymes which is safer and environment-friendly in nature. The process of biomining involves bioleaching (extraction of base metals), and biooxidation (pre-treatment of silver and gold bearing minerals). Both bioleaching and biooxidation utilize similar sulphur oxidizing and/or acidophilic microorgisms for solubilisation of metal containing sulphides (Gumulya et al. 2018). Different thermophilic (*Sulfolobus, Metallosphera*), and acidophilic (*Acidithiobacillus, Ferroplasma*) strains have been successfully employed in biomining techniques (Vera et al. 2013). The mesophilic and psychrophilic acidophiles strains may result in acid mine drainage (AMD) by generation of acidic water from oxidation of sulphides from mines; the most common sources of AMD are Cu, Zn, and Ni. The use of thermophiles is one of the possibilities for AMD reduction in a cost-effective manner (Coker et al. 2016). The acidophilic microorganisms involved in bioleaching for extraction of low-grade copper sulphide from Escondida mine (located 170 km south-east from Antofagasta, Chile) are *Acidithiobacillus ferroxidans, A. thioxidans, Leptospirillum ferriphilum*, and *Ferroplasma acidiphilum* (Demergasso et al. 2010, Acosta et al. 2017). The members of genus *Sulfobacillus* are involved in heap bleaching, which has been approved by prokaryotic acidophile microarray (PAM) studies (Remonsellez et al. 2009). Molecular studies conducted by 16S rRNA analysis, real time PCR, and metagenomics proved that various bacteria (*A. ferroxidans, A. thioxidans, A. caldus, A. ferrivorans, L. ferriphilum, Sulfobacillus thermosulfidooxidans, S. acidophilus, Acidiphilum* spp.), and archaea (*Ferroplasma acidiphilium, F. acidarmanus, Sulfolobus* spp.) are involved in heap leaching (Soto et al. 2013, Acosta et al. 2017). Halotolerant strains (able to tolerate up to 1.7 M NaCl concentration) were isolated from marine sediment samples of Comau Jjord, Northern Patagonia, Chile, and the isolated strains belong to phylum *Actinobacteria* (Undabarrena et al. 2016). Further, the genome sequencing of *Streptomyces* sp. K-KF8 showed the presence of 49 heavy metal resistance genes, and could tolerate different heavy metals including As (100 mM), Cd (1.5 mM), Co (6 mM), Cr (20 mM) Cu (0.75 mM), Hg (60μM), Ni (15 mM), Te (40 μM), and Zn (50 mM) (Undabarrena et al. 2017). There are various acidophilic and chemolithotrophic microbial strains (*Ferrimicrobium, Rhodococcus, Sulphobacillus*), which have been utilized to enhance copper and uranium bioleaching processes at industrial scale (Ijaz et al. 2017). Some genetic and metabolic engineering tools such as regulated gene expression (*tac* promoter, *cycA1* promoter, *tusA* promoter, *araS* promoter, *tetH* promoter) and protein expression (2-ketodecarboxylase, acyl-ACP reductase, aldehyde deformylating decarbonylase, ABCE1 protein), DNA delivery (electroporation, conjugation), shuttle vectors (pTMZ48, pKMZ51, pJRD215, pSDRA1, pEXS-series, pMJ03, pMSD2, pLAtc1), marker-less gene knockout (kanamycin mutated allele, insertion of *lacS* gene), and reporter genes (*gusA*, β-glucuronidase and *lacS*, β-galactosidase) have been used for altering the gene and protein expression of acidophiles (*Acidithiobacillus ferroxidans, A. caldus, Sulfolobus acidocaldarius, S. islandicus, S. solfataricus*) to enhance their biomining efficiency (Meng et al. 2013, Zhang et al. 2014, Gumulya et al. 2018).

3.9 Application in Chemical Synthesis

An extracellular, highly stable serine proteases has been reported from extreme halophilic bacterium *Halobacterium halobium*; this enzyme showed optimal activity in the presence of 4 M NaCl. The chemical synthesis process for industrial production of dipeptide aspartame (L-aspartyl-L-phenylalaninemethyl ester) uses thermophilic enzyme (Vielle 2001). The synthesis of organic compounds has been done by carbon-carbon bond forming enzymes such as transaldolases, transketolases and hydroxynitrile lyases, and transformation of nitriles has been achieved using nitrile-degrading enzymes (Resch et al. 2011). Extremozymes like γ-lacatamase, L-aminoacylase, pyroglutamyl carboxyl peptidase, carboxyl esterase, carbonic anhydrase, transaminase, hydrolase, and dehydrogenase are important in production of pharmacologically active compounds (Littlechild 2015). There are various thermophilic enzymes such as cytochrome P450, glycosyl hydrolase, hydantoinase, and secondary alcohol dehydrogenase with high degree of stereo- and region-selectivity which are highly used in synthetic chemistry (Pandey et al. 2016).

4. Conclusions and Future Prospects

Extremophiles are the microorganisms which can survive under adverse environmental conditions, and these microbes have gained attention in past few years due to their potential to produce valuable resources like extremozymes, polysaccharides, peptides, biopolymers, biosurfactants and extremolytes for developing bio-based economy. Extremozymes show optimal activity under extreme conditions with high degree of stability, and hence they are the source of novel enzymes having biotechnological potential in agriculture, biomining, bioremediation, detergents, food and beverages, leather, paper and pulp, pharmaceuticals, and textiles industries. Despite the current understanding of extremophiles to grow under different boundaries of life, there is lack of studies addressing their tolerance to multiple extremes. Hence, studies should be conducted to understand their tolerance to different habitats, and interaction factor between multiple habitats. Considering the emerging demand of extremozymes in biotechnological industries, extensive research is going on to provide bio-based economy. The actual number of extremozymes reported till date is limited to fulfil industrial need due to availability of non-cultivational microbial strains, and impracticability of conventional streaking methods for isolation of extremophiles. There is also a large technical gap between production of extremozymes at laboratory scale to commercialized product formation. Metagenome screening and genome mining are also conducted to overcome bottlenecks for exploring non-cultivable extremophilic microorganisms and their biocatalysts. Novel culturing technique like iChip can be used for identification of novel industrially important enzymes from uncultured bacteria. Some strategies like directed evolution through iterative process of mutagenesis, rational design by computational methods, and rational engineering approaches are used in enhancing conformational specificity, and enzyme stability to get industry quality results for novel enzymes. The rational design methods used for industrial applications of enzymes are rosetta, molecular d, structure-guided consensus analysis, flexible region stabilization, disulphide by design code, structure-guided consensus analysis, unfolding free energy, and B-factor analysis. Computational enzyme design lacked proper standardization, reliability of results, and clustered in small number of laboratories. It is expected that rational design will become a regular procedure in biochemical laboratories in coming years for computational enzyme design. Therefore, the development of novel microbial strains, efficient mass production of extremozymes, and genetic and protein engineering may help in advancement of extremozymes for different industrial processes. We strongly believe that vast research on exploration of novel extremophiles from diverse habitats, their exploitation for various metabolite synthesis using omics tools (genomics, proteomics, metabolomics), recombination studies and computational tools may give new directions to synthesis of novel, economic, efficient and environment- friendly industrial products.

Acknowledgement

Yogita Lugani acknowledges the Department of Science and Technology (DST), Ministry of Science and Technology, Government of India (SP/YO/385/2018) for the financial assistance.

References

Abomohra, A.E.F., M. El-Sheekh and D. Hanelt. 2017. Screening of marine microalgae isolated from the hypersaline Bardawil lagoon for biodiesel feedstock. Renew. Energy 101: 1266–1272. https://doi.org/10.1016/j. renene.2016.10.015.

Ahirwar, S., H. Soni, B.P. Prajapati and N. Kango. 2017. Isolation and screening of thermophilic and thermotolerant fungi for production of hemicellulases from heated environments. Mycology 8(3): 125–134. https://doi.org/10. 1080/21501203.2017.1337657.

Ai, L., Y. Huang and C. Wang. 2018. Purification and characterization of halophilic lipase of *Chromohalobacter* sp. From ancient salt well. J. Basic Microbiol. 58(8): 647–657. https://doi.org/10.1002/jobm.201800116.

Al-Bader, D., M.K. Kansour, R. Rayan and S.S. Radwan. 2013. Biofilm comprising phototrophic, diazotrophic, and hydrocarbon-utilizing bacteria: A promising consortium in the bioremediation of aquatic hydrocarbon pollutants. Environ. Sci. Pollut. Res. 20(5): 3252–3262. https://doi.org/10.1007/s11356-012-1251-z.

Alsafadi, D., F.I. Khalili, H. Juwhari and B. Lahlouh. 2018. Purification and biochemical characterization of photo-active membrane protein bacteriorhodopsin from *Haloarcula marismortui*, an extreme halophile from the Dead Sea. Int. J. Biol. Macromol. 118: 1942–1947. https://doi.org/10.1016/j.ijbiomac.2018.07.045.

Amils, R., C. Ellis-Evans and H. Hinghofer-Szalkay. 2007. Life in Extreme Environments. Springer. https://doi. org/10.1007/978-1-4020-6285-8.

Amoozegar, M.A., M. Siroosi, S. Atashgahi, H. Smidt and A. Ventosa. 2017. Systematics of haloarchaea and biotechnological potential of their hydrolytic enzymes. Microbiology 163(5): 623–645. https://doi.org/10.1099/ mic.0.000463.

Anupama, A. and G. Jayaraman. 2011. Detergent stable, halotolerant α-amylase from *Bacillus aquimaris* VITP4 exhibits revesible unfolding. Int. J. Appl. Biol. Pharm. Technol. 2: 366–376.

Asoodeh, A. and M. Lagzian. 2012. Purification and characterization of a new glucoamylopullulanase from thermotolerant alkaliphilic *Bacillus subtilis* DR8806 of a hot mineral spring. Process Biochem. 47(5): 806–815. https://doi.org/10.1016/j.procbio.2012.02.018.

Avanti, C., V. Saluja, E.L.P. Van Streun, H.W. Frijlink and W.L.J. Hinrichs. 2014. Stability of lysozyme in aqueous extremolyte solutions during heat shock and accelerated thermal conditions. PLoS ONE 9(1): 2–7. https://doi. org/10.1371/journal.pone.0086244.

Awazu, N., T. Shodai, H. Takakura, M. Kitagawa, H. Mukai and I. Kato. 2011. Microorganism-derived psychrophilic endonuclease. U.S. Patent # 8,034,597. https://patents.google.com/patent/US8034597.

Baker-Austin, C. and M. Dopson. 2007. Life in acid: pH homeostasis in acidophiles. Trends Microbiol. 15(4):165–171. https://doi.org/10.1016/j.tim.2007.02.005.

Barnard, D., A. Casanueva, M. Tuffin and D. Cowan. 2010. Extremophiles in biofuel synthesis. Environ. Technol. 31(8-9): 871–888. https://doi.org/10.1080/09593331003710236.

Barns, S.M., C.F. Delwichet, J.D. Palmert and N.R. Pace. 1996. Perspectives on archaeal diversity, thermophily and monophyly from environmental rRNA sequences. Proc. Natl. Acad. Sci. U.S.A. 93(17): 9188–9193. http://www. bio.indiana.edu/-nrpace/pacelab/pub-.

Blum, J.S., S. Han, B. Lanoil, C. Saltikov, B. Witte, F.R. Tabita, S. Langley, T.J. Beveridge, L. Jahnke and R.S. Oremland. 2009. Ecophysiology of "*Halarsenatibacter silvermanii*" strain SLAS-1 T, gen. nov., sp. nov., a facultative chemoautotrophic arsenate respirer from salt-saturated Searles Lake, California. Appl. Environ. Microbiol. 75(7): 1950–1960. https://doi.org/10.1128/AEM.02614-08.

Bosma, E.F., A.H.P. van de Weijer, M.J.A. Daas, J. van der Oost, W.M. de Vos and R. van Kranenburg. 2015. Isolation and screening of *Thermophilic bacilli* from compost for electrotransformation and fermentation: Characterization of *Bacillus smithii* ET 138 as a new biocatalyst. Appl. Environ. Microbiol. 81(5): 1874–1883. https://doi.org/10.1128/AEM.03640-14.

Bowers, K.J. and J. Wiegel. 2011. Temperature and pH optima of extremely halophilic archaea: A mini-review. Extremophiles 15(2): 119–128. https://doi.org/10.1007/s00792-010-0347-y.

Byrne, R.T., A.J. Klingele, E.L. Cabot, W.S. Schackwitz, J.A. Martin, J. Martin, Z. Wang, E.A. Wood, C. Pennacchio, L.A. Pennacchio and N.T. Perna. 2014. Evolution of extreme resistance to ionizing radiation via genetic adaptation of DNA repair. eLife2014(3): 1–18. https://doi.org/10.7554/eLife.01322.

Campos, V.L., G. Escalante, J. Yañez, C.A. Zaror and M.A. Mondaca. 2009. Isolation of arsenite-oxidizing bacteria from a natural biofilm associated to volcanic rocks of Atacama Desert, Chile. J. Basic Microbiol. 49(1): 93–97. https://doi.org/10.1002/jobm.200900028.

Cavicchioli, R., T. Charlton, H. Ertan, S.M. Omar, K.S. Siddiqui and T.J. Williams. 2011a. Biotechnological uses of enzymes from psychrophiles. Microb. Biotechnol. 4(4): 449–460. https://doi.org/10.1111/j.1751-7915.2011.00258.x.

Cavicchioli, R., R. Amils, D. Wagner and T. Mcgenity. 2011b. Life and applications of extremophiles. Environ. Microbiol. 13(8): 1903–1907. https://doi.org/10.1111/j.1462-2920.2011.02512.x.

Cencic, A. and W. Chingwaru. 2010. The role of functional foods, nutraceuticals, and food supplements in intestinal health. Nutrients 2(6): 611–625. https://doi.org/10.3390/nu2060611.

Charlesworth, J. and B.P. Burns. 2016. Extremophilic adaptations and biotechnological applications in diverse environments. AIMS Microbiol. 2(3): 251–261. https://doi.org/10.3934/microbiol.2016.3.251.

Chisti, Y. 2008. Biodiesel from microalgae beats bioethanol. Trends Biotechnol. 26(3): 126–131. https://doi.org/10.1016/j.tibtech.2007.12.002.

Choi, Y.J., J.M. Hur, S. Lim, M. Jo, D.H. Kim and J. Choi. 2014. Induction of apoptosis by deinoxanthin in human cancer cells. Anticancer Res. 34(4): 1829–1836.

Choudhary, S. and C. Schmidt-Dannert. 2010. Applications of quorum sensing in biotechnology. Appl. Microbiol. Biotechnol. 86(5): 1267–1279. https://doi.org/10.1007/s00253-010-2521-7.

Coker, J.A. 2016. Extremophiles and biotechnology: Current uses and prospects. F1000Res. 5: 1–7. https://doi.org/10.12688/f1000research.7432.1.

Colman, D.R., S. Poudel, T.L. Hamilton, J.R. Havig, M.J. Selensky, E.L. Shock and E.S. Boyd. 2018. Geobiological feedbacks and the evolution of thermoacidophiles. ISME J. 12(1): 225–236. https://doi.org/10.1038/ismej.2017.162.

Correa-Llantén, D.N., S.A. Muñoz-Ibacache, M.E. Castro, P.A. Muñoz and J.M. Blamey. 2013. Gold nanoparticles synthesized by *Geobacillus* sp. strain ID17 a thermophilic bacterium isolated from Deception Island, Antarctica. Microb. Cell Fact. 12(1): 2–7. https://doi.org/10.1186/1475-2859-12-75.

Corti Monzón, G., M. Nisenbaum, M.K. Herrera Seitz and S.E. Murialdo. 2018. New findings on aromatic compounds' degradation and their metabolic pathways, the biosurfactant production and motility of the *Halophilic Bacterium Halomonas* sp. KHS3. Curr. Microbiol. 75(8): 1108–1118. https://doi.org/10.1007/s00284-018-1497-x.

Czech, L., L. Hermann, N. Stöveken, A.A. Richter, A. Höppner, S.H. Smits, J. Heider and E. Bremer. 2018. Role of the extremolytes ectoine and hydroxyectoine as stress protectants and nutrients: Genetics, phylogenomics, biochemistry, and structural analysis. Genes 9(4): 1–58. https://doi.org/10.3390/genes9040177.

D'Amico, S., T. Collins, J.C. Marx, G. Feller and C. Gerday. 2006. Psychrophilic microorganisms: Challenges for life. EMBO Reports 7(4): 385–389. https://doi.org/10.1038/sj.embor.7400662.

Dalmasso, C., P. Oger, G. Selva, D. Courtine, S. L'haridon, A. Garlaschelli, E. Roussel, J. Miyazaki, J. Reveillaud, M. Jebbar and K. Takai. 2016. *Thermococcus piezophilus* sp. nov., a novel hyperthermophilic and piezophilic archaeon with a broad pressure range for growth, isolated from a deepest hydrothermal vent at the Mid-Cayman Rise. Syst. Appl. Microbiol. 39(7): 440–444. https://doi.org/10.1016/j.syapm.2016.08.003.

Datta, S., B. Holmes, J.I. Park, Z. Chen, D.C. Dibble, M. Hadi, H.W. Blanch, B.A. Simmons and R. Sapra. 2010. Ionic liquid tolerant hyperthermophilic cellulases for biomass pretreatment and hydrolysis. Green Chem. 12(2): 338–345. https://doi.org/10.1039/b916564a.

David, A., T. Govil, A.K. Tripathi, J. McGeary, K. Farrar and R.K. Sani. 2018. Thermophilic anaerobic digestion: Enhanced and sustainable methane production from co-digestion of food and lignocellulosic wastes. Energies 11(8): 1–13. https://doi.org/10.3390/en11082058.

De Maayer, P., D. Anderson, C. Cary and D.A. Cowan. 2014. Some like it cold: Understanding the survival strategies of psychrophiles. EMBO Rep. 15(5): 508–517. https://doi.org/10.1002/embr.201338170.

Del-Cid, A., P. Ubilla, M.C. Ravanal, E. Medina, I. Vaca, G. Levicán, J. Eyzaguirre and R. Chávez. 2014. Cold-active xylanase produced by fungi associated with Antarctic marine sponges. Appl. Biochem. Biotechnol. 172(1): 524–532. https://doi.org/10.1007/s12010-013-0551-1.

Demergasso, C., F. Galleguillos, P. Soto, M. Serón and V. Iturriaga. 2010. Microbial succession during a heap bioleaching cycle of low grade copper sulfides: Does this knowledge mean a real input for industrial process design and control? Hydrometallurgy 104(3-4): 382–390. https://doi.org/10.1016/j.hydromet.2010.04.016.

Dennett, G.V. and J.M. Blamey. 2016. A new thermophilic nitrilase from an antarctic hyperthermophilic microorganism. Front. Bioeng. Biotechnol. 4(5): 1–9. https://doi.org/10.3389/fbioe.2016.00005.

Depeursinge, A., D. Racoceanu, J. Iavindrasana, G. Cohen, A. Platon, P.A. Poletti and H. Müller. 2010. Fusing visual and clinical information for lung tissue classification in HRCT Data. Artif. Intell. Med. 50(1): 13–21. https://doi.org/10.1016/j.

Donati, E.R., R.K. Sani, K.M. Goh and K.G. Chan. 2019. Editorial: Recent advances in bioremediation/biodegradation by extreme microorganisms. Front. Microbiol. 10: 1–2. https://doi.org/10.3389/fmicb.2019.01851.

Dopson, M. and D.S. Holmes. 2014. Metal resistance in acidophilic microorganisms and its significance for biotechnologies. Appl. Microbiol. Biotechnol. 98(19): 8133–8144. https://doi.org/10.1007/s00253-014-5982-2.

Dornez, E., P. Verjans, F. Arnaut, J.A. Delcour and C.M. Courtin. 2011. Use of psychrophilic xylanases provides insight into the xylanase functionality in bread making. J. Agric. Food Chem. 59(17): 9553–9562. https://doi.org/10.1021/jf201752g.

Dumorné, K., D.C. Córdova, M. Astorga-Eló and P. Renganathan. 2017. Extremozymes: A potential source for industrial applications. J. Microbiol. Biotechnol. 27(4): 649–659. https://doi.org/10.4014/jmb.1611.11006.

Dumorné, K. 2018. Biotechnological and industrial applications of enzymes produced by extremophilic bacteria. A Mini Review. Preprints 2018 2018010198. https://doi.org/10.20944/preprints201801.0198.v1.

Fakruddin, M. 2017. Thermostable enzymes and their industrial application: A review. Discovery Publ. 53(254): 147–152.

Fröls, S. 2013. Archaeal biofilms: Widespread and complex. Biochem. Soc. Trans. 41(1): 393–398. https://doi.org/10.1042/BST20120304.

Fujiwara, S. 2002. Extremophiles: Developments of their special functions and potential resources. J. Biosci. Bioeng. 94(6): 518–525. https://doi.org/10.1016/S1389-1723(02)80189-X.

Gabani, P. and O.V. Singh. 2013. Radiation-resistant extremophiles and their potential in biotechnology and therapeutics. Appl. Microbiol. Biotechnol. 97(3): 993–1004. https://doi.org/10.1007/s00253-012-4642-7.

García, M.T., E. Mellado, J.C. Ostos and A. Ventosa. 2004. *Halomonas organivorans* sp. nov., a moderate halophile able to degrade aromatic compounds. Int. J. Syst. Evol. Microbiol. 54(5): 1723–1728. https://doi.org/10.1099/ijs.0.63114-0.

Georlette, D., Z.O. Jonsson, F. Van Petegem, J.P. Chessa, J. Van Beeumen, U. Hübscher and C. Gerday. 2000. A DNA ligase from the psychrophile *Pseudoalteromonas haloplanktis* gives insights into the adaptation of proteins to low temperatures. Europ. J. Biochem. 267(12): 3502–3512. https://doi.org/10.1046/j.1432-1327.2000.01377.x.

Ghosh, M., K.K. Pulicherla, V.P.B. Rekha, P.K. Raja and K.R.S. Sambasiva Rao. 2012. Cold active β-galactosidase from *Thalassospira* sp. 3SC-21 to use in milk lactose hydrolysis: A novel source from deep waters of Bay-of-Bengal. World J. Microbiol. Biotechnol. 28(9): 2859–2869. https://doi.org/10.1007/s11274-012-1097-z.

Golyshina, O.V. and K.N. Timmis. 2005. Ferroplasma and relatives, recently discovered cell wall-lacking archaea making a living in extremely acid, heavy metal-rich environments. Environ. Microbiol. 7(9): 1277–1288. https://doi.org/10.1111/j.1462-2920.2005.00861.x.

Gomes, J., I. Gomes, K. Terler, N. Gubala, G. Ditzelmüller and W. Steiner. 2000. Optimisation of culture medium and conditions for α-L-arabinofuranosidase production by the extreme thermophilic eubacterium *Rhodothermus marinus*. Enzyme Microb. Technol. 27(6): 414–422. https://doi.org/10.1016/S0141-0229(00)00229-5.

Gomes, J. and W. Steiner. 2004. The biocatalytic potential of extremophiles and extremozymes. Food Technol. Biotechnol. 42(4): 223–235.

Grant, W.D. 2004. Life at low water activity. Philos. Trans. R. Soc. Lond. B Biol. Sci. 359(1448): 1249–1267. https://doi.org/10.1098/rstb.2004.1502.

Green, S.J., O. Prakash, P. Jasrotia, W.A. Overholt, E. Cardenas, D. Hubbard, J.M Tiedje, D.B. Watson, C.W. Schadt, S.C. Brooks and J.E. Kostka. 2012. Denitrifying bacteria from the genus *Rhodanobacter* dominate bacterial communities in the highly contaminated subsurface of a nuclear legacy waste site. Appl. Environ. Microbiol. 78(4): 1039–1047. https://doi.org/10.1128/AEM.06435-11.

Guerra, M., K. González, C. González, B. Parra and M. Martínez. 2015. Dormancy in *Deinococcus* sp. UDEC-P1 as a survival strategy to escape from deleterious effects of carbon starvation and temperature. Int. Microbiol. 18(3): 189–194. https://doi.org/10.2436/20.1501.01.249.

Gumulya, Y., N.J. Boxall, H.N. Khaleque, V. Santala, R.P. Carlson and A.H. Kaksonen. 2018. In a quest for engineering acidophiles for biomining applications: Challenges and opportunities. Genes 9(2): 116. https://doi.org/10.3390/genes9020116.

Hajela, P., M. Patel and A. Soni. 2012. Exploring the biotechnological applications of halophilic archaea. J. Pure Appl. Microbiol. 6(3): 1185–1197.

Hill, J., E. Nelson, D. Tilman, S. Polasky and D. Tiffany. 2006. Environmental, economic, and energetic costs and benefits of biodiesel and ethanol biofuels. Proc. Natl. Acad. Sci. U.S.A. 103(30): 11206–11210. https://doi.org/10.1073/pnas.0604600103.

Horner, T.W., M.L. Dunn, D.L. Eggett and L.V. Ogden. 2011. β-Galactosidase activity of commercial lactase samples in raw and pasteurized milk at refrigerated temperatures. J. Dairy Sci. 94(7): 3242–3249. https://doi.org/10.3168/jds.2010-3742.

Ijaz, K., J.I. Wattoo, B. Zeshan, T. Majeed, T. Riaz, S. Khalid and S. Baig. 2017. Potential impact of microbial consortia in biomining and bioleaching of comnercial metals. Adv. Life Sci. 5(1): 13–18. https://doi.org/2310-5380.

Irwin, J.A. and A.W. Baird. 2004. Extremophiles and their application to veterinary medicine. Irish Veter. J. 57(6): 348–354. https://doi.org/10.1186/2046-0481-57-6-348.

Jackson, B.R., C. Noble, M. Lavesa-Curto, P.L. Bond and R.P. Bowater. 2007. Characterization of an ATP-dependent DNA ligase from the acidophilic archaeon *"Ferroplasma acidarmanus"* Fer1. Extremophiles 11(2): 315–327. https://doi.org/10.1007/s00792-006-0041-2.

Jebbar, M., B. Franzetti, E. Girard and P. Oger. 2015. Microbial diversity and adaptation to high hydrostatic pressure in deep-sea hydrothermal vents prokaryotes. Extremophiles 19(4): 721–740. https://doi.org/10.1007/s00792-015-0760-3.

Jiang, Z.Q., X.T. Li, S.Q. Yang, L.T. Li, Y. Li and W.Y. Feng. 2006. Biobleach boosting effect of recombinant xylanase B from the hyperthermophilic Thermotoga maritima on wheat straw pulp. Appl. Microbiol. Biotechnol. 70(1): 65–71. https://doi.org/10.1007/s00253-005-0036-4.

Jolivet, E., E. Corre, S. L'Haridon, P. Forterre and D. Prieur. 2004. *Thermococcus marinus* sp. nov. and *Thermococcus radiotolerans* sp. nov., two hyperthermophilic archaea from deep-sea hydrothermal vents that resist ionizing radiation. Extremophiles 8(3): 219–227. https://doi.org/10.1007/s00792-004-0380-9.

Joshi, S. and T. Satyanarayana. 2013. Biotechnology of cold-active proteases. Biology 2(2): 755–783. https://doi.org/10.3390/biology2020755.

Kotresh, K.R., S. Neelagund and D.M. Gurumurthy. 2019. Novel *Geobacillus thermoleovorans* KNG 112 *Thermophilic Bacteria* from bandaru hot spring: A potential producer of thermostable enzymes. Asian J. Pharm. Clin. Res. 13(1): 134–141. https://doi.org/10.22159/ajpcr.2020.v13i1.36008.

Kalita, D. and S.R. Joshi. 2017. Study on bioremediation of Lead by exopolysaccharide producing metallophilic bacterium isolated from extreme habitat. Biotechnol. Rep. 16: 48–57. https://doi.org/10.1016/j.btre.2017.11.003.

Kaur, A., N. Capalash and P. Sharma. 2019. Communication mechanisms in extremophiles: Exploring their existence and industrial applications. Microbiol. Res. 221: 15–27. https://doi.org/10.1016/j.micres.2019.01.003.

Khamson, A., P. Sumpavapol, P. Tangwatcharin and S. Sorapukdee. 2019. Optimization of microbial collagenolytic enzyme production by *Bacillus subtilis* subsp. subtilis S13 using Plackett-Burman and response surface methodology. Int. J. Agric. Technol. 15(6): 913–924.

Klein, J., T. Schwarz and G. Lentzen. 2007. Ectoine as a natural component of food: Detection in red smear cheeses. J. Dairy Res. 74(4): 446–451. https://doi.org/10.1017/S0022029907002774.

Kobori, H., C.W. Sullivan and H. Shizuya. 1984. Heat-labile alkaline phosphatase from Antarctic bacteria: Rapid 5' end-labeling of nucleic acids. Proc. Natl. Acad. Sci. U.S.A. 81(21): 6691–6695. https://doi.org/10.1073/pnas.81.21.6691.

Koch, R., P. Zablowski, A. Spreinat and G. Antranikian. 1990. Extremely thermostable amylolytic enzyme from the archaebacterium *Pyrococcus furiosus*. FEMS Microbiol. Lett. 71(1–2): 21–26. https://doi.org/10.1016/0378-1097(90)90026-M.

Krisko, A. and M. Radman. 2013. Biology of extreme radiation resistance: The way of *Deinococcus radiodurans*. Cold Spring Harb. Perspect. Biol. 5(7): 1–11. https://doi.org/10.1101/cshperspect.a012765.

Krzmarzick, M.J., D.K. Taylor, X. Fu and A.L. McCutchan. 2018. Diversity and niche of archaea in bioremediation. Archaea 2018: 1–17. https://doi.org/10.1155/2018/3194108.

Kuddus, M. 2013. Cold-active detergent-stable extracellular α-amylase from *Bacillus cereus* GA6: Biochemical characteristics and its perspectives in laundry detergent formulation. J. Biochem. Technol. 4(4): 636–644.

Kumar, S. and R. Nussinov. 2001. How do thermophilic proteins deal with heat? Cell. Mol. Life Sci. 58(9): 1216–1233. https://doi.org/10.1007/PL00000935.

Kunte, H., G. Lentzen and E. Galinski. 2014. Industrial production of the cell protectant ectoine: Protection mechanisms, processes, and products. Curr. Biotechnol. 3(1): 10–25. https://doi.org/10.2174/22115501113026660037.

Lanes, O., I. Leiros, A.O. Smalås and N.P. Willassen. 2002. Identification, cloning, and expression of uracil-DNA glycosylase from Atlantic cod (*Gadus morhua*): Characterization and homology modeling of the cold-active catalytic domain. Extremophiles 6(1): 73–86. https://doi.org/10.1007/s007920100225.

Lee, D.H., S.L. Choi, E. Rha, S.J. Kim, S.J. Yeom, J.H. Moon and S.G. Lee. 2015. A novel psychrophilic alkaline phosphatase from the metagenome of tidal flat sediments. BMC Biotechnol. 15(1): 1–13. https://doi.org/10.1186/s12896-015-0115-2.

Lévêque, E., S. Janeček, B. Haye and A. Belarbi. 2000. Thermophilic archaeal amylolytic enzymes. Enzyme Microb. Technol. 26(1): 3–14. https://doi.org/10.1016/S0141-0229(99)00142-8.

Liszka, M.J., M.E. Clark, E. Schneider and D.S. Clark. 2012. Nature versus nurture: Developing enzymes that function under extreme conditions. Annu. Rev. Chem. Biomol. Eng. 3(1): 77–102. https://doi.org/10.1146/annurev-chembioeng-061010-114239.

Littlechild, J.A. 2015. Enzymes from extreme environments and their industrial applications. Front. Bioeng. Biotechnol. 3: 1–9. https://doi.org/10.3389/fbioe.2015.00161.

Liu, L., N. Salam, J.Y. Jiao, H.C. Jiang, E.M. Zhou, Y.R. Yin, H. Ming and W.J. Li. 2016. Diversity of culturable thermophilic actinobacteria in hot springs in Tengchong, China and studies of their biosynthetic gene profiles. Microb. Ecol. 72(1): 150–162. https://doi.org/10.1007/s00248-016-0756-2.

Ma, Y., Y. Xue, W.D. Grant, N.C. Collins, A.W. Duckworth, R.P. Van Steenbergen and B.E. Jones. 2004. *Alkalimonas amylolytica* gen. nov., sp. nov., and *Alkalimonas delamerensis* gen. nov., sp. nov., novel alkaliphilic bacteria from soda lakes in China and East Africa. Extremophiles 8(3): 193–200. https://doi.org/10.1007/s00792-004-0377-4.

Madern, D., C. Ebel and G. Zaccai. 2000. Halophilic adaptation of enzymes. Extremophiles 4(2): 91–98. https://doi.org/10.1007/s007920050142.

Marasco, R., E. Rolli, B. Ettoumi, G. Vigani, F. Mapelli, S. Borin, A.F. Abou-Hadid, U.A. El-Behairy, C. Sorlini, A. Cherif and G. Zocchi. 2012. A drought resistance-promoting microbiome is selected by root system under desert farming. PLoS One7(10): 1–14. https://doi.org/10.1371/journal.pone.0048479.

Marhuenda-Egea, F.C., S. Piera-Velázquez, C. Cadenas and E. Cadenas. 2002. An extreme halophilic enzyme active at low salt in reversed micelles. J. Biotechnol. 93(2): 159–164. https://doi.org/10.1016/S0168-1656(01)00392-3.

Martinez, N., G. Michoud, A. Cario, J. Ollivier, B. Franzetti, M. Jebbar, P. Oger and J. Peters. 2016. High protein flexibility and reduced hydration water dynamics are key pressure adaptive strategies in prokaryotes. Sci. Rep. 6: 1–11. https://doi.org/10.1038/srep32816.

McKeown, R.M., C. Scully, A.M. Enright, F.A. Chinalia, C. Lee, T. Mahony, G. Collins and V. O'flaherty. 2009. Psychrophilic methanogenic community development during long-term cultivation of anaerobic granular biofilms. ISME J. 3(11): 1231–1242. https://doi.org/10.1038/ismej.2009.67.

Meng, J., H. Wang, X. Liu, J. Lin, X. Pang and J. Lin. 2013. Construction of small plasmid vectors for use in genetic improvement of the extremely acidophilic *Acidithiobacillus caldus*. Microbiol. Res. 168(8): 469–476. https://doi.org/10.1016/j.micres.2013.04.003.

Merino, N., H.S. Aronson, D.P. Bojanova, J. Feyhl-Buska, M.L. Wong, S. Zhang and D. Giovannelli. 2019. Living at the extremes: Extremophiles and the limits of life in a planetary context. Front. Microbiol. 10(780): 1–25. https://doi.org/10.3389/fmicb.2019.00780.

Mesbah, N.M. and J. Wiegel. 2017. A Halophilic, Alkalithermostable, ionic liquid-tolerant cellulase and its application in *in situ* saccharification of rice straw. Bioenergy Res. 10(2): 583–591. https://doi.org/10.1007/s12155-017-9825-8.

Michetti, D., B.O. Brandsdal, D. Bon, G.V. Isaksen, M. Tiberti and E. Papaleo. 2017. A comparative study of cold-and warmadapted Endonucleases a using sequence analyses and molecular dynamics simulations. PLoS One 12(2): 1–18. https://doi.org/10.1371/journal.pone.0169586.

Mijts, B.N. and B.K.C. Patel. 2002. Cloning, sequencing and expression of an α-amylase gene, *amyA*, from the thermophilic halophile *Halothermothrix orenii* and purification and biochemical characterization of the recombinant enzyme. Microbiology 148(8): 2343–2349. https://doi.org/10.1099/00221287-148-8-2343.

Mnif, S., M. Chamkha, M. Labat and S. Sayadi. 2011. Simultaneous hydrocarbon biodegradation and biosurfactant production by oilfield-selected bacteria. J. Appl. Microbiol. 111(3): 525–536. https://doi.org/10.1111/j.1365-2672.2011.05071.x.

Mykytczuk, N.C.S., S.J. Foote, C.R. Omelon, G. Southam, C.W. Greer and L.G. Whyte. 2013. Bacterial growth at −15°C; molecular insights from the permafrost bacterium *Planococcus halocryophilus* Or1. ISME J. 7(6): 1211–1226. https://doi.org/10.1038/ismej.2013.8.

Nakagawa, T., T. Nagaoka, S. Taniguchi, T. Miyaji and N. Tomizuka. 2004. Isolation and characterization of psychrophilic yeasts producing cold-adapted pectinolytic enzymes. Lett. Appl. Microbiol. 38(5): 383–387. https://doi.org/10.1111/j.1472-765X.2004.01503.x.

Navarro-González, R., E. Iñiguez, J. De La Rosa and C.P. McKay. 2009. Characterization of organics, microorganisms, desert soils, and mars-like soils by thermal volatilization coupled to mass spectrometry and their implications for the search for organics on mars by phoenix and future space missions. Astrobiology 9(8): 703–715. https://doi.org/10.1089/ast.2008.0284.

Navarro, C.A., D. Bernath and C.A. Jerez. 2013. Heavy metal resistance strategies of acidophilic bacteria and their acquisition: Importance for biomining and bioremediation. Biol. Res. 46(4): 363–371. http://doi.org/10.4067/S0716-97602013000400008.

Niehaus, F., C. Bertoldo, M. Kähler and G. Antranikian. 1999. Extremophiles as a source of novel enzymes for industrial application. Appl. Microbiol. Biotechnol. 51(6): 711–729. https://doi.org/10.1007/s002530051456.

Orellana, R., C. Macaya, G. Bravo, F. Dorochesi, A. Cumsille, R. Valencia, C. Rojas and M. Seeger. 2018. Living at the frontiers of life: extremophiles in chile and their potential for bioremediation. Front. Microbiol. 9: 1–25. https://doi.org/10.3389/fmicb.2018.02309.

Oren, A. 2010. Industrial and environmental applications of halophilic microorganisms. Environ. Technol. 31(8-9): 825–834. https://doi.org/10.1080/09593330903370026.

Paiardini, A., G. Gianese, F. Bossa and S. Pascarella. 2003. Structural plasticity of thermophilic serine hydroxymethyltransferases. Proteins: Struct. Funct. Genet. 50(1): 122–134. https://doi.org/10.1002/prot.10268.

Pandey, R.K., A. Barh, D. Chandra, S. Chandra, V. Pandey Pankaj and L. Tewari. 2016. Biotechnological applications of hyperthermophilic enzymes. Int. J. Curr. Res. Acad. Rev. 5(3): 39–47. https://doi.org/10.20546/ijcrar.2016.403.005.

Parihar, J. and A. Bagaria. 2019. The extremes of life and extremozymes: Diversity and perspectives. Acta Sci. Microbiol. 3(1): 107–119. https://doi.org/10.31080/asmi.2020.03.the-extremes-of-life-and-extremozymes-diversity-and-perspectives.

Park, J.T., H.N. Song, T.Y. Jung, M.H. Lee, S.G. Park, E.J. Woo and K.H. Park. 2013. A novel domain arrangement in a monomeric cyclodextrin-hydrolyzing enzyme from the hyperthermophile *Pyrococcus furiosus*. BBA- Proteins Proteom. 1834(1): 380–386. https://doi.org/10.1016/j.bbapap.2012.08.001.

Pettit, R.K. 2011. Culturability and secondary metabolite diversity of extreme microbes: Expanding contribution of deep sea and deep-sea vent microbes to natural product discovery. Mar. Biotechnol. 13(1): 1–11. https://doi.org/10.1007/s10126-010-9294-y.

Pikuta, E.V., R.B. Hoover and J. Tang. 2007. Microbial extremophiles at the limits of life. Crit. Rev. Microbiol. 33(3): 183–209. https://doi.org/10.1080/10408410701451948.

Preiss, L., D.B. Hicks, S. Suzuki, T. Meier and T.A. Krulwich. 2015. Alkaliphilic bacteria with impact on industrial applications, concepts of early life forms, and bioenergetics of ATP synthesis. Front. Bioeng. Biotechnol. 3: 1–16. https://doi.org/10.3389/fbioe.2015.00075.

Pulicherla, K.K., M. Ghosh, P.S. Kumar and K.R.S. Rao. 2011. Psychrozymes—The next generation industrial enzymes. J. Marine Sci. Res. Development 1(1): 1–7. https://doi.org/10.4172/2155-9910.1000102.

Purwasena, I.A., Y. Sugai and K. Sasaki. 2014. Estimation of the potential of an anaerobic thermophilic oil-degrading bacterium as a candidate for MEOR. J. Petrol. Explor. Prod. Technol. 4(2): 189–200. https://doi.org/10.1007/s13202-013-0095-5.

Qin, Y., Z. Huang and Z. Liu. 2014. A novel cold-active and salt-tolerant α-amylase from marine bacterium *Zunongwangia profunda*: Molecular cloning, heterologous expression and biochemical characterization. Extremophiles 18(2): 271–281. https://doi.org/10.1007/s00792-013-0614-9.

Quillaguamán, J., S. Hashim, F. Bento, B. Mattiasson and R. Hatti-Kaul. 2005. Poly(β-hydroxybutyrate) production by a moderate halophile, Halomonas boliviensis LC1 using starch hydrolysate as substrate. J. Appl. Microbiol. 99(1): 151–157. https://doi.org/10.1111/j.1365-2672.2005.02589.x.

Raddadi, N., A. Cherif, D. Daffonchio, M. Neifar and F. Fava. 2015. Biotechnological applications of extremophiles, extremozymes and extremolytes. Appl. Microbiol. Biotechnol. 99(19): 7907–7913. https://doi.org/10.1007/s00253-015-6874-9.

Rasoul-Amini, S., P. Mousavi, N. Montazeri-Najafabady, M.A. Mobasher, S.B. Mousavi, F. Vosough, F. Dabbagh and Y. Ghasemi. 2014. Biodiesel properties of native strain of *Dunaliella Salina*. Int. J. Renew. Energy Res. 4(1): 39–41. https://doi.org/10.20508/ijrer.36803.

Rastogi, G., G.L. Muppidi, R.N. Gurram, A. Adhikari, K.M. Bischoff, S.R. Hughes, W.A. Apel, S.S. Bang, D.J. Dixon and R.K. Sani. 2009. Isolation and characterization of cellulose-degrading bacteria from the deep subsurface of the Homestake gold mine, Lead, South Dakota, USA. J. Ind. Microbiol. Biotechnol. 36(4): 585–598. https://doi.org/10.1007/s10295-009-0528-9.

Reed, C.J., H. Lewis, E. Trejo, V. Winston and C. Evilia. 2013. Protein adaptations in archaeal extremophiles. Archaea 2013: 1–14. https://doi.org/10.1155/2013/373275.

Remonsellez, F., F. Galleguillos, M. Moreno-Paz, V. Parro, M. Acosta and C. Demergasso. 2009. Dynamic of active microorganisms inhabiting a bioleaching industrial heap of low-grade copper sulfide ore monitored by real-time PCR and oligonucleotide prokaryotic acidophile microarray. Microb. Biotechnol. 2(6): 613–624. https://doi.org/10.1111/j.1751-7915.2009.00112.x.

Resch, V., J.H. Schrittwieser, E. Siirola and W. Kroutil. 2011. Novel carbon-carbon bond formations for biocatalysis. Curr. Opin. Biotechnol. 22(6): 793–799. https://doi.org/10.1016/j.copbio.2011.02.002.

Rifaat, H.M., Z.A. Nagieb and Y.M. Ahmed. 2005. Production of Xylanases by *Streptomyces* species and their bleaching effect on rice straw. Appl. Ecol. Environ. Res. 4(1): 151–160. http://www.ecology.kee.hu.

Rigoldi, F., S. Donini, A. Redaelli, E. Parisini and A. Gautieri. 2018. Review: Engineering of thermostable enzymes for industrial applications. APL Bioeng. 2(1): 011501. https://doi.org/10.1063/1.4997367.

Rodríguez-Rojas, F., W. Díaz-Vásquez, A. Undabarrena, P. Muñoz-Díaz, F. Arenas and C. Vásquez. 2016. Mercury-mediated cross-resistance to tellurite in *Pseudomonas* spp. isolated from the Chilean Antarctic territory. Metallomics 8(1): 108–117. https://doi.org/10.1039/c5mt00256g.

Rolland, J.L., Y. Gueguen, C. Persillon, J.M. Masson and J. Dietrich. 2004. Characterization of a thermophilic DNA ligase from the archaeon *Thermococcus fumicolans*. FEMS Microbiol. Lett. 236(2): 267–273. https://doi.org/10.1016/j.femsle.2004.05.045.

Rolli, E., R. Marasco, G. Vigani, B. Ettoumi, F. Mapelli, M.L. Deangelis, C. Gandolfi, E. Casati, F. Previtali, R. Gerbino and F. Pierotti Cei. 2015. Improved plant resistance to drought is promoted by the root-associated microbiome as a water stress-dependent trait. Environ. Microbiol. 17(2): 316–331. https://doi.org/10.1111/1462-2920.12439.

Ruiz, D.M. and R.E. De Castro. 2007. Effect of organic solvents on the activity and stability of an extracellular protease secreted by the haloalkaliphilic archaeon *Natrialba magadii*. J. Ind. Microbiol. Biotechnol. 34(2): 111–115. https://doi.org/10.1007/s10295-006-0174-4.

Sarethy, I.P., Y. Saxena, A. Kapoor, M. Sharma, S.K. Sharma, V. Gupta and S. Gupta. 2011. Alkaliphilic bacteria: Applications in industrial biotechnology. J. Ind. Microbiol. Biotechnol. 38(7): 769–790. https://doi.org/10.1007/s10295-011-0968-x.

Sarmiento, F., R. Peralta and J.M. Blamey. 2015. Cold and hot extremozymes: Industrial relevance and current trends. Front. Bioeng. Biotechnol. 3: 148. https://doi.org/10.3389/fbioe.2015.00148.

Satyanarayana, T., S.M. Noorwez, S. Kumar, J.L.U.M. Rao, M. Ezhilvannan and P. Kaur. 2004. Development of an ideal starch saccharification process using amylolytic enzymes from thermophiles. Biochem. Soc. Trans. 32(2): 276–278. https://doi.org/10.1042/BST0320276.

Scambelluri, M., G. Pennacchioni, M. Gilio, M. Bestmann, O. Plümper and F. Nestola. 2017. Fossil intermediate-depth earthquakes in subducting slabs linked to differential stress release. Nature Geosci. 10(12): 960–966. https://doi.org/10.1038/s41561-017-0010-7.

Schiraldi, C., M. Giuliano and M. De Rosa. 2002. Perspectives on biotechnological applications of archaea. Archaea 1(2): 75–86. https://doi.org/10.1155/2002/436561.

Schleper, C., G. Puhler, H.P. Klenk and W. Zillig. 1996. *Picrophilus oshimae* and *Picrophilus torridus* fam. nov., gen. nov., sp. nov., two species of hyperacidophilic, thermophilic, heterotrophic, aerobic archaea. Int. J. Syst. Bacteriol. 46: 814–816.

Schmid, A.K., D.J. Reiss, M. Pan, T. Koide and N.S. Baliga. 2009. A single transcription factor regulates evolutionarily diverse but functionally linked metabolic pathways in response to nutrient availability. Mol. Syst. Biol. 5: 1–15. https://doi.org/10.1038/msb.2009.40.

Schuster, B. 2018. S-layer protein-based biosensors. Biosensors 8(2): 40. https://doi.org/10.3390/bios8020040.

Serour, E. and G. Antranikian. 2002. Novel thermoactive glucoamylases from thermoacidophilic archea thermoplasma acidophilum, *Picrophilus torridus* and *Picrophilus oshimae*. Antonie van Leeuwenhoek Int. J. Gen. Mol. Microbiol. 81(1–4): 73–83. https://doi.org/10.1023/A:1020525525490.

Shafiei, M., A.A. Ziaee and M.A. Amoozegar. 2011. Purification and characterization of an organic-solvent-tolerant halophilic α-amylase from the moderately halophilic *Nesterenkonia* sp. strain F. J. Ind. Microbiol. Biotechnol. 38(2): 275–281. https://doi.org/10.1007/s10295-010-0770-1.

Sharma, A., Y. Kawarabayasi and T. Satyanarayanak. 2012. Acidophilic bacteria and archaea: Acid stable biocatalysts and their potential applications. Extremophiles 16(1): 1–19. https://doi.org/10.1007/s00792-011-0402-3.

Shi, H., Y. Zhang, X. Li, Y. Huang, L. Wang, Y. Wang, H. Ding and F. Wang. 2013. A novel highly thermostable xylanase stimulated by Ca^{2+} from *Thermotoga thermarum*: Cloning, expression and characterization. Biotechnol. Biofuels 6(1): 1–9. https://doi.org/10.1186/1754-6834-6-26.

Shraddha., R. Shekher, S. Sehgal, M. Kamthania and A. Kumar. 2011. Laccase: Microbial sources, production, purification, and potential biotechnological applications. Enzyme Res. 2011: 1–11. https://doi.org/10.4061/2011/217861.

Singh, O.V. and P. Gabani. 2011. Extremophiles: Radiation resistance microbial reserves and therapeutic implications. J. Appl. Microbiol. 110(4): 851–861. https://doi.org/10.1111/j.1365-2672.2011.04971.x.

Sleytr, U.B. and M. Sára. 1997. Bacterial and archaeal S-layer proteins: Structure-function relationships and their biotechnological applications. Trends Biotechnol. 15(1): 20–26. https://doi.org/10.1016/S0167-7799(96)10063-9.

Soto, P., M. Acosta, P. Tapia, Y. Contador, A. Velásquez, C. Espoz, C. Pinilla, P.A. Galleguillos and C. Demergasso. 2013. From mesophilic to moderate thermophilic populations in an industrial heap bioleaching process. Adv. Mat. Res. 825: 376–379. https://doi.org/10.4028/www.scientific.net/AMR.825.376.

Suzuki, S., J.G. Kuenen, K. Schipper, S. Van Der Velde, S.I. Ishii, A. Wu, D.Y. Sorokin, A. Tenney, X. Meng, P.L. Morrill and Y. Kamagata. 2014. Physiological and genomic features of highly alkaliphilic hydrogen-utilizing Beta-proteobacteria from a continental serpentinizing site. Nat. Commun. 5: 1–12. https://doi.org/10.1038/ncomms4900.

Takai, K., M. Miyazaki, H. Hirayama, S. Nakagawa, J. Querellou and A. Godfroy. 2009. Isolation and physiological characterization of two novel, piezophilic, thermophilic chemolithoautotrophs from a deep-sea hydrothermal vent chimney. Environ. Microbiol. 11(8): 1983–1997. https://doi.org/10.1111/j.1462-2920.2009.01921.x.

Taylor, I.N., R.C. Brown, M. Bycroft, G. King, J.A. Littlechild, M.C. Lloyd, C. Praquin, H.S. Toogood and S.J.C. Taylor. 2004. Application of thermophilic enzymes in commercial biotransformation processes. Biochem. Soc. Trans. 32(2): 290–292. https://doi.org/10.1042/BST0320290.

Temsaah, H.R., A.F. Azmy, M. Raslan, A.E. Ahmed and W.G. Hozayen. 2018. Isolation and characterization of thermophilic enzymes producing microorganisms for potential therapeutic and industrial use. J. Pure Appl. Microbiol. 12(4): 1687–1702. https://doi.org/10.22207/JPAM.12.4.02.

Tortorella, E., P. Tedesco, F. Palma Esposito, G.G. January, R. Fani, M. Jaspars and D. De Pascale. 2018. Antibiotics from deep-sea microorganisms: Current discoveries and perspectives. Mar. Drugs 16(10): 1–16. https://doi.org/10.3390/md16100355.

Tuysuz, E., N. Gonul-Baltaci, M.A. Omeroglu, A. Adiguzel, M. Taskin and H. Ozkan. 2020. Co-production of amylase and protease by locally isolated thermophilic bacterium *Anoxybacillus rupiensis* T2 in Sterile and Non-sterile Media Using Waste Potato Peels as Substrate. Waste Biomass Valorization (in press). https://doi.org/10.1007/s12649-020-00936-3.

Undabarrena, A., F. Beltrametti, F.P. Claverías, M. González, E.R. Moore, M. Seeger and B. Cámara. 2016. Exploring the diversity and antimicrobial potential of marine actinobacteria from the comau fjord in Northern Patagonia, Chile. Front. Microbiol. 7: 1–16. https://doi.org/10.3389/fmicb.2016.01135.

Undabarrena, A., J.A. Ugalde, M. Seeger and B. Cámara. 2017. Genomic data mining of the marine actinobacteria *Streptomyces* sp. H-KF8 unveils insights into multi-stress related genes and metabolic pathways involved in antimicrobial synthesis. Peer J, 2017: 1–35. https://doi.org/10.7717/peerj.2912.

Usui, K., T. Hiraki, J. Kawamoto, T. Kurihara, Y. Nogi, C. Kato and F. Abe. 2012. Eicosapentaenoic acid plays a role in stabilizing dynamic membrane structure in the deep-sea piezophile *Shewanella violacea*: A study employing high-pressure time-resolved fluorescence anisotropy measurement. Biochim. Biophys. Acta Biomembrane 1818(3): 574–583. https://doi.org/10.1016/j.bbamem.2011.10.010.

Van De Werken, H.J., M.R. Verhaart, A.L. VanFossen, K. Willquist, D.L. Lewis, J.D. Nichols, H.P. Goorissen, E.F. Mongodin, K.E. Nelson, E.W. Van Niel and A.J. Stams. 2008. Hydrogenomics of the extremely thermophilic bacterium *Caldicellulosiruptor saccharolyticus*. Appl. Environ. Microbiol. 74(21): 6720–6729. https://doi.org/10.1128/AEM.00968-08.

Vera, M., A. Schippers and W. Sand. 2013. Progress in bioleaching: Fundamentals and mechanisms of bacterial metal sulfide oxidation-part A. Appl. Microbiol. Biotechnol. 97(17): 7529–7541. https://doi.org/10.1007/s00253-013-4954-2.

Vieille, C. and G.J. Zeikus. 2001. Hyperthermophilic Enzymes. Microbiol. Mol. Biol. Rev. 65(1): 1–43. https://doi.org/10.1128/MMBR.65.1.1.

Yano, J.K. and T.L. Poulos. 2003. New understandings of thermostable and peizostable enzymes. Curr. Opin. Biotechnol. 14(4): 360–365. https://doi.org/10.1016/S0958-1669(03)00075-2.

Yildiz, S.Y., N. Radchenkova, K.Y. Arga, M. Kambourova and E. Toksoy Oner. 2015. Genomic analysis of *Brevibacillus thermoruber* 423 reveals its biotechnological and industrial potential. Appl. Microbiol. Biotechnol. 99(5): 2277–2289. https://doi.org/10.1007/s00253-015-6388-5.

Young, S.Y. and N.L. Young. 2004. Purification and some properties of superoxide dismutase from *Deinococcus radiophilus*, the UV-resistant bacterium. Extremophiles 8(3): 237–242. https://doi.org/10.1007/s00792-004-0383-6.

Zhang, M.J., C.Y. Jiang, X.Y. You and S.J. Liu. 2014. Construction and application of an expression vector from the new plasmid pLAtc1 of *Acidithiobacillus caldus*. Appl. Microbiol. Biotechnol. 98(9): 4083–4094. https://doi.org/10.1007/s00253-014-5507-z.

Zhu, S., D. Song, C. Gong, P. Tang, X. Li, J. Wang and G. Zheng. 2013. Biosynthesis of nucleoside analogues via thermostable nucleoside phosphorylase. Appl. Microbiol. Biotechnol. 97(15): 6769–6778. https://doi.org/10.1007/s00253-012-4542-x.

2

Extremophiles for Sustainable Bio-energy Production
Diversity, Enzymology and Current Applications

Amit Verma,[1] *Tirath Raj,*[2,6] *Shulbhi Verma,*[3] *Varun Kumar*[4] and *Ruchi Agrawal*[2,5,*]

1. Introduction

Extremophiles are defined as organisms surviving under extreme habitats which have inhospitable condition of pH, temperature, salinity, water availability, pressure, etc. These environments have unique organisms, especially microorganisms having specialized metabolic machinery to tackle the harsh, extreme conditions. Extremophiles mainly come from archea and bacteria, but recently some are reported from eukaryotes (Zeldes et al. 2015, Coker 2019). Most of the present extremophiles are reported from archea and they inhabit the extreme sites on Earth like volcanic vents to hypersaline lakes. Extremophiles are gaining importance presently due to their wide range of hydrolytic enzymes which can be put to various biotechnological applications. Apart from this, these extremophiles have chaperones and heat shock proteins (HSPs) which escape their proteins from proteolysis and denaturation. This property makes these microorganisms interesting for study and utilization in different applications (Zeldes et al. 2015). These extremophiles are the source of unique biochemicals including extremozymes or enzymes with extreme stability and activity under extreme conditions of high and low temperatures, pressure, high alkaline and acidic conditions. Extremozymes are rugged proteins resisting denaturation as well as bear activity even at partial unfolded conditions (Coker 2019). Presently, extremozymes are also attracting the bioenergy researchers as biofuel production largely relies on utilization of agricultural and lignocellulosic wastes. This conversion requires robust enzymes and extremozymes present all the properties required for the bioenergy applications (Fig. 1).

Conventional fuels can only be replaced by biofuels if they are produced efficiently by the hydrolysis of agricultural and lignocellulosic wastes (Kaur et al. 2021, Agrawal et al. 2021a, Agrawal et al. 2021b). Based upon the utilization of raw material, biofuels are classified as 1st generation biofuels which are obtained through processing of materials like sugarcane, corn, beets,

[1] Department of Biochemistry, College of Basic Science & Humanities, S.D A.U, S.K Nagar, Gujarat-385506, India.
[2] DBT-IOC Centre for Advanced Bioenergy Research, Indian Oil R&D Centre, Sector 13. Faridabad, Haryana.
[3] Department of Biotechnology, College of Basic Science & Humanities, S.D A.U, S.K Nagar, Gujarat-385506, India.
[4] Nanotechnology and Advanced Biomaterials Group, Avantha Centre for Industrial Research and Development, Yamunanagar-135001, Haryana, INDIA.
[5] TERI-Deakin Nanobiotechnology Centre, The Energy and Resources Institute, TERI Gram, Gurugram, India.
[6] Environmental Processes and Materials Laboratory (EPML), School of Civil and Environmental Engineering, Yonsei University, Seoul, South Korea 03722.
* Corresponding author: dr.ruchiagrawal010@gmail.com

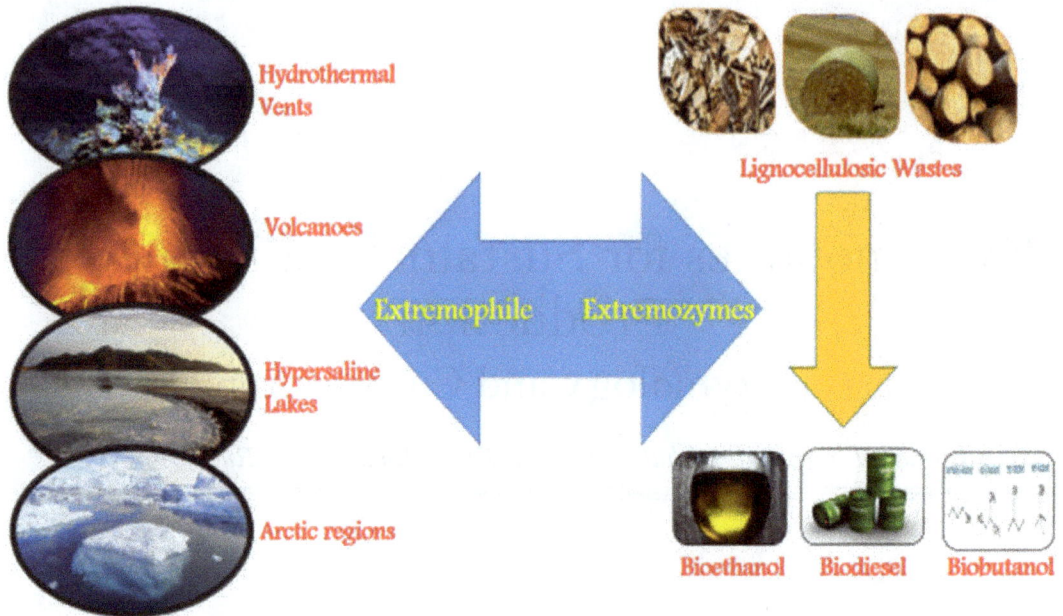

Fig. 1. Extremophiles and their extremozymes: Different habitats and bioenergy applications.

wheat, sorghum which have their own pros and cons. 2nd generation biofuels are produced through utilization of waste materials of various kinds, especially recalcitrant lignocellulosic materials (Robak and Balcerek 2018). However, utilization of such recalcitrant materials requires physical and chemical processing to make the material utilizable under fermentation conditions (Agrawal et al. 2021c, Satlewal et al. 2018a, Negi et al. 2019). Thus, challenge lies in the use of such materials and maintaining a continuous system for bioenergy production from such substrates (Satlewal et al. 2018b, Agrawal et al. 2018a). Extremophiles, due to their robustness, solve many of these obstacles in use of such substrates and they present a new hope for developing "*Single Step Fermentation*" procedures, replacing the various pretreatment steps. The present applications of extremozymes as well as extremophiles in bioenergy sector are therefore gaining momentum and present chapter focuses on the recent details of such studies.

2. Extremophile Diversity for Bioenergy

Microorganisms (archea, bacteria, fungi and microalgae) having adaptations to survive and flourish under harsh or extreme environments are known as extremophiles. Extreme environmental conditions are the conditions in which most of the living organisms are unable to survive, like extreme temperatures, pH, pressure, salt concentration, lack of water and presence of heavy metals and radiations. The microorganisms which are normally grown at extreme high and low temperatures are denoted as thermophiles and psychrophiles, respectively. The microorganisms having tolerance for extremes of pH and pressure are known as acidophiles-alkaliphiles and barophiles, respectively. High salt concentration tolerating microorganisms are called halophiles. Some of the microorganisms are found living under high scarcity of water and these are known xerophiles. Microorganisms growing in high radiation environments are designated as radiotolerants. These amazing microorganisms have developed various strategies to adapt to such harsh living conditions. These microorganisms successfully maintain the structure and functions of cellular membranes and enzymes, regulate the metabolism and intracellular environment and use new or modified energy transduction mechanisms to cope with such extreme habitats. Being able to adapt to extreme environmental conditions, these microorganisms and their enzymes ("extremozymes") are of great value for application in industrial processes, especially bioenergy domain. Extremophiles have a great diversity according to their living environmental conditions and following is the brief description about these major groups.

2.1 Thermophiles

Microorganisms which are grown and thrive at temperature range of 50°C to 120°C are known as thermophiles. On the basis of temperature range, these microorganisms are further divided into thermophiles (50–85°C) and hyperthermophiles (85–120°C) (Bala and Singh 2018). Thermophiles mainly belong to prokaryotes, like spore forming anerobic *Clostridium* and aerobic and anerobic *Bacilli*, photosynthetic bacteria, blue green algae (micro algae), sulphur reducing and oxidizing bacteria, and methane producing and oxidizing bacteria. Besides, some eukaryotic microorganism such as filamentous fungi and algae are also thermophilic in nature. The main microbial members of hyperthermophiles group belong to archaea and bacteria (Zeldes et al. 2015). These microorganisms mainly have been isolated from manure, compost, piles of agricultural and forestry plant biomass, sands of beach, hot springs, desert soils, coal mine soils, nuclear reactor effluents, Dead Sea valley soils and natural water heaters (Singh et al. 2016, Bhagia et al. 2021).

These thermophiles and hyperthermophiles cope with elevated temperature by stabilization of various proteins through structural modification (high compact) and sequence modification to inhibit their denaturation (Berezovsky and Shakhnovich 2005, Dominy et al. 2004). Besides, these microorganisms also have heat shock proteins (chaperons) to stop the structural modifications in proteins at high temperature (Sterner and Liebl 2001). Thermophilic archaea use modified Entner-Doudoroff and Embden Meyerhof pathways along with non–oxidative pentose phosphate pathway to cope with high temperature (Van der Oost and Siebers 2007).

2.2 Psychrophiles

Microorganisms that colonized in cold environments (cold adapted) under 15–20°C (from deep sea to mountains and Polar Regions) are designated as psychrophiles. Besides natural environments, these microorganisms are also colonized in human-made low temperature places like deep freezers. Psychrophiles living in ocean depth are known as Piezo-psychrophiles and psychrophiles inhabiting elevated salt concentrations are designated as Halo-psychrophiles (Kumar et al. 2011). Psychrophiles include a variety of microorganisms like Gram-positive and Gram-negative bacteria, fungi, yeast and microalgae. Not only cold environment, the Polar Regions also have other extreme conditions like strong ultraviolet radiation, robust winds and dryness. In such harsh conditions, psychrophiles support their life by producing special antifreeze proteins, UV protecting compounds, polyunsaturated fatty acids and various antioxidants (Kim et al. 2020, Jung et al. 2014, Nogueira et al. 2015). In cold temperature, psychrophiles also faced the challenges of lack in thermal energy and problem of high viscosity and to overcome these problems, protein modification is induced by chaperones and cold-shock proteins in cells (D'Amico et al. 2006). The enzyme proteins remain active by flexible structure and by lowering the enthalpy-driven interactions (Violot et al. 2005, Berger et al. 1996). The fluidity of membranes is maintained by high amount of polyunsaturated, unsaturated and methylbranched fatty acids (Chintalapati et al. 2004). From metabolism point of view, psychrophiles maintain ATP generation and cofactors production by enhancing the activities or up-regulating enzymes of central metabolic pathways (Russell 2000).

2.3 Acidophiles and Alkaliphiles

Microorganisms that easily survive and proliferate in acidic habitats (below pH 4.0) are recognized as acidophiles. Acidophiles is a diverse group of microorganisms including bacteria, archaea, fungi (yeast), algae and protozoa. Usually, acidophiles are found in sulphur rich geysers, pools, solaphataric areas and mines of coal and metals. These microorganisms can also be isolated from industrial effluent treatment plants. Acidophiles have cellular level (structural and physiological) specific adaptations (pH homeostasis) to tolerate the extreme low pH (Golyshina et al. 2000, Crossman et al.

2004). The most common mechanism to survive in highly acidic environment is to pump out the acid from cell and maintaining the neutral pH or weak acidic conditions (5.0–7.0 pH) within cell for proper functioning of various enzymes and biochemicals (Matin 1999, Jolivet et al. 2004, Sharma et al. 2011). To stop the proton influx, a reverse membrane potential is developed having influx of potassium ions (Baker-Austin and Dopson 2007). Proton impermeable cellular membranes also play an important role to reduce the influx of protons (Konings et al. 2002). The cell membranes of archaea have tetraether lipids instead of ester bonds (Shimada et al. 2002, Batrakov et al. 2002).

Alkaliphiles are microorganisms that are optimally colonized at pH 9.0 and above 9.0 usually between pH 10.0 and 12.0. They do not grow near neutral pH (6.5–7.0). These microorganisms are naturally found in neutral or high alkaline environments like soda lakes, alkaline soils of sub ground level, carbonate springs and deep sea. Sometimes, alkaliphiles are also found in acidic soils. Alkaliphiles have two main groups on the basis of physiology, alkaliphiles and haloalkaliphiles. Haloalkaliphiles need high salinity (~ 33% NaCl) with alkaline environment for their optimal growth. This group of extremophiles comprises several types of microorganisms including bacteria (spore forming aerobic and non-spore forming anaerobic), archaea and fungi (yeast) (Chinnathambi 2015, Sarethy et al. 2011). Alkaliphiles tolerate strong alkaline environment by maintaining the pH homeostasis (Padan et al. 2005). They balance or decrease cytoplasmic pH by enhanced production of acids from deamination of amino acids and sugar fermentation (Richard and Foster 2004). They also modify (acidic) their cell surfaces for attraction and attachment of cations for pH homeostasis through activation of monovalent cation/proton antiporters (Wang et al. 2004, Kitada et al. 2000, Swartz et al. 2005).

2.4 Halophiles

Halophiles are the microorganisms that have a requirement of high salt concentration, especially of sodium chloride for their growth. These are found in hypersaline habitats such as salt marshes, salty lakes, salt pans, deep salt mines and coastal and submarine pools (Setati 2010, DasSarma and DasSarma 2015). On the basis of salt concentration tolerance, they are categorized into slight halophiles (2–5% NaCl), moderate halophiles (5–15% NaCl) and extreme halophiles (15–30% NaCl) (Yin et al. 2015). The microorganisms of all three main life domains (archaea, bacteria and eukarya) have the halo tolerance (Quillaguamán et al. 2010, Yin et al. 2015). To survive in hyper saline environments, these microorganisms have some specific physiological strategies. Haloarchaea or halophilic archaea apply a "salt-in" tactic by KCl accumulation (equal to surrounding NaCl) within cells' cytoplasm to counter the high saline environment. The halophilic bacteria and eukaryotes use a "salt-out" strategy to cope with high salt stress. These microorganisms accumulate or synthesize glycine betaine glycerol, ectoine, trehalose and sucrose to balance the high salt concentration of surrounding environments (Roberts et al. 2005, Oren and Mana 2003).

2.5 Barophiles

Microorganisms able to survive under extreme pressure (upto 120 MPa) are referred to as barophiles or piezophiles (Eisenmenger and Reyes-De-Corcuera 2009). Barophiles have the adaptation for tolerating the high pressure in cold environments, while piezophiles have the adaptation for high pressure and all allowable temperatures. They are usually naturally found in deep sea sediments and submarine hydrothermal vents with variety of microorganisms like archaea, bacteria and fungi. At high pressure, these microorganisms face a great challenge to maintain their membrane fluidity and they manage it by increasing unsaturated fatty acids in membranes (Valentine and Valentine 2004). At high pressure condition, cells often loose transporter activities and these organisms upregulate their transporters to cope with the problem (Abe 2007). Owing to their occurrence in diverse temperature range,these microorganisms have heat and cold-shock proteins to overcome the problem of protein denaturation (Martin et al. 2002).

2.6 Xerophiles

Xerophiles are the microorganisms that can survive and proliferate in nearly absence or extreme lack of water (desiccation, below 0.85 water activity) or in environments of high solute concentrations (hypertonicity) (Lebre et al. 2017). Usually, bacteria and fungi of arid and desert habitats show this capacity. In desiccation or hypertonicity, microorganisms survive through applying various modifications in their membranes and cytoplasm. To cope with extreme fluctuations in water availability, they have increased ratio of unsaturated fatty acids (cis-mono-unsaturated fatty acids for maintaining the liquid crystalline phase) and cyclopropane fatty acids (for stability of intracellular proteins) (Lebre et al. 2017). At low water activity, to maintain osmotic equilibrium and protein function,xerophiles apply a "bi-phasic salting-out" process in which charged solutes like glutamate and potassium accumulate in cytoplasm to balance the osmotic stress and after it these solutes are replaced with organic solutes for long term osmotic balance and protein functions (Riedel and Lehner 2007, Sleatorand Hill 2002).

3. Metabolic Features and Biocatalysis

Extremophile microorganisms established specific molecular arrangement in order to establish life in extreme harsh environment, acidic or alkaline conditions, i.e., pH (> 8 and < 4), high pressure (> 500 atm), high or low temperatures (55°C to 121°C and –2°C to 20°C), high concentrations of pollutants, and salts (2–5 M NaCl), and geographical barriers (UVR resistance > 600 J/m) (Shrestha et al. 2018). Biocatalyst prepared from these microorganisms, i.e., extremozymes,possess specific physiological, enzymatic characteristics, i.e., salt allowance, thermo stability, etc. Hence, may be used for various industrial biocatalytic processes and opens a way for future research (Agrawal et al. 2018b, Agrawal et al. 2018c).

Thus, on the basis of their biological activity, these microorganism can be characterized as acidophile (growth at ≤ 3.0 pH), alkaliphile (grow > 9 pH), endolith, hyperthermophile, hypolith, oligotrop (obtained with concerted media composition with 1–15 mg organic carbon/liter), piezophile (survive at high-pressure atmospheric conditions, nearly 50 MPa), psychrophile (growth of ≤ 15°C and a maximum temperature for growth of < 20°C), radio resistant, thermophile, toxitolerant, and xerophil, hyperthermophiles (grow above ≥ 80°C). For instance, thermophilic proteins, i.e., *cellulases, amylases, xylanases, mannoses, pectinases, proteases*, etc., can be stable between 41 to 122°C (Dumorné et al. 2017, Kaur et al. 2019). The growth features, enzymes and industrial applications of different types of extremophiles are tabulated in Table 1.

Table 1. Growth features, enzymes and industrial applications of different types of extremophiles.

Extremophiles	Growth features	Isolated enzyme	Industrial applications
Thermophiles	Growth temp. > 80°C (Hyperthermophile; 60–80°C) (Thermophiles)	Proteases, Glycosyl hydrolases (e.g., amylases, glucoamylases, xylanases, lipases, esterases, DNA polymerases, Xylanases, Lipases, esterases)	Hydrolysis of starch, cellulose, chitin, pectin processing, textiles, baking, brewing, detergents, biosensors, molecular biology, peptide synthesis, biocatalysts in organic media
Psychrophiles	Grow below < 15°C	Lipases, Proteases, Dehydrogenases. Cellulases, DNA polymerase	Starch processing
Halophiles	High salt conc. (e.g., 2–5 M NaCl)	Proteases, Dehydrogenases	Fed component
Alkaliphiles	pH > 9	Proteases, Cellulases, Oxidases	Desulfurization of coal
Acidophiles	pH < 2–3	Amylases, Glucoamylases, Proteases, Cellulases	Food processing and antibiotic applications
Piezophiles	High pressure > 120 MPa	Amylases, Glucoamylases, DNA polymerase	Food processing and antibiotic applications
Mesophiles	Growth 15–55°C	Archaea, Bacteria	Food industry, pharmacy industry

Thermophiles

For instance, thermopiles contain chaperons (Hsp60 complex), heat shock proteins, which prevent thermopiles from folding and denaturation. Thermopiles mainly consist of two different mechanisms which stabilize proteins at high temperature (Verma et al. 2016, Abe and Horikoshi 2001). In one way, protein structures are modified significantly to more compact as compared to mesospheric homologous through various interactions. In second way, sequential build up of new strong interactions are developed at high temperature and the structure remains same as mesophillic structure. In addition to that, thermophillic archaea consume sugars and follow modified Embden-Meyerhof (EM), Entner-Doudoroff (ED) pathways with replacement of oxidative to non-oxidative PP pathway. For example, *Thermococcuszilligii* and *Thermococcusstetteri* strains have ADP-dependent phosphofructokinase and unique glyceraldehyde-3-phosphate: ferredoxinoxidoreductase and follows modified EM pathway.

Piezophiles

In sea depth, pressure reached to 70–140 MPa, which make life highly critical where a group of extremophiles, i.e., Piezophiles are found under high hydrostatic pressure condition. Piezophiles adapted their microscopic structure by gene regulations and cellular structure. It was proposed that cell division, osmolytes and polyunsaturated fatty acids' productions, high pressure flexibility and multimeric antioxidant proteins may help their existence. For instance, the presence of extended helices in the 16S ribosomal RNA (rRNA) genes helps to grow these bacteria at high pressures (Abe and Horikoshi 2001).

Halophiles

Halophiles are capable of surviving in hypersaline (high salt conc., i.e., 5 M NaCl and 4 M KCl) conditions and can be found in deep salt lakes, sea floor, e.g., Red Sea, Mediterranean Sea, etc. They are able to maintain osmotic balance inside cell membrane and maintain isotonic state in cytoplasm. They also posses high thermal stability and are tolerant to a wide range of pHs, etc. For example, an extracellular protease from *Halobacterium halobium* has been used for peptide synthesis in water/N0-dimethylformamide media with high selectivity. Enzymes produced from halophiles are highly stable in salt conditions with low water activity as well as in organic solvents. Characteristic and microscopic study reveals that these halophilic enzymes are made up of archaea and have greater amino acids functionalities, i.e., serine, threonine, aspartic and glutamic acids on surface. This high concentration of amino acid groups with low hydrophobic characters results in enhanced salt bridges to co-operate with electrostatics interactions, which prevent the halophytic proteins, aggregation and precipitation. *Halophilicarchaea,* the *Halobacteriaceae,* accumulate K^+ intracellularly for osmoregulations, requiring adaptation to intra and extra cellular protein activity and stability, while other bacteria accumulate compatible solutes (e.g., glycine, betaine, sugars, polyols, amino acids and ectoines), which help them to maintain an environment isotonic with the growth medium. The production of halophilic enzymes, such as *xylanases, amylases, proteases* and *lipases,* has been reported for some halophiles belonging to the genera *Acinetobacter, Haloferax, Halobacterium, Halorhabdus, Marinococcus, Micrococcus, Natronococcus, Bacillus, Halobacillus* and *Halothermothrix*. These halophilicesterases and lipases may be used for flavoring in food industry. For instance, *halophilic lipase LipBL* from *Marinobacter lipolyticus SM19* has been explored for hydrolysis of olive and fish oil, which produced eicosapentanoic acid, which has high commercial value (Madigan and Orent 1999, Amoozegar et al. 2019).

Psychrophiles

Nowadays, biotechnologies are more concerned about energy consumption with minimum energy use. In sea and at poles, temperature drops to minus. Therefore, it's highly challenging to survive at extreme low temperature conditions. Therefore, psychrophiles organisms are certain low temperature extremophiles which could be used for industrial application where energy consumption is low. Psychrophiles are adapted and achieved an active cellular metabolic rate at –25°C. Additionally, DNA synthesis can be performed at –20°C (Siddiqui et al. 2013). Psychrophiles can be further characterized in two extremophiles based upon the growing temperature. Eurypsychrophiles lie in deep sea level from broad temperature ranges, and stenopsychrophiles (true psychrophiles) are those which can't grow above 20°C. Psychrophiles are able to retain their existence at low temperature due to presence of cold-shock proteins and RNA chaperones, enhanced tRNA flexibility, enhanced membrane fluidity for maintaining the semi-fluid state of the membranes, and the production of cold-active secondary metabolites, enzymes, pigments and antifreeze protein and have flexible structures. The isolated enzyme, i.e., psychrophilic proteases, amylases or lipases, have great applications for cellulose-degradation in pulp and paper industry, starch and cellulose hydrolysis in beverage and biofuel industries as these enzymes may provide economic benefits with energy consumption and production costs. Besides that, several low temperature enzymes like L-glutaminase and L-asparaginase could be used in food processing industries (Siddiqui et al. 2013). Low temperature adapted enzymes may have significant importance in molecular biology in sequential reactions, etc. For instance, xylanases isolated from deep sea psyhrophilic, i.e., *Zunongwangia profunda* and *Flammeovirga Pacifica*, can be used as additive in food industries.

In context, there are some lipolytic enzymes from thermopiles, i.e., triglycerol lipases and carboxulesterases, which could be used for hydrolysis of long chain fatty esters and short chain Carboxylesterases may help in catalysis of short acyl esters < 10 carbon) (Amoozegar et al. 2019). All these enzymes have α/β fold and triad catalytic residues, i.e., histidine, serine, aspartic or glutamic acids, etc. Screening and characteristics study reveals that these enzymes poses high stability under various acidic conditions and are stable at high temperature. For instance, Esterases from *S. acidophilus* and, *Pyrobaculum* sp. strains have high acidic and thermal stability and thus represent unique industrial applicability.

Alkaliphiles/Acidophile

These organisms are adapted to live under variable pH range, which can be useful for highly alkaline and highly acidic reaction conditions (Takeuchi et al. 2001). Due to pH adaptation, these organisms exclude any specific protein to maintain the peripasmatic space and neutral pH internally. Hence, by virtue of this, the lipases and enzyme isolated from alkaliphiles and acidophile are used in food and detergent preparation. Certain proteases, amylases, lipases and other enzymes are resistant to high pH variations and chelate concentration. These organisms, alkaliphiles and acidophiles, follow different strategies to survive at varying pH. For instance, alkaliphiles are associated with negatively charged structured polymers with peptidoglycan, which reduces the overall charge density on bacterial outer cell surface, which stabilizes the cell membrane. Additionally, fatty acids present in alkaliphilic bacterial strains also contain saturated and mono-unsaturated straight-chain fatty acid chains (Elleuche et al. 2015).

On the other side, acidophiles follow combinations of different mechanisms since a positively charged membrane surface having high internal buffer capacity, by virtue of over-expression of H⁺ exporting enzymes and unique transport systems, makes them a great biocatalytic material for laundry and dishwashing processes. For instance, *cellulase* free *xylanases* are used for bio-bleaching of pulp and paper for paper industry. Similarly, *pectinases* are also used in degumming of ramie fibers and *catalase* and *peroxidase* or *oxidoreductase* could serve for removal of residual H_2O_2 from textile waste streams of the textile (Agrawal et al. 2017). Proteolytic enzymes act as biocatalyst for hydrolysis of proteinaceous materials. For instance, serine *proteases* proteolysin isolated from a thermophilic bacterium *Coprothermobacter proteolyticus* is highly preferred in soap industry due to high thermal stability and high pH tolerance (Van Den Burg 2013).

4. Applications in Bioenergy Production

Extremophiles microorganism can be potentially considered as potent bioenergy producer. They have the ability to survive in extreme condition. Abilities to thrive in extreme condition have changed microbes at the genetic level, and their adaption is reflected in its metabolic system. Due to non-competition in terms of food, they can be utilized as bioenergy source in a broad way (Table 2).

Table 2. Bioenergy Production using Extremophiles Microorganism.

S. No.	Microorganism	Process	Biofuel
1.	*Saccharomyces cerevisiae*	fermentation	Ethanol
2.	*Clostridia* sp.	cellulases	Butanol
3.	*Thermoanaerobacterium saccharolyticum* ALK2	fermentation	Ethanol
4.	*Methanococcusjannaschii*	fermentation	1-butanol and propanol
5.	*Spirulina*	photobioreactor system	biodiesel
6.	*Chlorella* spp.	photobioreactor system	biodiesel
7.	*Dunaliella*	photobioreactor system	biodiesel
8.	*Caldicellulosiruptor saccharolyticus*	fermentation	hydrogen
9.	*Thermotogomaritima*	fermentation	hydrogen
10.	Methanosarcina thermophila	fermentation	biogas

4.1 Extremophiles Microorganism in Production of Ethanol

Lignocellulose is the part of plant cell wall utilized for biofuel through deconstruction of plants' cell wall. Generally, it involves three main steps: first is the pretreatment which breaks recalcitrant lignocelluloses, second is the enzymatic hydrolysis of the polysaccharides to convert into fermentable sugar, and finally, the fermentation process converts these sugars into ethanol, e.g., *Ruminococcus flavefaciens, Clostridium thermocellum, Clostridium cellulolyticum, Geobacillus* and *Humicolabrevis*, etc. (Barnard et al. 2010).

4.2 Extremophiles Microorganism in Production of Biodiesel

Biodiesel obtained from plant residues is quite similar to diesel in terms of properties. The biodiesel production depends on catalyst driven chemical reaction between vegetable oils and alcohol such as methanol, ethanol, propanol, isopropanol, etc. Several extremophiles microbes which participate in the production of biodiesel include *Botryococcus braunii, Cyanidium caldarium* and *Galdieria sulphuraria*, etc. (Barnard et al. 2010).

4.3 Extremophiles Microorganism in Production of Biobutanol

Biobutanol production is carried through *Clostridium* spp. by using raw biomass of corn, sugar beet, and sorghum. In the initial stage, it is partially converted to hydrogen and butyric acid, which is converted into butanol. Apart from *Clostridium* spp., *Pseudomonas* genus and *Bacillus* spp. are utilized for the production of biobutanol. Commercially, biobutanolis produced through the use of a mixture of thermophiles as well as thermostable enzymes for the production from waste biomass. The thermophilic fermentations are conducted with genetically modified microbial strains optimized to produce butanol which is commercially sold as Butafuel.

4.4 Extremophiles Microorganism in Production of Hydrogen

Hydrogen is a very important element on Earth and is considered as a viable biofuel because it has the ability to convert electrical energy in fuel cells, transportation power, and heat power. So generating hydrogen from microbes can be effective in terms of cost, pollution and energy saving alternative. In the production of hydrogen, microbes act on substrate and sequentially convert raw slurries and wastes to amino acid, sugars and fatty acid. Then they are anaerobically oxidized and fermented into organic acids and alcohols, which may then be converted to hydrogen and methane by methanogens. Thermophilic bacterial hydrogen producers belong to the genus *Thermoanaerobacterium, Thermoanaerobacter, Thermotoga, Caldicellulosiruptor saccharo-lyticus* and archaeal genera *Thermococcus* and *Pyrococcus*. Extremophiles such as *Acidithiobacillus, Arthrobacter, Bacillus, Caldicellulosiruptor, Clostridium, Coprothermobacter*, and *Enterobacter* are used in biorefinery (Zhu et al. 2020).

4.5 Extremophiles Microorganism in Production of Biogas

Biogas is a product of anaerobic degradation of organic substrates and can be easily produced in small industrial units. It produces heat and electric power, which can be used in house and for running small industries' engine. It's pollution free, cost effective, and environmentally friendly. Biogas can be produced from animals manure, organic households, garden wastes, municipal refuse, and frying oil. Generating methane from plant material through microbial technologies usually occurs in a three-step process. At last step, it yields up to 70% CH_4 and CO_2, NH_3 and H_2S as by-products. More recently, the use of thermophilic organisms for these purposes has gained momentum, based largely on their ability to speed up the process (Maus et al. 2016).

5. Conclusion and Future Prospects

Studies related to metabolic features of extremophiles and unique properties of their enzymes and protein increased their utilization in many industrial applications. Owing to the limitations of mesophilic organisms and their enzymes to fit into industrial applications, many of these were replaced by the extremophiles and extremozymes due to their property of robustness and versatility. Similar is the case with bioenergy sector which depends upon hydrolysis of various substrates. Bioenergy production can only replace the conventional fuels if they are produced using low cost waste materials in place of food grade or high cost raw materials. One of the probable solutions for this is utilization of extremophiles with low cost waste materials under fermentation conditions for production of biofuels. This is further supported by increase in knowledge of embolic and biochemical features of these extremophiles. Thus, present bioenergy researches are carried with an aim to scale up the biofuel production using extremophiles and extremozymes.

Acknowledgement

RA is thankful to TERI (The Energy and Resources Institute), Gurugram India and DBT -IOC Centre for providing all infrastructural, analytical, and financial assistance. The SERB (Science and Engineering Research Board) and DBT (Department of Biotechnology), Ministry of Science and Technology, Govt. of India is duly acknowledged for providing the funding and other economic assistance (SRG/2020/001306).

References

Abe, F. and K. Horikoshi. 2001. The biotechnological potential of piezophiles. Trends in Biotechnol. 19(3): 102–108.
Abe, F. 2007. Exploration of the effects of high hydrostatic pressure on microbial growth, physiology and survival: Perspectives from piezophysiology. Biosci. Biotechnol. Biochem. 71(10): 2347–2357.

Agrawal, R., A. Satlewal, B. Sharma, A. Mathur, R. Gupta, D. Tuli and M. Adsul. 2017. Induction of cellulases by disaccharides or their derivatives in *Penicillium janthinellum* EMS-UV-8 mutant. Biofuels 8(5): 615–622. doi: 10.1080/17597269.2016.1242692.

Agrawal, R., A. Verma and A. Satlewal. 2018a. Bioprospecting PGPR microflora by novel immunobased techniques. pp. 465–478. *In*: Prasad, R., S.S. Gill and N. Tuteja (eds.). Crop Improvement Through Microbial Biotechnology. Elsevier.

Agrawal, R., S. Semwal, R. Kumar, A. Mathur, R.P. Gupta, D.K. Tuli and A. Satlewal. 2018b. Synergistic enzyme cocktail to enhance hydrolysis of steam exploded wheat straw at pilot scale. Frontiers in Energy Research 6(122). doi: 10.3389/fenrg.2018.00122.

Agrawal, R., B. Bhadana, A.S. Mathur, R. Kumar, R.P. Gupta and A. Satlewal. 2018c. Improved enzymatic hydrolysis of pilot scale pretreated rice straw at high total solids loading. Frontiers in Energy Research 6(115). doi: 10.3389/fenrg.2018.00115.

Agrawal, R., P. Kumari, P. Sivagurunathan, A. Satlewal, R. Kumar, R.P. Gupta and S.K. Puri. 2021a. Chapter 8— Pretreatment process and its effect on enzymatic hydrolysis of biomass. pp. 145–169. *In*: Tuli, D.K. and A. Kuila (eds.). Current Status and Future Scope of Microbial Cellulases. Elsevier.

Agrawal, R., A. Verma, R.R. Singhania, S. Varjani, C. Di Dong and A. Kumar Patel. 2021b. Current understanding of the inhibition factors and their mechanism of action for the lignocellulosic biomass hydrolysis. Bioresource Technology 332: 125042. doi: https://doi.org/10.1016/j.biortech.2021.125042.

Agrawal, R., A. Verma, S. Verma and A. Varma. 2021c. Industrial methanogenesis: biomethane production from organic wastes for energy supplementation. pp. 99–115. *In*: Prasad, R., Kumar, V., Singh, J. and Upadhyaya, C.P. (eds.). Recent Developments in Microbial Technologies. Singapore: Springer Singapore.

Amoozegar, M.A., A. Safarpour, K. AkbariNoghabi, T. Bakhtiary and A. Ventosa. 2019. Halophiles and their vast potential in biofuel production. Frontiers in Microbiol. 10: 1895.

Baker-Austin, C. and M. Dopson. 2007. Life in acid: pH homeostasis in acidophiles. Trends Microbiol. 15(4): 165–171.

Barnard, D., A. Casanueva, M. Tuffin and D. Cowan. 2010. Extremophiles in biofuel synthesis. Environmental Technol. 31(8-9): 871–888.

Batrakov, S.G., T.A. Pivovarova, S.E. Esipov, V.I. Sheichenko and G.I. Karavaiko. 2002. β-D-Glucopyranosylcaldarchaetidylglycerol is the main lipid of the acidophilic, mesophilic, ferrous iron-oxidisingarchaeon *Ferroplasma acidiphilum*. Biochim. Biophys. Acta 1581(1-2): 29–35.

Berezovsky, I.N. and E.I. Shakhnovich. 2005. Physics and evolution of thermophilic adaptation. Proc. Natl. Acad. Sci. 102(36): 12742–12747.

Berger, F., N. Morellet, F. Menu and P. Potier. 1996. Cold shock and cold acclimation proteins in the psychrotrophic bacterium *Arthrobacter globiformis* SI55. J. Bacteriol. 178(11): 2999–3007.

Bhagia, S., K. Bornani, R. Agrawal, A. Satlewal, J. Ďurkovič, R. Lagaňa, M. Bhagia, C.G. Yoo, X. Zhao, V. Kunc, Y. Pu, S. Ozcan and A.J. Ragauskas. 2021. Critical review of FDM 3D printing of PLA biocomposites filled with biomass resources, characterization, biodegradability, upcycling and opportunities for biorefineries. Applied Materials Today 24: 101078.

Chinnathambi, A. 2015. Industrial important enzymes from alkaliphiles—an overview. Biosci. Biotechnol. Res. Asia 12(3): 2007–2016.

Chintalapati, S., M.D. Kiran and S. Shivaji. 2004. Role of membrane lipid fatty acids in cold adaptation. Cellul. Mol. Biol. 50(5): 631–642.

Choudhary, J., S. Singh and L. Nain. 2016. Thermotolerant fermenting yeasts for simultaneous saccharification fermentation of lignocellulosic biomass. Electron. J. Biotechnol. 21: 82–92.

Coker, J.A. 2019. Recent advances in understanding extremophiles. F1000Research. 8: F1000 Faculty Rev-1917. https://doi.org/10.12688/f1000research.20765.1.

Crossman, L., M. Holden, A. Pain and J. Parkhill. 2004. Genomes beyond compare. Nat. Rev. Microbiol. 2(8): 616–618.

D'Amico, S., T. Collins, J.C. Marx, G. Feller and C. Gerday. 2006. Psychrophilic microorganisms: Challenges for life. EMBO Rep. 7(4): 385–389.

DasSarma, S. and P. DasSarma. 2015. Halophiles and their enzymes: Negativity put to good use. Curr. Opin. Microbiol. 25: 120–126.

Dominy, B.N., H. Minoux and C.L. Brooks 3rd. 2004. An electrostatic basis for the stability of thermophilic proteins. Proteins 57(1): 128–141.

Dumorné, K., D.C. Córdova, M. Astorga-Eló and P. Renganathan. 2017. Extremozymes: A potential source for industrial applications. J. Microbiol. Biotechnol. 27(4): 649–659.

Eisenmenger, M.J. and J.I. Reyes-De-Corcuera. 2009. High pressure enhancement of enzymes: A review. Enzyme Microb. Technol. 45(5): 331–347.

Elleuche, S., C. Schäfers, S. Blank, C. Schröder and G. Antranikian. 2015. Exploration of extremophiles for high temperature biotechnological processes. Current Opinion in Microbiology 25: 113–119.

Golyshina, O.V., T.A. Pivovarova, G.I. Karavaiko, T.F. Kondratéva, E.R. Moore, W.R. Abraham, H. Lünsdorf, K.N. Timmis, M.M. Yakimov and P.N. Golyshin. 2000. *Ferroplasma acidiphilum* gen. nov., sp. nov., an acidophilic, autotrophic, ferrous-iron-oxidizing, cell-wall-lacking, mesophilic member of the *Ferroplasmaceae fam. nov.*, comprising a distinct lineage of the Archaea. Int. J. Syst. Evol. Microbiol. 50(3): 997–1006.

Jolivet, E., E. Corre, S. L'Haridon, P. Forterre and D. Prieur. 2004. *Thermococcus marinus* sp. nov. and *Thermococcus radiotolerans* sp. nov., two hyperthermophilic archaea from deep-sea hydrothermal vents that resist ionizing radiation. Extremophiles 8(3): 219–227.

Jung, W., Y. Gwak, P.L. Davies, H.J. Kim and E. Jin. 2014. Isolation and characterization of antifreeze proteins from the antarctic marine microalga *Pyramimonas gelidicola*. Mar. Biotechnol. 16(5): 502–512.

Kaur, A., N. Capalash and P. Sharma. 2019. Communication mechanisms in extremophiles: Exploring their existence and industrial applications. Microbiol. Res. 221: 15–27.

Kaur, P., N. Sharma, M. Munagala, R. Rajkhowa, B. Aallardyce, Y.N. Shastri and R. Agrawal. 2021. Nanocellulose: resources, physio-chemical properties, current uses and future applications. Frontiers in Nanotechnology.

Kim, E.J., S. Kim, H.G. Choi and S.J. Han. 2020. Co-production of biodiesel and bioethanol using psychrophilic microalga *Chlamydomonas* sp. KNM0029C isolated from Arctic sea ice. Biotechnol. Biofuels 13(1): 1–13.

Kitada, M., S. Kosono and T. Kudo. 2000. The Na+/H+ antiporter of alkaliphilic *Bacillus* sp. Extremophiles 4(5): 253–258.

Konings, W.N., S.V. Albers, S. Koning and A.J. Driessen. 2002. The cell membrane plays a crucial role in survival of bacteria and archaea in extreme environments. Antonie Van Leeuwenhoek 81(1-4): 61–72.

Kumar, P.S., M. Ghosh, K.K. Pulicherla and K.R.S.S. Rao. 2011. Cold active enzymes from the marine psychrophiles: Biotechnological perspective. Adv. Biotechnol. 10: 16–20.

Lebre, P.H., P. De Maayer and D.A. Cowan. 2017. Xerotolerant bacteria: surviving through a dry spell. Nat. Rev. Microbiol. 15(5): 285–296.

Madigan, M.T. and A. Orent. 1999. Thermophilic and halophilic extremophiles. Current Opinion in Microbiology 2(3): 265–269.

Martin, D., D.H. Bartlett and M.F. Roberts. 2002. Solute accumulation in the deep-sea bacterium *Photobacterium profundum*. Extremophiles 6(6): 507–514.

Matin, A. 1999. pH homeostasis in acidophiles. Novartis Found Sym. 221: 152–163.

Maus, I., D.E. Koeck, K.G. Cibis, S. Hahnke, Y.S. Kim, T. Langeret, J. Kreubel, M. Erhard, A. Bremges, S. Off, Y. Stolze, S. Jaenicke, A. Goesmann, A. Sczyrba, P. Scherer, H. König, W.H. Schwarz, V.V. Zverlov, W. Liebl, A. Pühler, A. Schlüter and M. Klocke. 2016. Unraveling the microbiome of a thermophilic biogas plant by metagenome and metatranscriptome analysis complemented by characterization of bacterial and archaeal isolates. Biotechnol. Biofuels 9: 171. https://doi.org/10.1186/s13068-016-0581-3.1.

Negi, H., R. Agrawal, A. Verma and R. Goel. 2019. Chapter 14—Municipal solid waste to bioenergy: current status, opportunities, and challenges in indian context. pp. 191–203. *In*: Singh, J.S. and D.P. Singh (eds.). New and Future Developments in Microbial Biotechnology and Bioengineering. Elsevier.

Nogueira, D.P.K., A.F. Silva, O.Q. Araújo and R.M. Chaloub. 2015. Impact of temperature and light intensity on triacylglycerol accumulation in marine microalgae. Biomass Bioenergy 72: 280–287.

Oren, A. and L. Mana. 2003. Sugar metabolism in the extremely halophilic bacterium *Salinibacterruber*. FEMS Microbiol. Lett. 223(1): 83–87.

Padan, E., E. Bibi, M. Ito and T.A. Krulwich. 2005. Alkaline pH homeostasis in bacteria: new insights. Biochim. Biophys. ActaBiomembr. 1717(2): 67–88.

Quillaguamán, J., H. Guzmán, D. Van-Thuoc and R. Hatti-Kaul. 2010. Synthesis and production of polyhydroxyalkanoates by halophiles: Current potential and future prospects. Appl. Microbiol. Biotechnol. 85(6): 1687–1696.

Richard, H. and J.W. Foster. 2004. *Escherichia coli* glutamate-and arginine-dependent acid resistance systems increase internal pH and reverse transmembrane potential. J. Bacterial. 186(18): 6032–6041.

Riedel, K. and A. Lehner. 2007. Identification of proteins involved in osmotic stress response in *Enterobacter sakazakii* by proteomics. Proteomics 7: 1217–1231.

Robak, K. and M. Balcerek. 2018. Review of second generation bioethanol production from residual biomass. Food Technol. Biotechnol. 56(2): 174–187. https://doi.org/10.17113/ftb.56.02.18.5428.

Roberts, M.F. 2005. Organic compatible solutes of halotolerant and halophilic microorganisms. Saline Syst. 1: 5.

Russell, N.J. 2000. Toward a molecular understanding of cold activity of enzymes from psychrophiles. Extremophiles 4(2): 83–90.

Sarethy, I.P., Y. Saxena, A. Kapoor, M. Sharma, S.K. Sharma, V. Gupta and S. Gupta. 2011. Alkaliphilic bacteria: Applications in industrial biotechnology. J. Ind. Microbiol. Biotechnol. 38(7): 769.

Satlewal, A., R. Agrawal, S. Bhagia, J. Sangoro and A.J. Ragauskas. 2018a. Natural deep eutectic solvents for lignocellulosic biomass pretreatment: Recent developments, challenges and novel opportunities. Biotechnology Advances. doi: DOI 10.1016/j.biotechadv.2018.08.009.

Satlewal, A., R. Agrawal, S. Bhagia, P. Das and A.J. Ragauskas. 2018b. Rice straw as a feedstock for biofuels: Availability, recalcitrance, and chemical properties. Biofuels, Bioproducts and Biorefining 12(1): 83–107. doi: 10.1002/bbb.1818.

Setati, M.E. 2010. Diversity and industrial potential of hydrolase-producing halophilic/halotolerant eubacteria. Afr. J. Biotechnol. 9(11): 1555–1560.

Sharma, A., Y. Kawarabayasi and T. Satyanarayana. 2012. Acidophilic bacteria and archaea: Acid stable biocatalysts and their potential applications. Extremophiles 16(1): 1–19.

Shimada, H., N. Nemoto, Y. Shida, T. Oshima and A. Yamagishi. 2002. Complete polar lipid composition of *Thermoplasma acidophilum* HO-62 determined by high-performance liquid chromatography with evaporative light-scattering detection. J. Bacteriol. 184(2): 556–563.

Shrestha, N., G. Chilkoor, B. Vemuri, N. Rathinam, R.K. Sani and V. Gadhamshetty. 2018. Extremophiles for microbial-electrochemistry applications: A critical review. Bioresour. Technol. 255: 318–330.

Siddiqui, K.S., T.J. Williams, D. Wilkins, S. Yau, M.A. Allen, M.V. Brown, F.M. Lauro and R. Cavicchioli. 2013. Psychrophiles. Annual Review of Earth and Planetary Sciences 41: 87–115.

Singh, B., M.J. Poças-Fonseca, B.N. Johri and T. Satyanarayana. 2016. Thermophilic molds: biology and applications. Crit. Rev. Microbiol. 42(6): 985–1006.

Sleator, R.D. and C. Hill. 2002. Bacterial osmoadaptation: The role of osmolytes in bacterial stress and virulence. FEMS Microbiol. Rev. 26: 49–71.

Sterner, R.H. and W. Liebl. 2001. Thermophilic adaptation of proteins. Crit. Rev. Biochem. Mol. Biol. 36(1): 39–106.

Swartz, T.H., S. Ikewada, O. Ishikawa, M. Ito and T.A. Krulwich. 2005. The Mrp system: A giant among monovalent cation/proton antiporters? Extremophiles 9(5): 345–354.

Takeuchi, F., K. Iwahori, K. Kamimura, A. Negishi, T. Maeda and T. Sugio. 2001. Volatilization of mercury under acidic conditions from mercury-polluted soil by a mercury-resistant *Acidithiobacillus ferrooxidans* SUG 2-2. Biosci. Biotechnol. Biochem. 65: 1981–1986. https://doi.org/10.1271/bbb.65.1981.

Valentine, R.C. and D.L. Valentine. 2004. Omega-3 fatty acids in cellular membranes: A unified concept. Prog. Lipid Res. 43(5): 383–402.

Van Den Burg, B. 2003. Extremophiles as a source for novel enzymes. Current Opinion in Microbiology 6(3): 213–218.

Van Der Oost, J. and B. Siebers. 2007. The glycolytic pathways of Archaea: Evolution by tinkering. Archaea Evol. Physiol. Mol. Biol. 22: 247–260.

Verma A., H. Singh S. Anwar, A. Chattopadhyay, S. Kaur and G.S. Dhilon. 2017. Microbial keratinases: Industrial enzymes with waste management potential. Critical Reviews in Biotechnology 37(4): 476–491. https://doi.org/10.1080/07388551.2016.1185388.

Wang, Z., D.B. Hicks, A.A. Guffanti, K. Baldwin and T.A. Krulwich. 2004. Replacement of amino acid sequence features of a-and c-subunits of ATP synthases of alkaliphilic *Bacillus* with the *Bacillus* consensus sequence results in defective oxidative phosphorylation and non-fermentative growth at pH 10.5. J. Biol. Chem. 279(25): 26546–26554.

Yin, J., J.C. Chen, Q. Wu and G.Q. Chen. 2015. Halophiles, coming stars for industrial biotechnology. Biotechnol. Adv. 33(7): 1433–1442.

Zeldes, B.M., M.W. Keller, A.J. Loder, C.T. Straub, M.W. Adams and R.M. Kelly. 2015. Extremely thermophilic microorganisms as metabolic engineering platforms for production of fuels and industrial chemicals. Front. Microbiol. 6: 1209.

Zhu, D., W.A. Adebisi, F. Ahmad, S. Sethupathy, D. Blessing and J. Sun. 2020. Recent development of extremophilic bacteria and their application in biorefinery. Front. Bioeng. Biotechnol. 8: 843. https://doi.org/10.3389/fbioe.2020.00483.

3

Extremophiles
Life of Microorganisms in Extreme Environments

Rahul Kumar,[1,*] *Ramchander Merugu,*[2] *Swati Mohapatra,*[3]
Sneha Sharma[4] and *Hemlata*[4]

3.1 Introduction

An extreme environment is a habitat that is considered as very harsh or difficult to sustain because of its impressively extraordinary conditions, for example, temperature, pH, and availability to various vitality sources or under high pressure (Rampelotto 2013). Pressure conditions might be amazingly maximum or minimum; immersing or depressing content of oxygen or CO_2 level in the troposphere or Earth's atmosphere; elevated levels of harmful radiation, acidity, or alkalinity; the nonappearance of water; water containing a high convergence of salt or sugar; the existence of sulphur, crude oil, and other poisonous substances (M.L.E.R.). Extreme environment symbolizes an elite or unprecedented environment that is home for novel microorganisms including archaea, bacteria and fungi (Sayed et al. 2019). The diversity of microorganisms can survive nearly in each and every habitat available on the planet Earth (Fig. 3.1). They can take nutrition even from the extreme environmental circumstances for their reproduction, growth and development (Kumar and Sharma 2020b). The condition such as impartial range of pH, temperature limits between 4 to 40°C, atmospheric pressure ranging around one atm, availability of water, satisfactory amount of supplements and salts, low frequency of hydrostatic pressure as well as of ionizing radiations are essential to promote human living systems (Rothschild and Mancinelli 2001). Thus, ecological frameworks, for example, thermal springs, deserts, lakes having saline and alkaline nature, sea bottom areas and profound warm vent-holes that are not perfect for the growth and development as well as for the existence of human being refers to be "extreme" (Brock 1978, Satyanarayana et al. 2005). Extremophilic microorganisms can grow and develop in extreme or very harsh environmental circumstances (Oarga 2009). The characteristics and features of extremophilic microorganisms are fascinating microbiologists since early times. The microorganisms can be delineated as acidophilic (idealistic growth under pH range between 1.0 to 5.0) and alkaliphilic (ideal survival at pH more than 9.0); halophilic (ideal survival under maximum saline conditions); thermophilic (ideal growth and survival under temperature range of 60 to 80°C); hyperthermophilic (ideal survival and growth

[1] Chair of Hydrobiology and Fishery, Institute of Agricultural and Environmental Sciences, Estonian University of Life Sciences (EMU), Fr. R. Kreutzwaldi 1, Tartu 51006, Estonia.
[2] Department of Biochemistry, Mahatma Gandhi University, Anneparthy, Nalgonda-508254, Telangana State, India.
[3] Department of Infection Biology, School of Medicine, Wankwong University, Iksan, South Korea.
[4] Department of Environmental Sciences, H.N.B. Garhwal University (A Central University), Srinagar Garhwal- 246174, Uttarakhand, India.
* Email: rahul.khadwalia@gmail.com
ORCID ID: 0000-0003-4085-6148

at temperatures above 80°C) and psychrophilic (ideal survival at temperature of 10°C or below than this); Piezophilic, earlier known as barophilic (ideal growth in a hydrostatic pressure of high range); oligotrophic (development in nutrition deficient conditions); endolithic (grows within tiny pores of a stone) and xerophilic (grows in less water conditions or in deficiency of water). Few extremophilic organisms at the same time are adjusted to different growth conditions (polyextremophile), which includes thermoacidophiles and haloalkaliphiles (Encyclopaedia Britannica 2014). Among extremophiles, the thermophiles or heat-loving microbes can survive at a very high temperature (Beg et al. 2000, Akmar et al. 2011). Thermophiles are found in distinct biotopes like hot water ponds, geohot water silts and coastal solfataras (Rothschild and Manicineli 2001). Hot water ponds contain water temperature undoubtedly higher than the air temperature of the region where the hot water pond is located (Sen et al. 2010). Hot water ponds are the manifestations of tectonic activity and represent extreme environmental conditions, which are available across the Himalayan region (Kumar et al. 2004). Hot water ponds are considered to be an important source of heat-loving microbes that can be tapped for miscellaneous operations in distinct areas. Heat-loving or thermophilic microorganisms are having enzymes that are useful and effective even at the extreme high range of temperature, which makes these thermophilic microorganisms valuable for various industries including the medical or pharmaceutical sector (Tekere et al. 2015).

The investigation of extremophiles gives an inclusion of the physico-chemical boundaries characterizing existence on the planet Earth and can provide understanding about how life started on Earth. The term "extremophiles" refers to organisms that live under "extreme" environments. For every specific extreme or harsh environmental condition examined, an assortment of microorganisms demonstrated that the microorganisms not exclusively endure such growth circumstances but these circumstances are must for their growth and development (Pikuta and Hoover 2007). Extremophiles or extremophilic microorganisms are characterized by the environmental conditions wherein they grow and develop ideally. These hardy organisms are momentous not just as a result of the situations in which they live but in addition, a large number of them couldn't survive in evidently typical, moderate environmental conditions. Extreme environmental conditions might be viewed as normal, which become harder, for the growth and development of most living organisms (Merino et al. 2019). Extremophilic life forms are essentially prokaryotic (archaea and bacteria), and very few eukaryotes. Extremophiles can be found under each of the three life domains: Archaea, Bacteria and Eukarya (Fig. 3.2).

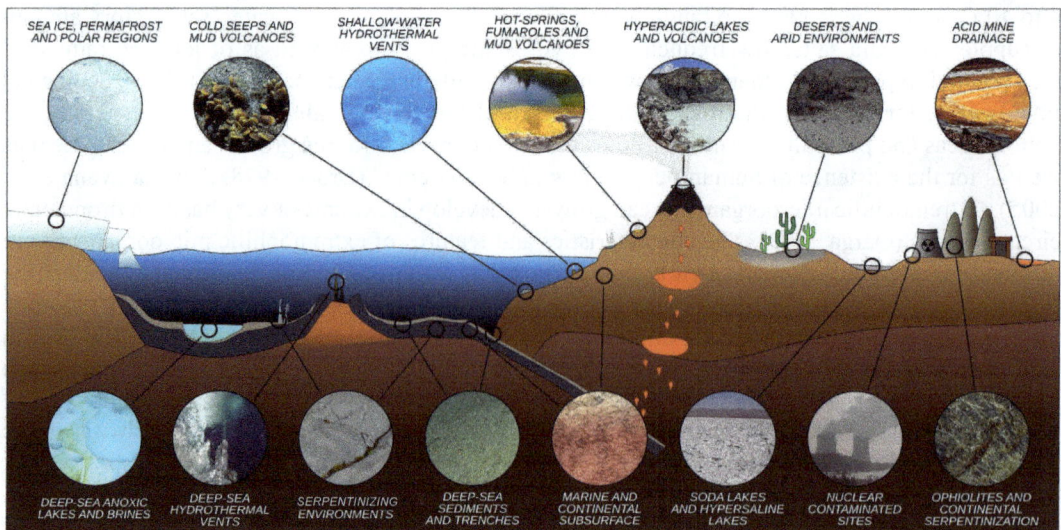

Fig. 3.1. Depicting a glorified cross-section of Earth's crust demonstrating the variety of extreme environments and their approximate location (Source: Merino et al. 2019).

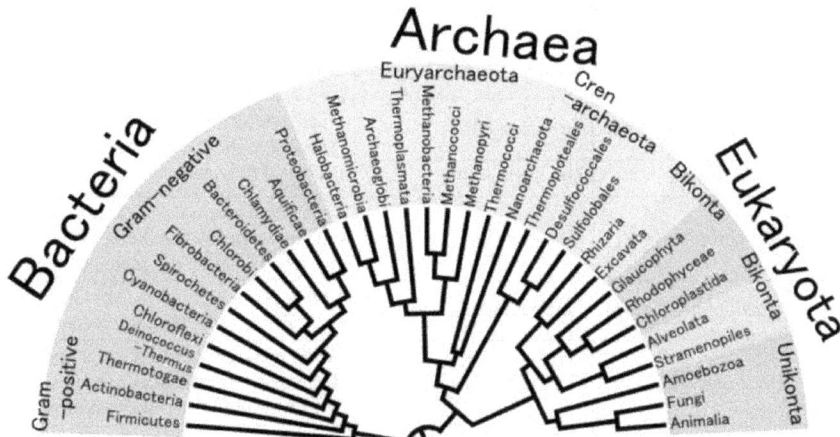

Fig. 3.2. The three domains of cellular life as indicated by metagenomic analysis (Source: Otohinoyi and Omodele 2015).

A lot bigger range of variety of organisms is realized that can endure harsh environmental circumstances and develop further, but it is not mandatory, ideally in harsh circumstances; such life forms are characterized to be extremotrophs (Mueller et al. 2005). The difference among extremophiles and extremotrophs isn't just a semantic one, and it features various crucial issues identifying with the exploratory investigation, for example, (1) wrong techniques or methods that have been utilized for separating presumptive extremophilic microorganisms, (2) presumptive extremophilic microorganisms might be undermined by serial dilution cultivation in artificial growth parameters or conditions, (3) lacking endeavors to decide if life forms or organisms are versatile to only little variations in natural factors. A lot of species can survive and sustain under extreme environmental conditions in an inactive form; unfortunately, they are not able to thrive or replicate endlessly in harsh or extreme growth circumstances. Since numerous extremophiles are a part of the domain 'Archaea' and mostly known archaeas are extremophiles, hence, occasionally, the terms have been utilized reciprocally. Most of the extremophilic organisms are microbes, such as the immediately recognized upper range of ideal survival temperature is 113°C in case of archaea, which becomes 95°C in case of bacteria, while 62°C for unicellular eukaryotes rather than multicellular eukaryotes that rarely appeared to grow over 50°C (Rothschild and Manicineli 2001). Microorganisms belonging to the Domain Archaea are extraordinarily hyperthermophilic, as they show ideal growth and survival at a temperature range of 80°C or more than this. The investigations of extremophilic and extremotrophic eukaryotic organisms have been moderately ignored in contrast with their prokaryotic partners. Nevertheless, extremophily is progressively revealed among algae and fungi, such as Dunaliella species (well known since very long time) and *Cyanidiales algae* (Toplin et al. 2008).

Extremophilic microorganisms or extremophiles are named as well as characterized by the conditions under which they live and survive (Table 3.1). The exploration of extreme environmental situations or conditions and investigations of extremophilic microorganisms not just offered significant experiences into science and development but additionally gave a piece of key information to extraterrestrial life since these tiny organisms are generally acknowledged to symbolise the earliest life forms available on the planet Earth. In addition, studies related to the taxonomy of extremophiles enable the Archaea to find its place as the third domain under the evolutionary or universal tree of life (Rainey and Oren 2006, Rothschild and Manicineli 2001). Higher organisms were accepted to be not able to survive in extreme environmental conditions most probably due to their cell multifaceted nature and compartmentalization. In this way, different research on the extremophiles mainly concentrated on organisms belonging to the Archaeal and Bacterial domains. However, a large number of eukaryotes, including both unicellular and multicellular, have emerged

Table 3.1. Classification of extremophiles.

Environmental Parameter	Type	Characteristics	Adaptation Mechanisms
Temperature	Hyperthermophiles	Growth > 80°C	Heat-stable biomolecules (proteins, nucleic acids, lipids). Production of long-chained, saturated, branched fatty acids and cyclic lipids
	Thermophiles	Growth 45–80°C	
	Psychrophiles	Growth < 10°C	Production of unsaturated fatty acids to counteract decreased membrane fluidity; Reduced cell size; Increased fraction of ordered cellular water
Radiation	Radiophiles	Nuclear radiation loving	Strong pigmentation (melanoids, carotenoids, etc.). Efficient DNA repair mechanisms
Pressure	Piezophiles	Pressure loving	When pressure increases, or temperature decreases, the molecules in lipid membranes pack tighter, resulting in decreased membrane fluidity
Dessication	Xerophiles	Anhydrobiotic (Mpa) −4 to −10 (bacteria); −10 to −70 (actinomycetes and fungi)	Accumulation of inorganic or organic osmolytes; Production of extracellular polysaccharides; They live inside rocks to use water condensation (spores)
Salinity	Halophiles	Salt loving (2–5 M)	Accumulating osmolytes. Salt-tolerant (and salt-dependent) enzymes
	Osmophiles	Survive in high sugar environments	
pH	Alkaliphiles	pH > 9	Membrane impermeability for OH-ions. Efficient proton uptake mediated by membrane antiporters. Negatively charged cell wall polymers.
	Acidophiles	Low pH loving	Proton exclusion, efficient efflux system. Acid stable membrane. Proton-driven secondary transporters
Chemicals	Metallophiles	Able to tolerate high concentrations of metal	Efflux pump. Sequestering and/or detoxification of metals (reduction, alkylation)
	Toxitolerant	Tolerates toxic and xenobiotic chemicals like benzene	Efficient efflux pump. Detoxification or decomposition of toxins and xenobiotics

Source: Rothschild and Mancinelli 2001, Horikoshi and Grant 1998, Torsvik and Ovreas 2007

to thrive and flourish under extreme environmental conditions (for example, *Alvinella pompejana, Tetrahymena thermophila* and *Dunaliella salina*). Hence, extremophilic microorganisms span Archaea, Bacteria and Eukarya (Andrade et al. 1999, Rothschild and Mancineli 2001). A range of heterotrophic and methanogenic archaea; photosynthetic, lithotrophic, and heterotrophic bacteria; and photosynthetic and heterotrophic eukaryotes are a part of halophilic microorganisms. There exists a range of microorganisms capable of enduring extreme dehydration, including fungi, yeast, bacteria, tardigrades, worms, etc. Microorganisms that can thrive under significant level of harmful radiation includes two Rubrobacter species, *Deinococcus radiodurans*, and the green algae *Dunaliella bardawil* (Rothschild and Mancineli 2001, Margesin and Schinner 2001, Rainey and Oren 2006). A large variety of higher organisms survives in normal or ordinary growth circumstances, i.e., these organisms develop and flourish in neural conditions like acidity, salt concentration, water requirement, oxygen availability, temperature, pressure, etc. (Grant 1988, Aguilar 1996, Aguilar et al. 1998, Antranikian et al. 2005)

The title "extremophile" was first coined by MacElroy in the year 1974 to depict these living organisms, and keeping in mind that, most of the extremophiles are prokaryotic including some species of protozoa, algae, and fungi (MacElroy 1974, Townsend 2012, Horikoshi and Grant 1998, Rainey and Oren 2006). "As conditions become more demanding, extreme environments become exclusively populated by prokaryotes" (Horikoshi 1998). A specific habitat with its unique environment probably contains microorganisms. An important example is the Dead Sea which was considered inert or without life, but in reality contains prokaryotes and even eukaryotes (Arahal et al. 2000, Buchalo et al. 1998). Extremophiles can be categorised under two vast or extensive categories: obligate extremophilic organisms that need at least one harsh environmental condition so as to develop and reproduce; the other one is facultative extremophilic organisms that can easily tolerate the harsh growth circumstances in which most of the microorganisms can't survive. During the present time, a few genera or orders include only organisms which have the capability to survive in extreme circumstances, while other genera or orders include both extremophilic and non-extremophilic organisms (Rampelotto 2013). Under few circumstances, the extremophiles are pretended to be phylogenetically more established ones (Wiegel and Adams 1998). The extremophilic organisms are best delineated by the min., max., and opt. attributes of the harsh environmental conditions that are required for growth and development, i.e., T_{min}, T_{opt}, and T_{max} in case of thermophilic microorganisms or thermophiles (Andrade et al. 1999). The review focuses on examining different classifications of microorganisms (Fig. 3.1).

Besides intellectual interest, enthusiasm for contemplating extremophiles originates out of their conceivable usefulness for industrial procedures, and the possible origins of their lives on Earth (Villar and Edwards 2006, Lentzen and Schwarz 2006, Javaux 2006, Wiegel and Adams 1998, Shock 1997, Litchfield 1998, Stetter 1996). A few instances of industrial utilizations of extremophiles are given in Table 3.4. Ferdinand Julius Cohn (1828–1898) is recognized as one of the founders of modern bacteriology. He represented the broad discussions on the blue-green algae (BGA) or algal mat near hot water springs in the 19th century itself (Cohn 1897). The most recent two decades demonstrated the interest associated with the microbes, especially those who can survive under harsh environmental circumstances. Many different extreme environments that were not explored before under the view of scientific and technical limitations are presently being explored, which has prompted the confinement of an abundance of novel microorganisms. One reason for the interest in extremophilic organisms is the possibility to examine the associated microorganisms or their processes performed in financially significant processes. Remarkably, the thermophilic microorganisms or thermophiles are responsible for yielding a range of proteins or enzymes widely utilized in the field of biotechnology (Hough and Danson 1999). A well-known example of extremophile-based biotechnology is the Taq polymerase enzymes used for DNA amplification by using PCR machine. The following enzyme obtained from a thermophilic microorganism *Thermus aquaticus*, which is isolated from a hot water spring located in the Yellowstone National Park (Brock and Freeze 1969), is the one and only enzyme obtained from an extremophile that has found various important applications in this field. The term "Superbugs" was coined various years back (Horikoshi and Grant 1991) for extremophilic organisms for their unique properties, signifies both the abilities of these organisms to live in unusual environments and the possibilities for their economic exploitation.

3.2 Types of Extremophiles

Extremophile word is derived from two different words, "*extremus*" and "*philia*". *Extremus* is a Latin word which means "extreme" and *philia* is a Greek word which means "love". So the meaning of extremophiles is those organisms who love extreme conditions or those organisms who love to grow or reproduce in extreme environmental conditions (Arora and Panosyan 2019). Extremophiles are primarily prokaryotes (archaea and bacteria) with few instances of eukaryotes like algae, fungi and protozoa (Rampelotto 2013). Extremophiles can be defined on the basis of environmental conditions. Extremophiles are divided into two broad categories. These categories are extremophilic and extremotolerant organisms (Gupta et al. 2014). Extremophilic organisms need at least one extreme environmental condition for their existence by all means; for example, thermophile and psychrophile. However, extremotolerant organisms survive ideally in moderate environmental circumstances but they can also survive in extreme environmental circumstances (Rampelotto 2010); for example, tardigrades (also known as water bears, they are small aquatic animals). Some species of tardigrades tolerate complete dehydration (Weronika and Lukasz 2017). On the basis of environmental conditions required for the growth and development of microorganisms, the extremophiles can be categorized into various groups (Table 3.1).

3.2.1 Thermophiles

Among all the requirements, temperature is one of the basic and physical factors of the environment that performs an important function in the growth and distribution of microorganisms on Earth. Temperature is considered as one of the major environmental factor that controls the activities and survival of microbes (Kumar and Sharma 2020a, Kumar et al. 2013, Takacs-Vesbach et al. 2008, Abou-Shanab 2007). Thermophilic microbes are those that can survive under temperature ranging within 45°C to 80°C (Table 3.2). Thermophiles mostly inhabit hot springs. Thermophiles possess physically and chemically stabilized enzymes, and higher metabolism (Princeteejay 2013). Most of the thermophiles belong to the domain Archaea. *Methanopyrus kandleri* is a thermophilic microorganism that can grow and reproduce at temperature in excess of 250°F (Vieille and Zeikus 2001). The other instances of thermophilic microorganisms are *Thermus aquaticus, Thermococcus litoralis, Bacillus stearothermophilus,* etc. A broad range of thermophilic microorganisms have been isolated from thermal springs, mud pots, geysers, hot spring algal mat, and active volcanoes (Kumar and Sharma 2020b). A wide range of thermal springs or hot water springs are available across the world. Few of them are Kamchatka in Russia and Gauri Kund, Draupadi Kund, Gangnani Kund, Tapovan, Narad Kund, Tapt Kund, Tattapani, etc., in India. These thermophilic habitats are the home of wide range of thermophilic and thermo tolerant microorganisms.

Table 3.2. "Record-holding" extremophiles and their environmental limits.

Environmental factors	Low/High	Microorganisms	Habitat	Phylogenetic affiliation	Tolerance to stress
Temperature	High	*Pyrolobus fumarii*	Hot undersea hydrothermal vents	Crenarchaeota	Maximum 113°C, Optimum 106°C, Minimum 90°C
	Low	*Polaromonas vacuolata*	Sea-Ice	Bacteria	Minimum 0°C Optimum 4°C, Maximum 12°C
pH	Low	*Picrophilus oshimae*	Acidic hot springs	Euryarchaeota	Minimum pH −0.06, Optimum pH 0.7, Maximum pH 4 (is also thermophilic)
	High	*Natronobacterium gregoryi*	Soda lakes	Euryarchaeota	Maximum pH 12, Optimum pH 10, Minimum pH 8.5 (is also halophilic)
Hydrostatic pressure	-	Strain MT41	Mariana Trench	Bacteria	Maximum > 100 MPa, Optimum 70 Mpa, Minimum 50 Mpa
Salt Concentration	-	*Halobacterium salinarum*	Salt lakes, salted hides, salted fish	Euryarchaeota	Maximum NaCl saturation, Optimum 250 g l^{-1} salt, Minimum 150 g l^{-1} salt
Ultraviolet and ionizing radiation	-	*Deinococcus radiodurans*	Isolated from ground meat, radioactive waste, nasal secretion of elephant; true habitats unknown	Bacteria	Resistant to 1.5 kGy gamma radiation and to 1500 J m^{-2} of ultraviolet radiation

Source: Ferreira et al. 1997, Madigan 2000, Seckbach and Oren 2004, Rainey and Oren 2006

Mechanism of Survival

i. **Protein Stability:** A highly hydrophobic core, that reduces the tendency of the proteins to unfold. Chaperonins are also present which functions to refold partially denatured proteins (Motojima 2015).

ii. **DNA Stability:** High temperature is lethal to DNA as it melts the DNA (Grogan 2002). A variety of mechanisms are involved in stabilizing it and these are: Thermostabilizers, Reverse Gyrase and DNA binding proteins.

iii. **Lipid Stability:** Lipid bilayers consist of branched chain hydrocarbons that are linked through ether linkages to glycerol (Chong et al. 2012). All thermophiles produce lipids constructed upon the diphytanyl tetraether model which is more thermotolerant.

3.2.2 Hyperthermophiles

A hyperthermophilic microorganism is one that can live and thrive at very high or extreme high temperature range between 80°C to 110°C (Brock 1986, Singh et al. 2012, Kumar and Sharma 2019a,b) (Table 3.2). Hyperthermophiles are one of the three types of groups of thermophiles. Many hyperthermophiles are from the domain Archaea found in thermal springs and active volcanoes (Princeteejay 2013). A wide range of hot water springs are available across the world. Few of them are Yellowstone thermal spring at Yellowstone National Park in the USA, Surya Kund, Saldhar, Ringigad, etc., in India. These hyperthermophilic microorganisms also follow similar mechanism of survival as that of thermophilic microorganisms.

3.2.3 Psychrophiles

Among the extremophiles, the psychrophilic bacteria, fungi and actinomycetes are the organisms which can grow, survive and produce some enzymes and compounds having an economical value at very low temperatures, mainly between 10°C to –10°C and less than this temperature (Morita 1975). The term psychrophile means cold loving or the organisms that love to grow in cold temperature. The word psychrophile was initially used by Schmidt-Nielsen in 1992 (Ingraham and Stokes 1959). Psychrophilic microorganisms are those microbes that grow well at temperature near to water's freezing point (Table 3.2). For example: *Moritella profunda,* a psychropiezophilic microorganism adapted to cold and living in deep sea that shows maximum growth at 2°C (Princeteejay 2013). The first reporting of the existence of bacteria that are capable of growing at the temperature of 0°C was done in the year 1887 by Forster. Microorganisms that are cold loving can be termed as psychrophiles. The psychrophiles have cardinal growth temperatures (Amico et al. 2006). Microorganisms that survive and reproduce better at temperature from 10 to –10°C or less than this are termed as psychrophiles. Example: *Psychtobacter* spp., *Candida Antarctica, Ralstonia eutropha, Acinetobacter lwoffii, Pseudomonas aeruginosa, Microbacterium paraoxydans, Arthrobacterium sulfonivorans* and *Microbacterium scheliferi.* Most of the strictly characterized psychrophilic microorganisms can be isolated from high altitude glacier-fed lakes, snow-fed lakes, Arctic and Antarctica regions or Polar Regions, glaciers and their snouts, etc. (Foster and Winans 1975, Kumar and Sharma 2020c, 2021). Such extreme cold habitats are available across the world, such as Satopanth Lake, Neel Tal, Roop Kund, Dev Tal, Chorabari Tal, Kedar Tal, etc., which are glacier-fed lakes. Hemkund Lake, Maldaru Tal, Lamb Tal, Masuri Tal, Lingam Tal, Brahmi Tal, etc., are snow-fed lakes. Satopanth, Bhagirathi-Kharak, Gangotri, Yamunotri, Banderpoonch, Chaukhamba, Khatling, Milam, Kafni, Dronagiri, etc., are the famous glaciers. These psychrophilic habitats are the home of wide range of psychrophilic and psychrotolerant microorganisms or psychrotrophs.

Mechanism of Survival

The smoothness of membranes diminishes with declining temperature. Accordingly, living organisms enhance the proportion of unsaturated to saturated fatty acids (Brenchly 2008). Furthermore, the ability to withstand temperatures below freezing point depends on two main strategies: protection of the cells from ice formation or deposition of ice crystals by the process of freezing evasion, and whether ice crystal creates protection from any possible harm or damage during defrosting or melting process (Bialkoska et al. 2020). The proteins utilized for both the procedures are misleadingly named 'antifreeze' molecules—molecules that really permits the occurrence of hysteresis (Cavicchioli et al. 2011a). Hysteresis brings down the freezing point of water up to 9–18°C, which enables their survival (Rothschild and Mancinelli 2001).

3.2.4 Alkaliphiles

Alkaliphiles generally refer to microorganisms that grow well at high pH 9 (Table 3.2). Some of the alkaliphiles strains grow at pH 10–13 (Horikoshi 1999). Obligate alkaliphiles are those alkaliphiles that grow only at pH 9 or above. Facultative akaliphiles are those that can grow in optimally alkaline condition and can also survive in neutral pH. Alkaliphiles are useful source of stable cells and enzymes that are used in high pH range for biotechnological and other applications (Princeteejay 2013). Example: *Natronomonas pharaonis, Thiohalospira alkaliphila, Natronococcus gregory, Deinococcus thermus,* and *Pyrolobus fumarii.*

Mechanism of Survival

Under the conditions of high alkalinity, the concentration of H^+ ions is extremely low and the cells experience difficulty in the utilization of ATP-synthase in order to generate energy. Alkaliphiles or high pH loving microorganisms circumvent such issues by actively pumping these ions and transporting others in order to manage their internal procedures at neutral pH. Also, the cell wall of the alkaliphiles serves like a safeguard of harsh environmental circumstances (Preiss et al. 2015).

3.2.5 Acidophiles

Acidophiles are the living microbes that can grow well in pH 0 to 5.5 (Table 3.2). Fungi are more tolerant to acidic condition than bacteria (Princeteejay 2013). Many bacteria are acidophilic whereas, some of them are acidophilic. Some of them are obligate acidophiles (unable to grow in neutral pH 7). Acidophiles are able to sustain in this acidic habitat because its cells pump out noxious hydrogen ions quickly enough not to harm the DNA inside the nucleus. If they couldn't do this, then acidophilic microorganisms would not be able to survive (Reed et al. 2013). Examples: *Acidithiobacillus, Laptospirillum,* several genera of archaebacteria such as Ferroplasma, thermoplasma. Majority of the acidophiles have been identified from magma or lava and acidic mine drainage (Okibe et al. 2003). Acidophiles are not simply present in fascinating conditions, for example, Yellowstone National Park or deep-sea hydrothermal vents.

Mechanism of Survival (Baker-Austin and Dopson 2007, Princeteejay 2013) (Fig. 3.3)

 i. The cell membrane is profoundly impermeable to protons
 ii. Membrane channels have a decreased pore size
 iii. Excess protons are siphoned out of the cell
 iv. Proton uncoupling by organic acid
 v. DNA and protein harm caused by low pH can be fixed by chaperones
 vi. Intracellular compounds may be settled by 'iron rivets'

3.2.6 Halophiles

Halophiles are the living organisms that can survive in an environment having high salt concentration (Ma et al. 2010). The name "halophiles" signifies "salt-adoring" in Greek. Most of the halophiles are considered as bacteria, while some are categorized as primitive eukaryotes. Eukaryotes are very

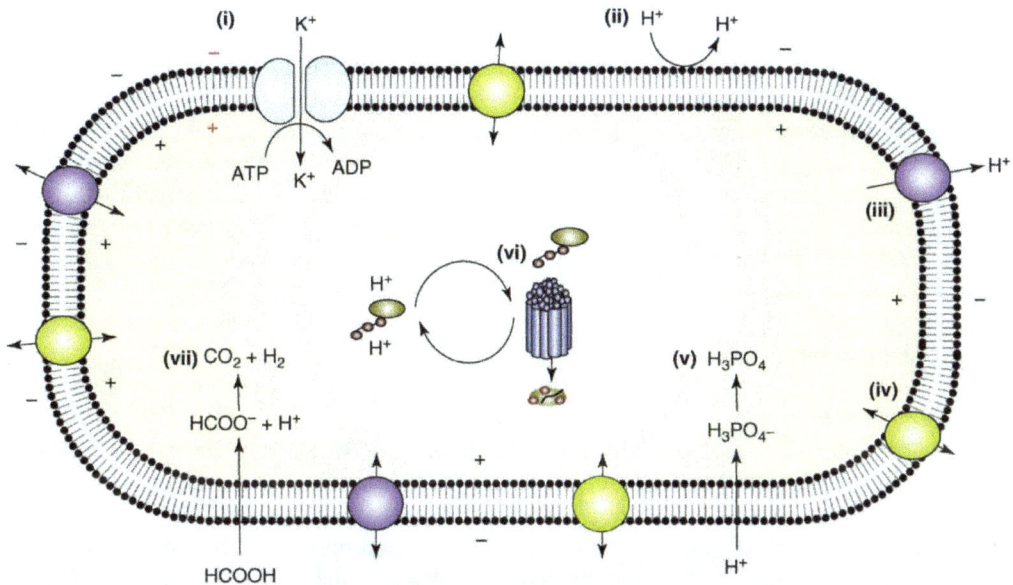

Fig. 3.3. Processes associated with pH homeostasis in acidophiles. (i) Acidophiles reverse the Dc to partially deflect the inward flow of protons. One potential mechanism of generating a reversed Dc is by potassium transport—a predominance of potassium-transporting ATPases is found in acidophile genomes. (ii) Many acidophiles have evolved highly impermeable cell membranes to retard the influx of protons into the cell. (iii) DpH is maintained through active proton export by transporters (blue). (iv) The sequencing of several acidophile genome sequences has indicated that there is a higher proportion of secondary transporters (green) than in neutralophiles. Overall, they reduce the energy demands associated with pumping necessary solutes and nutrients into the cell. (v) The presence and availability of enzymes and/or chemicals capable of binding and sequestering protons might help to maintain pH homeostasis. (vi) Comparative genome analysis suggests that a larger proportion of DNA and protein repair systems might be present in acidophiles compared with neutralophiles and that this could be associated with the cellular demands of life at low pH. (vii) Organic acids that function as uncouplers in acidophiles might be degraded by heterotrophic acidophiles (Source: Baker-Austin and Dopson 2007).

complex and unpredictable living organisms having a nucleus and several organelles conjugated with the cell membranes (Princeteejay 2013). These extremophiles flourish in environmental conditions that have an extreme level of salinity (Table 3.2). They possess ability to balance osmotic pressure and resist denaturing impact of salts. Halophiles are found in places having high salinity, for example, the Great Salt Lake and the Dead Sea (Ma et al. 2010). They are one of a kind since they require significant levels of salt that would be deadly to most of the living organisms. Example: *Halobacterium halobium, Dunaliella salina* (Oren 2008).

Classification: Halophiles are classified as Slightly halophile: 2–5%; Moderately halophile: 5–10%; Extremely halophile: 20–30%; Non halophile: less than 2% and Halotolerant—those that doesn't need high salt to grow but are not affected by it (Ventosa et al. 1998). The two biggest and best-examined hypersaline water bodies, i.e., the pH of both the Great Salt Lake and the Dead Sea range near 7.0, that is, neutral, in spite of the fact that the Great Salt Lake is somewhat basic while the Dead Sea is near to acidic. The Sambhar Salt Lake is also a perfect example of hypersaline lake that is located in the Jaipur city of Rajasthan in India along the National Highway 8.

Mechanism of survival

Osmoregulation: Osmosis is the process of flow of solvent from the region of higher to the lower concentration (Janacek and Sigler 2000). Halophiles maintain high solute concentration within their cells (Oren 1999). Few variants of halophiles have unusual chloride pumps that transfer chloride ions from the environment to the cells.

3.2.7 Piezophiles

Piezophiles, previously termed as barophiles. These are those microorganisms that can survive under great or extreme pressures (Table 3.2). The pressure often exceeds 380 atm (38 MPa). Microorganisms that prefer high pressure conditions above 10 mpa for their growth are called piezophiles (Kato 2011). Deep-sea, deep-sea hydrothermal vents, deep-oceanic crust, sedimentary sub-seafloor, deep-sea cold seep and continental deep-biosphere are the home to these microorganisms (Kato 2011). Enzymes isolated from piezophiles are stable at high pressure and do not need specific pressure related adaptations; for example, *Pyrococcus yayanosii* can survive at pressures ranges to 150 MPa (Abe and Horikoshi 2001, Zeng et al. 2009, Ichiye 2018). The majority of piezophiles are bacteria, and can be grouped under any one of the five genera—Shewanella, Colwellia, Photobacterium, Moritella, and Psychromonas.

Mechanism of survival

Protein-protein interactions are delicate to enhanced pressure, which can become the explanation behind enzyme dissociation (Sharma et al. 2002). The pressure is also known to alter gene expression. Furthermore, when pressure increases, or temperature decreases, the molecules in lipid membranes that are packed more tightly result in diminished membrane fluidity (Bartlett 2002). Living organisms frequently circumvent this issue by enhancing the extent of unsaturated fatty acids under their membranes (Aertsen et al. 2009). Pressure influences chemical equilibria and rates of reaction, on the basis of involved reaction and activation volumes (Oger and Jebbar 2010). Other mechanisms applied in piezophilic organisms are:

Fatty Acids: Deep-ocean microorganisms are considered to preserve membrane functionality at extreme high pressure and low temperature by expanding the ratio of unsaturated fatty acids in their lipids (Simonato et al. 2006, Bartlett 2002).

Respiratory Chain: The transmembrane proteins play a key role in cell stabilization. High-pressure consequences on such proteins could emerge from the impact of pressure on proteins in a direct way or on the lipid environment in which they work (Simonato et al. 2006).

Membrane Proteins: Outer membrane protein high pressure is accurate at 28 MPa, whereas outer membrane protein low pressure is accurate at 0.1 MPa (Bartlett 2002).

3.2.8 Radiophiles

These are those living microbes that are able to live within the sight of high radiation levels or these microorganisms are radiation lovers (Rampelotto 2013). Radiophiles are the microorganisms that can thrive in conditions with high level of radiations including ultraviolet and nuclear radiation (Rasuk et al. 2016). These microorganisms are highly resistant to high level of ionizing and UV radiations (Table 3.2). For example: *Deinococcus radiodurans.* The bacterium *Deinococcus radiodurans* can grow continuously, without any effect on its growth rate, in the presence of 6,000 rad/hr (Lange et al. 1998).

Mechanism of survival

Radiophiles are highly resistant to high level of ionizing and ultraviolet radiations (Daly 2011). They show high radioresisteance and have growth of > 400 J/m^2 UV. Examples of radiophiles are *Deinococcus radiodurans* (Sandigursky et al. 2004), *Deinococcus radiophilus* (Yun and Lee 2004), *Thermococcus marinus* and *Thermococcus radiotolerans* (Jolivet et al. 2004).

   These microorganisms thrive in such extreme environments due to their defensive mechanisms provided by primary and secondary metabolic products, usually characterized by extremolytes and extremozymes (Daly et al. 2007). These products play a role in absorbing wide spectrum of radiation, thereby preserving the organism's DNA. Examples of these metabolites include biopterin, phlorotannin, porphyra-334, palythine, shinorine, mycosporine-like amino acids and scytonemin (Gabani and Singh 2013). Other mechanisms employed by these microorganisms include the use of polyploidy strategy (presence of identical duplicates of chromosomes); this prevents total chromosome damaged in such a way that there is an extra chromosome available, which should be the one to be destroyed as a result of radiation, resulting in the survival of the microorganisms (Singh and Gabani 2011, Gabani and Singh 2013). In addition to the polyploidy strategy, microorganisms also have a conventional DNA repair mechanism; an example is *Deinococcus radiodurans* that utilizes RecA-independent double strand break repair mechanism identified as synthesis-dependent single strand annealing (Seckbach et al. 2013, Biello 2006).

Furthermore, studies have also reveal that manganese (Mn2+) and iron (Fe2+) intracellular level play a role in making radioresistant microorganisms have resistance to radiation (Gabani et al. 2014). It was found that in *Deinococcus radiodurans*, ionizing radiation leads to oxidative modification of proteins, consequently destroying them (Singh and Gabani 2011). However, it was shown that Mn2+ protects these proteins by forming a complex with phosphate, reducing the reactive superoxides to peroxides, while iron Fe2+ helps in the repair and reannealing of the damaged DNA fragments (Daly et al. 2007). Further study on *Deinococcus radiodurans* shows that after exposure to extreme radiation, the shattered DNA each functions as a template, the damaged ends are removed and fragments are joined with corresponding nucleotide sequence by the help of the enzyme PolA. The newly formed single strand then utilizes the Waston and Crick theory of base pairing, and thereby nucleotide thymine is annealed with 5-bromodeoxyuridine specifically in order to have a functional set of chromosomes (Biello 2006).

3.2.9 Metallophiles

The extremophilic microorganisms are able to live under the conditions with high metal ion concentration (Table 3.2). For example: *Ferroplasma* species. Metallophytes have unique characteristics, which allow them to develop biological mechanisms in order to survive and reproduce on metalliferous conditions without affecting from their toxicity. Numerous metallophytes are known in temperate territories of Northern Hemisphere as they act as bioindicators against high heavy metals concentration (Midhat et al. 2019).

Mechanism of survival

Metallophiles have excellent tolerance for high concentration of heavy metals such as cadmium, arsenic, lead, mercury, and copper (Tchounwou et al. 2012). Metallophilic microorganisms have their origin from volcanic regions, geothermal and hydrothermal vents as well as from industrially contaminated locations (Arora 2019). These microorganisms implicitly possess characteristics that can influence the biomineralization process by altering the physicochemical behaviour of surroundings, metal speciation and toxicity (Gadd and Pan 2015). They have an endogenous ability to delicately modulate their physiology in order to overcome the harmful effects of external toxic metal conditions.

Microorganisms able to tolerate heavy metals more than other microorganisms in extreme conditions have led to the emergence of heavy metals tolerant species and are referred to as metallophiles (Reed et al. 2013). These tolerant microorganisms handle heavy metals by either carrying out homeostasis on essential heavy metals like every other microorganism or detoxify non-essential heavy metals, as well as excess essential heavy metals (Rajakaruna and Boyd 2013). They detoxify by using an efflux pump mechanism, which pumps out these heavy metals via essential metal transporters and/or by biotransformation or precipitation of heavy metals using their cellular proteins (Rothschild and Mancinelli 2001).

3.2.10 Xerophiles

These microorganisms grow in extremely dry circumstances or the conditions having deficiency of water (Table 3.2). For example: *Trichosporonoides nigrescens*. A xerophile (in Greek, xēros stands for 'dry', and philos stands for 'loving') is an extremophilic organism capable of growing and reproducing in conditions with low availability of water (Rampelotto 2013, Arora and Panosyan 2019). The xerophiles are also known as "xerotolerant", i.e., tolerant of dry conditions. They are capable of surviving in an environment with the water activity below 0.8, above which it is typical for most of the living organisms on the planet Earth (Grant 2004). Typically, the activity of xerotolerance is utilized with respect to matric drying, where a substance has lesser water concentration. These environments incorporate parched desert soil. The term osmotolerance is generally applicable to living organisms that survive in solutions with high concentration of solutes, for example, halophiles (Stratford et al. 2019). Xeric stress caused due to deficiency of water known as desiccation and high solute concentration known as hypertonicity are responsible for excess loss of water from the cells, leading to pressure on components such as metabolic, biochemical, physiological and physical. The physiological and molecular adaptations of xerophiles enable them to flourish under stringent xeric condition (Lebre et al. 2017). Other than desert soil, these microorganisms can survive in hypersaline aquatic conditions; for example, *Halobacterium salinarum* NRC-1 (Arora and Panosyan 2019).

Mechanism of Survival

Xerophiles survive desiccation by entering a state of metabolism called anhydrobiosis, thereby inhibiting water loss until they are rehydrated (Thiel 2014). They do this by retaining intracellular moisture against a concentration gradient by accumulating or synthesizing compatible solutes internally to high concentration, in so doing maintaining cell turgor (Seckbach et al. 2013). These compatible solutes are small organic molecules such as polyols, amino acids and sugars, protecting enzymes from unfolding during the intense conditions such as dehydration, allowing usual cellular activities to continue (Pitt and Hocking 1980). However, survival depends on the water activity of the environment during this period and the rate of moisture loss by organism to the environment (Seckbach et al. 2013). Xerophiles on rehydration undergo tissue repair in order to achieve optimal metabolism, enabling them to grow and reproduce effectively on rehydration (Thiel 2014).

3.2.11 Poly-extremophiles

The term polyextremophile describes organisms which can thrive in more than one extreme. Example includes an archaea, *Sulfolobus acidocaldarius*, that flourishes under pH 3 and 80°C (Rothschild and Mancinelli 2001). They are able to develop under concomitant extreme conditions required for growth and development, such as the high concentration of sodium salt, high pH and inflated temperature. However, only a very few extreme halophilic microorganisms have the capacity to survive under extreme growth conditions like high pH and high temperature. Microorganisms which are able to grow at high pH temperature are termed as extremophiles (Table 3.3), and those that can sustain all the above conditions are called "polyextremophiles" (Wiegel and Kevbrin 2003, Roberts 2005, Bowers et al. 2009).

3.3 Importance of Extremophiles

Extremophiles are useful in the field of biotechnology, as they produce extremozymes. Extremozymes can be explained as those which can catalyze a reaction in extreme conditions of temperature, pH and pressures and hence can be used in industries (Coker 2016) (Table 3.4). In order to thrive under extreme environmental conditions, extremophiles establish a variety of molecular strategies. The extremophilic microorganisms survive under extreme conditions by acclimatizing in different ranges of environmental conditions. This includes high temperature ranging from 45°C to 121°C and lowest ranging temperature of –2°C to 20°C. The pressure tolerance range of extremophiles is more than 500 atmospheres. The microorganisms have high tolerance for alkaline pH (pH > 8) as well as for acidic pH (pH < 4). They can exist in radiations (UVR resistance > 600 J/m), diverse saline conditions (2–5 M KCl or NaCl), utmost heavy metals (arsenic, zinc, cadmium), and absence of nutrients (e.g., water, ice, air, rock, or soil) (Cavicchioli et al. 2011b, Deppe et al. 2005, Navarro-

Table 3.3. Some examples of "polyextremophilic" microorganisms, adapted to life at a combination of several environmental extremes.

Environmental factors	Microorganisms	Habitat	Phylogenetic affiliation	Tolerance to stress
Low pH and High temperature	*Picrophilus oshimae*	Acidic hot springs	Euuyachaeota	Minimum pH: 0.06; Max. temperature: 65°C
	Cyanidium caldarium	Acidic hot springs	Eukarya	Max. temperature: 57°C; Minimum pH: 0.2
High pH and High Salt Concentration	*Natronobacterium gregoryi*	Soda Lakes	Archaea-Euryarchaeota	Max. pH: 12; NaCl saturation
High temperature and High pH	*Thermococcus alkaliphilus*	Shallow marine hydrothermal springs	Archaea-Euryarchaeota	Max. temperature: 90°C; Max. pH: 10.5
High temperature and High pressure	*Thermococcus barophilus*	Thermal vent; Mid-Atlantic Ridge	Archaea-Euryarchaeota	Max. temperature: 100°C; Requires 15-17.5 Mpa pressure at highest temp.
High temperature and High radiation	*Deinococcus geothermalis*	Thermal springs	Bacteria	Max. temperature 50°C; Resistant to at least 10 kGy gamma radiations
Low temperature and High pressure	All deep-sea bacteria	Deep Sea Environments	Bacteria	Live at 2–4°C and Pressure of 50-110Mpa

Source: Seckbach and Oren 2004, Rainey and Oren 2006

Table 3.4. Extremophiles and some examples of their biotechnological products and applications.

Extremophiles	Applications
Thermophiles and Hyperthermophiles	
DNA polymerases	DNA amplification by PCR or Genetic Engineering
Lipases, Pullulanases and Proteases	Detergents, Baking, Brewing
Amylases	Glucose, Fructose for sweeteners
Xylanases	Paper bleaching
Halophiles	
Bacteriorhodopsin	Optical switches and photocurrent generators
Lipids	Liposomes for drug delivery and cosmetics
Compatible solutes, e.g., Ectoin	Protein, DNA and Cell Protectants
α-Linoleic acid, β-Carotene and Cell extracts, e.g., *Spirulina* and *Dunaliella*	Health foods, dietary supplements, food colouring and feedstock
Psychrophiles	
Alkaline phosphates	Molecular biology
Proteases, lipases and cellulases	Detergents, Cheese maturation, Dairy production
Amylases	Polymer degradation in detergents
Dehydrogenases	Biosensors
Polyunsaturated fatty acids	Food additives, dietary supplements
Ice nucleating proteins	Artificial snow, food industry, e.g., ice cream
Alkaliphiles and Acidophiles	
Proteases, cellulases, lipases and pullulanases	Detergents
Elastases, Keritinases	Hide de-hairing
Cyclodextrins	Foodstuffs, chemicals and pharmaceuticals
Acidophiles	Fine papers, waste treatment and degumming
Sulphur oxidizing acidophiles	Recovery of metals and de-sulphurication of coal
Chalcopyrite concentrate	Valuable metals' recovery
Acidophiles	Organic acids and solvents
Piezophiles	
Whole microorganisms	Formation of gels and starch granules
Matallophiles	
Whole microorganisms	Ore-bleaching, bioremediation, biomineralization
Radiophiles	
Whole microorganisms	Bioremediation of radionuclide contaminated sites

Source: Demirjian et al. 2001, Kasar 2013

González et al. 2009, Seitz et al. 1997). The extremophilic microorganisms are popular under studies since decades; however, thermophilic, halophilic, acidophilic, and piezophilic proteins, due to their applications in biotechnology and industries, are increasingly getting attention (Cárdenas et al. 2010, López-López et al. 2014, Yildiz et al. 2015). Other than this, the classification of extremophiles also includes alkaliphile, hyperthermophile, metalotolerant, psychrophile, radioresistant, toxitolerant, and xerophile (Cowan et al. 2015, Qin et al. 2009). The application of more than 3,000 enzymes in the field of biotechnology and industry has been identified but the supply of enzymes is still not meeting the demand of industries (Demirjian et al. 2001, Van den Burg 2003).

3.3.1 Applications of Thermophilic Microorganisms

Thermophilic enzymes are highly useful in biotechnological application and show high degree of tolerance at extreme temperatures (Singh et al. 2011a). Thermozymes such as cellulases, xylanases, amylases, proteases have various uses in the field of biotechnology and industries. The thermostable cellulases have potential to enable the competent and cost-effective types of the saccharification and fermentation method in order to transform cellulosic biomass into biofuels. The xylanases, a hemicellulose part of agro-industrial dregs, contribute in derivation of hexose and pentose sugars having various biotechnological uses. Microbial xylanasesis has been getting popularity since few years as it signifies one of the largest industrial enzyme group. Various sectors such as food, feed, textile, fuel, detergents, paper and pulp industries and waste treatment sectors use the biotechnological applications for xylanases (Azeri et al. 2010). The applications in the paper bleaching, and in treatment of environmental contaminants are significant features of cellulase, hemicellulases, and xylanases (De Pascale et al. 2008, Unsworth et al. 2007). Amylases are one of the most important enzymes used in the field of biotechnology, mainly in process of starch hydrolysis. There are different sources of amylases like plants, animals and microorganisms but the microbial amylases, due to their productivity and thermostability, are widely useful in the industrial sectors (Burhan et al. 2003). Different strains of microorganisms producing amylase have applications in food industry (fermented food media), while soil microorganism originating from wastes and mud are useful mainly in chemical industry. Amylase derived from *Pyrococcus furiosus* has huge scope for usefulness in mutational studies. Maltoheptaose produced from mutation in pancreatic fistula amylase can be used for industrial applications (Rosenbaum et al. 2012).

The proteolytic enzymes of thermophiles show a peculiar feature due to their stability and activeness even at high temperature. Another characteristic includes resistance to organic solvents, detergents, tolerance to high and low pH and other denaturants. The following features allow its use in technological processes. Reduced risk of microbial contamination, lower viscosity and better susceptibility to some proteins as well as enzyme molecules are some of the advantages of Proteolytic thermozymes (Synowiecki 2010). The production of dipeptides and starch-processing is among the different applications of proteases (Bruins et al. 2001, Jayakumar et al. 2012). Thermostable enzymes such as lipases are used in biotechnological and industrial processes such as esterification, interesterification, transesterification, grease hydrolysis and organic biosynthesis (Staiano et al. 2005). The other application of lipase involves the paper industry, leather industry, milk industry, dyed products manufacturing, and in pharmaceuticals (Eichler 2001, Haki and Rakshit 2003). Biodetergents retain enzymes such as amylase, cellulase, protease and lipase having high degree of resistance to the extreme conditions. Thermophilic DNA polymerases study can help in developing innovative technologies for human health, e.g., methylation specific PCR (R. L. Momparler, Universite´ de Montre´al). The following technique is used for diagnoses of methylated cancer related genes at premature stages; the applicability of the method is enhanced due to specifically amplified methylated targets of thermostable DNA polymerases (Rossi et al. 2003).

a. **Textile dyes remediation:** The bacterium *Geobacillus thermocatenulatus* produces laccases which show very higher catalytic activity and this enzyme is higly stable at varying temperature and pH levels and can be used widely for the removal of dyes that cause environmental pollution (Verma and Shirkot 2014).

b. **Dairy processing:** Thermophilic *bacilli* such as *Anoxybacillus flavithermus* and *Geobacillus* spp. are used as hygiene indicator in dairy processing (Burgess et al. 2010).

c. **Medical application:** Lactobacillus, enterococci and yeasts can be used as probiotics and can be used for treating different kinds of gastrointestinal problems (Fijan 2014).

d. **Agricultural residues saccharification:** *Sporotricum thermophile* LAR5 is able to use agricultural waste to produce cellulose titre which is very stable at different pH and high temperature range (Bajaj et al. 2014).

3.3.2 Applications of Hyperthermophilic Microorganisms

Enzymes obtained from extreme heat-loving microorganisms are of paramount importance as they act as biocatalysts for various industrial applications. Such enzymes can work effectively even at extreme high temperatures (Lopez et al. 2013). The bacterial species such as *Bacillus subtilis* have antidiarrhoeal properties. This species is also used to produce fermented soybeans, yogurt, ice cream, milk and cheese. The bacterial species *Streptococcus thermophilus* has a property to boost the immunity of a human body and can also be used in the fermentation of cheese. *Brevibacillus borstelensis* can be used for the degradation of polyethylene and long-chain hydrocarbons (Hadad et al. 2005, Khalil et al. 2018). *Geobacillus stearothermophilus* showed significant esterase producing ability, hence it can be used for the production of thermostable esterase (Ghati et al. 2013, Pathak and Rathod 2014). Other applications of hyperthermophiles are:

a. **Bioethanol production:** Lignocellulose is a non-edible plant which is mostly used in the production of bioenergy or biofuel (Hood 2016). Cellulose is the most abundant and renewable non fossil carbon source. Hyperthermophilic cellulose enzyme produced by *Pyrococcus horikoshii* is used for cellulose degradation (Wang et al. 2011). Hyperthermophiles are used in cellulytic reaction for sustainable bioethanol production.

b. **Starch processing:** Starch bioprocessing usually involves high temperature dependent two steps (Pervez et al. 2014):

i. Liquefaction

ii. Saccharification

Thermophiles and hyperthermophiles play important role in starch degradation (Turner et al. 2007). They produce starch hydrolyzing enzymes like α-amylase, and glucoamylase for degradation of starch under high temperature (Elleuche and Antranikian 2013).

3.3.3 Applications of Psychrophilic Microorganisms

(a) **Bioremediation:** Excess use of fossil fuels and other human activities are increasing environmental pollution day by day (Perera 2018). Psychrophiles play important role as environmental cleaners. They successfully degrade pollutants of petroleum hydrocarbons in extreme cold temperature (Morita 1975).

(b) **Food Industry:** Cold active enzymes produced by psychrophiles are extremely used in the food industry. These enzymes are used in order to avoid detrimental effects on taste, texture and nutritional value (Feller and Gerday 2003). It is used to produce lactose free milk derived foods for lactose intolerant people who represent approx. 30% of world population (Silanikove et al. 2015).

It can be used to convert lactose into D-tagatose (a natural high added value sweetener with low calorie and gycemic index) which has number of health benefits. For example: reducing blood sugar levels, and lowering risk of heart diseases.

(c) **Medicinal application:** Psychrophilic bacteria may constitute an economic elementary source for aquaculture industry with favourable activities such as cholesterol and triglyceride transport for nervous system and cardiovascular health. At industrial level, polar yeast *Candida antarctica* is used to produce two cold active lipase A and B, which is involved in large number of applications related to pharmaceutical products and cosmetic products (Sisinthy and Prasad 2009, Andualema and Gessesse 2012).

(d) **Detergent and fabric industry:** Cold adapted enzymes produced by psychrophiles are used in detergent formulations like proteases lipases, and α-enzymes are used to improve the efficiency of detergent and also reduce the amount of chemicals used in order to protect the texture and colour of fabrics (Hasan et al. 2010).

3.3.4 Applications of Alkaliphilic Microorganisms

(a) **Pharmaceutical compounds:** Alkaliphiles produce various types of antibiotics. The alkaliphilic actinomycetes, Nocardiopsis strain, is producer of phenazine, which is used to prevent nausea and vomiting related certain conditions. It is also used to treat severe allergic reactions (Horikoshi 1999).

(b) **Lignin degradation:** Lignin is a constituent of plant cell wall. It is complex polymer of phenyl propane (Liu et al. 2018). It is difficult to degrade. Filamentous bacteria (actinomycetes) can decompose lignin to some extent (Kirby 2005). Alkaliphilic species are mostly important for degradation of phenolic lignin compounds (Zhu et al. 2018).

(c) **Food:** Spirulina a cyanobacterium, commonly called single celled protein, has long been a staple food in Africa (Sarma et al. 2008). Spirulina grows naturally in alkaline lakes. Spirulina is commercially produced in many countries for health food in the form of powder granules or flakes and as tablets or capsules (Soni et al. 2017).

3.3.5 Applications of Acidophilic Microorganisms

Acidophilic microorganisms' enzymes have been proven to have great potential for different applications such as biofuel and ethanol production in the field of industries and biotechnology (Golyshina and Timmis 2005, Jackson et al. 2007).

a. **Recovery of metals:** Sulfolobus (aerobic) and Acidianus (facultative anaerobic), these are two archaebacteria that have the potential to recover metal from its ores through microbial leaching (Schippers 2007). These microbes easily adapt to the low pH condition and are useful in bioleaching processing.

b. **Industry:** Acid stable enzyme is mainly used in the industry for fruit processing, baking, animal feed and also used in pharmaceutical industry (Singh et al. 2016).

3.3.6 Applications of Halophilic Microorganisms

Different sources of halophilic extremozymes such as xylanases, amylases, proteases, and lipases are Halobacterium, Marinococcus, Bacillus, Acinetobacter, Natronococcus, Haloferax, Halorhabdus, Micrococcus, Halobacillus and Halothermothrix (Sutrisno et al. 2004, Taylor et al. 2004, Woosowska and Synowiecki 2004). The industrial uses of lipases and esterases include manufacturing of food, biodiesel, and polyunsaturated fatty acids (Litchfield 2011, Schreck and Grunden 2014). Halophilic extremozymes from microorganisms provide good amount of opportunities for the food industries, as well as industries of bioremediation and biosynthetic processes (Oren 2010). The utilization of halophilic enzymes in the field of biotechnology is due to their ability to tolerate high salinity and high organic solvent conditions (Zaccai 2004).

a. **Glycerol production by Dunaliella:** Glycerol is produced by the alga Dunaliella. Cells grown in near-saturated NaCl solutions may contain over 6–7 M intracellular glycerol (Jovor 1989). The mass cultivation of Dunaliella for commercial production of glycerol is being attempted.

b. **β-carotene production using Dunaliella:** The isolation of green algae *Dunaliella salina* and *Dunaliella bardawil* is mostly used for the β-carotene production that is the highest success story of halophiles (Raja et al. 2007). During mid-1960s, the first plant was established in Ukraine for the mass culture of Dunaliella. There is a high demand of pigment β-carotene (Oren 2010). It has antioxidant property and is also a major source of Vitamin A.

c. **Ectoine production:** Ectoine is a multipurpose osmotic solute produced from haloalkaliphilic photosynthetic sulfur bacterium *Ectothiorhodospira halochloris*, which protects enzymes and nucleic acids under extreme conditions of temperature, pH and pressure (Oren 2010).

3.3.7 Applications of Piezophilic Microorganisms

The reaction of piezophilic proteins offers great industrial and biotechnological potential, particularly in the food industry (Cannio et al. 2004, Giuliano et al. 2004, Dumorné et al. 2017). Piezophilic proteins have shown high efficiency in the detergent and food industries and chemical products (Cavicchioli et al. 2011b).

Industrial process: Piezophiles microorganisms and their enzymes have considerable potential for use in many applications. Piezophiles produce that type of enzymes which are stable at high pressure condition. Piezophile is mainly used in food industry; here, high pressure is applied for the processing and sterilization of food material (Abe and Horikoshi 2001, Ichiye 2018).

3.3.8 Applications of Radiophilic Microorganisms

a. **Biotechnology:** These radiation resistant organisms are mainly used in new and adaptive mechanism of repair of DNA, antioxidant and enzymatic defense system (Krisko and Radman 2013). Fast as well as accurate repair of genomes such as *Deinococcus radiodurans* cells attain various oxidative stress prevention properties and tolerance mechanisms such as cell cleaning by the removal of oxidized macromolecules, selective protection of protein against oxidative damage and suppression in the production of reactive oxygen species (Slade and Radman 2011).

b. **Treatment of radioactive waste:** Radiophiles are mainly used in the treatment of radioactive waste because of their ability to survive in high ionizing radiation (Rampelotto 2013).

3.3.9 Applications of Metallophilic Microorganisms

a. **Removal of heavy metals:** Pollution by heavy metals like (Cr, Cu, Zn, As, Hg) poses threat to public health, fisheries, wildlife, etc. Heavy metals give rise to many diseases in human beings; for example: arsenic causes black foot disease, cadmium causes itai-itai disease, and mercury causes minamata disease. Metallophiles are mainly used in removal of heavy metals from soil, sediments and wastewater (Kapahi and Sachdeva 2019). Metallophiles also show high potential in biomining of expensive metals from effluents of industrial processes (Zhuang et al. 2015).

3.3.10 Applications of Xerophilic Microorganisms

Xerophiles accumulate glycerol as a good solute to adjust the internal and external osmotic pressure (Rampelotto 2013). The common food preservation technique of reducing water activities may not prevent the development of xerophilic living microorganisms, often resulting in food decay. Some species of mold and yeast are xerophilic. Mold growth on bread is an example of food deterioration by xerophilic living microbes. Instances of xerophiles include *Trichosporonoides nigrescens* (Hocking and Pitt 1981).

3.4 Conclusion

Extremophiles have pushed our understanding of the limits of life in all directions since they were first identified. Extremophiles have the capability to survive in extremely harsh environmental conditions. These microorganisms can be separated into different classifications, for example, acidophiles, alkaliphiles, halophiles, thermophiles, hyperthermophiles, psychrophiles, barophiles, endoliths, and xerophiles. Extremophiles are basically prokaryotic (archaea and microorganisms), and very few are eukaryotes. Extremophiles can be found under each of the three life domains: Archaea, Bacteria, and Eukarya. Extremophiles are useful in the field of biotechnology, as they produce extremozymes. These enzymes have great economic potential in many industrial processes, including agricultural, chemical, and pharmaceutical applications. Many consumer products will increasingly benefit from the addition or exploitation of extremozymes. These extremozymes will be used in novel biocatalytic processes that are faster, more specific, and ecofriendly. Simultaneous advancements of protein designing and coordinated development advances will result in further tailoring and improving biocatalytic characteristics, which will enhance the utilization of extremozymes obtained from extremophiles in the industry.

References

Abe, F. and K. Horikoshi. 2000. Tryptophan permease gene TAT2 confers high-pressure growth in *Saccharomyces cerevisiae*. Mol. Cell. Biol. 20(21): 8093–8102.
Abe, F. and K. Horikoshi. 2001. The biotechnological potential of piezophiles. Trends Biotechnol. 19(3): 102–108.
Abou-Shanab, R.A.I. 2007. Characterization and 16S rDNA identification of thermotolerant bacteria isolated from hot springs. Res. J. Appl. Sci. 3: 994–1000.
Aertsen, A., F. Meersman, M.E.G. Hendrick, R.F. Vogel and C.W. Michiels. 2009. Biotechnology under high pressure: Applications and implications. Trends Biotechnol. 27: 434–441.
Aguilar, A. 1996. Extremophile research in the European Union: From fundamental aspects to industrial expectations. FEMS Microbiol. Rev. 18: 89–92.
Aguilar, A., T. Ingemansson and E. Magniea. 1998. Extremophile microorganisms as cell factories: Support from the European Union. Extremophiles 2: 367–373.
Akmar, H.N., I. Asma, B. Venugopal, L.Y. Latha and S. Sasidharan. 2011. Identification of appropriate sample and culture method for isolation of new thermophilic bacteria from hot pond. Afr. J. Microbiol. Res. 5(3): 217–221.
D'Amico, S., T. Collins, J.C. Marx, G. Feller and C. Gerday. 2006. Psychrophilic microorganisms: Challenges for life. EMBO Rep. 7(4): 385–389.
Andrade, Carolina M.M.C., Nei. Jr. Pereira and G. Antranikian. 1999. Extremely thermophilic microorganisms and their polymer-hidrolytic enzymes. Rev. de. Microbiol. 30(4): 287–298.
Andualema, B. and A. Gessesse. 2012. Microbial lipases and their industrial applications: Review. Biotechnol. J. 11: 100–118.
Antranikian, G., C.E. Vorgias and C. Bertoldo. 2005. Extreme environments as a resource for microorganisms and novel biocatalysts. Adv. Biochem. Engin. 96: 219–262.
Arahal, D.R., M.C. Gutierrez, B.E. Volcani and A. Ventosa. 2000. Taxonomic analysis of extremely halophilic archaea isolated from 56-yearsold Dead Sea brine samples. Syst. Appl. Microbiol. 23: 376–385.
Arora, N.K. and H. Panosyan. 2019. Extremphiles: Appications and roles in environmental sustainability. Environmental Sustainability 2: 217–218.
Azeri, C., A.U. Tamer and M. Oskay. 2010. Thermoactive cellulase-free xylanase production from alkaliphilic *Bacillus* strains using various agro-residues and their potential in biobleaching of kraft pulp. Afr. j. biotechnol. 9(1): 63–72.
Baker-Austin, C. and M. Dopson. 2007. Life in acid: pH homeostasis in acidophiles. Trends Microbiol. 15(4): 165–171.
Bajaj, B.K., M. Sharma and R.S. Rao. 2014. Agricultural residues for production of cellulase from *Sporotrichum thermophile* LAR5 and its application for saccharification of rice straw. J. Mater. Environ. Sci. 5(5): 1454–1460.
Bartlett, D.H. 2002. Pressure effects on *in vivo* microbial processes. Biochim. Biophys. Acta 1595(1-2): 367–381.
Battista, J.R. 1997. Against all odds: The survival strategies of *Deinococcus radiodurans*. Annu. Rev. Microbiol. 51: 203–224.
Beg, Q.K., B. Bhushan, M. Kapoor and G.S. Hoondal. 2000. Production and characterization of thermostable xylanase and pectinase from *Streptomyces* sp. QG-11-3. J. Ind. Microbiol. Biotechnol. 24: 396–402.

Bialkowska, A., E. Majewska, A. Olczak and A. Twarda-Clapa. 2020. Ice binding proteins: Diverse biological roles and applications in different types of industry. Biomolecules 10(2): 274.

Biello, D. 2006. Cheating DNA Death: How an Extremophile Repairs Shattered Chromosomes. http://www. scientificamerican.com/article/cheating-dna-death-how-an/.

Bowers, K.J., N.M. Mesbah and J. Wiegel. 2009. Biodiversity of poly-extremophilic Bacteria: Does combining the extremes of high salt, alkaline pH and elevated temperature approach a physico-chemical boundary for life? Saline Systems 5: 1–9.

Brenchly, Jean E. 2008. The Characterization of Psychrophilic Microorganisms and their potentially useful Cold-Active Glycosidases. Final Progress Report, United States. doi: 10.2172/959114.

Brock, T.D. and H. Freeze. 1969. Thermus aquaticus gen. n. and sp. n., a non-sporulating extreme thermophile. J. Bacteriol. Res. 98: 289–297.

Brock, T.D. 1978. Thermophilic Microorganisms and Life at High Temperatures, New-York, USA, Springer-Verlag.

Brock, T.D. 1986. An overview of the thermophiles. pp. 2–15. *In*: Brock, T. (eds.). Thermophiles: General, Molecular and Applied Microbiology. John Wiley and Sons, New York.

Bruins, M.E., A.E. Janssen and R.M. Boom. 2001. Thermozymes and their applications: A review of recent literature and patents. Appl. Biochem. Biotechnol. 90: 155–186.

Buchalo, A.S., E. Nevo, S.P. Wasser, A. Oren and H. Molitoris. 1998. Fungal life in the extremely hypersaline water of the Dead Sea: First records. Proc. Royal Soc. BIOL SCI 265: 1461–1465.

Burgess, S.A., D. Lindsay and S.H. Flint. 2010. *Thermophilic bacilli* and their importance in dairy processing. Int. J. Food Microbiol. 144(2): 215–25.

Burhan, A., U. Nisa, C. Gökhan, C. Ömer, A. Ashabil and G. Osman. 2003. Enzymatic properties of a novel thermostable, thermophilic, alkaline and chelator resistant amylase from an alkaliphilic *Bacillus* sp. isolate ANT-6. Process Biochem. 38: 1397–1403.

Cannio, R., N. Di Prizito, M. Rossi and A. Morana. 2004. A xylan degrading strain of Sulfolobus solfataricus: Isolation and characterization of the xylanase activity. Extremophiles 8: 117–124.

Cárdenas, J.P., J. Valdés, R. Quatrini, F. Duarte and D.S. Holmes. 2010. Lessons from the genomes of extremely acidophilic bacteria and archaea with special emphasis on bioleaching microorganisms. Appl. Microbiol. Biotechnol. 88: 605–620.

Cavicchioli, R., D. Amils and T. McGenity. 2011. Life and applications of extremophiles. Environ. Microbiol. 13: 1903–1907.

Cavicchioli, R., T. Charlton, H. Ertan, S. Mohd Omar, K.S. Siddiqui and T.J. Williams. 2011. Biotechnological uses of enzymes from psychrophiles. Microb. Biotechnol. 4(4): 449–460.

Chong, P.L.G., U. Ayesa, V.P. Daswani and E.C. Hur. 2012. On physical properties of tetraether lipid membranes: Effects of Cyclopentane Rings. Lipid Biology of Archaea, Article ID 138439 | https://doi.org/10.1155/2012/138439.

Cohn, F. 1897. Die Pflanze. Vortra¨ge aus dem Gebiete der Botanik, 3rd edn., vol. 2. J. U. Kern's Verlag (Max Mu¨ ller), Breslau.

Coker, J.A. 2016. Extremophiles and biotechnology: Current uses and prospects. F1000 Research, 5, F1000 Faculty Rev-396. https://doi.org/10.12688/f1000research.7432.1.

Cowan, D.A., J.B. Ramond, T.P. Makhalanyane and P. De Maayer. 2015. Metagenomics of extreme environments. Curr. Opin. Microbiol. 25: 97–102.

Daly, M.J., E.K. Gaidamakova, V.Y. Matrosova, A. Vasilenko, M. Zhai, R.D. Leapman, B. Lai, B. Ravel, S.M. Li, K.M. Kemner and J.K. Fredrickson. 2007. Protein oxidation implicated as the primary determinant of bacterial radioresistance. PLoS Biol. 5(4): 769–779.

Daly, M.J., E.K. Gaidamakova, V.Y. Matrosova, J.G. Kiang, R. Fukumoto, D.Y. Lee, N.B. Wehr, G.A. Viteri, B.S. Berlett and R.D. Levine. 2010. Small-molecule antioxidant proteome-shields in *Deinococcus radiodurans*. PLoS One 5(9): e12570.

Daly, M.J. 2011. Deinococcus radiodurans: revising the molecular basis for radiation effects on cells. *In*: Horikoshi, K. (eds.). Extremophiles Handbook. Springer, Tokyo. https://doi.org/10.1007/978-4-431-53898-1_53.

De Pascale, D., A.M. Cusano, F. Author, E. Parrilli, G. di Prisco, G. Marino and M.L. Tutino. 2008. The cold-active Lip1 lipase from the Antarctic bacterium *Pseudoalteromonas haloplanktis* TAC125 is a member of a new bacterial lipolytic enzyme family. Extremophiles 12: 311–323.

Demirjian, D.C., F. Morís-Varas and C.S. Cassidy. 2001. Enzymes from extremophiles. Curr. Opin. Chem. Biol. 5: 144–151.

Deppe, U., H.H. Richnow, W. Michaelis and G. Antranikian. 2005. Degradation of crude oil by an arctic microbial consortium. Extremophiles 9: 461–470.

Dumorné, K., D. Camacho Córdova, M. Astorga-Eló and P. Renganathan. 2017. Extremozymes: A Potential Source for Industrial Applications. J. Microbiol. Biotechnol. 27(4): 649–659.

Eichler, J. 2001. Biotechnological uses of archaeal extremozymes. Biotechnol. Adv. 19: 261–278.

62   *Extremophiles: General and Plant Biomass Based Biorefinery*

Elleuche, S. and G. Antranikian. 2013. Starch-hydrolyzing enzymes from thermophiles. pp. 509–533. *In*: Thermophilic Microbes Environmental and Industrial Biotechnology.

Englander, J., E. Klein, V. Brumfeld, A.K. Sharma, A.J. Doherty and A. Minsky. 2004. DNA toroids: Framework for DNA repair in *Deinococcus radiodurans* and in germinating bacterial spores. J. Bacteriol. 186: 5973–5977.

Extremophile. 2014. Encyclopædia Britannica. Encyclopædia Britannica Ultimate Reference Suite. Chicago: Encyclopædia Britannica.

Feller, G. and C. Gerday. 2003. Psychrophilic enzymes: Hot topics in cold adaptation. Nat. Rev. Microbiol. 1: 200–2008.

Ferreira, A.C., M.F. Nobre, F.A. Rainey, M.T. Silva, R. Waite, J. Burghardt, A.P. Chung and M.S. da Costa. 1997. *Deinococcus geothermalis* sp. nov. and *Deinococcus murrayi* sp. nov., two extremely radiation-resistant and slightly thermophilic species from hot springs. Int. J. Syst. Bacteriol. 47: 939–947.

Fijan, S. 2014. Microorganisms with claimed probiotic properties: An overview of recent literature. Int. J. Environ. Res. Public Health 11(5): 4745–4767.

Foster, T.L. and L. Jr. Winans. 1975. Psychrophilic microorganisms from areas associated with the Viking spacecraft. Appl. Microbiol. 30(4): 546–550.

Gabani, P. and O.V. Singh. 2013. Radiation-resistant extremophiles and their potential in biotechnology and therapeutics. Appl. Microbiol. Biotechnol. 97(3): 993–1004.

Gabani, P., D. Prakash and O.V. Singh. 2014. Bio-signature of ultraviolet-radiation-resistant extremophiles from elevated land. Am. J. Microbiol. Res. 2(3): 94–104.

Gadd, G.M. and X. Pan. 2016. Biomineralization, bioremediation and biorecovery of toxic metals and radionuclides. Geomicrobiol. J. 33(3-4): 175–178.

Ghati, A., K. Sarkar and G. Paul. 2013. Isolation, characterization and molecular identification of Esterolytic bacteria from an Indian Hot pond. Curr. Res. Microbiol. Biotechnol. 1(4): 196–202.

Giuliano, M., C. Schiraldi, M.R. Marotta, J. Hugenholtz and M. De Rosa. 2004. Expression of *Sulfolobus solfataricus* α- glucosidase in *Lactococcus lactis*. Appl. Microbiol. Biotechnol. 64: 829–32.

Golyshina, O. and K.N. Timmis. 2005. Ferroplasma and relatives, recently discovered cell wall-lacking archaea making a living in extremely acid, heavy metal-rich environments. Environ. Microbiol. 7: 1277–1288.

Grant, W.D. 1988. Bacteria from alkaline, saline environments and their potential in biotechnology. J. Chem. Technol. Biotechnol. 42: 291–294.

Grant, W.D. 2004. Life at low water activity. Philos. Trans. R. Soc. Lond., B, Biol. Sci. PHILOS T R SOC B 359(1448): 1249–1267.

Grogan, D.W. 2002. Hyperthermophiles and the problem of DNA instability. Mol. Microbiol. 28(6): 1043–1049.

Gupta, G.N., S. Srivastava, S.K. Khare and V. Prakash. 2014. Extremophiles: An overview of microorganism from extreme environment. Int. J. Agric. Environ. Biotechnol. 7(2): 371–380.

Hadad, D., S. Geresh and A. Sivan. 2005. Biodegradation of polyethylene by the thermophilic bacterium *Brevibacillus borstelensis*. J. Appl. Microbiol. 98(5): 1093–1100.

Haki, G.D. and S.K. Rakshit. 2003. Developments in industrially important thermostable enzymes: A review. Bioresour. Technol. 89: 7–34.

Hasan, F., A.A. Shah, S. Javed and A. Hameed. 2010. Enzymes used in detergents: Lipases. Afr. J. Biotechnol. 9(31): 4836–4844.

Hocking, A.D. and J.I. Pitt. 1981. *Trichosporonoides nigrescens* sp. nov., a new xerophilic yeast-like fungus. Antonie Leeuwenhoek 47(5): 411–421.

Hood, E.E. 2016. Plant-based biofuels. F1000. https://doi.org/10.12688/f1000research.7418.1.

Horikoshi, K. and W.D. Grant (eds.). 1991. Superbugs. Microorganisms in Extreme Environments, Japan Scientific Societies Press, Tokyo – Springer-Verlag, Berlin.

Horikoshi, K. 1998. Introduction. *In*: Horikoshi, K. and W.D. Grant (eds.). Extremophiles: Microbial Life in Extreme Environments. Wiley-Liss, New York.

Horikoshi, K. and W.D. Grant. 1998. Extremophiles—Microbial Life in Extreme Environments, Wiley-Liss (eds.), New York.

Horikoshi, K. 1999. Alkaliphiles: Some applications of their products for biotechnology. Microbiol. Mol. Biol. Rev. 63: 735–750.

Hough, D.W. and M.J. Danson. 1999. Extremozymes. Curr. Opin. Chem. Biol. 3: 39–46.

Ichiye, T. 2018. Enzymes from piezophiles. Semin. Cell Dev. Biol. 84: 138–146.

Ingraham, J.L. and J.L. Stokes. 1959. Psychrophilic bacteria. Bacteriol. Rev. 23: 97–108.

Jackson, B.R., C. Noble, M. Lavesa-Curto, P.L. Bond and R.P. Bowater. 2007. Characterization of an ATP-dependent DNA ligase from the acidophilic archaeon "*Ferroplasma acidarmanus*" Extremophiles 11: 315–327.

Janacek, K. and K. Sigler. 2000. Osmosis: Membranes impermeable and permeable for solutes, mechanism of osmosis across porous membranes. Physiol. Res. 49(2): 191–195.

Javaux, E.J. 2006. Extreme life on Earth—past, present and possibly beyond. Res. Microbiol. 157: 37–48.

Javor, B. 1989. *Dunaliella* and Other Halophilic, Eucaryotic Algae. *In*: Hypersaline Environments. Brock/ Springer Series in Contemporary Bioscience. Springer, Berlin, Heidelberg. https://doi.org/10.1007/978-3-642-74370-2_10.

Jayakumar, R., S. Jayashree, B. Annapurna and S. Seshadri. 2012. Characterization of thermostable serine alkaline protease from an alkaliphilic strain *Bacillus pumilus* MCAS8 and its applications. Appl. Biochem. Biotechnol. 168: 1849–1866.

Jolivet, E., E. Corre, S. L'Haridon, P. Forterre and D. Prieur. 2004. *Thermococcus marinus* sp. nov., and *Thermococcus radiotolerans* sp. nov., two hyperthermophilic archaea from deep-sea hydrothermal vents that resist ionizing radiation. Extremophiles 8: 219–227.

Kapahi, M. and S. Sachdeva. 2019. Bioremediation options for heavy metal pollution. J. Health Pollut. 9(24): 191203.

Kasar, H. 2013. Medical and biotechnological applications of extremophiles. Microbiol. Today, pp. 10–18.

Kato, C. 2011. Distribution of piezophiles. *In*: Horikoshi, K. (eds.). Extremophiles Handbook. Springer, Tokyo. https://doi.org/10.1007/978-4-431-53898-1_29.

Khalil, A.B., N. Sivakumar, M. Arslan, H. Saleem and S. Qarawi. 2018. Insights into *Brevibacillus borstelensis* AK1 through Whole Genome Sequencing: A Thermophilic Bacterium Isolated from a Hot Spring in Saudi Arabia. Biomed Res. Int., https://doi.org/10.1155/2018/5862437.

Kirbi, R. 2005. Actinomycetes and lignin degradation. Adv. Appl. Microbiol. 58C: 125–168.

Krisko, A. and M. Radman. 2013. Biology of extreme radiation resistance: The way of *Deinococcus radiodurans*. Cold Spring Harb. Perspect. Biol. CSH PERSPECT BIOL 5(7): a012765.

Kumar, B., P. Trivedi, A.K. Mishra, A. Pandey and L.M.S. Palni. 2004. Microbial diversity of soil from two hot ponds in Uttaranchal Himalaya. Microbiol. Res. 159: 141–146.

Kumar, N., A. Singh and P. Sharma. 2013. To study the Physico-Chemical properties and Bacteriological examination of Hot Spring water from Vashisht region in Distt. Kullu of Himachal Pradesh, India. Int. J. Environ. Sci. 2: 28–31.

Kumar, R. and R.C. Sharma. 2019a. Microbial diversity and physico-chemical attributes of two hot water springs in the Garhwal Himalaya, India. J. Microbiol., Biotechnol. Food Sci. 8(6): 1249–1253.

Kumar, R. and R.C. Sharma. 2019b. Determination of microbial diversity and physico-chemical characteristics of two hot water ponds near Badrinath Shrine in the Uttarakhand, India. IWRA (India) Journal 9(1): 16–23.

Kumar, R. and R.C. Sharma. 2020a. Thermophilic microbial diversity and physicochemical attributes of thermal springs in the Garhwal Himalaya. Environ. Exp. Biol. 18: 143–152.

Kumar, R. and R.C. Sharma. 2020b. Microbial diversity in relation to physico-chemical properties of hot water ponds located in the Yamunotri landscape of Garhwal Himalaya. Heliyon 6: e04850.

Kumar, R. and R.C. Sharma. 2020c. Exploration of psychrophilic microbial diversity and physicochemical environmental variables of Glacier-Fed Lakes in the Garhwal Himalaya, India. Taiwan Water Conserv. 68(3): 38–51.

Kumar, R. and R.C. Sharma. 2021. Psychrophilic microbial diversity and physicochemical characteristics of glaciers in the Garhwal Himalaya, India. J. Microbiol., Biotechnol. Food Sci. 10(5): e2096.

Lange, C., L. Wackett, K. Minton and M.J. Daly. 1998. Construction and characterization of recombinant *Deinococcus radiodurans* for organopollutant degradation in radioactive mixed waste environments. Nat. Biotechnol. 16: 929.

Lebre, P., P.D. Maayer and D.A. Cowan. 2017. Xerotolerant bacteria: Surviving through a dry spell. Nat. Rev. Microbiol. 15(5): 10.1038/nrmicro.2017.16.

Lentzen, G. and T. Schwarz. 2006. Extremolytes: Natural compounds from extremophiles for versatile applications. Appl. Microbiol. Biotechnol. 72: 623–634.

Litchfield, C.D. 1998. Survival strategies for microorganisms in hypersaline environments and their relevance to life on early Mars. Meteorit. Planet Sci. 33: 813–819.

Litchfield, C.D. 2011. Potential for industrial products from the halophilic Archaea. J. Ind. Microbiol. Biotechnol. 38: 1635–1647.

Liu, Q., L. Luo and L. Zheng. 2018. Lignins: Biosynthesis and biological functions in plants. Int. J. Mol. Sci. 19(2): 335.

Lopez, O.L., M.E. Cerdan and M.I. Gonzalez-Siso. 2013. Hot spring metagenomics. Life 2: 308–320.

López-López, O., M.E. Cerdán and M.I. González-Siso. 2014. New extremophilic lipases and esterases from metagenomics. Curr. Protein Pept. Sci. 15: 445–455.

M.L.E.R. "Types of Extreme Environments". Extreme Environments. https://serc.carleton.edu/microbelife/extreme/about.html.

Ma, Y., A.E.A. Galinski, W.D. Grant, A. Oren and A. Ventosa. 2010. Halophiles 2010: Life in saline environments. Appl. Environ. Microbiol. 76(21): 6971–6981.

MacElroy, M. 1974. Some comments on the evolution of extremophiles. Biosystems–6: 74–75.

Madigan, M.T. 2000. Bacterial habitats in extreme environments. pp. 61–72. *In*: Seckbach, J. (ed.). Journey to Diverse Microbial Worlds. Adaptation to Exotic Environments. Kluwer Academic Publishers, Dordrecht.

Margesin, R. and F. Schinner. 2001. Potential of halotolerant and halophilic microorganisms for biotechnology. Extremophiles 5: 73–83.

Merino, N., H.S. Aronson, D.P. Bojanova, J. Feyhl-Buska, M.L. Wong, S. Zhang and D. Giovannelli. 2019. Living at the Extremes: Extremophiles and the Limits of Life in a Planetary Context. Front. Microbiol. 10: 780.

Midhat, L., N. Ouazzani, A. Hejjaj, A. Ouhammou and L. Mandi. 2019. Accumulation of heavy metals in metallophytes from three mining sites (Southern Centre Morocco) and evaluation of their phytoremediation potential. Ecotoxicol. Environ. Saf. 169: 150–160.

Morita, R. 1975. Psychrophilic bacteria. Bacteriol. Rev. 39: 144–167.

Motojima, F. 2015. How do chaperonins fold protein? Biophysics 11: 93–102.

Mueller, D.R., W.F. Vincent, S. Bonilla and I. Laurion. 2005. Extremotrophs, extremophiles and broadband pigmentation strategies in a high arctic ice shelf ecosystem. FEMS Microbiol. Ecol. 53: 73–87.

Navarro-González, R., E. Iniguez, J. de la Rosa and C.R. McKay. 2009. Characterization of organics, microorganisms, desert soil, and Mars-like soils by thermal volatilization coupled to mass spectrometry and their implications for the search for organics on Mars by Phoenix and future space missions. Astrobiology 9: 703–711.

Oarga, A. 2009. Life in extreme environments. Rev. Biol. Ciênc. Terra 9(1): 1–9.

Oger, P.M. and M. Jebbar. 2010. The many ways of coping with pressure. Res. Microbiol. 161(10): 799–809.

Okibe, N., M. Gericke, K.B. Hallberg and D.B. Johnson. 2003. Enumeration and characterization of acidophilic microorganisms isolated from a pilot plant stirred-tank bioleaching operation. Appl. Environ. Microbiol. 69(4): 1936–1943.

Oren, A. 1999. Bioenergetic aspects of halophilism. Microbiol. Mol. Biol. Rev. 63(2): 334–348.

Oren, A. 2008. Microbial life at high salt concentrations: Phylogenetic and metabolic diversity. Saline syst. 4: 2.

Oren, A. 2010. Industrial and environmental applications of halophilic microorganisms. Environ. Technol. 31(8-9): 825–834.

Otohinoyi, D.A. and I. Omodele. 2015. Prospecting microbial extremophiles as valuable resources of biomolecules for biotechnological applications. Int. J. Sci. Res. 4(1): 1042–1059.

Pathak, A.P. and M.G. Rathod. 2014. Cultivable bacterial diversity of terrestrial hot water pond of Unkeshwar, India. J. Biochem. Technol. 5(4): 814–818.

Perera, F. 2017. Pollution from fossil-fuel combustion is the leading environmental threat to global pediatric health and equity: Solutions exist. Int. J. Environ. Res. Public Health 15(1): 16.

Pervez, S., A. Aman, S. Iqbal, N.N. Siddiqui and S.A.U. Qader. 2014. Saccharification and liquefaction of cassava starch: an alternative source for the production of bioethanol using amylolytic enzymes by double fermentation process. BMC Biotechnol. 14: 49.

Pikuta, E.V. and R.B. Hoover. 2007. Microbial extremophiles at the limits of life. Crit. Rev. Microbiol. 33: 183–209.

Pitt, J.I. and A.D. Hocking. 1980. Dichloran-glycerol medium for enumeration of xerophilic fungi from low-moisture foods. Appl. Environ. Microbiol. 39(3): 488–492.

Preiss, L., D.B. Hicks, S. Suzuki, T. Meier and T.A. Krulwich. 2015. Alkaliphilic bacteria with impact on industrial applications, concepts of early life forms, and bioenergetics of ATP synthesis. Front. Bioeng. Biotechnol. 3: 75.

Princeteejay. 2013. Extremophiles Life under extreme conditions. Microbiol. Today, pp. 3–9.

Qin, J., B. Zhao and X. Wang. 2009. Non-sterilized fermentative production of polymer-grade L-lactic acid by a newly isolated thermophilic strain *Bacillus* sp. PLoS One 4: 43–59.

Rainey, F.A. and A. Oren. 2006. Methods in Microbiology, vol. 35, Extremophiles. New York: Academic Press.

Raja, R., S.H. Iswarya, D. Balasubramanyam and R. Rengasamy. 2007. PCR-Identification of *Dunaliella salina* (Volvocales, Chlorophyta) and its growth characteristics. Microbiol. Res. 162(2): 168–76.

Rajakaruna, N. and R.S. Boyd. 2013. Oxford bibliographies in ecology "heavy metal tolerance". Oxford University Press, Madison Avenue, New York.

Rampelotto, P.H. 2010. Resistance of microorganisms to extreme environmental conditions and its contribution to astrobiology. Sustainability 2: 1602–1623.

Rampelotto, P.H. 2013. Extremophiles and extreme environments. Life (Basel, Switzerland) 3(3): 482–485.

Rasuk, M.C., G.M. Ferrer, J.R. Moreno, M.E. Farías and V. Albarracin. 2016. The diversity of microbial extremophiles. In book: Molecular Diversity of Environmental Prokaryotes. pp. 87–126.

Reed, C.J., H. Lewis, E. Trejo, V. Winston and C. Evilia. 2013. Protein adaptations in archaeal extremophiles, Archaea, 2013: 1–14.

Roberts, M.F. 2005. Organic compatible solutes of halotolerant and halophilic microorganisms. Saline Syst. 1: 1–5.

Rosenbaum, E., F. Gabel, M.A. Durá, S. Finet, C. Cléry-Barraud, P. Masson and B. Franzetti. 2012. Effects of hydrostatic pressure on the quaternary structure and enzymatic activity of a large peptidase complex from *Pyrococcus horikoshii*. Arch. Biochem. Biophys. 517: 104–110.

Rossi, M., M. Ciaramella, R. Cannio, F.M. Pisani, M. Moracci and S. Bartolucci. 2003. Extremophiles. J. Bacteriol. 185(13): 3683–3689.

Rothschild, L.J. and R.L. Mancinelli. 2001. Life in extreme environments. Nature 409: 1092–1101.

Sandigursky, M., S. Sandigursky, P. Sonati, M.J. Daly and W.A. Franklin. 2004. Multiple uracil-DNA glycosylase activities in *Deinococcus radiodurans*. DNA Repair (Amst.) 3(2): 163–169.

Sarma, A.P., P. Petar and S.D.S. Murthy. 2008. Spirulina as a source of single cell protein. Int. J. Plant Res. 21(1): 35–45.

Satyanarayana, T., C. Raghukumar and S. Shivaji. 2005. Extremophilic microbes: Diversity and perspectives. Curr. Sci. 89(1): 78–90.

Sayed, A.M., M.H.A. Hassan, H.A. Alhadrami, H.M. Hassan, M. Goodfellow and M.E. Rateb. 2019. Extreme Environments: Microbiology leading to specialized metabolites. J. Appl. Microbiol. 128: 630–657.

Schippers, A. 2007. Microorganisms involved in bioleaching and nucleic acid-based molecular methods for their identification and quantification. In book: Microbial Processing of Metal Sulfides, pp. 3–33.

Schreck, S.D. and A.M. Grunden. 2014. Biotechnological applications of halophilic lipases and thioesterases. Appl. Microbiol. Biotechnol. 98: 1011–1021.

Seckbach, J. and A. Oren. 2004. Introduction to the extremophiles. pp. 373–393. *In*: Seckbach, J. (ed.). Origins. Kluwer Academic Publishers, Dordrecht.

Seckbach, J., A. Oren and L. Stan-Lotter. 2013. Polyextremophiles: Life Under Multiple Forms of Stress. http://books. google.com.ng/books?id=vRGvZmALC3YC&pg=PT34&lpg=PT34&dq=metabolism+of+xerophile&source= bl&ots=yWfNUMe5zP&sig=1YO2xnZcZqMu5nrZppt6bZ1ubiE&hl=en&sa=X&ei= mUcrVPybEpPhaNr1gqgN&ved=0CE8Q6AEwBg#v=onepage&q=metabolism%20of%20xerophile&f=false.

Seitz, K.H., C. Studdert, J. Sanchez and R. de Castro. 1997. Intracellular proteolytic activity of the haloalkaliphilic archaeon *Natronococcus occultus*. Effect of starvation. J. Basic Microbiol. 7: 313–322.

Sen, S.K., S.K. Mohapatra, S. Satpathy and G.T.V. Rao. 2010. Characterization of hot water pond source isolated clones of bacteria and their industrial applicability. Int. J. Chem. Res. 2: 1–7.

Sharma, A., J.H. Scott, G.D. Cody, M.L. Fogel, R.M. Hazen, R.J. Hemley and W.T. Huntress. 2002. Microbial activity at gigapascal pressures. Science 295: 1514–1516.

Shock, E.L. 1997. High temperature life without photosynthesisias a model for Mars. J. Geophys. Res. 102: 23687–23694.

Silanikove, N., G. Leitner and U. Merin. 2015. The interrelationships between lactose intolerance and the modern dairy industry: Global perspectives in evolutional and historical backgrounds. Nutrients 7(9): 7312–7331.

Simonato, F., S. Campanaro, M.F. Lauro, V. Alessandro, M. D'Angelo, N. Vitulo, G. Valle and D. Bartlett. 2006. Piezophilic adaptation: A genomic point of view. J. Biotechnol. 126: 11–25.

Singh, G., A. Bhalla and P.K. Ralhan. 2011a. Extremophiles and extremozymes: Importance in current biotechnology. ELBA Bioflux 3(1): 1–9.

Singh, G., A. Bhalla, Ralhan and K. Paramjit. 2011b. Extremophiles and extremozyrnes: Importance in current biotechnology. Extreme Life, Biospeology and Astrobiology 3(1): 46–54.

Singh, O.V. and P. Gabani. 2011. Extremophiles: Radiation resistance microbial reserves and therapeutic implications. J. Appl. Microbiol. 110: 851–861.

Singh, R., M. Kumar, A. Mittal and P.K. Mehta. 2016. Microbial enzymes: Industrial progress in 21st century. 3 Biotech 6(2): 174.

Singh, S.P., R.J. Shukla and B.A. Kikani. 2012. Molecular diversity and biotechnological relevance of thermophilic actinobacteria. *In*: Satyanarayana, T., J. Littlechild and Y. Kawarabayasi (eds.). Thermophiles in Environmental and Industrial Biotechnology. Springer Publication, U.K.

Sisinthy, S. and G.S. Prasad. 2009. Antarctic yeasts: Biodiversity and potential applications. In book: Yeast Biotechnology: Diversity and Applications, pp. 3–18.

Slade, D. and M. Radman. 2011. Oxidative stress resistance in *Deinococcus radiodurans*. Microbiol. Mol. Biol. Rev. 75(1): 133–191.

Soni, R.A., K. Sudhakar and R. Rana. 2017. Spirulina—From growth to nutritional product: A review. Trends Food Sci. Technol. 69: 157–171.

Staiano, M., P. Bazzicalupo, M. Rossi and S. D'Auria. 2005. Glucose biosensors as models for the development of advanced protein-based biosensors. Mol. Biosyst. 1: 354–362.

Stetter, K.O. 1996. Hyperthermophilic prokaryotes. FEMS Microbiol. Rev. 18: 149–158.

Stratford, M., H. Steels, M. Novodvorska, D.B. Archer and S.V. Avery. 2019. Extreme osmotolerance and halotolerance in food-relevant yeasts and the role of glycerol-dependent cell individuality. Front. Microbiol. 9: 3238.

Sutrisno, A., M. Ueda, Y. Abe, M. Nakazawa and K. Miyatake. 2004. A chitinase with high activity toward partially Nacetylated chitosan from a new, moderately thermophilic, chitin-degrading bacterium, *Ralstonia* sp. A-471. Appl. Microbiol. Biotechnol. 63: 398–406.

Synowiecki, J. 2010. Some applications of thermophiles and their enzymes for protein processing. Afr. J. Biotechnol. 9(42): 7020–7025.

Takacs-Vesbach, C., K. Mitchell, O. Jackson-Weaver and A. Reysenbach. 2008. Volcanic calderas delineate biogeographic provinces among Yellowstone thermophiles. Environ. Microbiol. 10: 1681–1689.

Taylor, I.N.R., C. Brown, M. Rycroft, G. King, J.A. Littlechild, M.C. Lloyd, C. Praquin, H.S. Toogood and S.J.C. Taylor. 2004. Application of thermophilic enzymes in commercial biotransformation processes. Biochem. Soc. Trans. 32: 290–292.

Tchounwou, P.B., C.G. Yedjou, A.K. Patlolla and D.J. Sutton. 2012. Heavy metal toxicity and the environment. Experientia Suppl. 101: 133–164.

Tekere, M., A. Lötter, J. Olivier and S. Venter. 2015. Bacterial diversity in some South African Hot water Ponds: A Metagenomic analysis. Proceedings World Geohot water Congress, Melbourne, Australia, pp. 1–8.

Thiel, V. 2014. Extreme environments. Accessed from: http://www.springerreference.com/docs/html/chapterdbid/187271.html.

Toplin, J.A., T.B. Norris, C.R. Lehr, T.R. McDermott and R.W. Castenholz. 2008. Biogeographic and phylogenetic diversity of thermoacidophilic Cyanidiales in Yellowstone National Park, Japan, and New Zealand. Appl. Environ. Microbiol. 74: 2822–2833.

Torsvik, V. and L. Ovreas. 2007. Microbial diversity, life strategies, and adaptation to life in extreme soils. *In*: Dion, P. and C.S. Nautiyal (eds.). Microbiology in Extreme Soils, Springer 13: 3–14.

Townsend, T. 2012. Extremophiles Lecture Retrieved from http://zuserver2.star.ucl.ac.uk/~rhdt/diploma/lecture_6/ Date 20/12/2014.

Turner, P., G. Mamo and E.N. Karlsson. 2007. Potential and utilization of thermophiles and thermostable enzymes in biorefining. Microb. Cell Fact. 6: 9.

Unsworth, L.D., O.J. Van Der and S. Koutsopoulos. 2007. Hyperthermophilic enzymes-stability, activity and implementation strategies for high temperature applications. FEBS J. 274: 4044–4056.

Van den Burg, B. 2003. Extremophiles as a source for novel enzymes. Curr. Opin. Microbiol. 6: 213–218.

Verma, A and P. Shirkot. 2014. Purification and characterization of thermostable laccase from thermophilic *Geobacillus thermocatenulatus* MS5 and its applications in removal of Textile Dyes. Scholars Academic Journal of Biosciences 2(8): 479–485.

Ventosa, A., J.J. Nieto and A. Oren. 1998. Biology of moderately halophilic aerobic bacteria. Microbiol. Mol. Biol. Rev. 62(2): 504–544.

Vieille, C. and G.J. Zeikus. 2001. Hyperthermophilic enzymes: Sources, uses, and molecular mechanisms for thermostability. Microbiol. Mol. Biol. Rev. 65(1): 1–43.

Villar, S.E. and H.G. Edwards. 2006. Raman spectroscopy in astrobiology. Anal. Bioanal. Chem. 384: 100–113.

Wang, H., F. Squina, F. Segato, A. Mort, D. Lee, K. Pappan and R. Prade. 2011. High-temperature enzymatic breakdown of cellulose. Appl. Environ. Microbiol. 77(15): 5199–5206.

Weronika, E. and K. Łukasz. 2017. Tardigrades in space research—past and future. Origins of life and evolution of the biosphere. Orig. Life Evol. Biosph. 47(4): 545–553.

Wiegel, J. and M.W.W. Adams. 1998. Thermophiles—The keys to Molecular Evolution and the Origin of Life? Taylor and Francis, London.

Wiegel, J. and V.V. Kevbrin. 2003. Alkalithermophiles. Biochem. Soc. Trans. 32: 1–2.

Woosowska, S. and J. Synowiecki. 2004. Thermostable glucosidase with broad substrate specificity suitable for processing of lactose-containing products. Food Chem. 85: 181–187.

Yildiz, S.Y., N. Radchenkova, K.Y. Arga, M. Kambourova and O.E. Toksoy. 2015. Genomic analysis of *Brevibacillus thermoruber* 423 reveals its biotechnological and industrial potential. Appl. Microbiol. Biotechnol. 99: 2277–2289.

Yun, Y.S. and Y.N. Lee. 2004. Purification and some properties of superoxide dismutase from *Deinococcus radiophilus*, the UV-resistant bacterium. Extremophiles 8(3): 237–242.

Zaccai, G. 2004. The effect of water on protein dynamics. Philos. Trans. R. Soc. Lond., B, Biol. Sci. 359: 1269–1275.

Zeng, X., J.L. Birrien, Y. Fouquet, G. Cherkashkov, M. Jebbar, J. Querellou, P. Oger, M.A. Cambon-Bonavita, X. Xiao and D. Preur. 2009. *Pyrococcus* CH1, an obligate piezophilic hyperthermophile: Extending the upper pressure-temperature limits for life. International Society for Microb. Ecol. 3(7): 873–876.

Zhu, D., P. Zhang, C. Xie, W. Zhang, J. Sun, W.J. Qian and B. Yang. 2017. Biodegradation of alkaline lignin by *Bacillus ligniniphilus* L1. Biotechnol. Biofuels 10: 44.

Zhuang, W.Q., J.P. Fitts, C.M. Ajo-Franklin, S. Maes, L. Alvarez-Cohen and T. Hennebel. 2015. Recovery of critical metals using biometallurgy. Curr. Opin. Biotechnol. 33: 327–335.

4

Understanding the Role of a Thermophilic Genus *Geobacillus* in the Hydrolytic and Oxidative Degradation of Plant Biomass
Insights from Genomics and Metagenomic Analyses

Tanvi Govil,[1,4] *David R. Salem*[1,2,3,4]* and *Rajesh K. Sani*[1,4, 5,]*

1. Introduction

Cellulose, hemicellulose, and lignin are major constituents of lignocellulose-containing raw materials, but a small amount of pectin, nitrogen compounds, and mineral residues are also present in these feedstocks (Yang 2007). Depending on the origin, the amounts of the indicated constituents differ and generally the lignocellulosic feedstock, that are rich in hemicellulose and cellulose, are ideal raw materials to produce biofuel. The structure of the lignocellulose is compact with different bonding among cellulose, hemicellulose, and lignin that makes lignocellulose a very complex substrate for enzymes (Andlar et al. 2018). Hemicellulose and cellulose are wrapped in a lignin matrix, which prohibits the microorganisms' access to them. Moreover, bacteria that are good at cellulase production generally possess poor machinery for producing enzymes that degrade hemicelluloses and lignin, and vice versa. These factors are major barriers for hydrolysis and fermentation of lignocellulose biomass. Therefore, identification of unique microbial strains having efficient lignin-degrading capability together with polysaccharide-hydrolyzing capabilities is vital to the realization of industrial-scale biofuel production from lignocellulose biomass (Tsegaye et al. 2019).

In nature, fungi are known to produce a broad set of synergistically acting enzymes with diverse catalytic activities for the hydrolysis of renewable lignocellulose-containing raw materials. The enzymatic degradation of lignocellulose-containing raw materials is achieved through the multiple carbohydrate-active enzymes, usually acting together with complementary, synergistic activities, and modes of action. Their extracellular enzymatic system includes two types of enzymes: hydrolytic,

---
[1] Department of Chemical and Biological Engineering, South Dakota School of Mines and Technology, Rapid City, SD 57701, USA.
[2] Department of Materials and Metallurgical Engineering, South Dakota School of Mines and Technology, Rapid City, SD 57701, USA.
[3] Department of Nanoscience and Nanoengineering, South Dakota School of Mines and Technology, Rapid City, SD 57701, USA.
[4] Composite and Nanocomposite Advanced Manufacturing - Biomaterials Center, Rapid City, SD 57701, USA.
[5] BuG ReMeDEE consortium, Rapid City, SD 57701, USA.
* Corresponding authors: David.salem@sdsmt.edu; Rajesh.Sani@sdsmt.edu

responsible for polysaccharide degradation; and oxidative, which degrade lignin and open phenyl rings (Andlar et al. 2018). Using these enzymes, fungi decompose and assimilate all lignocellulose constituents: lignin, cellulose, and hemicellulose. Degradation of lignin indeed is more rapid, efficient, and extensive than hemicellulose and cellulose in white rot fungi (e.g., *Phanerochaete chrysosporium, Phanerochaete carnosa, Pleurotus ostreatus, Pycnoporus cinnabarinus, Botrytis cinerea, Stropharia coronilla,* and *Trametes versicolor*) than other microorganisms (Andlar et al. 2018). Comparatively, knowledge of bacterial lignin-degradation in environmental contexts is limited, and only certain bacteria groups are known to rapidly metabolize cellulose and hemicellulose while only slightly modifying lignin.

Fungal microbes are more active than bacteria to co-degrade lignin and other lignocellulosic polymers. However, members of Actinobacteria, Firmicutes, α-proteobacteria, and γ-proteobacteria are known to have the capabilities to degrade lignin and lignin model compounds, and hence are capable of degrading and co-metabolizing most of the carbohydrates in plant biomass (Tsegaye et al. 2019). This dual capability towards delignification with simultaneous hydrolysis of lignocellulosicis is an advantage for the ultimate improvement of the lignocellulosic biorefinery that would involve a one-step conversion of lignocellulose to product. Heterotrophs belonging to group Firmicutes that thrive at these elevated temperatures are of special interest for their robust and temperature-stable carbohydrate active enzymes (CAZymes) capable of deconstructing even the most recalcitrant parts of plant biomass. Recent work with thermophilic strains belonging to *Geobacillus* has produced exciting results. *Geobacilli* are active in environments such as hot plant composts, and examination of their genome sequences reveals that they are endowed with a battery of enzymes dedicated to hydrolyzing plant polysaccharides (Brumm et al. 2015, De Maayer et al. 2014, Zeigler 2014). In further sections, we will summarize the current state of knowledge about the thermophilic genus *Geobacillus*, including genomic, metagenomic, and other omics analysis to highlight the extraordinary functional diversity of members of *Geobacillus* as a rich source of biomass-degrading enzymes of potential biotechnological significance.

2. *Geobacillus*: The Plant Biomass Deconstructing genus

Geobacillus bacteria are rod-shaped, aerobic or facultatively anaerobic, endospore-forming microbes, with known temperature optima for most isolates located between about 45 and 70°C. This temperature range for *Geobacillus* growth classifies them as "thermophiles", which are widely distributed and readily isolated from natural and man-made thermophilic biotopes like hot springs, geothermal soils, hot subterranean oilfields and natural gas wells, and hydrothermal vents (Zeigler 2014). Interestingly, due to their ability to sporulate and their catabolic versatility, their existence has even been verified in cool soils and cold ocean sediments in anomalously high numbers, given that the ambient temperatures of these environments are significantly below their minimum requirement for growth (Zeigler 2014).

A closer look into the literature reveals that the niche this particular bacterial group is associated with, at the respective places of isolation, involves their active participation in a complex, dynamic process termed composting, i.e., biodegradation of organic matter, usually derived from plants. *Geobacillus* are indeed opportunistic decomposers of plant-derived organic matter, capable of rapid growth under transient thermophilic conditions, but endowed with mechanisms to survive long periods of time when growth is impossible. With the power of comparative genetics, which involves scanning the *Geobacillus* dispensable genomes for genes encoding known enzymes for hydrolysis, uptake, and utilization of these polysaccharides and their components, their involvement in a thermophilic plant biomass hydrolysis process has been fully confirmed (Zeigler 2014).

3. Maturation of *Geobacillus* genomics

Following the details on the genome online database (JGI GOLD), at the time of writing (7 April 2020), there are 27 *Geobacillus* genome sequencing projects that are complete and published;

permanent drafts are available for another 62 *Geobacillus* genomes, 36 genome drafts are incomplete, while 1 genome project is under progress. This brings the total number of genome sequencing projects involving *Geobacillus* to 126, with GenBank ID's publicly available for 89 of those projects involving representatives of 17 species with validly published name: *G. thermodenitrificans, G. stearothermophilus, G. thermoleovorans, G. thermocatenulatus, G. kaustophilus, G. zalihae, G. caloxylosilyticus, G. gargensis, G. toebii, G. proteiniphilus, G. uzenensis, G. yumthangensis, G. lituanicus, G. icigianus, G. vulcani, G. jurassicus,*and *G. subterraneus*, along with some strains that have not been assigned to named species, which remain uncertain (Table 1). The availability of complete *Geobacillus* genome sequences has enabled or accelerated the research on the genomic analysis of their capability to code for and express many industrially important enzymes, including lignocellulose degrading enzymes such as cellulases, xylanases, and ligninases.

Table 1. *Geobacillus* genome projects reported on the GOLD online database https://gold.jgi.doe.gov/project?id= Gp0005586.

Geobacillus project status with JGI	Number of genomes
Complete and published genomes	27
Complete but unpublished genomes	62
Partial genomes	36
In progress	1
Total representative species	17

List: G. thermodenitrificans, G. stearothermophilus, G. thermoleovorans, G. thermocatenulatus, G. kaustophilus, G. zalihae, G. caloxylosilyticus, G. gargensis, G. toebii, G. proteiniphilus, G. uzenensis, G. yumthangensis, G. lituanicus, G. icigianus, G. vulcani, G. jurassicus, and G. subterraneus, along with some strains that have not been assigned to named species, which remain uncertain

3.1 Comparative Geobacillus genomics to Identify the Xylan Degrading Cluster

Representing hemicellulose, xylan is a heteropolysaccharide with a basic backbone of β-1,4-linked xylose residues that is heavily modified with a range of (a) neutral sugars such as arabinose, mannose, mannitol, galactose, (b) charged sugars such as methylglucuronic acid, and (c) acetate groups (Shulami et al. 2014). These modifications result in a remarkable diversity of xylan chemical compositions and structures and the necessity for synergistic action by specific glycoside hydrolases (GHs), including endo-β-1,4-xylanase (EC 3.2.1.8), which hydrolyze the main chain backbone, and accessory enzymes that act on the side chains, such as α-L-arabinofuranosidase (EC 3.2.1.55), α-glucuronidase (EC 3.2.1.139), β-xylosidase (EC 3.2.1.37), and acetyl xylan esterase (EC 3.2.1.72) to essentially degrade the xylan, the arabinan, and the galactan systems (Fig. 1) (Brumm et al. 2015, Shulami et al. 2014).

Amongst thermophiles, *Geobacillus stearothermophilus* is a very well-studied bacteria encoding many xylan-degrading enzymes. In this bacterium, a 23.5 kb xylan utilizing genomic DNA fragment was recognized in 1999 (Shulami et al. 1999). It is organized in at least three transcriptional units: unit 1 coding for the extracellular GH10 xylanase, xylanase T-6; unit 2 encoding an intracellular xylanase and three β-xylosidase; and Unit 3 encoding 12 genes involved in transport and metabolism of galacturonate and glucuronate (GlcUA), including an intracellular α-glucuronidase (*aguA*), 5 genes (*kdgK, kdgA, uxaC, uxuA,* and *uxuB*) for GlcUA catabolism, and a potential regulatory gene, *uxuR* (resembles repressors of the GntR family). Based on the identified genes, this gene cluster was found good enough to transport and metabolize branched xylo-oligosaccharides substituted with glucuronic acid and/or methyl D-glucuronic acid (MeGlcUA) (Shulami et al. 1999).

In 2011, a 38 kb gene cluster was identified in the same bacterium, lying contiguous to the xylan utilization cluster, for the hydrolysis and utilization of arabinofuranosyl substituents and for the main chain deacetylation (Shulami et al. 2011). This fragment consists of an extracellular GH43 endo α-1,5-arabinanase (AbnA), a unique three-component regulatory system (*araPST*)

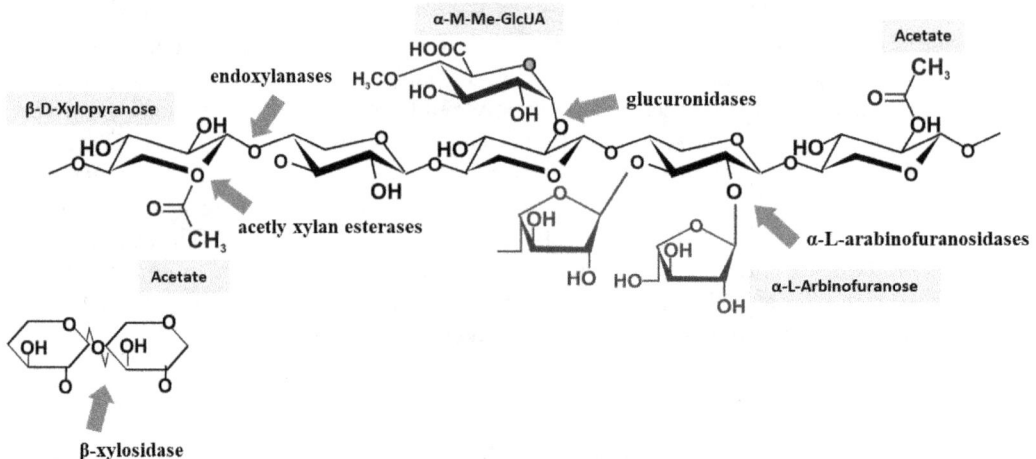

Fig. 1. Structure of xylan and the xylanolytic enzymes involved in hemicellulose degradation.

to sense extracellular arabinose; a specific ABC transporter for arabinose (AraEGH); a second putative ABC sugar transporter (AbnEFJ) to interact specifically with linear and branched arabino-oligosaccharides; an arabinose repressor (AraR); and a repertoire of intracellular enzymes, including two α-L-arabinofuranosidases (AbfA and AbfB), a α-L-arabinopyranosidase (Abp), and two xylanacetylesterases (CE4). This gene cluster within the same hemicellulose utilization locus of *G. stearothermophilus* T-6 was also shown to play a role in the degradation of the pectin-associated L-arabinan polymerin to shorter arabino-saccharides and arabinose, further highlighting the capacity of *Geobacillus* spp. to degrade and utilize polymers in plant biomass (De Maayer et al. 2014). In 2014, a group led by Shulami further identified multiple regulator sites regulated by XylR, CodY, and XynX, catabolite repression (CcpA), and quorum sensing, bringing the single locus (NCBI Accession # DQ868502) to approximately 76 kb in size, which encodes 60 proteins, divided into 13 distinct gene clusters (De Maayer et al. 2014).

Of these, seven clusters (Clusters K, F, H, D, G, I, and L) can be considered as central to hemicellulose degradation, with the proteins encoded in these clusters driving the extracellular degradation (cluster K), transport (cluster F and H) and internal cleavage of arabinoglucuronoxylan into metabolizable monosaccharides (clusters D, G, I and L). Four additional gene clusters (B,E,J, and M) are not essential for the degradation of the hemicellulose polymer, but rather encode enzymes involved in pathways for the metabolism of the end-product pentose sugars arabinose and xylose and for uronic acids. A gene cluster A encodes for an L-arabinose transporter, and additional cluster C has been known for the degradation and utilization of the pectin-associated polymer L-arabinan (Fig. 2) (De Maayer et al. 2014).

These studies, with *Geobacillus stearothermophilus* T-6 as the model strain, indicated that this strain is equipped with the necessary enzymes required for complete xylan hydrolysis. Later, with the help of comparative genomics, and using the *G. stearothermophilus* T-6 hemicellulose utilization locus as genetic marker, orthologous hemicellulose utilization (*HUS*) loci were identified in the complete and partial genomes of 17 out of 24 *Geobacillus* strains by De Maayer et al., but with extensive variability (De Maayer et al. 2014). 12 out of 13 orthologous gene clusters were present in *G. thermodenitrificans* NG80-2, *G. thermodenitrificans* DSM465, *Geobacillus* sp. G11MC16, *Geobacillus* sp. WSUCF1, *Geobacillus* sp. Y412MC52, *Geobacillus* sp. Y412MC61, *Geobacillus* sp. C56-T3, *Geobacillus* sp. CAMR12739, *Geobacillus* sp. A8, *G. kaustophilus* GBlys, and *Geobacillus* sp. CAMR5420. 11 gene clusters were found in *G. thermopakistaniensis* MAS1, *G. caldoxylolyticus* CIC9, and *Geobacillus* sp. GHH01. 8 gene clusters were found to be contained within *Geobacillus* sp. JF8, and *G. thermoglucosidasius* C56YS93. These 17 strains, possessing between 8 and 12 gene clusters, including between 5 and 7 of the clusters with a role in hemicellulose

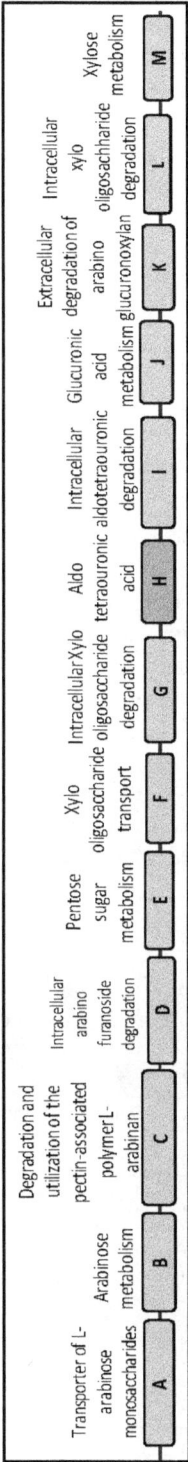

Fig. 2. Schematic diagram of the *G. stearothermophilus* T-6 hemicellulose utilization locus. The *G. stearothermophilus* T-6 HUS (hemicellulose utilization system) locus was subdivided into thirteen gene clusters on the basis of their predicted function (De Maayer et al. 2014).

degradation plus extracellular GH10 family xylanase XynA1, indicate that both the organization and the individual genes of carbohydrate metabolism are fairly conserved throughout the genus. Further three strains, namely *G. thermoglucosidasius* CCB_US3_UF5, *G. thermoglucosidasius* B23, and *G. kaustophilus* HTA426, only carry 3 orthologous gene clusters, namely those for xylose metabolism, arabinose transport and L-arabinose metabolism, suggesting that these three strains do not have the capacity to degrade hemicellulose, and rather make use of L-arabinose and D-xylose monomers that may be present in the environment (De Maayer et al. 2014).

Somewhat similar findings have been reported by Brumm et al. (Brumm et al. 2015), who carried out comparative genomics on six *Geobacillus xylanolytic* strains (*G. thermoglucosidasius* strain YS93, *G. thermodenitrificans* strain 1 MC16, *G. thermocatenulatus* strain 56T3, 2 strains belonging to *G. stearothermophilus* (MC52, MC61), and *Geobacillus* sp. strain 56T2), and identified an approximately 200 kb unique supercluster in all six strains, containing 5to 8 distinct carbohydrate degradation clusters in a single genomic region, a feature not seen in genera outside *Geobacillus* (Brumm et al. 2015). Indeed, this genome region contains clusters for the utilization of not only xylan, xylose, arabinose, glucuronic acid (the four key features of a hemicellulose), but also has the fructose, mannitol, inositol, α-mannoside, α-1,6, glucosides, and cellobiose carbohydrate utilization clusters (Brumm et al. 2015). According to them, while the gene clusters for xylan, xylose, glucuronic acid, fructose, mannitol, and cellobiose were present in all six strains, domains for the utilization of arabinan (and arabinose) were found in 5of the 6 strains. After more in-depth analysis of these domains in their study, the authors claimed that the gene clusters for utilization of these 7 substrates (arabinan, fructose, mannitol, inositol, α-mannoside, α-1,6, glucosides, and cellobiose) have identical organization across the genus and that the individual proteins have a high percentage identity to their homologs (Brumm et al. 2015), indicating high conservation for utilization of these seven carbohydrates in *Geobacillus*. In contrast, the inositol, α-1,4-glucosides, and α-mannoside utilization clusters were present in this region in just one strain, indicating extensive variability for the last of these three sugars among the *Geobacillus* hemicellulose utilization systems (Brumm et al. 2015). These findings have been summarized in Table 2.

A similar arrangement of gene clusters and gene orthologs distributed in the *Geobacillus* genus has been reported in many other studies, a summary of which can be found in Hussein et al. (Hussein et al. 2015). Altogether, the classical works by groups led by Shulami (Shulami et al. 1999, 2011, 2014), DeMaayer (De Maayer et al. 2014), and Brumm (Brumm et al. 2015) suggest that a single large chromosomal hemicellulose utilization system (HUS) is a common feature encoded across all major branches of the *Geobacillus* genus phylogeny, but is not observed in any other genus. This unique system is organized as a super-cluster encoding, which not only includes xylan, arabinan, and glucuronic acid domains, but also cellobiose, mannose, fructose, mannitol, and inositol gene clusters embedded within the HUS.

Table 2. Summary of the 10 major functional clusters found in the 200 kb conserved regions of *Geobacillus* spp. (Brumm et al. 2015).

	Xyn	Xyl	Cell	Fruc	Glucn	Mtl	Inos	Ara	α-Man	α-gluc
MC52	+	+	+	+	+	+	-	+	-	+
MC61	+	+	+	+	+	+	-	+	-	+
56T2	+	+	+	+	+	+	-	+	-	-
YS93	+	+	+	+	+	+	-	-	-	-
56T3	+	+	+	+	+	+	-	+	-	-
1MC16	+	+	+	+	+	+	+	+	+	+

Here: Xyn: xylan; Xyl: xylose; Cell: Cellobiose; Fruc: Fructose: Mtl: Manitol; Glucn: Glucuronic acid; Inos: Inositol; Ara: Arabinose; α-Man: Mannosidase; α-gluc: α-1,4-glucosides. Strains MC52, MC61: belong to *G. stearothermophilus;* Strain1MC16: represents *G. thermodenitrificans;* strain56T3: represents *G. thermocatenulatus;* and 56T2: represents a new *Geobacillus* sp.

3.2 Cellulose Cluster in Geobacillus spp.

In nature, the synergistic action between at least three classes of glycoside hydrolases (GHs), collectively called 'cellulases', are responsible for releasing a physiologically relevant amount of sugar from the crystalline cellulose present in biomass. Amongst these, 1, 4-β-endoglucanase (EC 3.2.1.4) hydrolyzes internal bonds in the cellulose, generating poly oligo-saccharides; 1, 4-β-exoglucanaseacts on the exposed chain ends in a unidirectional manner, either from non-reducing or reducing ends of cellulose polysaccharide chains, liberating cellobiose as the major product; and, lastly, β-glucosidases (EC 3.2.1.86) (β-D-glucoside glucohydrolase or cellobiase) convert cellobiose into glucose (Fig. 3).

Today, scientists in the field agree on a stringent definition in which enzymes that can hydrolyze a purified crystalline cellulose or a crystalline product of partially depolymerized cellulose (e.g., Avicel®) are cellulases. In comparison, an enzyme that hydrolyzes any of a variety of non-physiological cellulose surrogates, including carboxymethyl cellulose, phosphoric acid swollen cellulose or 4-methylumbelliferyl-β-D-cellobioside, but are inactive on crystalline cellulose, is not a cellulase (but can be termed putative cellulase), even if they may be involved in cellulose degradation (Brumm 2013).

Among the Gram-positive organisms, there are a significant number of cellulose degraders, including the model bacterium *Acidothermus cellulolyticus* (Brumm 2013), *Ruminiclostridium cellulolyticum* (Fosses et al. 2017), *Clostridium thermocellum* (Chow and Wu 2017), members of *Cellulolomonas* spp. (Kenyon et al. 2005), a number of *Streptomyces* species (Brumm 2013), and *Caldicellulosiruptor* species (Blumer-Schuette 2020), in which cellulose degradation is achieved through a variety of secreted or complexed carbohydrate active enzymes (CAZymes) with different substrate specificities, such as endoglucanases, cellobiohydrolases, xylanases, and pectinases, sometimes organized in an extracellular organelle called cellulosome.

In contrast, despite major efforts to find cellulolytic *Geobacillus* spp., significant evidence of a strain capable of efficient cellulose conversion has only recently emerged (Daas et al. 2016, 2018a,b). While most of the *Geobacillus* spp. can effectively degrade hemicellulose, they are unable to break down crystalline or amorphous cellulose (Bashir et al. 2019). Despite their possession of β-glucosidases or endoglucanases like the (a) GH5 endoglucanase CelA5 from *Geobacillus* sp. 70PC53 (Ng et al. 2009), (b) endoglucanase from *G. thermoleovorans* T4 (Tai et al. 2004), (c) endoglucanase from *Geobacillus* sp. WSUCF1 (Rastogi et al. 2010), (d) endoglucanase from *Geobacillus* sp. HTA426 (Potprommanee et al. 2017), (e) β-1,4-glucosidase from the *G. thermoglucosidasius* NCIMB 11955 (Bashir et al. 2019), to name a few, the strains were only

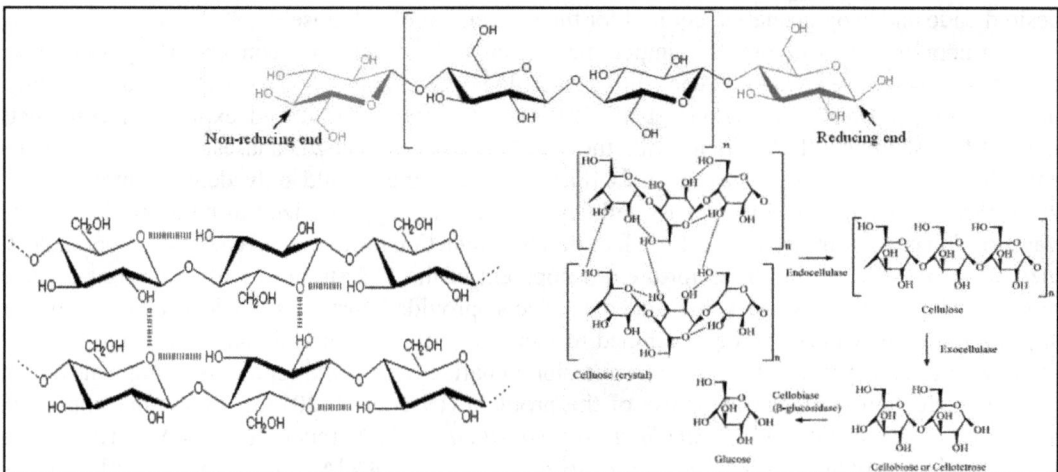

Fig. 3. Molecular structure of a cellulose molecule.

shown to be active on synthetic substrates such as carboxy-methyl cellulose (CMC), and phosphoric acid-swollen cellulose (PASC). None of these strains was able to efficiently degrade cellulose even when isolated from microcrystalline cellulose or composted plant biomass. This had placed a question mark on the utility of members of this genus as prominent hosts for a consolidated bioprocessing process, despite the existence of a hemicellulose complex in *Geobacillus* across the genus.

However, in a 2016 study on the isolation of *Geobacillus* strains from compost samples, several strains of *G. thermodenitrificans* were shown to grow on carboxymethyl cellulose, and for some of these strains clear degradation of cellulose was demonstrated by using the Congo red assay (Daas et al. 2016). In their follow up work in 2018, the same group further performed a metagenome screen on 73 of those previously isolated *G. thermodenitrificans* strains using Hidden Markov Models (HMM) profiles of all known CAZy families that contain endo and/or exoglucanases, to retrieve potential cellulases (Daas et al. 2018a). Out of the total of 82 hits for potential endoglucanases or exoglucanases, they reported an endoglucanase GE40 with 55% sequence identity and similar protein architecture to the well characterized GH5 family endoglucanases Cel5b from *Bacillus halodurans*. This GE40 is comprised of a GH5_4 catalytic domain, an immunoglobulin (Ig)-like module and a carbohydrate binding module (CBM) belonging to family 46, with high activity towards cellulose and barley derived β-glucan (Daas et al. 2018a). The Ig-like module is believed to act as a structural hinge, thereby holding the GH5_4 catalytic domain and the CBM family 46 in position for optimal enzymatic activity.

The CBM46 domain indeed has been shown to contribute to xyloglucan hydrolysis in the intact plant cell walls (Liberato et al. 2016, Venditto et al. 2015), and hence the presence of Ig like module, CBM 46, and GH5_4 catalytic domain in GE40 qualify it to be considered a true cellulase. Hence, in their further work, the authors expressed GE40 cellulase in *E. coli*, and found high activity of the recombinant *E. coli* towards cellulose, further confirming the *G. thermodenitrificans* strain T81's GE40 endoglucanase as a potential cellulase. This important work by Daas et al. (Daas et al. 2016, 2018a,b) shows the potential of metagenome mining for the discovery of novel cellulases, and also provides a starting point for further development of *Geobacillus thermodenitrificans* spp. as potential hosts for consolidated bioprocessing (Daas et al. 2018a). Previously, one member of this species viz. *G. thermodenitrificans* strain T12, was already shown to possess excellent xylanolytic and pectinolytic activity (Daas et al. 2016). It was reported that stain T12 has all the genes that are present in the *G. stearothermophilus* strain T-6 hemicellulose utilization (HUS) locus except for the arabinan degradation cluster. Instead, the HUS locus of strain T12 contains genes for both an inositol and a pectate degradation pathway (Daas et al. 2018b). Nevertheless, the *G. thermodenitrificans* strain T81's GE40 endoglucanase was not found in strain T12's HUS locus, making it lack the desired endo and exoglucanases required for the conversion of cellulose.

An approach to overcome the impediment from lack of cellulose conversion is to engineer the required cellulose encoding genes into a suitable host of the genus *Geobacillus*. Consequently, in the work by Daas et al. (Daas et al. 2018a), the authors introduced expression constructs into T12 containing GE40, along with the *C. thermocellum* exoglucanasescel Kgene and the endoglucanase celC gene (Daas et al. 2018a). However, they could only detect minor activity in *G. thermodenitrificans* T12 against cellulase, which they hypothesized to be a problem at the transcription or early translational stage, hampering production of functional enzymes. Nonetheless, their work provides a solid basis for the development of the T12 strain in a host for consolidated bioprocessing of biomass to value added products, provided further suitable promoters and/or signal peptide sequences can be developed to increase protein yield and thereby the extracellular activity required for efficient cellulose conversion in native *Geobacillus* spp., assuming insufficient protein production is indeed the cause of the problem (Daas et al. 2018a). In another approach, heterologous production of *Clostridium thermocellum* CtCelA (endoglucanase), Thermobifda *fusca* Cel6B (exoglucanase), *Caldicellulosiruptor bescii* CbCelA (endo and exo-glucanase), and Thermoanaerobacter *brockii* CglT (β-glucosidases) has been demonstrated in an otherwise non-cellulolytic, *G. thermoglucosidasius* NCIMB 11955 wild-type strain, with subsequent production of sufficient simple sugars from pre-treated wheat straw by the recombinant strain to make it suitable for CBP (Bashir et al. 2019). Altogether, these studies provide a starting point for further development of *Geobacillus* spp. as potential new host for biotechnological production of environmentally benign chemicals from renewable resources. Furthermore, success of these studies demonstrates genetic accessibility of *Geobacillus* spp. towards genetic manipulation, which is frequently challenging.

3.3 Ligninases Cluster in Geobacillus spp.

Bacterial laccases are the least studied plant-biomass degrading enzymes, but they are attracting new interest. Although not common throughout the bacterial domain, they are widespread within specific genera such as *Streptomyces, Azospirillum, Bacillus* and *Geobacillus.* Within the phylum Firmicutes, the only known enzyme with laccase activity that has been fully characterized and studied at the structural level is the outer endospore coat component of *Bacillus subtilis* (Chauhan et al. 2017, Enguita et al. 2002). In *Geobacillus*, many species have now been reported with laccase activity (Basheer et al. 2017, Castagnaro 2014, Moon et al. 2018, Rai et al. 2019, Singh et al. 2019, Verma and Shirkot 2014) and, based on the results from genome sequence comparisons, a consensus is building in the research community that the laccase activities reported from strains of *Geobacillus* are most similar to proteins belonging to the multicopper oxidase family, which includes polyphenol oxidases (Moon et al. 2018). For example, a recent genome sequencing study of the thermophilic *Geobacillus thermoleovorans* strain RL, isolated from a hot water spring, revealed the presence of genes encoding multicopper polyphenol oxidase, in addition to hydrolytic enzymes of glycoside hydrolase, a-and b-glucosidase, xylanase, amylase, neopullulanase, pullulanase and lipases (Singh et al. 2019).

Genome analysis of laccases from several *Geobacillus* species has revealed the presence of a 26 amino acid long signal peptide for secretion, along with four copper binding regions (Basheer et al. 2017). Beyond this, there does not exist, currently, any crystal structures or a common characterization of laccases (multicopper oxidases) from *Geobacillus*. Moreover, there is a dearth of knowledge and understanding of the general mechanisms for transcriptional regulation of laccase-like multicopper oxidases from this group in response to metal ions, various aromatic compounds related to lignin or lignin derivatives, nitrogen and carbon sources. Using tools of comparative modelling, it will be very interesting to elucidate the structure of laccase-like multicopper oxidases from *Geobacillus* spp. On a similar note, by using the power of comparative genomics, analyses of both the sequences and putative functions of laccase will be helpful towards postulating the molecular mechanisms underlying the regulation of laccases by different stimuli in *Geobacillus* spp.

4. Secretomics of *Geobacillus* spp.

In general, Gram-positive bacteria lack the outer membrane and periplasmic space where many exported proteins would otherwise be retained, and therefore they can secrete a large number of proteins that play a significant metabolic role in the adaptation to the ecological niches that they occupy (Lebre et al. 2018). For example, *Bacillus subtilis* str. 168, one of the most fully characterized Gram-positive bacteria, has been shown to possess excellent capacity to secrete around 200 extracellular proteins (Anné et al. 2014). However, within the Gram positive world, three general schemes for degradation of plant cell wall polysaccharides have been evolved by nature and can be described as follows. Aerobic *Streptomycetes* are largely tapped for decomposition of lignocelluloses, and produce a large number of extracellular free cellulases, hemicellulases, and ligninases (approximately 10% of their total proteins) that work synergistically to completely degrade the polymers into mono- and disaccharides (Ramachandra et al. 1987, Saini et al. 2015). Anaerobic bacteria, such as *Clostridia*, have evolved unique multienzyme complexes, named cellulosomes, that integrate many cellulolytic and hemicellulolytic enzymes on the cell surface and mediate both the attachment of the cell to the crystalline polymer and its controlled hydrolysis at the cell-substrate interface, while maintaining the sugar products close to the cells at relatively high concentrations. In contrast to these observations, members of *Geobacillus* spp. secrete only a limited number of endo-type enzymes that commence the extracellular degradation of lignocelluloses into Oligosaccharides, which are relatively large, and their final breakdown is carried out by intracellular enzymes, after transporting these Oligosaccharides into the cell via dedicated sugar permeases. For instance, in a study where the secretomes of 49 *Geobacillus* sp. were determined, the average percentage of secreted proteins across the compared genomes was calculated as 3.82% of total genome protein content, with just a small core of 51 distinct proteins with glycoside hydrolase (GH) domains, which include enzymes involved in the degradation of complex polysaccharides from plant cell wall, i.e., hemicellulose, cellulose and pectin (Lebre et al. 2018). According to the authors, the protein families that dominated this *Geobacillus* secretome dataset included β-galactosidases (GH2), α-amylases (GH13), chitinases (GH18), and lytic transglycolyses (GH23). One other protein family that was detected in 23 *Geobacillus* secretome included a GH10 family XynA1 (WP_044731438) endoxylanase, which degrades the xylan backbone into xylooligosaccharides before transport across the cell membrane. Some more protein families that represented this dataset, although in minor quantities, included: GH1 and GH3 family β-glucosidases, GH5-cellulases, GH10-endo-β-1,3-xylanases, GH43 and GH52-β-D-xylosidases, GH3-α-L arabinofuranosidases, GH43-endo-α-L-arabinanases, GH1-β-galactosidases, GH27-α-galactosidases, GH53-β-1,4-galactanases, GH70-transglucosylases, GH27-α-N-acetylgalactosaminidases, and GH73-β-N-acetylglucosaminidases (Lebre et al. 2018). The functional analysis of the global secretome of *Geobacillus*, combined with evidences from comparative genome models (Brumm et al. 2015, De Maayer et al. 2014), therefore reveal yet another feature unique to the *Geobacillus* spp., viz their dependence on a minimum number of secreted enzymes for utilization of carbohydrates. As stated by Hussein et al. [10], an undeniable pattern emerges that suggests *Geobacillus* spp. fully secrete only a small number of glycosidehydrolases that degrade non-crystalline polymeric substrates to short oligomers, which are then further hydrolyzed to monomers by non-secreted glycoside hydrolases and glycosidases after being transported inside the cell. This metabolic setup is highly conserved throughout the *Geobacillus* genome for a majority of the lignocellulosic degrading clusters. Indeed, this catabolic strategy reveals a notable metabolic efficiency, employing a minimal set of secreted enzymes together with enhanced energy gain through transporting (then internally hydrolyzing) oligomers rather than monomers (Hussein et al. 2015).

4.1 Hemicellulose Hydrolysis

Since the major part of the xylan backbone is comprised of xylose units, the first of the major CAZyme enzyme involved in xylan and xylose utilization includes (a) a single secreted xylanase (XynA) that degrades xylan into oligosaccharides; (b) a two component system (xynDC) and three protein (xynEFG) ABC permeases that transport xylose and xylooligosaccharides; (c) further degradation of the transported oligosaccharides into monosaccharides within the cell by intracellular xylanase (XynA2), xylosidases (XynB and XynB2) a-glucuronidase (AguA) (De Maayer et al. 2014). Likewise, for the utilization of arabinan, an extracellular GH43 endo α-1,5-arabinanase (AbnA) has been shown to first hydrolyze pectin-associated L-arabinan polymers into shorter arabino-saccharides and arabinose in *G. stearothermophilus* T-6, which is then finally metabolized after transporting inside the cells (with a dedicated AguEFG transporter) by a group of intracellular hydrolases comprising α-L-arabinofuranosidases (AbfA and AbfB), β-L-arabinopyranosidase, and an arabinanase (AbnB) (Shulami et al. 2011) (Fig. 4). GH10 xylanase (XynA) and GH43 endo α-1,5-arabinanase (AbnA) are, so far, the only two singly secreted GHs in *Geobacillus* spp. that have been found to be used in the extracellular degradation of xylan and arabinan (Daas et al. 2017, Huang et al. 2017, Hussein et al. 2015) [All the rest of the enzymes, involved in pentose and hexose utilization, are intracellular, and none of the *Geobacillus* spp. secretexylosidases or arabinofuranosidases (Brumm et al. 2015, De Maayer et al. 2014, Shulami et al. 1999, 2011, 2014).

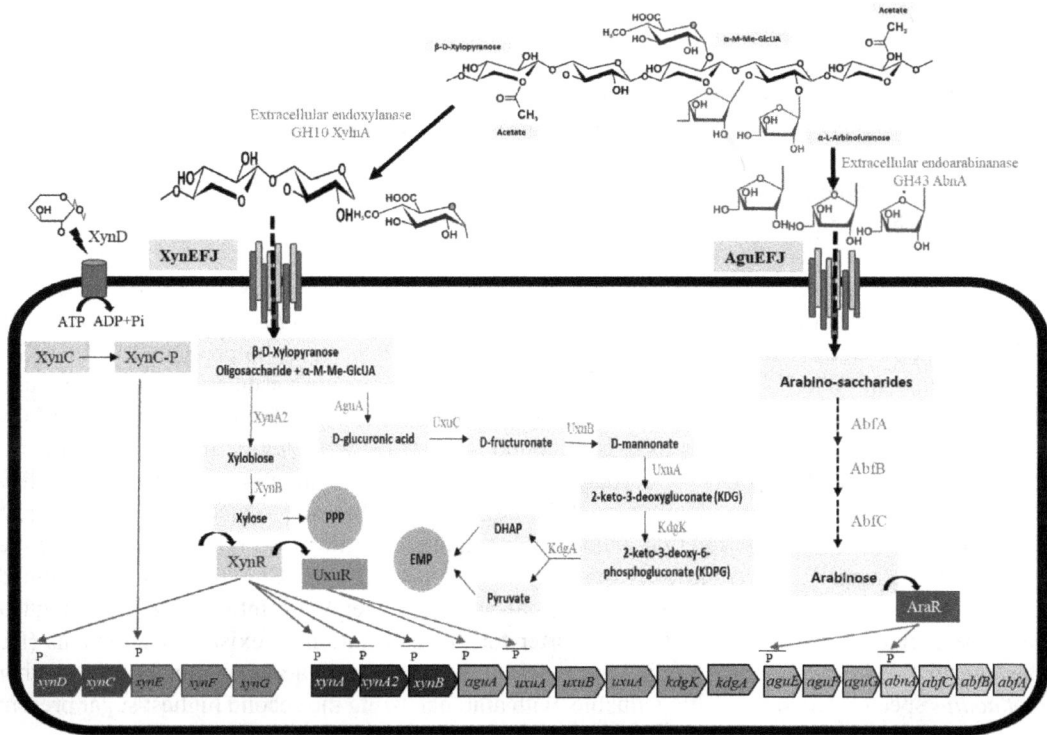

Fig. 4. Brief representation of hemicellulose utilization system from *Geobacillus* spp. The scheme depicted shows that the utilization of xylan is initiated by just two of the secreted extracellular enzymes: an extracellular GH10 endo-β-1,4-xylanase (XylnA), and an extracellular endo-α-1,5-arabinanase, into xylo-oligosaccharides, which are subsequently, transported inside the cell via two dedicated ABC sugar transporters, XynEFG for xylooligosaccharides and methylglucuronic acids, and AguEFG for arabinoligosachharides. Here, before getting transported inside the cells, the sugars interact with the extracellular domain of the class I histidine kinase sensor protein XynD, which triggers phosphorylation of the response regulator XynC. It is the phosphorylated XynC that then activates the expression of XynEFG, and AguEFG transporters. Once inside the cell, majority of the enzymes are present to carry out the final breakdown, with the whole machinery tightly regulated by master regulators XylR, UxuR, and AraR, which regulates the transcriptional units of intracellular enzymes. Here, XynA2 = intracellular xylanase, XynB = β-xylosidase, AguA = α-glucuronidase, UxuC = uronate isomerase, UxuB = D-mannonate oxidoreductase, UxuA = D-mannonate hydrolase, KdgK = KDG kinase, KdgA = KDGP aldolase..

For utilizing the third component, normally associated with a hemicellulose, i.e., glucuronic acid and sometimes galactose, the only secreted proteins in *Geobacillus* are from the extracellular solute-binding protein family 1, with none of the strains possessing the gene for extracellular α-galactosidase, or α-L-glucuronidase, or acetylxylanesterase (axe2). This suggests a limited ability to utilize exogenous galactose or galactans (Brumm et al. 2015). Indeed, for the transport of galactan/glucuronic acid, *Geobacillus* spp. seems to lack a dedicated transport system, and it has been hypothesized that the members utilize galactose and galactans linked to xylan or arabinan that was transported into the cell via xylan or arabinan transporter systems (Brumm et al. 2015). Indeed, the presence of solute-binding protein family 1, along with multiple genes for intracellular β-galactosidases, α-galactosidase and also α-glucuronidase (agu) tell-tales the ability of *Geobacillus* spp. to utilize a wide range of galactose-containing oligosaccharides (De Maayer et al. 2014).

These observations and arguments are valid for the *G. stearothermophilus* strain T6, whose genome has largely been elucidated for various polysaccharides degradation, transport and metabolism. However, in other *Geobacillus* species and strains, despite the presence of a hemicellulose utilization system as a common feature, comparison of the hemicellulose utilization loci at the genes and proteins level reveals extensive variation. For example, orthologs of a thermostable extracellular xylanase XynA1 and endo α-1,5-arabinanase (AbnA) that degrades arabinoxylan to its component xylooligosaccharides and arabinose have been identified in a number of *Geobacillus* spp. (De Maayer et al. 2014, Hussein et al. 2015). By comparative analysis of genomes, in some species of *Geobacillus*, two of the specific carbohydrate uptake transporters (CUT) operons, specifically XynEFG and AguEFG, have been found to be either disrupted by transposon insertions or replaced by orthologs of CUT transporters. For instance, a distinct CUT1 transporter (cutJKL) is encoded within the genome of *Geobacillus* sp. WSUCF1 in the location occupied by abnEFG in *G. stearothermophilus* T-6. The translated products for these genes share only 43.7% average amino acid identity to *G. stearothermophilus* T-6 AbnEFJ, but share greater orthology (71.7% average amino acid identity) with an ABC transporter of an unknown sugar substrate in *Bacillus halodurans* C-125 (BH1864–1866). Because of the shared orthology with abnEFG, it can be presumed that cutKJL in *Geobacillus* sp. WSUCF1 can potentially encode an arabinosaccharide transporter, although whether this CUT system facilitates the uptake of linear or branched saccharides is unknown. Many such orthologous CUT transporters have been found in other *Geobacillus* sp. too, more details for which can be found in the work by De Maayer et al. (De Maayer et al. 2014), suggesting that they likely code for a transporter for the uptake of the degraded saccharide products. However, for rest of the CUT transporters, experimental validation is required. Similarly, in other strains where the orthologs of the aguEFG are completely missing, as in *Geobacillus* sp. C56-T2, it can be assumed that the strain can also not utilize aldotetraouronicacid as a metabolic intermediate, although it cannot be excluded that an alternative transporter for this substrate may exist in such strains (De Maayer et al. 2014). In the future, finding the genes or its orthologs for arabinan metabolism in other *Geobacillus* species would be highly valuable, with arabinan being the second highest sugar present in xylan, after xylose.

4.2 Cellulose Hydrolysis

After the xylose/xylan gene cluster, the genes that are conserved amongst 97% of the *Geobacillus* spp. belong to the cellobiose, fructose, mannose, and sucrose metabolism cluster. It is believed that in *Geobacillus* spp., a dedicated three component phosphotransferase system (PTS) is present as a transporter system which enables the intake of cellobiose and fructose. The enzymes of this PTS transporter uses phosphoenolpyruvate to transport the sugar into the cell and phosphorylate it, generating intracellular fructose-1-phosphate or cellobiose-6-phosphate or mannitol-1-phosphate. Once inside the cells, a gene encoding for 6-phospho-β-glucosidase converts cellobiose-6-phosphate to glucose and glucose-6-phosphate, another gene coding for 1-phosphofructokinase converts fructose-1-phosphate to fructose-1,6-diphosphateand, similarly, a gene coding for mannitol-1-phosphate5-dehydrogenase converts the mannitol-1-phosphate to fructose-1-phosphate. Hence, although mannose is subsequently assimilated in the bacteria via the fructose utilization pathway, each of the pathways has its separate regulator. While a Deoxyribonucleoside regulator (DeoR) family transcriptional regulator controls fructose uptake, a Mercuric resistance operon regulatory protein (MerR) family transcriptional regulator controls cellobiose uptake, and a Mannitol operon repressor (MtlR) family transcriptional regulator controls mannitol uptake in all *Geobacillus* spp. (Brumm et al. 2015). With this analysis, it can be safely assumed that machinery to transport and execute intracellular hydrolysis of cellulose and related oligosaccharides in *Geobacillus* spp. is less variant compared to that observed with HUS loci in this genus. However, more studies that are focused exclusively on annotation of cellulose utilization loci in *Geobacillus* using the tools of comparative genomics, can potentially serve as a backbone for the better understanding of this genus, and consequent improvement of its cellulolytic capacities or its capacity to degrade structurally diverse cellulose substrates.

Indeed, from the work of Lebre et al. on global secretomics of *Geobacillus* (Lebre et al. 2018), it is worth noting that the authors reported that 46.07% of proteins in the global secretome were hypotheticalor had no significant homology to domains and sequences in the databases used in the study. The presence of these secreted "dark matter" proteins highlights the fact that *Geobacillus* biology is still largely unexplored and has the capacity to reveal novel traits, functions and products of biotechnological value (Lebre et al. 2018).

5. Insights into Mechanisms used to Maintain Cell-substrate Proximity

Plant cell walls are recalcitrant to degradation because of their complex configurations of polysaccharides. To tackle this intricacy, noncatalytic carbohydrate-binding modules (CBMs) are being recognized as discretely folded units within the multi-modular structures of the glycoside hydrolases, where they play critical roles in the recognition of plant cell wall components and potentiating the activity of the enzymes (Gilbert et al. 2013). CBMs enhance the hydrolysis of lignocelluloses by targeting and increasing the local concentration of either individual glycoside hydrolases or multienzyme cellulosome complexes on the surfaces of carbohydrate substrates or bacterial cell wall (Rincon et al. 2007, Sainz-Polo et al. 2015). In addition to targeting the catalytic modules to their substrates, CBM's enabling role in catalytic depolymerization of polysaccharides, including crystalline and noncrystalline cellulose, soluble and insoluble xylan, chitin, β-glucans, mannan, galactan, and starch, has already been recognized. In Gram positive bacteria, multiple CBMs have been discovered in different genera, each with a particular specificity. For example, while the major ligand recognized by CBM3 is crystalline cellulose (Hernandez-Gomez et al. 2015), CBM46 is a Gram-positive bacterial CBM family that plays a role in recognizing decorated and unsubstituted forms of the xylans (Venditto et al. 2015). Family 6 CBMs (CBM6) are distinct from other CBM families in that these protein modules contain multiple distinct ligand binding sites and have binding specificities towards a number of substrates, including both branched and debranchedxylan, ß-1,4-glucan and agarose (van Bueren et al. 2005). In *Geobacillus* spp., the only CBM that has been reported to date is a CBM6 non-catalytic module at the N-terminal end (amino acid positions 32–285) of the GH5 protein orthologs of the intracellular β-xylosidasesof most of the species (De Maayer et al. 2014). Since GE40, a putative cellulase found in *G. thermodenitrificans* species (Daas et al. 2018a), also belongs to a GH5 protein family, it is possible that CBM6 can be found in its domain. This claim needs thorough characterization and detailed crystallographic study of the putative cellulase. Nevertheless, from these findings, another unique feature of the *Geobacillus* enzymology emerges: the lack of targeting by attached carbohydrate binding modules (CBM) (Brumm et al. 2015, Lombard et al. 2014). CBM modules are believed to improve enzyme efficiency by providing specific non-catalytic binding to the correct substrate (Gilbert et al. 2013). CBM modules are present in many of the xylanases produced by thermophilic Gram-positive organisms including *Clostridium thermocellum* and *Caldicellulosiruptor* species (http://www.cazy. org/) (Brumm et al. 2015, Lombard et al. 2014). The lack of CBM modules may indicate that the *Geobacillus* enzymes predate the evolution of CBM modules. Alternately, the lack of CBM modules provides the given *Geobacillus* enzymes with the ability to utilize a broader range of substrates at the cost of a lower rate of hydrolysis.

Furthermore, there is a dearth of any comparative genomics on the mechanisms used by *Geobacillus* spp. to permit elaboration on how each species differ in the number of genes encoding for proteins and enzymes involved in attachment to lignocellulosic substrates. Like any other Gram-positive bacteria, the S-layer of *Geobacillus* comprises protein and carbohydrate polymers that helps them to maintain the required cell-substrate specificity. In one study, where *Geobacillus stearothermophilus* NRS 2004/3a served as model organism, WsaF, a rhamnosyltransferase, was proposed to function as both an enzyme and to promote adherence to a polysaccharide, a rhamnolipidin this case (Steiner et al. 2010). In the future, comparative genomics and proteomics studies aimed at identifying the proteins involved in attachment to cellulose, xylan, or any other plant biomass in their environment, and unique to various *Geobacillus* species, can provide important new insights into the ligninolytic capacity of this genus.

6. Future perspective

Coupled with the previously established genomics, transcriptomics and proteomics data available for the *Geobacillus* genus, the stage is set for further advancements towards targeted strain development. Initial reports of multiple regulatory mechanisms, with inducible promoters that control the expression of hemicellulose degrading machinery in this genus, are supporting the nascent field of *Geobacillus* synthetic biology (Shulami et al. 2014). Moving forward, the use of established 'omics tools' to identify additional regulatory mechanisms occurring for cellulose and lignin degradation will be critical for continued advancements in the understanding and utilization of this genus. Further improvements in microcrystalline cellulose hydrolysis may be forthcoming through metabolic engineering or through GH5 domains with enhanced activity.

7. Conclusion

Geobacillus spp. are active in environments such as hot plant composts, and examination of their genome sequences reveal that they are endowed with a battery of genomic clusters encoding for sensors, transporters and enzymes dedicated to hydrolyzing plant polysaccharides. Comparative genomic analyses have divulged the existence of a large, conserved ~ 200 kb centralized hemicellulose utilization in almost all the members of the group, making this locus a unique feature of *Geobacillus* spp. The evolutionary advantages of a single cluster versus a random distribution are unclear but suggest a single cluster may be an adaptation to life under extreme conditions. So far, any potential genomic cluster or multidomain for cellulases has not been located in *Geobacillus* genomes to date, nor have the mechanisms underlying lignin degradation by multicopper oxidases in *Geobacillus* been identified. It is possible that, in their natural environments, *Geobacillus* strains form part of consortia, whereby strains deficient in genes or pathways required for hemicellulose, cellulose, or lignin degradation and utilization may be complemented by other members of the consortia. More than ever, understanding the genome structure of these species, by taking advantage of the availability of genome sequencing and omics tools, is vital for enhancing the functionality of this non-model environmental genus and its high value utilization in promising biotechnology processes, including consolidated bioprocessing schemes.

Acknowledgments

We acknowledge the support from the CNAM-Bio Center, supported by the South Dakota Governor's Office of Economic Development. Authors also gratefully acknowledge the support from the National Science Foundation (Award #1736255, #1849206, and #1920954) and the Department of Chemical and Biological Engineering at the South Dakota School of Mines and Technology.

References

Andlar, M., T. Rezić, N. Marđetko, D. Kracher, R. Ludwig and B. Šantek. 2018. Lignocellulose degradation: An overview of fungi and fungal enzymes involved in lignocellulose degradation. Eng. Life Sci. 18(11): 768–778.

Anné, J., K. Vrancken, L. Van Mellaert, J. Van Impe and K. Bernaerts. 2014. Protein secretion biotechnology in Gram-positive bacteria with special emphasis on *Streptomyces lividans*. Biochimica et Biophysica Acta (BBA) - Molecular Cell Research 1843(8): 1750–1761.

Basheer, S., N. Rashid, R. Ashraf, M.S. Akram, M.A. Siddiqui, T. Imanaka and M. Akhtar. 2017. Identification of a novel copper-activated and halide-tolerant laccase in *Geobacillus thermopakistaniensis*. Extremophiles 21(3): 563–571.

Bashir, Z., L. Sheng, A. Anil, A. Lali, N.P. Minton and Y. Zhang. 2019. Engineering *Geobacillus thermoglucosidasius* for direct utilisation of holocellulose from wheat straw. Biotechnology for Biofuels 12: 199–199.

Blumer-Schuette, S.E. 2020. Insights into thermophilic plant biomass hydrolysis from *caldicellulosiruptor* systems Biology. Microorganisms 8(3).

Brumm, P.J. 2013. Bacterial genomes: What they teach us about cellulose degradation. Biofuels 4(6): 669–681.

Brumm, P.J., P. De Maayer, D.A. Mead and D.A. Cowan. 2015. Genomic analysis of six new *Geobacillus* strains reveals highly conserved carbohydrate degradation architectures and strategies. Frontiers in Microbiology 6(430).

Castagnaro, A. 2014. Overexpression and Characterization of a Laccase from *Geobacillus Thermoglucosidasius*, Vol. Master's Theses and Capstones. University of New Hampshire, Durham.

Chauhan, P.S., B. Goradia and A. Saxena. 2017. Bacterial laccase: Recent update on production, properties and industrial applications. 3 Biotech. 7(5): 323–323.

Chow, N. and J.H.D. Wu. 2017. Chapter 10—The cellulosome: A supramolecular assembly of microbial biomass-degrading enzymes. pp. 243–266. *In*: Brahmachari, G. (ed.). Biotechnology of Microbial Enzymes. Academic Press.

Daas, M.J., A.H. van de Weijer, W.M. de Vos, J. van der Oost and R. van Kranenburg. 2016. Isolation of a genetically accessible thermophilic xylan degrading bacterium from compost. Biotechnol. Biofuels 9: 210.

Daas, M.J.A., P.M. Martínez, A.H.P. van de Weijer, J. van der Oost, W.M. de Vos, M.A. Kabel and R. van Kranenburg. 2017. Biochemical characterization of the xylan hydrolysis profile of the extracellular endo-xylanase from *Geobacillus thermodenitrificans* T12. BMC Biotechnology 17(1): 44–44.

Daas, M.J.A., B. Nijsse, A.H.P. van de Weijer, B.W.A.J. Groenendaal, F. Janssen, J. van der Oost and R. van Kranenburg. 2018a. Engineering *Geobacillus thermodenitrificans* to introduce cellulolytic activity; expression of native and heterologous cellulase genes. BMC Biotechnology 18(1): 42.

Daas, M.J.A., B. Vriesendorp, A.H.P. van de Weijer, J. van der Oost and R. van Kranenburg. 2018b. Complete genome sequence of *Geobacillus thermodenitrificans* T12, A potential host for biotechnological applications. Curr. Microbiol. 75(1): 49–56.

De Maayer, P., P.J. Brumm, D.A. Mead and D.A. Cowan. 2014. Comparative analysis of the *Geobacillus hemicellulose* utilization locus reveals a highly variable target for improved hemicellulolysis. BMC Genomics 15: 836.

Enguita, F.J., P.M. Matias, L.O. Martins, D. Plácido, A.O. Henriques and M.A. Carrondo. 2002. Spore-coat laccase CotA from *Bacillus subtilis*: Crystallization and preliminary X-ray characterization by the MAD method. Acta Crystallogr. D Biol. Crystallogr 58(Pt 9): 1490–3.

Fosses, A., M. Maté, N. Franche, N. Liu, Y. Denis, R. Borne, P. de Philip, H.-P. Fierobe and S. Perret. 2017. A seven-gene cluster in *Ruminiclostridium cellulolyticum* is essential for signalization, uptake and catabolism of the degradation products of cellulose hydrolysis. Biotechnology for Biofuels 10(1): 250.

Gilbert, H.J., J.P. Knox and A.B. Boraston. 2013. Advances in understanding the molecular basis of plant cell wall polysaccharide recognition by carbohydrate-binding modules. Current Opinion in Structural Biology 23(5): 669–677.

Hernandez-Gomez, M.C., M.G. Rydahl, A. Rogowski, C. Morland, A. Cartmell, L. Crouch, A. Labourel, C.M.G.A. Fontes, W.G.T. Willats, H.J. Gilbert and J.P. Knox. 2015. Recognition of xyloglucan by the crystalline cellulose-binding site of a family 3a carbohydrate-binding module. FEBS Letters 589(18): 2297–2303.

Huang, D., J. Liu, Y. Qi, K. Yang, Y. Xu and L. Feng. 2017. Synergistic hydrolysis of xylan using novel xylanases, β-xylosidases, and an α-L-arabinofuranosidase from *Geobacillus thermodenitrificans* NG80-2. Appl. Microbiol. Biotechnol. 101(15): 6023–6037.

Hussein, A.H., B.K. Lisowska and D.J. Leak. 2015. The genus *Geobacillus* and their biotechnological potential. Adv. Appl. Microbiol. 92: 1–48.

Kenyon, W.J., S.W. Esch and C.S. Buller. 2005. The curdlan-type exopolysaccharide produced by *Cellulomonas flavigena* KU forms part of an extracellular glycocalyx involved in cellulose degradation. Antonie Van Leeuwenhoek 87(2): 143–8.

Lebre, P.H., H. Aliyu, P. De Maayer and D.A. Cowan. 2018. *In silico* characterization of the global *Geobacillus* and *Parageobacillus secretome*. Microbial Cell Factories 17(1): 156–156.

Liberato, M.V., R.L. Silveira, É.T. Prates, E.A. de Araujo, V.O.A. Pellegrini, C.M. Camilo, M.A. Kadowaki, M.d.O. Neto, A. Popov, M.S. Skaf and I. Polikarpov. 2016. Molecular characterization of a family 5 glycoside hydrolase suggests an induced-fit enzymatic mechanism. Scientific Reports 6(1): 23473.

Lombard, V., H. Golaconda Ramulu, E. Drula, P.M. Coutinho and B. Henrissat. 2014. The carbohydrate-active enzymes database (CAZy) in 2013. Nucleic Acids Res, 42(Database issue), D490-5.

Moon, S.J., H.W. Kim and S.J. Jeon. 2018. Biochemical characterization of a thermostable cobalt- or copper-dependent polyphenol oxidase with dye decolorizing ability from *Geobacillus* sp. JS12. Enzyme Microb. Technol. 118: 30–36.

Ng, I.S., C.-W. Li, Y.-F. Yeh, P.T. Chen, J.-L. Chir, C.-H. Ma, S.-M. Yu, T.-h.D. Ho and C.-G. Tong. 2009. A novel endo-glucanase from the thermophilic bacterium *Geobacillus* sp. 70PC53 with high activity and stability over a broad range of temperatures. Extremophiles 13(3): 425–435.

Potprommanee, L., X.-Q. Wang, Y.-J. Han, D. Nyobe, Y.-P. Peng, Q. Huang, J.-y. Liu, Y.-L. Liao and K.-L. Chang. 2017. Characterization of a thermophilic cellulase from *Geobacillus* sp. HTA426, an efficient cellulase-producer on alkali pretreated of lignocellulosic biomass. PLOS ONE 12(4): e0175004.

Rai, R., M. Bibra, B.S. Chadha and R.K. Sani. 2019. Enhanced hydrolysis of lignocellulosic biomass with doping of a highly thermostable recombinant laccase. International Journal of Biological Macromolecules 137: 232–237.

Ramachandra, M., D.L. Crawford and A.L. Pometto. 1987. Extracellular enzyme activities during lignocellulose degradation by *Streptomyces* spp.: A comparative study of wild-type and genetically manipulated strains. Applied and Environmental Microbiology 53(12): 2754–2760.

Rastogi, G., A. Bhalla, A. Adhikari, K.M. Bischoff, S.R. Hughes, L.P. Christopher and R.K. Sani. 2010. Characterization of thermostable cellulases produced by Bacillus and *Geobacillus* strains. Bioresource Technology 101(22): 8798–8806.

Rincon, M.T., T. Cepeljnik, J.C. Martin, Y. Barak, R. Lamed, E.A. Bayer and H.J. Flint. 2007. A novel cell surface anchored cellulose-binding protein encoded by the Gene Cluster of Ruminococcus flavefaciens. Journal of Bacteriology 189(13): 4774–4783.

Saini, A., N.K. Aggarwal, A. Sharma and A. Yadav. 2015. Actinomycetes: A source of lignocellulolytic enzymes. Enzyme Research 2015: 279381–279381.

Sainz-Polo, M.A., B. González, M. Menéndez, F.I.J. Pastor and J. Sanz-Aparicio. 2015. Exploring multimodularity in plant cell wall deconstruction: STRUCTURAL AND FUNCTIONAL ANALYSIS OF Xyn10C CONTAINING THE CBM22-1–CBM22-2 TANDEM. Journal of Biological Chemistry 290(28): 17116–17130.

Shulami, S., O. Gat, A.L. Sonenshein and Y. Shoham. 1999. The glucuronic acid utilization gene cluster from *Bacillus stearothermophilus* T-6. Journal of Bacteriology 181(12): 3695–3704.

Shulami, S., A. Raz-Pasteur, O. Tabachnikov, S. Gilead-Gropper, I. Shner and Y. Shoham. 2011. The Arabinan Utilization System of Geobacillus stearothermophilus. Journal of Bacteriology 193(11): 2838–2850.

Shulami, S., O. Shenker, Y. Langut, N. Lavid, O. Gat, G. Zaide, A. Zehavi, A.L. Sonenshein and Y. Shoham. 2014. Multiple regulatory mechanisms control the expression of the *Geobacillus stearothermophilus* Gene for Extracellular Xylanase. Journal of Biological Chemistry 289(37): 25957–25975.

Singh, D.N., U. Sood, A.K. Singh, V. Gupta, M. Shakarad, C.D. Rawat and R. Lal. 2019. Genome sequencing revealed the biotechnological potential of an obligate *Thermophile Geobacillus* thermoleovorans Strain RL isolated from hot water spring. Indian Journal of Microbiology 59(3): 351–355.

Steiner, K., G. Hagelueken, P. Messner, C. Schäffer and J.H. Naismith. 2010. Structural basis of substrate binding in WsaF, a rhamnosyltransferase from *Geobacillus stearothermophilus*. Journal of Molecular Biology 397(2): 436–447.

Tai, S.-K., H.-P.P. Lin, J. Kuo and J.-K. Liu. 2004. Isolation and characterization of a cellulolytic *Geobacillus thermoleovorans* T4 strain from sugar refinery wastewater. Extremophiles 8(5): 345–349.

Tsegaye, B., C. Balomajumder and P. Roy. 2019. Microbial delignification and hydrolysis of lignocellulosic biomass to enhance biofuel production: An overview and future prospect. Bulletin of the National Research Centre 43(1): 51.

van Bueren, A.L., C. Morland, H.J. Gilbert and A.B. Boraston. 2005. Family 6 carbohydrate binding modules recognize the non-reducing end of beta-1,3-linked glucans by presenting a unique ligand binding surface. J. Biol. Chem. 280(1): 530–7.

Venditto, I., S. Najmudin, A.S. Luís, L.M.A. Ferreira, K. Sakka, J.P. Knox, H.J. Gilbert and C.M.G.A. Fontes. 2015. Family 46 carbohydrate-binding modules contribute to the enzymatic hydrolysis of xyloglucan and β-1,3–1,4-Glucans through distinct mechanisms. Journal of Biological Chemistry 290(17): 10572–10586.

Verma, A.P. and P. Shirkot. 2014. Purification and characterization of thermostable laccase from thermophilic *Geobacillus thermocatenulatus* MS 5 and its applications in removal of Textile Dyes. Scholars Academic Journal of Biosciences (SAJB) 2(8): 479–485.

Yang, S.-T. 2007. Chapter 1—Bioprocessing—from biotechnology to biorefinery. pp. 1–24. *In*: Yang, S.-T. (ed.). Bioprocessing for Value-Added Products from Renewable Resources. Elsevier. Amsterdam.

Zeigler, D.R. 2014. The *Geobacillus paradox*: Why is a thermophilic bacterial genus so prevalent on a mesophilic planet? Microbiology 160(Pt 1): 1–11.

5

Bioprospecting Extremophiles for Sustainable Biobased Industry

Neha Basotra,[1] *Yashika Raheja,*[1] *Gaurav Sharma,*[1] *Kumud Ashish Singh,*[2] *Diksha Sharma,*[3] *Rohit Rai*[4],* and *Bhupinder Singh Chadha*[1]

1. Introduction

The majority of enzymes used to date have been produced by mesophilic micro-organisms such as bacteria, fungi, archea and these mesophilic enzymes are often inhibited under the extreme conditions of many industrial processes. Therefore, extensive research is being carried out to discover novel sources of isolation, experimental techniques, and analytical methods in search of robust biocatalysts (Lugani et al. 2020, Rai et al. 2016a,b). A number of microbes (extremophiles) that flourish in extreme environments are the excellent alternative source of replacement enzymes of mesophilic counterparts currently used in these procedures (Sharma and Rai 2019, Rai et al. 2020). Extremophiles are specifically acquiring much attention and a number of these microbes have been isolated in pure culture, their genomes have been studied, and their enzymes are characterized by either academic or industrial laboratories (López-López et al. 2014, Yildiz et al. 2015). Various biotechnological products like biofuels (Barnard et al. 2010), bioinsecticides (Rubio-Infante and Moreno-Fierros 2015), cellulases for making 'stone-washed' jeans (Miettinen-Oinonen and Suominen 2002) are being manufactured through different reaction processes. Enzymes like lipases and proteases are widely used in detergent industry but they have their own limitation for usage at extreme conditions of temperature, pressure, pH and salinity due to their mesophilic origin (Herbert 1992). Several efforts are being made to improve the efficiency of these enzymes by employing genetic and/or chemical modification (Gatti-Lafranconi et al. 2010, Siddiqui et al. 2009) as well as immobilization approaches (Mukhopadhyay et al. 2015), in order to produce biocatalysts with amended properties such as enhanced activity/stability to use further in specific industrial processes. However, this can be time consuming and, more importantly, expensive strategy. In recent past, alternative approaches adopted the use of extremophilic organisms and their enzyme products, due to their robustness and versatility. Dearth of knowledge of the physiology and metabolic pathways of these extremophiles, as well as the lack of genetic tools for enzyme/organism improvement by genetic engineering are the major impediments in this approach. With the advance of genomics and other molecular biology tools, value-added extremophilic-derived products are envisaged.

---

[1] Department of Microbiology, Guru Nanak Dev University, Punjab, India – 143005.
[2] Department of Microbiology, Lovely Professional University, Punjab, India – 144411.
[3] Department of Biotechnology, CT Group of Institutions, Shahpur, Punjab, India – 144410.
[4] Faculty of Applied Medical Sciences, Lovely Professional University, Punjab, India – 144411.
* Corresponding author: rohitraisharma44@gmail.com

Extremozymes are well adapted to work under the extreme conditions of many industrial processes. Until now, few extremophiles/extremozymes have found their way into large-scale use in the field of biotechnology (Elleuche et al. 2014); however, their potential is undeniable in many applications. Few examples of extremozymes being used successfully in biotechnological industries are the thermostable DNA polymerases used in the polymerase chain reaction (PCR) (Ishino and Ishino 2014), various enzymes used in the generation of biofuels (Barnard et al. 2010), microbes employed in the mining process (Johnson 2014), and carotenoids used in the cosmetic and food industries (Oren 2010). Additionally, extremozymes are also used in making lactose-free milk (Coker and Brenchley 2006); the production of antibiotics, anticancer, and antifungal drugs (Herbert 1992); and lastly, the generation of electricity or electrons leaching for current production which can be used or stored (Dopson et al. 2016). Extremophiles/extremozyme could replace currently used mesophilic enzymes to make reactions more efficient and/or cost effective. The discovery of new extremophilic microorganisms and their enzymes have a great impact on the field of biocatalysis (Rai et al. 2019).

This chapter discusses the bioprospecting of extremophiles/extremozymes, which is the hotly pursued area of research, by employing various techniques such as omics analysis for the identification of countless novel enzymes and their application for improving sustainability of existing biotechnological processes towards a greener biobased industry. Furthermore, existing technologies for the production of value-added products and their limitations and strategies to overcome those limitations are also discussed. A graphical overview of this chapter has been presented in Fig. 1.

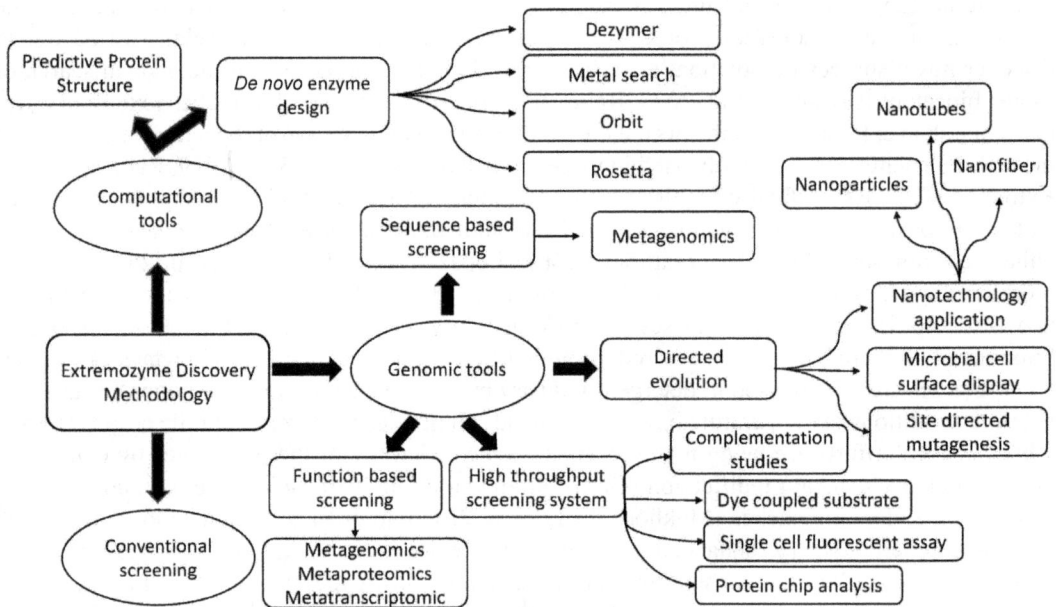

Fig. 1. An overview of the extremozyme discovery tools.

2. Ecology and Classification of Extremophiles

Extremophiles are organisms that have the potential to thrive in extremes of temperature (as high as 122°C and as low as −12°C), pressure (as high as 1000 atm), salinity (up to and including saturating levels), and pH (from 0 to 6 and 8 to 12) (10–15). These organisms have adapted to survive in ecological niches that are uninhabitable to others. Some of the examples are hot springs, solfataric fields, deep-sea hydrothermal vents, soda lakes, inland saline systems, hot and cold deserts, solar salterns, highly contaminated atmospheres with nuclear waste or heavy metals, as well as lithic or rock environments. Extremophiles, that have evolved to exist in a variety of extreme environments, fall into a number of different classes including thermophiles, halophiles, acidophiles, alkaliphiles, psychrophiles, and barophiles (piezophiles). Polyextremophiles are those that can survive in more than one of these extreme conditions. Psychrophiles and thermophiles are the class of extremophiles that inhabit extreme cold and hot environments, respectively, whereas halophiles are capable of thriving in the presence of high salt concentrations; these extremophiles employ different survival strategies to survive in their respective environment (Oren 2013). Psychrophilic prokaryotes mostly belong to bacteria and archaea and can be found within the genera *Halobacterium, Alteromonas, Shewanella, Pseudoalteromonas, Psychrobacter, Arthrobacter, Gelidibacter, Colwellia, Marinobacter, Psychroflexus, Pseudomonas, Methanococcoides* and *Methanolobus* (De Maayer et al. 2014). Acidophiles (evolved to survive in acidic environments) are also typically adapted to ecological niches with high temperatures, high salinity, or heavy metal concentrations since these conditions often co-occur, for example, in areas of acid drainage (Dopson and Holmes 2014). Alkalophiles thrive in alkaline environments such as soda lakes or gypsum-based soils and are often halophiles. They include bacteria from different genera encompassing among others *Bacillus, Halomonas*, and *Pseudomonas* (Sarethy et al. 2011) as well as archaea belonging to the genera *Halobiforma, Halalkalicoccus, Halorubrum, Natrialba, Natronococcus*, and *Natronorubrum* (Bowers and Wiegel 2011). Piezophiles (barophiles) are a group of extremophiles that dominates the deep-sea and deep subsurface environments and produce compatible solutes and polyunsaturated fatty acids and are capable of forming multimeric and antioxidant proteins for surviving under extremely high hydrostatic pressures (Zhang et al. 2015a). Piezophiles are mostly psychrophilic Gram-negative bacterial species belonging to the genera *Shewanella, Photobacterium, Colwellia, Psychromonas, Thioprofundum*, and *Moritella*, but some derived from archea can be found among the genera *Sulfolobus, Thermococcus*, and *Pyrococcus* (Zhang et al. 2015b). Metalophiles are also acidophiles and are adapted to high concentrations of heavy metals, thereby allowing metalophiles to thrive in metal-polluted sites (Orell et al. 2013). Metalophiles include both bacteria from the genera *Acidithiobacillus, Alicyclobacillus, Acidiphilium, Leptospirillum, Acidimicrobium, Ferrimicrobium*, and *Sulfobacillus* and archaea from the genera *Sulfolobus, Metallosphaera, Ferroplasma, Acidiplasma*, and *Acidianus* (Dopson and Holmes 2014). Radiophiles thrive in environments of high oxidative stress and radiation (UV, gamma, and X-rays) because of their ability to repair extensive DNA damage. Radiophiles are found among various microbial groups and species including bacteria from the genera *Deinococcus, Bacillus, Rubrobacter*, and *Kineococcus*, and the family Geodermatophilaceae and cyanobacteria including the genera *Nostoc* and *Chroococcidiopsis* (Gtari et al. 2012, Gabani and Singh 2013).

The protein stability mechanism under extreme environmental conditions varies in different microbial species and level of adaption required for survival. For the acidophiles or alkaliphiles, the intracellular proteins are not tolerant at extreme pH since the intracellular pH is maintained at pH 5.0–6.0 and the cell membrane proteins are more tolerant to extreme pH. During industrial processes, esterase has to function in solvent containing media and thus it should be solvent tolerant. The important functional characteristic that helps esterases to be stable in organic solvents include the presence of large number of negatively charged amino acid residues on their surface. This mechanism is also utilized by the halophiles to cope with high salt salinity, and hence, enzymes from such bacteria are expected to work in the presence of organic solvents. Similarly, cold active enzymes have tremendous applications in textile and detergent industry, fruit juice clarification, environmental bioremediation, etc., and can be prospected from psychrophiles, which are normally adapted to function in extremely cold temperatures (Santiago et al. 2016). So, detailed knowledge about how the extremophilic biocatalysts could achieve their special properties and their structure function relationship is extremely important for developing novel enzymes with desirable properties and last decades witnessed tremendous developments in this area.

3. Bioprospecting of Extremophiles/Extremozymes using various Techniques

The massive number of already discovered enzymes possessing different activities represents only a tiny fraction of the natural diversity. More than 99% of microbes existing in nature are unculturable under laboratory conditions. Various approaches have been devised to prospect these extreme environmental niches and uncultivable microbes. The identification of novel enzymes from such extreme habitats would involve the use of specialised techniques like microbial culturing of uncultivable and high throughput screening approaches for desired activities/properties (Wang et al. 2016). Some of the approaches currently in use for the prospection of these extremophiles are described below:

3.1 Genomics for Extremozyme Discovery

The unexplored genetic diversity of extreme habitats of temperature, pH, alkalinity, and salinity can be exploited for the discovery of new and potent enzymes for industrial use. The most successful story in this regard is the discovery of biotechnologically important Taq DNA polymerase used for PCR, which was purified from the extreme thermophilic bacterium *Thermus aquaticus* (Chien et al. 1976) with a commercial market of about $500 million in 2009 (Adrio and Demain 2014). Sequencing of genomes obtained from extremophiles offers us with a massive sum of genes encoding extremozymes, which are of interest for basic as well as applied research. By exploiting genomics tools, the mechanism of action of robust enzymes can be explored and can deliver information on the enigmas behind the 3D structure of extremozymes. Novel extremozymes with great potential for industrial applications have been identified in the last few years by using a combination of computational and structure-based analysis with evolutionary driven approaches, such as synthetic biology or directed evolution (Bornscheuer et al. 2012, Ibrahim et al. 2016). With the advent of various sequence-based or activity-based (function-based) screening approaches, development in this field has been accelerated. In activity- or function-based screening approaches, the identification of novel enzymes is performed by high-throughput screening system using protein chip analysis, dye-coupled substrates, single cell fluorescent assays or complementation studies of process-specific mutants (Adam and Perner 2017, Maruthamuthu et al. 2016, Park et al. 2012, Santoro et al. 2002). In this post-genomic era, the sharing of a number of sequences in the public domain, metagenomics, metatranscriptomics or metaproteomics (sequence-based approaches) will dramatically accelerate the discovery of novel biological systems (Prosser 2015, Wilmes et al. 2015, Basner and Antranikian 2014). In order to produce these extremozymes in bulk, their heterologous expression by employing various mesophilic or thermophilic hosts including *Escherichia coli*, *Thermus thermophilus* or *Sulfolobus solfataricus*, *Saccharomyces cerevisiae*, *Pichia pastoris* have been successfully used (Morana et al. 1995, Cava et al. 2009, Latiffi et al. 2013). The production in large quantities is crucial to implement a sustainable bioeconomy strategy.

3.2 Metagenomics for the Analysis of Microbial Communities

Since more than 95% of microorganisms cannot be cultured in lab conditions (unculturable), therefore, the progress in the field of genomics has significant impact on understanding microbial physiology in extreme habitats. Using the metagenomics tool which is defined as a genomic analysis of consortium of micro-organisms present in an environmental sample under study, it is now possible to directly clone the DNA from extreme habitats into various kinds of cloning plasmids such as fosmids, phagemids or bacterial artificial plasmids (BACs) and therefore analyze the microbial communities in these environments and study the impact of environmental factors like changes in temperature, pH, salt concentration on the diversity of extremophilic archaea and bacteria (Handelsman 2004). Metagenomics studies of these extremophiles have led to new visions into microbial metabolism and evolutionary relationships in a microbial community (Benson et al. 2011, Pearce et al. 2012, Venter et al. 2004, Ventosa et al. 2015). In addition to the analysis of type of microbes present in a metagenome sample, in further studies, type of changes or shifts that microbial communities undergo during fermentation processes or what kind of starting cultures provide an optimal distribution and adaption to the corresponding industrial process, were also investigated by metagenomics. Such strategies have been reported for various industrial applications including food fermentation processes or biogas production (Roske et al. 2014, Bora et al. 2016). In further advancement, the profound knowledge of the impact of environmental conditions on microbial communities, the microbial adaptation and the underlying regulation processes can be gained by combining the metagenomics tool with expression studies which include metatranscriptomics or metaproteomics. Metatranscriptomics and metaproteomics are the emerging molecular biology tools that enable the study of transcribed genes and translated proteins, respectively, at a specific time point and in response to a given environment. These studies are possible with the development of NGS (Next Generation Sequencing) technology which has brought these studies to the next level, involving the direct sequencing of RNA without requiring the previously needed laborious cDNA synthesis or cloning steps and resulting in growing amount of datasets (Frias-Lopez et al. 2008, Gilbert et al. 2008). These breakthroughs have been possible due to advances in Next Generation Sequencing (NGS) technology which resulted not only in elimination of tedious and time consuming cDNA synthesis/cloning steps during RNA sequencing but availability of genomic sequences information for greater applications as well (Frias-Lopez et al. 2008, Gilbert et al. 2008). Large number of genomes of extremophiles have been sequenced by next-generation sequencing which delivers rich resources for prospecting of novel enzymes with potent activities. Hence, by exploiting the above mentioned technologies, a clear picture of functional and regulating networks in the cells of each individual microorganism existing within the microbial consortium of interest can be obtained (Martinez et al. 2016). Furthermore, the information gained on gene, transcript or protein level can be exploited in the search of novel targets or for metabolic engineering of strains of industrial relevance (Wilmes et al. 2006). Therefore, in order to improve the production of a specific product, pre-existing pathways can be modified by protein engineering or novel ones can be built from scratch, which are most suitable for the corresponding industrial process. Moreover, high-throughput screening methods for the identification of mutated variants or improved strains have also been successfully developed (Brugger et al. 2014).

3.3 Metagenomics for Discovery of Novel Extremozymes

Unculturable microorganisms represent the major fraction of all existing microorganisms and screening of metagenomic libraries from these non-cultivable microbes inhabiting extreme ecological niches enables the discovery of robust enzymes possessed by these extremophiles (Pace 1997, Sogin et al. 2006). Activity or function-based approaches offer the identification of promising active enzymes without requiring any prior knowledge of their coding sequences. However, reliance on cultivable microbes and resulting low expression levels in the heterologous hosts are some of the disadvantages posed by function-based approaches. On the contrary, shotgun metagenomics facilitates the sequencing of environmental DNA directly from the sample, skipping any cloning steps and resulting in whole sequences of protein encoding genes. This permits a functional depiction of large parts or complete genomes of each microorganism present in the consortium under study. These new metagenomic methods depend on the possibility to store, process and analyze vast amounts of data, which could only be achieved by digitalization in sciences. Therefore, metagenomics is a great example of how the digitalization has accelerated scientific progress in the field of extremophiles. On the other hand, it is interesting to note that extremozymes, e.g. DNA polymerases from the thermophiles *Thermus aquaticus* and *Pyrococcus furiosus*, have revolutionized life sciences during the last three decades. Hence, various industrially relevant extremozymes like DNA-modifying enzymes, lipases, esterases, proteases, amylases, cellulases have been discovered by using metagenomics, thus showing its importance in the search for novel biocatalysts and, consequently, for a sustainable industry (Warnecke et al. 2007, Chow et al. 2012).

Moreover, a huge number of "commercial useful enzymes" (CUEs), identified using metagenomics, have been made publicly accessible by the "MetaBioMe" database since 2010. These CUEs were classified into nine application categories, specified as agriculture, energy, food and nutrition, biotechnology, biosensor, environment and health (Sharma et al. 2010). Crucial requirements for shotgun metagenomics are bioinformatic tools to assemble the short reads generated to continuous sequences (contigs or scaffolds) after sequencing. The contigs thereby enable the gene prediction and functional assignments of the metagenomic sequences. Especially, the reconstruction of each individual genome sequence within a metagenome requires the complex bioinformatics workflows successfully developed in the last years (Sharma et al. 2010, Milshteyn et al. 2014). Altogether, the adoption of the reported trends into industrial applications offers a promising way to switch from conventional processes to more biobased processes leading towards an industrial revolution and into a sustainable technology landscape.

4. Industrial Applications of Extremophiles

Extremophilic biocatalysts exhibit potential biotechnological applications due to their unique enzymatic characteristics and physiological properties ranging from the bioremediation of noxious pollutants to the generation of biomolecules of medical and industrial use. Extremozymes have developed molecular mechanisms (Hough and Danson 1999) in order to adapt to extreme physicochemical conditions that have found relevant applications as biocatalysts in industrial biotransformation processes. Extremozymes derived from psychrophiles have been observed to display high catalytic efficiency in the detergent and food industries and for the production of fine chemicals (Cavicchioli et al. 2011). Halophilic enzymes are capable of being active and stable under low water activity and in many cases in presence of organic solvents, and therefore have the potential for industrial applications (Raddadi et al. 2013). Examples of these extremozymes include polysaccharide-degrading enzymes for the degradation of cellulose, xylan, and starch (Raddadi et al. 2013, Elleuche et al. 2014). For example, the extremotolerant cellulases from *Paenibacillus tarimensis* L88, isolated from the Sahara Desert in southern Tunisia, were observed to have high functionality across a broad pH range (3.0 to 10.5), at high temperatures (80°C) and high salt concentrations (up to 5-M NaCl) (Raddadi et al. 2013). Carboxymethyl cellulase activity has been detected in the presence of 40% (v/v) 1-butyl-3-methylimidazolium chloride or 20% (w/v) 1-ethyl-3-methylimidazolium acetate ionic liquids and was maintained after exposure to organic solvents, detergents, heavy metals, and even under high alkalinity. *Paenibacillus tarimensis* is a potent source of cellulases with applications in detergent, textile, and pulp and paper industries; it also has potential for simultaneous ionic liquid treatment and saccharification of lignocellulose in biorefinery processes (Raddadi et al. 2013). Some halophilic enzymes are lipolytic, such as lipases and esterases, and have the ability to hydrolyze long-chain acylglycerols ($\geq$ C10) and short-chain fatty esters ($\leq$ C10), respectively. Enzymes derived from bacterial alkalophiles are widely applied in the detergent and laundry industries (Sarethy et al. 2011). Piezophiles may also be valuable to the food industry in processes that require high pressures (Zhang et al. 2015) and piezophilic bacteria could also be a source of essential fatty acids like omega-3-polyunsaturated fatty acids since these compounds are produced by the bacteria to stabilize the cell membrane under high pressure (Zhang et al. 2015). Enzymes produced by radiotolerant microorganisms have been shown to be resistant to other stresses. For example, Shao et al. (2013) characterized lipases from the radiation tolerant bacterium *Deinococcus radiodurans* expressed in *Escherichia coli*. Thermozymes are extremozymes produced by thermophilic and hyperthermophilic microorganisms and are stable at high temperature. These enzymes are also often able to tolerate proteolysis and harsh conditions like the presence of denaturing agents and organic solvents as well as high salinity. Benefits of using thermozymes include reduced risk of contamination, lower viscosity, and higher solubility of substrates.

A thermostable nucleoside phosphorylase has been characterized from hyperthermophilic aerobic crenarchaeon *Aeropyrum pernix* K1 and has been used for the synthesis of nucleoside analogues used in antiviral therapies as an alternative to chemical synthesis (Zhu et al. 2013). Other thermozymes also include proteases like thermolysin used in the synthesis of dipeptides, pretaq protease used to clean DNA prior to PCR amplification, and starch-processing and DNA-processing enzymes (Jayakumar et al. 2012). In addition to the above mentioned extremozymes, other enzymes are also suitable for use in further industrial processes. For example, alcohol dehydrogenases can be used to synthesize building blocks for the chemical industry, such as optically active alcohols, or to synthesize cofactors such as NAD and NADP. Meanwhile, nitrile-degrading enzymes are of interest for the transformation of nitriles and carbon-carbon bond forming enzymes like aldolases, transketolases and hydroxynitrile lyases are useful in organic synthesis (Resch et al. 2011, Demirjian et al. 2001).

4.1 Extremolytes

Extremophiles, when exposed to stressful environmental conditions, tend to accumulate organic compounds known as extremolytes that can account for up to 25% of dry cell weight. Several compounds of polyol derivatives (betaine, echoin and hydroxyectoin), carbohydrates such as trehalose and mannose derivatives (mannosylglycerate [firoin] and mannosylglyceramide [firoin-A]), glucosylglucosylglycerate (GGG) and various amino acids are examples of extremolytes (Lentzen and Schwarz 2006, Alarico et al. 2013, Bougouffa et al. 2014). The reliable reference of UV radiation-protective compounds such as bacterioruberin, ectoines, mycosporin-like amino acids (MAAs), melanin and scytonemin are the UV-resistant extremophilic bacteria (Gabani and Singh 2013). Several archaea accumulate negatively charged inositol and glycerol derivatives, such as phosphodiester di-myoinositol-1,1'-phosphate and α-diglycerol phosphate or cyclic 2,3-diphosphoglycerate and trianionic pyrophosphate (Lentzen and Schwarz 2006). Extremolytes have the ability to be used in the pharmaceutical industry and have been used mainly in cosmetics. The UV safety sunscreens in the cosmetics industry use MAAs, which have also been proposed as possible protective agents for cancers caused by UV radiation, such as the melanoma (de la Coba et al. 2009) and may be specifically involved as therapeutic candidates in the future. As a possible candidate for the development of a novel pharmacophore to produce protein kinase inhibitors such as antiproliferative and anti-inflammatory drugs, Scytonemin, a component in sunscreens (Soule et al. 2009) was also proposed. Bacterioruberin is produced from radiation resistant microbes such as *Halobacterium* and *Rubrobacterium* which aid in repairing damaged DNA strands from ionizing UV radiation to prevent cancer in the human skin (Singh and Gabani 2011). Choi et al. (2014) studied the function of deinoxanthin (carotenoid) in forestalling malignancy. It was isolated from the radio resistant bacterium D. *radiodurans* which incited apoptosis of cancer cells, proposing that this carotenoid might be valuable as a chemopreventive operator. Extremolytes can likewise be utilized to stabilize proteins and nucleic acids. Protein instability is a key challenge for the use of therapeutic protein-based medicines, especially in aqueous formulations. In the absence of other protein stabilizers due to their ability to stabilize proteins *in vivo* and *in vitro*, extremolytes provide an attractive alternative for stabilization and survival of the responsive proteins (Avanti et al. 2014). Extremolytes have also been reported to inhibit the misfolding and/or aggregation of proteins, and are therefore promising candidates for drugs' production for several diseases (Faria et al. 2013). In order to minimize signal-dependent events resulting from *in vitro* and *in vivo* exposure to carbon nanoparticles, firoin and ectoine have been observed, expanding the fields of application for these compatible solutes. Indeed, such events include the activation of mitogen-activated protein kinases or the upregulation of proinflammatory cytokines, apoptosis and proliferation in lung epithelial cells, which could lead to lung cancer, chronic obstructive pulmonary disease, and fibrosis (Autengruber et al. 2014). In addition, in the food industry, extremolytes have found application for the production of functional foods, food items that have an additional positive health benefit by improving short-term well-being/performance potential or by reducing such diseases in the long-term (Cencic and Chingwaru 2010). For example, ectoine has been documented to accumulate (up to 89 mg/100 g of product) in some cheeses that have been treated with *Brevibacterium linens* for surface ripening of the product (Klein et al. 2007).

4.2 Biofuels

Agricultural crop residues (e.g., corn straw, beets, wheat straw, and sugar cane bagasse) represent the largest renewable lignocellulosic biomass for the production of bioethanol to counter act the adverse effects fossil fuel consumption (Basotra et al. 2016, Rai et al. 2020, Raheja et al. 2020). Biofuels can be divided into first and second generation, depending on the particular source. Biofuels of the first generation are those derived from the 'readily' hydrolyzed sugars, starches and oils of available crops, while biofuels of the second generation are derived from lignocellulosic content, which is more hydrolysis-resistant. Biofuels may also be categorized as final products: butanol, ethanol, hydrogen, methane and biodiesel. The use of a chemical method accompanied by the use of mesophilic microorganisms as *Saccharomyces cerevisiae* and *Clostridium* species (Lee et al. 2008) is a common way for the production of biobutanol and bioethanol. Hydrogen production historically relies upon a chemical/catalytic process (Das and Veziroglu 2001) but more large micro-organism-based systems have recently been established with thermophiles *Caldicellulosiruptor saccharolyticus* and *Thermotoga elfii* (de Vrije et al. 2002). Unlike the other products, methane has only been manufactured by a consortium of methanogens (extremophiles, the sole biological producers of methane) (Barnard et al. 2010, Dhiman et al. 2018).

Many steps in the development of biofuels require high temperatures and extreme pH levels and are thus ideal candidates in the replacement of mesophilic species used in conventional methods. For instance, hemicellulose and pentose sugars such as xylose can be used as a starting material for the production of ethanol (Mahajan et al. 2016, Demain et al. 2005) by *Thermoanaerobacterium saccharolyticum*. This thermophile's engineered models have shown great promise in producing significant amount of ethanol and minimizing other side reactions/products (Basen et al. 2014). There is enormous potential for the use of strains of *Caldicellulosiruptor*, *Thermoanaerobacterium* (Ren et al. 2008), *Pyrococcus* (Baker et al. 2009), and *Aeropyrum* (Nishimura and Sako 2009) in the production of hydrogen through anaerobic fermentation. The research is still quite preliminary, but recent developments such as hyperthermophile engineering are very promising (Lipscomb et al. 2014). The two most commercially effective products produced by microorganisms are biodiesel and butanol. Biodiesel harvests the energy of algae with a high lipid content (> 75% dry weight), most of which produce long-chain hydrocarbons, including those found in petroleum. There are many extremophilic algae that fulfil these criteria, e.g., *Cyanidium caldarium* (Luca et al. 1981) and *Galdieria sulphuraria* (Pulz and Gross 2004). Engineered halophilic algae are also very promising as they can be grown in open containers because the high salinity needed for their growth inhibits other microbes. This implies that they can be cultivated in underused habitats such as the oceans and arid/desert environments (DasSarma et al. 2009). Compared to ethanol, butanol is very inhibitory to the growth of microorganisms (most species do not withstand more than 2%). As such, organisms need to be modified to overcome inhibition of the product and withstand large amounts of butanol. Green Biologics is currently using thermophilic *Clostridium* to manufacture biobutanol from corn stocks. Different organizations, for example, Gevo, Joule Unlimited, and Solazyme are likewise ready to deliver huge scope volumes of bioethanol and biodiesel just as fly fuel for both regular citizen and military use. Also, Sapphire Energy has moved one stage back simultaneously and creates what it calls 'Green Crude', which can go about as a substitution for raw petroleum in the current oil framework.

4.3 Bioethanol

Ethanol, a renewable substitute for current hydrocarbon fuels, is readily produced through the fermentation of sugars. Ethanol production is dominated by first generation feedstocks, e.g., corn, sugar cane, sugar beet. *Saccharomyces cerevisiae*-based fermentation is industrially proven and well developed for the production of ethanol. One of the major constituents of lignocellulosic biomass is pentose (C5) sugars, but lack of C5 catabolism by *S. cerevisiae* reduces the efficiency of the conversion process. This issue has been addressed by xylose and arabinose utilizing recombinant strains of *S. cerevisiae* and *Zymomonas mobilis* (Barnard et al. 2010). On the other hand, several wild-type thermophilic microorganisms are able to ferment C5 as well as C6 sugars. They also typically have a higher pH tolerance and maintain their activities through environmental fluctuations. Thermophilic *Clostridia* sp. operating at 60–65°C has cellulosomes, containing cellulases that are adept at digesting crystalline cellulose. *C. thermocellum* JW20 has broad substrate specificity, and can grow on cellulose, cellobiose, xylooligomers, glucose, fructose, and xylose. Elevated temperatures also reduce the risk of contaminations, as there are fewer indigenous bacteria that are competitive at higher temperatures (Taylor et al. 2008). Currently, one of the leading contaminants is *Lactobacilli*, which competes for carbohydrate substrates and produces inhibitors. Antibiotics will control contamination, but with an additional cost. *Lactobacilli* do not grow above 50°C and thus cannot compete with a thermophilic fermentative organism. In addition, elevated temperatures are good for distillation and ethanol recovery; however, the two major hurdles of poor ethanol yield and low ethanol tolerance in thermophiles remain. Novel organisms from extreme conditions provide a good industrial source of genes for enzymes because of their ability to deal with harsh industrial conditions (Blumer-Schuette et al. 2008). This has prompted the collection of novel microorganisms from an assortment of spots and conditions including the underground, mechanical waste streams, and warm vents (Blumer-Schuette et al. 2008, Bai et al. 2010, Zambare et al. 2011). In ethanol production, the sustained operation of the extremophilic enzymes is of interest because it allows for further variation in the conditions of pretreatment and hydrolysis while maintaining normal saccharification rates.

Saccharolytic fermentative thermophiles can produce lignocellulose deconstruction enzymes. They ferment pentose and hexose and are desirable for efficient bioethanol production and cost-effective operation. In fermentation, the focus needs to remain on increasing ethanol tolerance and yields. Adaptations or mutations that lead to higher ethanol tolerance and higher yields would make these organisms comparable to the current yeast-dominated process. Extremophiles have been engineered for higher ethanol tolerance, and co-fermentation of the major pentose and hexose sugars has also been demonstrated. *Thermoanaerobacterium saccharolyticum* ALK2 produced 33 g/L ethanol by fermentation of mixed sugars (xylose, glucose, mannose, and galactose) (Barnard et al. 2010). Some extremophile fermenters have shown cellulase activity that can help promote simultaneous saccharification and fermentation. In summary, novel microorganisms need to be compared directly with industrial organisms (e.g., *Saccharomyces* and *Zymomonas* spp.) for accurate assessment. The high yield of ethanol (> 20%, v/v), no side fermentation byproduct, reduced inhibitor sensitivity and elevated tolerance to ethanol, close to *Saccharomyces* and *Zymomonas* spp., should be used to achieve the optimal microbial strain (e.g., a thermophile). These thermophiles should have fast growth rates and carry out fermentation of pentose and hexose sugars at elevated temperatures (e.g., > 80°C, for easier recovery of ethanol). These microbial strains should also have high resistance to microbial contaminants. For the purpose of developing various metabolic engineering approaches to improve ethanol yield and tolerance, genome sequencing of thermophiles is desirable. Finally, if extremophiles could produce thermostable lignocellulosic-deconstruction-enzymes, hydrolyze the untreated biomass, and ferment the sugars in a single reactor, then this could add a new dimension to the available methods of replacing fossil fuels with biofuels. Such extremophiles may produce new engineering solutions to biofuels production, which are more efficient and more economical.

4.4 Biodiesel

Biodiesel, a clean-burning renewable energy source, is a fatty acid ester compound produced from the transesterification of triglycerides and fatty acids from vegetable oils, waste oils, animal fats, or microalgae with alcohol, usually methanol or ethanol. It contains long-chain alkyl (methyl, ethyl, or propyl) esters and has higher energy content (about 35 MJ/L) compared to ethanol or butanol. It increases fuel lubricity, is not detrimental to engine performance, and is cost comparable with current diesel. More importantly, it is biodegradable, non-toxic, and free of sulfur and aromatics (Yang et al. 2009). Third generation biodiesel, in particular using microalgae as the oil source, is a viable option as a biofuel. Biodiesel is currently produced from oil crops such as soybeans and rapeseed using a base catalyst for the transesterification process. As the focus is switching from second generation to third generation biofuels, microalgae are an excellent source of oil compared to crops such as soybean. They are able to produce 15- to 300-times more oil than conventional oil crops, are grown on non-arable land, do not conflict with the food versus fuel issue, and are continually harvested (Dragone et al. 2010). Extremophilic algae are being discovered and reviewed as an alternative oil source along with an extremophilic source of lipase enzyme as a biocatalyst in the transesterification process. This section will focus on extremophiles for biodiesel production, mainly the lipid (algal oil) source and extremophilic sources of lipase enzymes for the transesterification process.

4.5 Algae (Lipid Source)

Biodiesel from algal oil is considered to be more sustainable than from oil crops. Algae produce more oil and do not have to be grown on arable land. Algae can be mass cultivated by two different methods, in an open air pond system or closed system such as a photobioreactor. To prevent contamination in an open air system, such as a raceway pond, extremophilic strains of algae are preferred that can grow at high temperatures, either low or high pH values, and/or high salinity. *Spirulina*, *Dunaliella*, and *Chlorella* spp. are different types of algae that have been shown to grow well in highly alkaline, halophilic, nutrient rich media environments, respectively, and are good options for an open air system. In a closed photobioreactor system, there is no limitation on the type of strain since contamination is not an issue, and higher cell densities can be achieved (Dragone et al. 2010). In addition, after extraction of the lipids and oils for transesterification, the algal biomass can be used as an animal feed, biofertilizer, or for human nutrition. Value-added products may also be produced simultaneously with the oil, such as polyunsaturated fatty acids, polysaccharides, antioxidants, colors, food-coloring products, toxins, and stable isotopes. Another extremophilic source for oil is the psychrophilic yeast *Rhodotorula glacialis* DBVPG 4785. This yeast was isolated from an alpine glacial environment and can grow at zero and sub-zero temperatures but has optimum growth at 20°C. It was shown that under nutrient limitation with carbon in excess, larger amounts of lipids are produced by the yeast cell, up to 68% (lipids/biomass). The majority of the fatty acids produced is in the range of C14 to C18 with traces of C12 and includes saturated, mono-unsaturated, and polyunsaturated forms. The high glucose concentration increases mono-unsaturated forms at the expense of the poly-unsaturated forms. This lipid content is similar to soybean oil and rapeseed oil, and thus has high potential as a biodiesel precursor (Amaretti et al. 2010).

4.6 Lipases for Transesterification for Biodiesel Production

Lipases, part of the α/β hydrolase superfamily (E.C.3.1.1.3), are carboxyl ester hydrolases that catalyze hydrolysis and synthesis of long chain ester compounds. These enzymes perform transesterification of triglycerides to biodiesel, instead of using a typical chemical transesterification. In a chemical transesterification, either an acid or base catalyst is used such as sulfuric acid, sodium hydroxide, or potassium hydroxide. These chemicals are corrosive and have large negative impact on the purification process after transesterification but have a high yield of biodiesel with a low reaction time. As a biocatalyst in transesterification, lipases are advantageous due to their synthesis of specific alkyl esters, easy recovery of the biodiesel and glycerol by-product, and transesterification of solutions with a high free fatty acid composition, whereas chemical transesterification has high energy consumption and difficulty in converting triglycerides with high free fatty acid content. Lipases typically have low yields over extended periods of time and are intolerant to the organic solvents typically used in transesterification (Lam et al. 2010). The desired characteristics of lipases for biodiesel production include high alcohol (ethanol or methanol) tolerance, high organic solvent tolerance, high yield using mono, di, and triglycerides along with free fatty acids, low reaction time, temperature resistance, low product inhibition, and reusability of free or immobilized enzyme (Yang et al. 2009). Extremophilic sources of lipases such as psychrophiles, thermophiles, and organic solvent tolerant microorganisms may have more desirable characteristics for the transesterification of triglycerides and free fatty acids into biodiesel. An extracellular lipase that has been shown to be thermo and solvent-tolerant has been obtained from *Bacillus* sp., isolated from a soil sample near a furnace in the metallurgical industry. It has an optimum temperature of 55°C and an optimum pH of 7. The activity of the lipase in polar solvents, such as methanol or ethanol, was low since they disrupt the protective water layer around the enzyme's active site. The polar organic solvents have no such massive effect on the active site of the lipase enzyme in low water reactions. Non-polar organic solvents such as *n*-hexane and *t*-butanol also do not have this effect, and the enzyme remains stable. This lipase is also activated by Ca2+, Mg2+, and K+ but inhibited by ethylenediaminetetraacetic acid (EDTA), Fe2+, and Zn2+ (Sivaramakrishnan and Muthukumar 2012). *Bacillus pumilus* B106 also produces a thermo-tolerant lipase that has an optimum temperature of 50°C, optimum pH of 8.0, and salinity tolerance of 0–150% psu (practical salinity units). *B. pumilus* B106 lipase is active at low concentrations of organic solvents, i.e., it is activated at methanol concentrations from 10–20%, then is inhibited at higher concentrations (Zhang et al. 2009). Psychrophilic microorganisms are another unique source for lipase production for a less energy intensive biodiesel production process. *Psuedomonas fluorescens* B68 expresses an extracellular lipase with a temperature optimum psychrophilic lipase at a slightly higher temperature at 25°C and optimum pH of 9.0. This lipase is active at low concentrations of organic solvents such as dimethyl sulfoxide (DMSO), ethyl acetate, and acetonitrile but decreases slowly in activity as the concentration of solvent is increased, whereas in methanol or ethanol the enzyme is active in a concentration of solvent up to 30%, then is inhibited at higher concentrations (Yang et al. 2009). The thermo- and organo-tolerant lipase from *Bacillus* sp. can hydrolyze a wide variety of oils effectively including olive oil, palm oil, sunflower oil, castor oil, ground nut oil, and coconut oil. The yield of fatty acid methyl esters (FAMES) was 76% by *Oedogonium* sp. oil after 40 hours. The stability of the lipase in *t*-butanol allowed the higher yield due to the reduction of inhibitory effects of methanol and glycerol on the enzyme (Sivaramakrishnan and Muthukumar 2012). The psychrophilic lipase from *P. fluorescens* B68 most efficiently hydrolyzed *p*Np-caprate (C10 acyl group) and had a yield of 92% of biodiesel after 12 hours on soybean oil at 20°C. As the temperature was increased, the activity of the lipase decreased. Using olive oil as the substrate for biodiesel, the *P. lipolyticum* M37 lipase had 90, 90, and 70% yield after 48 hours using a 3-step, 2-step, and 1-step methanol feed, respectively, in comparison to the commercial CalB which had 90, 60, and 0% yield, respectively, under the same conditions. Using waste oil as the substrate, biodiesel was produced with the free fatty acids and the higher water content (22.6%) did not have an effect on production (Yang et al. 2009). Some of the drawbacks of using lipases instead of the traditional chemical transesterification are the lower reaction rates, low yield, and sensitivity to the methanol co-reactant. Methanol can be used as both the solvent and reactant in the transesterification process such as for the lipase from *P. lipolyticum* M37 and *B. pumilus* B106. Ratio of 1:3 (oil to methanol) is commonly used for the transesterification reaction, which is determined to be a methanol concentration of 10% in the reaction, which activates the *P. lipolyticum* M37 and *B. pumilus* B106 lipases (Yang et al. 2009, Zhang et al. 2009). The psychrophilic lipase enzymes have higher yields and reaction rates using industrial substrates for biodiesel production in comparison to the thermo- and organo-tolerant lipases (Sivaramakrishnan and Muthukumar 2012, Yang et al. 2009, Zhang et al. 2009). A psychrophilic lipase would lower the energy consumption needed in the transesterification process. This activity at low temperatures is due to the flexibility of the enzyme structure to enzyme-substrate complexes. Extremophiles, in short, may play a major role in biodiesel production. Extremophilic sources of algal oils are only necessary if an open air cultivation system is used. Higher biomass yields, thus larger lipid yields, can be achieved in a photobioreactor but the costs of building and maintaining a photobioreactor are higher than a simple raceway pond design. Biodiesel known as 'Soladiesel' produced by Solazyme (www.solazyme.com) is well established using algae under dark fermentation conditions. Another option is using a psychrophilic yeast such as *R. glacialis* DBVPG 4785 which can produce up to 68% (lipids/biomass) under optimum C:N ratios, whose oil is very similar to soybean and rapeseed oil. Extremophilic sources of lipase enzymes for the transesterification of the oil to biodiesel have a much greater potential for use in biodiesel production. A lipase that is methanol tolerant and has high yields and reaction rates at lower temperatures would replace the need for a catalyst in a chemical transesterification process. Lipases have broader substrate specificities, have lower energy requirements, and are more environmentally friendly than traditional catalysts.

4.7 Glycosyl Hydrolases and Sugars

The class of enzymes that hydrolyze the glycosidic bond between a carbohydrate and another moiety are glycosyl hydrolases and are divided into well over 100 families. The hydrolysis of the glycosidic bond generally takes place with the use of only two amino acids—a proton donor and a nucleophile/base and results in retention or inversion of the anomeric configuration of the resulting carbohydrate. Lactase (β-galactosidase) derived from organisms like *Kluyveromyces lactis* (Messia et al. 2007) is being used to make lactose-free milk and other dairy products in preventing lactose intolerance. Roughly 70% of the world's population suffers from lactose intolerance resulting from a lack or loss of β-galactosidase activity. However, for the enzyme to be active, the temperature of the dairy product must be raised (from about 5°C to 25°C). This elevation in temperature creates the potential for pathogens to grow as well as for altering the flavor profile of the milk. A simple solution to both issues is to use a β-galactosidase from a psychrophile (Coker and Brenchley 2006). This enzyme would be active at low temperature and hydrolyze lactose throughout the entire process from production to shipment and storage by the consumer (Coker 2004). This approach could save significant amounts of money by eliminating the heating step as well as achieve a high percentage of lactose hydrolysis.

Currently, several cold-adapted enzymes have been characterized and developed that perform on par with the currently used mesophilic enzymes when compared at their respective temperature optima (i.e., 15°C and 37°C) (Coker et al. 2003, Coker 2004). Similar to the industrial-scale hydrolysis of lactose, that of starch traditionally uses mesophilic enzymes. Starch-hydrolyzing enzymes comprise about 25% of the worldwide enzyme market; however, several adjustments in temperature and pH are needed for most of the reactions to ensure optimal conditions. Since the industrial processes involved in hydrolyzing starch require high temperatures (95°C for one step and 60°C for the other) and high pH, polyextremophilic (thermophilic and alkaliphilic) enzymes would be ideal. Currently, a glucoamylases from *Picrophilus* (Serour and Antranikian 2002), α-amylase from *Bacillus acidicola* (Sharma and Satyanarayana 2012), and a pullulanase from *Thermococcus kodakarensis* (Han et al. 2013) show great promise in replacing their mesophilic counterparts. However, amylases have also been isolated from halophiles, such as *Halomonas meridian* and *Natronococcus amylolyticus*, which could be useful in the process of producing high-fructose corn syrup, which is produced by hydrolyzing corn starch (Schiraldi et al. 2002). In addition to sugar hydrolysis, another promising application for extremophiles is the production of carbohydrates like trehalose and ectoine, which can be used as stabilizers for products like antibodies and vaccines (Argüelles 2000, Guo et al. 2000). Another example is ectoine, which has been shown to protect skin from UVA-induced damage. RonaCare™ Ectoin, produced by Merck KGaA (Darmstadt,

Germany), is used as a moisturizer and comes from halophilic microorganisms (DasSarma et al. 2009). In addition to trehalose and ectoine, several other carbohydrates are produced by halophiles as compatible solutes that can also be employed as preservatives (Roberts 2005).

4.8 Biomining

In addition to biofuels, another important application of extremophiles and their enzymes can be found in the mining sector (Podar and Reysenbach 2006). This process, also known as bioleaching, is the removal of insoluble metal sulfides or oxides by using microorganisms. It is a safer and more environmentally friendly way to extract metals compared with traditional heap leaching, which involves the use of several chemicals, including cyanide, to bind and separate specific minerals/ metals from others. In 1992, biomining accounted for 10% of worldwide copper production (Herbert 1992), but current estimates place it at around 15% for copper and 5% for gold (Vera et al. 2013). Extraction rates are around 90% from biomining compared with 60% for traditional heap leaching (Vera et al. 2013). Biomining techniques have successfully been employed to mine metals such as gold, silver, copper, zinc, nickel, and uranium. The organisms used in this process are acidophiles such as *Acidithiobacillus* and *Ferroplasma*. However, depending on the conditions, more thermophilic strains, like *Sulfolobus* and *Metallosphaera* (Podar and Reysenbach 2006, Vera et al. 2013), may have to be employed. Although biomining is generally safe, it does need to be tightly controlled, since it can result in acid mine drainage (AMD), which occurs when acidic water, generated by the oxidation of sulphides from the mine, begins flowing or leaching out of the mine. Since the acidophiles employed in biomining thrive in acidic and usually heavy-metal environments, AMD results in an environment that is not only very acidic but also rich in heavy metals. Copper, zinc, and nickel mines are the most common sources of AMD (Vera et al. 2013). Interestingly, mesophilic and sometimes psychrophilic acidophiles are the main culprits of AMD (Vera et al. 2013). However, when thermophiles are used in biomining, the possibilities of AMD are reduced and costs associated with the cooling of processing tanks are kept to a minimum.

4.9 Proteases/lipases

Proteases and lipases, combined with the gylcosyl hydrolases, account for more than 70% of all enzymes sold (Li et al. 2012), while proteases alone are the most widely used class of enzyme. Proteases have numerous applications in diverse fields; however, the largest application is in laundry detergents, where they have been a standard component since 1985 and are used to break apart and remove protein-based stains (Vojcic et al. 2015). The other major uses for proteases are in the fields of cheese making, brewing, and baking. Typically, the microbial proteases used are mesophilic and derived from *Bacillus* species and produced by companies such as Novozymes and Genencor. However, explorations using psychrophilic proteases to enhance cold water washing have taken place. Unfortunately, most psychrophilic enzymes have proven to be unusable due to low stability at room temperature. However, through directed evolution, a chimeric psychrophilic/ mesophilic protease was generated that improved performance during cold water washing (Tindbaek et al. 2004). Lipases are a billion-dollar industry (Jaeger et al. 1999) and very attractive for use in industrial settings because of their broad range of substrates, high degree of specificity, and stability (Hasan et al. 2006). Although their applications in laundry detergents (i.e., low temperatures and alkaline conditions) and organic synthesis (i.e., low water activity) require lipases to be active under extreme conditions, most lipases used are mesophilic. Many mesophilic lipases, which typically come from organisms like *Bacillus* and *Aspergillus* species, are active at high temperatures. As a result, extremophilic lipases are often overlooked; however, lipases from thermophilic *Bacillus* species have been shown to be more efficient than currently used enzymes (Imamura and Kitaura 2000).

5. Existing Technologies, Limitations and Strategies to Overcome limitations Deploying Metagenomics for Bio-prospecting

Screening of biomolecules from environmental samples is done by deploying two basic metagenomic approaches: (1) sequence-based screening and (2) function-based (Escobar-Zepeda et al. 2015). These two methods are accomplished by the construction of metagenomic libraries where genomic DNA is fragmented and then cloned in suitable expression vectors such as fosmids, cosmids (40 kb), or bacterial artificial chromosomes (40 kb) depending on the size of the target gene to get the expression of the required genes. *Escherichia coli* are used as host for the heterologous expression of metagenomic clones which are acquired after the construction of metagenomic library. Small sized insert clonal libraries are employed for the identification of novel enzymes which are encoded by a single gene or a small sized operon, whereas insert libraries which are large in size are indispensable for the isolation of large biosynthetic gene clusters involved in complex pathways comprising multiple genes (Daniel 2005).

(1) Sequence based Metagenomic Screening

Another approach for screening of metagenomic clones is the widely employed sequence based method. This strategy involves the use of oligonucleotide primer or probes using the colony hybridisation technique. The desired gene can be cloned in suitable expression vectors after amplifying by PCR with specific or degenerate primers. The main downside of using this method is that only those enzymes with known function and gene sequence can be identified which lessens the chances of isolating novel enzyme having entirely new sequences. Sequence based strategy plays an important role in identifying complex metagenomic community where new sequencing techniques such as 454-pyrosequencing accompanying newly developed computational tools and softwares have to be used in case of highly complex community, which easily analyzes the bulk metagenomic sequence information (Brito and Alm 2016).

(2) Function based Screening of Metagenomic Libraries

Diagnosis of enzyme activity (hemicellulase assay for the detection of hemicellulases such as xylanase, mannase, etc., starch-iodine test for amylase, cellulase screening assay for cellulase, etc.), heterologous complementation of host strains and induced gene expression and phenotypic detection are the commonly used function based strategies in metagenomics. For heterologous complementation, presence of target gene for growth under selective condition is required for complemented host strains. Donato and co-workers in 2010 in their study obtained 10–13 restriction clones expressing novel resistance genes during the screening of 446000 clones of soil metagenomic library which showed resistance to b-lactam and aminoglycoside antibiotic. For phenotypic detection, target enzyme's specific dyes and substrates are incorporated into the culture plate which primarily detects the presence of specific biocatalyst. In functional screening, prior knowledge of the gene sequence is not required to identify the novel enzymes encoding novel genes (Ferrer et al. 2009). However, the risks for gene expression failure mainly due to problems in promoter recognition, misfolded proteins, inappropriate post translational modification of target proteins, etc., are some of the flaws of using function based screening method. This can be overcome by using vectors that accommodate large insert size, with broad host range and allow expression in multiple hosts, employing rossetta *E. coli* strains having tRNA for rare amino acid codons (Perner et al. 2011). The third category of functional screening is a high throuput screening method, which is mainly based on substrate induced gene expression using GFP as reporter gene. Clones produce specific metabolite which induce the expression of GFP and is under the control of tightly regulated promoter. Gene sequences were successfully isolated from metagenomic library that was derived from ground water using hydrocarbon induction (Uchiyama et al. 2005).

(3) Directed Evolution

The main benefit of directed evolution is its remarkable pace compared to natural evolution which takes millions and millions of years (Wang et al. 2012). In directed evolution, mutations are introduced randomly into the genome followed by screening and selection of the desired function. Direction evolution involves three main steps: (1) Mutant library formation (2) Screening/selection of mutants with improved function, and (3) Isolation of improved genes. Successful mutations mostly accumulate in substrate binding pockets and catalytic active site. Random mutagenesis, chemical mutagenesis, saturation mutagenesis and DNA shuffling are major mutagenesis techniques. Error prone PCR is the most widely employed technique for generation of genetic diversity due to the easy experimental procedure for introducing mutations in target gene.

5.1 Selection Methods

5.1.1 Directed Evolution as a Powerful Approach for Industrial Enzymes Development

Directed evolution has been recently employed to overcome the limitations of natural enzymes and to improve the stability, temperature tolerance, pH tolerance and increase in affinity for enzyme to bind with the substrate by alteration at several sites (Cheriyan et al. 2011, Garcia-Ruiz et al. 2012). Several methods such as site directed mutagenesis, genome reshuffling along with the rational screening approaches pave a way for the development of enzymes for industrial purpose. Mutant strain of halophilic bacteria *Haloarcula marismortui* has high stability under extreme halophilic conditions by alteration in protein solvent interaction (Madern et al. 2000). Further, the catalytic activity of alcohol dehydrogenase enzyme from *Pyrococcus furiosus* mutant strain was higher at lower temperature production of hexanediol (Machielsen et al. 2008).

5.1.2 Site Directed Mutagenesis

Site-directed mutagenesis has been proven to be a novel technique for the development of proteins which serve as efficient biocatalysts for its application in industry. Several alkaline xylanase enzyme majorly used in biorefineries are subjected to site directed mutation to improve the thermostability of enzyme (Xiang et al. 2019). Site-directed mutagenesis involves the specific changes in the plasmid DNA, thereby leading to formation of altered protein. The advent of bioinformatic tools for the prediction of protein structure and targeting particular expression site has accelerated the research in order to obtain the mutants producing high yield of product with desired function. a-galactosidase gene from deep sea bacteria *Bacillus megaterium* has been improved in order to increase the enzyme activity that paves way for its industrial use. Replacement of glutamate residue at 12 position in the protein of hyperthermophilic archaea *Pyrococcus woesei* (Pw) has resulted in selective removal of cation binding sites, thereby making this protein more mesophile-like. This proposes an evolutionary advantage over the *de novo* synthesized proteins as they are able to function in different environment (Simmon et al. 2001). Engineered strain of *Sulfolobus tokodaii* has ability to encode both glucosamine-1-phosphate (GlcN-1-P) AcTase activity and galactosamine-1-phosphate (GalN-1-P) AcTase activity by changes in amino acid His308 (Zhang et al. 2015). Other studies report that replacement in Cys22 of Phi-class glutathionine S-transferase with alanine from *Oryzasativa* resulted in 2.2-fold increases in km value in comparison to wild type (Jo et al. 2012). Similarly, when the basic amino acid residues His275, His293, and His310 of a-amylase from *B. subtilis* were replaced with aspartic acid by site-directed mutation, 16.7-fold increase in kcat/km value of mutant was observed (Yang et al. 2013).

5.1.3 Microbial cell Surface Display

Microbial cell surface display allows the display of peptides and proteins on the microbial cell surface that endows them with unique feature. It majorly involves the fusing of peptides/proteins of interest with the anchoring motifs. Cell surface display system consists of three factors, passenger protein (protein to be displayed) fused to carrier protein (anchoring motif) and fusion protein (N-terminal fusion, C-terminal fusion) (Lee et al. 2003). Based on these proteins, the efficiency of protein display depends on microbial cell. Earlier the phage display method was known which was exploited for vaccine, but now a days bacterial and yeast display has provided its use in whole cell biocatalysis, biosensors and bio-adsorption. The recombinant protein inoculated in sheep had led to increased antibody response and reduced infectivity in *in vitro* assay. The engineered protein was prepared by genetically fusing antigen EG95 from *Echinococcus granulosus* on the host cell surface protein (Tan et al. 2012). In the field of medical microbiology, the live oral vaccine against coccidiosis was developed by fusing *Eimeriatenella* EtMic2 protein with *Saccharomyces cerevisiae* as host strain (Sun et al. 2014).

Several biorefineries using bioconversion platform for lignocellulosic biomass conversion to bioethanol involves the heterologous expression of cellulolytic enzymes in the *Pichia pastoris* using glycosylphosphatidylinositol inositol anchor. With the advancement in research recently, combined strategy for cell surface display of cellulolytic genes has been used where the endoglucanase (EG) and cellobiohydrolase I (CBHI) were produced in a β-glucosidase displaying *S. cerevisiae* (Inokuma et al. 2014). It was observed that display of gene cassette EG-D-CBHI-D and EG-SCBHI-S resulted in higher ethanol production (2.9 and 2.6 g/L from 10 g/L phosphoric acid swollen cellulose, respectively) (Liu et al. 2015). Mannans, which consist of major fraction of hemicelluloses, are also the source of carbohydrates which can be exploited for its subsequent conversion to ethanol. So the use of cell surface display helps the co-display of β-mannanase and β-mannosidase on the yeast cell surface, which has resulted in efficient hydrolysis of recalcitrant lignocellulosiscs (Inokuma et al. 2014, Ishii et al. 2016).

5.2 Nanotechnology Enabled Enhancement of Enzyme Activity

5.2.1 Nanoparticles

Nano-biocatalytic applications have recently emerged as simple, competent and reliable methods for enzyme immobilization. Nanoparticles are becoming increasingly important for immobilization purpose as their adaptability is improved. The non-toxic, biocompatible synthetic polymer (Lactic-Co-Glycolic Acid) polymer (PLGA) has been regularly used to help the immobilization of enzymes, providing enhanced catalytic stability, viable continuous operation and fast recycling of catalysts, thus reducing operating costs for different processes. For biocatalysis and biotransformation, enzyme immobilization on nanoporous gold is quite important. Enzymes immobilized by adsorption or covalent binding on chitosan coated magnetic nanoparticles have been reported to enhance their repeated re-usability. Due to their native high specific surface area, inter fibers porosity, low interference for mass transfer, simple management, and strong mechanical power, nanofibrous polymers have many advantages over other nanostructure supports (nanotubes, nanoparticles, and mesoporoussilica) (Johnson et al. 2011). α-amylase, immobilized on recently synthesized magnetic nanoparticles coated with silica and gold, increased the optimum temperature from 60°C to 80°C and also retained 60% of the enzyme activity after recycling for ten times (Singh et al. 2016). Burkholderia cepacia lipase (BCL) was immobilized to obtain a novel immobilized lipase on phenyl-modified ordered mesoporous silica (Ph-OMMs) and was evaluated as a 1-phenylethanol resolution catalyst, obtaining up to 50% conversion with more than 99% enantiomeric surplus within 25 minutes, which is around 65 times faster than that of free lipase (Zheng et al. 2017).

5.2.2 Nanofibers

Nanofibers have various unique properties such as acting as support to immobilize enzymes, more specific surface area leading to high enzyme filling, finer porous structures leading to higher active sites, lesser resistance to diffusion and ease in recovery, which can lead to continuous operations resulting in its wider application (Nair et al. 2007). Huang et al. (2003) used modified nanofibers with phospholipid moieties to immobilize lipase via adsorption. Lipase immobilization resulted in higher activity and Km with better thermal stability and lower Vmax as compared to free lipase. Similarly, Ye et al. (2006) used poly electrospun nanofibers made from acrylonitrile-co-maleic acid containing reactive carboxyl groups for lipase immobilization. Effect of temperature, pH and additives concentration was studied by Wang (2006) on adsorption capacity of polysulfone nanofibers.

5.2.3 Nanotube

Carbon nanotubes (CNTs) have gained wider attention in research for application in biosensors because of effective enzyme immobilization owing to its unique structure, mechanical, thermal and biocompatible properties (Asuri et al. 2007). Typical CNTs may have diameters ranging from one to tens of nanometres with length ranging to several hundred micrometres for single and multi-walled CNTs (Lee et al. 2010). Single-walled carbon nanotubes (SWNTs) have been found more effective in biocatalytic performance as compared to enzymes containing composites lacking SWNTs, as the former has larger specific surface area, enhanced adsorption and retention of enzyme molecules in aqueous solution whereas the latter loses enzyme molecules due to leaching (Rege et al. 2003).

5.3 Computational Tool for Designing Improved Enzymes

As discussed earlier, mesophilic enzymes which are commercially available do not have optimal properties for industrial process. Hence, in the past decades, with advancements in computation tools, it has been effectively leveraged to analyse desired enzymes from various available natural pool (Damborsky and Brezovsky 2009). Most of the enzymes isolated from natural pool do not have all the required properties like higher activity, higher selectivity, and higher stability in presence of organic solvents or higher accumulation of product and substrate with broad spectrum of substrate specificity for artificial substrates in order to be used for industrial production. This scenario has been changed by computational tools which supports in combining all the desired properties or features through engineering of proteins (Mak and Siegel 2014). Success rate is high in majority of those studies where original frame is maintained as observed in original crystal structure of enzyme with modification of side chains only. Majority of the computational tools used for designing of proteins employ two strategies: assessing strength of a particular amino acid sequence suitable for a specific framework using score function; and analysis of sequence and framework along with conformations of side chains. Energy functions for protein-design often contain a combination of physically-based and knowledge based terms (Huang et al. 2016).

5.3.1 Protein Structure Predictions

The lack of desired properties of stability, specificity, selectivity, etc., in natural enzymes is fulfilled by modelling of protein structures. Protein Data Bank act as a repository of structural data of protein (Berman et al. 2000). During evolution of proteins, their structures have been found to be more conserved as compared to their amino acid sequences. Hence, if there is similarity in amino acid sequence for two or more proteins, then there will be similarity in their structure as well. This principle forms the basis of structure model or homology modelling tool which is used to derive structure of various proteins. Examples of homology modelling tool include Modeller by Marti-

Renom et al. (2000), Swiss-Model by Arnold et al. (2006) and Rosetta by Mariani et al. (2011). For the homology tool to be effective, there should be at least 25% similarity in amino acid sequence for protein structure to be modelled (Dolan et al. 2012). Thus, for newly discovered protein if the sequence similarity is higher than 25%, the structure of this newly discovered protein can be modelled by comparing sequence to similar protein.

5.3.2 De novo Design of Enzymes

The initial focus area for *de novo* design is the active sites of the enzymes. It involves designing of catalytic activity by focusing on the smallest part of the basic structure, i.e., the set of atoms followed up to specific amino acid residues necessary for stability which is achieved by lowering of activation energy, thus providing stability during transition state of catalytic reaction. Designing of active sites requires initial knowledge of the involved enzymatic reactions, its molecular mechanism and other variables necessary for effective catalytic process. Quantum mechanics simulations have been proved to be effective computational tools which lead to accurate positioning of amino acid residues in appropriate geometrical structure, thus providing stability during transition state (Quin and Schmidt-Dannert 2011). Such designing of enzymes is referred to as theozymes (Kiss et al. 2013). Post availability of structural data encompassing target active sites, compatible protein framework is selected from Protein Data Bank. This is used as a template for RosettaMatch or Gess, which are types of molecular computational modelling tools (Smith et al. 2014). Post completion of template of protein framework including newly designed active sites, further relevant mutations are introduced into active sites to enhance molecular interactions and stability during transition state. Significant advancements have occurred in *de novo* designing of biocatalytic enzymes due to development of various computational tools by researchers, which in turn has supported designing and engineering of biocatalysts with desirable features, thus improving its efficiency. Few examples of such advanced computational tools are DEZYMER by Hellinga and Richards (1991), METAL SEARCH by Clarke and Yuan (1995), ORBIT by Dahiyat and Mayo (1996) and ROSETTA by Zanghellini et al. (2006). DEZYMER is a computational program that develops new ligand binding sites into protein of known 3D structure and it alters only the amino acid sequence and side chains without any alteration in the original protein framework. It helps in designing active sites with known geometric structures and provides general technique to design ligand binding sites and enzyme active sites for experimental testing whereas METAL SEARCH focuses on designing known protein structures with tetrahedral coordinated metal binding sites. Significant examples of enzymes designed by *de novo* are retro-aldol, Diels-Alder reactions and Kemp elimination (Kiss et al. 2011). Table 1 summarizes enzyme technologies currently used for the improvement of industrially important biocatalysts.

6. Future Directions

With the growing emphasis on greener technologies, use of biocatalysts in enzyme industry is being looked upon with great hope due to its environment friendly process. This is also being supported by increase in enzyme demand, thus forcing for advanced enzyme technology for industrialized process with the support from biological and chemical technologies. In order to have better avenues, extreme environments and advanced computational tools are being looked upon to identify or develop new enzymes for biocatalytic process. Genetically engineered enzymes, artificial enzymes, use of metagenomics, nanotubes like CNTs, proteome analysis, whole genome sequencing, Protein Data Banks, conventional culture methods, analysis of genetic materials, *de novo* design, *in silico* modelling, etc., are few of the areas used individually or in combinations to identify unexplored areas with the ultimate goal of having effective and optimal industrialized process for mass production.

Table 1. Enzyme technologies currently used for the improvement of industrially important biocatalysts.

Improvement method	Industrial enzyme with example	Reference
Fusion enzymes	N-terminal fusion of a yeast homolog of SUMO protein-Smt3 could confer elevated optimal temperature and improved operational stability to D-psicose 3-epimerase.	Patel et al. (2016)
	Optimal temperature and improved operational stability to D-psicose 3-epimerase.	
	The Smt3-D-psicose 3-epimerase conjugate System showed relatively better catalytic efficiency, and improved productivity Two hybrid cellulases (BaCel5127 and BaCel5167) from Bisporaantennata with replacement of the N-terminal (ba)3 (127 residues) or (ba)4 (167 residues)-barrel with the corresponding sequences of TeEgl5A from *Talaromyces emersonii* were produced in *Pichia pastoris* and biochemically characterized and showed improved catalytic performance compared to wild type	Zheng et al. (2016)
Metagenomic approach	Cellulase; Example : Identification of novel cellulase genes from metagenomic library	Yang et al. (2016)
	β-Glucosidase; Example: Identified novel thermotolerant b glucosidase from metagenomic library	Del Pozo et al. (2012)
Directed Evolution	Xylanase; Example: Xyn A from *B. subtilis* has been engineered through directed evolution for high thermal stability and pH tolerance (pH 12 and temperature 55 C) Engineered GH11 xylanase through directed evolution	Ruller et al. (2014)
Site directed mutagenesis	a-Amylase; Example: a-Amylase from *B. subtilis* was improved for pH stability by mutation in histidine residue	Yang et al. (2013)
Surface display	Mannase and chitosanase; Example: ManB, a mannanase from Bacillus licheniformis DSM13, and CsnA, a chitosanase from Bacillus subtilis ATCC 23857 were fused to different anchoring motifs of L. plantarum for covalent attachment to the cell surface for the development of whole cell biocatalyst	Nguyen et al. (2016)
Molecular Modelling	Cellulase; Example: Developed Stochartic cellulase automata based modelling approach to describe degradation of cellulosic material by a cellulase system at single molecule resolution	Eibinger et al. (2016)

References

Acevedo-Rocha, C.G., M.G. Hoesl, S. Nehring, M. Royter, C. Wolschner, B. Wiltschi, G. Antranikian and B. Nediljko. 2013. Non-canonical amino acids as a useful synthetic biological tool for lipase-catalysed reactions in hostile environments. Catal. Sci. Technol. 3: 1198–1201.

Adam, N. and M. Perner. 2017. Activity-based screening of metagenomic libraries for hydrogenase enzymes. Methods. Mol. Biol. 1539: 261–270.

Adrio, J.L. and A.L. Demain. 2014. Microbial enzymes: Tools for biotechnological processes. Biomolecules 4: 117–139.

Akeroyd, M., M. Olsthoorn, J. Gerritsma, D. Gutker-Vermaas, L. Ekkelkamp, T. van Rij, P. Klaassen, W. Plugge, Ed. Smit, K. Strupat, T. Wenzel, M. van Tilborg and R. van der Hoeven. 2013. Searching for microbial protein over-expression in a complex matrix using automated high throughput MS-based proteomics tools. J. Biotechnol. 164: 112–120.

Alarico, S., N. Empadinhas and M.S. da Costa. 2013. A new bacterial hydrolase specific for the compatible solutes α-D-mannopyranosyl-(1→2)-Dglycerate and α-D-glucopyranosyl-(1→2)-D-glycerate. Enzym. Microb. Technol. 52: 77–83.

Amaretti, A., S. Raimondi, M. Sala, L. Roncaglia, A. De Lucia, A. Leonardi and M. Rossi. 2010. Single cell oils of the cold-adapted oleaginous yeast *Rhodotorula glacialis* DBVPG 4785. Microb. Cell Factories 9: 73–79.

Antoni, D., V.V. Zverlov and W.H. Schwarz. 2007. Biofuels from microbes. Appl. Microbiol. Biotechnol. 77: 23–35.

Argüelles, J.C. 2000. Physiological roles of trehalose in bacteria and yeasts: A comparative analysis. Arch. Microbiol. 174: 217–224.

Arnold, K., L. Bordoli, J. Kopp and T. Schwede. 2006. The SWISS-MODEL workspace: A web-based environment for protein structure homology modelling. J. Bioinform. 22: 195–201.

Asuri, P., S.S. Bale, R.C. Pangule, D.A. Shah, R.S. Kane and J.S. Dordick. 2007. Structure, function, and stability of enzymes covalently attached to single-walled carbon nanotubes. Langmuir 23: 12318–12321.

Auman, A.J., J.L. Breezee, J.J. Gosink, P. Kämpfer and J.T. Staley. 2006. *Psychromonas ingrahamii* sp. nov., a novel gas vacuolate, psychrophilic bacterium isolated from Arctic polar sea ice. Int. J. Syst. Evol. Microbiol. 56: 1001–1007.

Autengruber, A., U. Sydlik, M. Kroker, T. Hornstein, N. Ale-Agha, D. Stöckmann et al. 2014. Signalling-dependent adverse health effects of carbon nanoparticles are prevented by the compatible solute mannosylglycerate (firoin) *in vitro* and *in vivo*. PLoS One 9: e111485.

Avanti, C., V. Saluja, E.L.P. Van Streun, H.W. Frijlink and W.L.J. Hinrichs. 2014. Stability of lysozyme in aqueous extremolyte solutions during heat shock and accelerated thermal conditions. PLoS One 9: e86244.

Bai, Y., J. Wang, Z. Zhang, P. Yang, P. Shi, H. Luo, K. Meng, H. Huang and B. Yao. 2010. A new xylanase from thermoacidophilic *Alicyclobacillus* sp. A4 with broad-range pH activity and pH stability. J. Ind. Microbiol. Biotechnol. 37: 187–194.

Baker, S.E., R.C. Hopkins, C.D. Blanchette, V.L. Walasworth, R. Sumbad, N.O. Fischer, A.E. Kuhn, M. Coleman, B.A. Chromy, S.E. Létant, P.D. Hoeprich, M.W.W. Adams and P.T. Henderson. 2009. Hydrogen production by a hyperthermophilic membrane-bound hydrogenase in water-soluble nanolipoprotein particles. J. Am. Chem. Soc. 131: 7508–7509.

Barnard, D., A. Casanueva, M. Tuffin and D. Cowan. 2010. Extremophiles in biofuel synthesis. Environ. Technol. 31: 871–888.

Basen, M., G.J. Schut, D.M. Nguyen, G.L. Lipscomb, R.A. Benn, C.J. Prybol, B.J. Vaccaro, F.L. Poole, R.M. Kelly and M.W.W. Adams. 2014. Single gene insertion drives bioalcohol production by a thermophilic archaeon. Proc. Natl. Acad. Sci. U.S.A. 111: 17618–17623.

Benson, C.A., R.W. Bizzoco, D.A. Lipson and S.T. Kelley. 2011. Microbial diversity in nonsulfur, sulfur and iron geothermal steam vents. Fems. Microbiol. Ecol. 76: 74–88.

Berman, H.M., J. Westbrook, Z. Feng, G. Gilliland and T.N. Bhat. 2000. The protein data bank. Nucleic Acids Res. 28: 235–242.

Blumer-Schuette, S.E., I. Kataeva, J. Westpheling, M.W. Adams and R.M. Kelly. 2008 Extreme thermophilic microoranisms for biomass conversion: Status and prospects. Curr. Opin. Biotechnol. 19: 210–217.

Bora, S.S., J. Keot, S. Das, K. Sarma and M. Barooah. 2016. Metagenomics analysis of microbial communities associated with a traditional rice wine starter culture (*Xaj-pitha*) of Assam, India. 3 Biotech. 6: 153.

Bornscheuer, U.T., G.W. Huisman, R.J. Kazlauskas, S. Lutz, J.C. Moore and K. Robins. 2012. Engineering the third wave of biocatalysis. Nature 485: 185–194.

Bougouffa, S., A. Radovanovic, M. Essack and V.B. Bajic. 2014. DEOP: A database on osmoprotectants and associated pathways. Database 2014: bau100.

Bowers, K.J. and J. Wiegel. 2011. Temperature and pH optima of extremely halophilic archaea: A mini-review. Extremophiles 15: 119–128.

Brito, I.L. and E.J. Alm. 2016. Tracking strains in the microbiome: Insights from metagenomics and models. Front. Microbiol. 7: 712.

Brown, C.T. 2015. Strain recovery from metagenomes. Nat. Biotechnol. 33: 1041–1043.

Brugger, D., I. Krondorfer, K. Zahma, T. Stoisser, J.M. Bolivar, B. Nidetzky, C.K. Peterbauer and D. Haltrich. 2014. Convenient microtiter plate-based, oxygen-independent activity assays for flavin-dependent oxidoreductases based on different redox dyes. Biotechnol. J. 9: 474–482.

Cava, F., A. Hidalgo and J. Berenguer. 2009. Thermus thermophilus as biological model. Extremophiles 13: 213–231.

Cavicchioli, R., T. Charlton, H. Ertan, S.M. Omar, K.S. Siddiqui and T.J. Williams. 2011. Biotechnological uses of enzymes from psychrophiles. Microb. Biotechnol. 4: 449–460.

Cencic, A. and W. Chingwaru. 2010. The role of functional foods, nutraceuticals, and food supplements in intestinal health. Nutrients 2: 611–625.

Chao, F.A., A. Morelli, J.C. Haugner, L. Churchfield, L.N. Hagmann, L. Shi, L.R. Masterson, R. Sarangi, G. Veglia and B. Seelig. 2013. Structure and dynamics of a primordial catalytic fold generated by *in vitro* evolution. Nat. Chem. Biol. 9: 81–83.

Chartrain, M., P.M. Salmon, D.K. Robinson and B.C. Buckland. 2000. Metabolic engineering and directed evolution for the production of pharmaceuticals. Curr. Opin. Biotechnol. 11: 209–214.

Chen, I., B.M. Dorr and D.R. Liu. 2011. A general strategy for the evolution of bond forming enzymes using yeast display. Proc. Natl. Acad. Sci. U.S.A. 108: 11399–11404.

Cheriyan, M., M.J. Walters, B.D. Kang, L.L. Anzaldi, E.J. Toone and C.A. Fierke. 2011. Directed evolution of a pyruvate aldolase to recognize a long chain acyl substrate. Bioorg. Med. Chem. 19: 6447–6453.

Chien, A., D.B. Edgar and J.M. Trela. 1976. Deoxyribonucleic acid polymerase from the extreme thermophile *Thermus aquaticus*. J. Bacteriol. 127: 1550–1557.

Choi, Y.J., J.M. Hur, S. Lim, M. Jo, D.H. Kim and J.I. Choi. 2014. Induction of apoptosis by deinoxanthin in human cancer cells. Anticancer. Res. 34: 1829–1835.

Chow, J., F. Kovacic, Y.D. Antonia, U. Krauss, Fr. Fersini, C. Schmeisser, B. Lauinger, P. Bongen, J. Pietruszka, M. Schmidt, I. Menyes, U.T. Bornscheuer, M. Eckstein, O. Thum, A. Liese, J. Mueller-Dieckmann, K.E. Jaeger and W.R. Streit. 2012. The metagenome-derived enzymes LipS and LipT increase the diversity of known lipases. PLoS One 7: e47665.

Clarke, N.D. and S.M. Yuan. 1995. Metal search: A computer program that helps design tetrahedral metal-binding sites. Proteins 23: 256–263.

Coker, J.A., P.P. Sheridan, J. Loveland-Curtze, K.R. Gutshall, A.J. Auman and J.E. Brenchley. 2003. Biochemical characterization of a beta-galactosidase with a low temperature optimum obtained from an Antarctic *Arthrobacter* isolate. J. Bacteriol. 185: 5473–5482.

Coker, J.A. 2004. Structure and function relationships in the cold-active betagalactosidase, BgaS, examining theories of enzyme cold-adaptation. Doctoral dissertation. Penn State. 2004.

Coker, J.A. and J.E. Brenchley. 2006. Protein engineering of a cold-active beta-galactosidase from *Arthrobacter* sp. SB to increase lactose hydrolysis reveals new sites affecting low temperature activity. Extremophiles 10: 515–24.

Dahiyat, B.I. and S.L. Mayo. 1996. Protein design automation. Protein Sci. 5: 895–903.

Damborsky, J. and J. Brezovsky. 2009. Computational tools for designing and engineering biocatalysts. Curr. Opin. Chem. Biol. 13: 26–34.

Daniel, R. 2005. The metagenomics of soil. Nat. Rev. Microbiol. 3: 470–478.

Das, D. and T.N. Veziroglu. 2001. Hydrogen production by biological processes. A survey of literature. Int. J. Hydrogen. Energy 26: 13–28.

DasSarma, P., J.A. Coker, V. Huse and S. DasSarma. 2009. Halophiles, Industrial Applications. *In*: Encyclopedia of Industrial Biotechnology. Hoboken, NJ, USA: John Wiley & Sons, Inc. 2009.

De la Coba, F., J. Aguilera, M.V. De Galvez, M. Alvarez, E. Gallego, F.L. Figueroa and E. Herrera. 2009. Prevention of the ultraviolet effects on clinical and histopathological changes, as well as the heat shock protein-70 expression in mouse skin by topical applications of algal UV absorbing compounds. J. Dermatol. Sci. 55: 161–169.

De Vrije, T., G.G. De Haas, G.B. Tan, E.R. Keijsers and P.A. claassen. 2002. Pretreatment of *Miscanthus* for hydrogen production by *Thermotoga elfii*. Int. J. Hydrog. Energy 27: 1381–1390.

Del Pozo, M.V., L. Fernandez-Arrojo, J. Gil-Martinez, A. Montesinos, T. Chernicova N., T.Y. Nechitaylo, A. Waliszek, M. Tortajada, A. Rojas, A.S. Huws, O.V. Golyshina, C.J. Newbold, J. Polaina, M. Ferrer and P.N. Golyshin. 2012. Microbial bglucosidases from cow rumen metagenome enhance the saccharification of lignocellulose in combination with commercial cellulase cocktail. Biotechnol. Biofuels 5: 1–13.

DeMaayer, P., D. Anderson, C. Cary and D.A. Cowan. 2014. Some like it cold: Understanding the survival strategies of psychrophiles. EMBO Rep. 15: 508–517.

Demain, A.L., M. Newcomb and J.H. Wu. 2005. Cellulase, clostridia, and ethanol. Microbiol. Mol. Biol. Rev. 69: 124–54.

Demirjian, D.C., F. Morís-Varas and C.S. Cassidy. 2001. Enzymes from extremophiles. Curr. Opin. Chem. Biol. 5: 144–151.

Dolan, M.A., J.W. Noah and D. Hurt. 2012. Comparison of common homology modelling algorithms: Application of user-defined alignments. Methods Mol. Biol. 857: 399–414.

Donato, J.J., L.A. Moe, B.J. Converse, K.D. Smart, F.C. Berklein, P.S. McManus and J. Handelsman. 2010. Metagenomic analysis of apple orchard soil reveals antibiotic resistance genes encoding predicted bifunctional proteins. Appl. Env. Microbiol. 76: 4396–4401.

Dopson, M. and D.S. Holmes. 2014. Metal resistance in acidophilic microorganisms and its significance for biotechnologies. Appl. Microbiol. Biotechnol. 98: 8133–8144.

Dopson, M., G. Ni and T.H. Sleutels. 2016. Possibilities for extremophilic microorganisms in microbial electrochemical systems. FEMS Microbiology Reviews 40(2): 164–181.

Dragone, G., B. Fernandes, A.A. Vicente and J.A. Teixeira. 2010. Third generation biofuels from microalgae. pp. 1355–1367. *In*: Current Research, Technology, and Education Topics in Applied Microbiology and Microbial Biotechnology. Formatex Research Center.

Eibinger, M., T. Zahel, T. Ganner, H. Plank and B. Nidetzky. 2016. Cellular automata modeling depicts degradation of cellulosic material by a cellulase system with single-molecule resolution. Biotechnol. Biofuels 9: 56.

Elleuche, S., C. Schröder, K. Sahm and G. Antranikian. 2014. Extremozymes—biocatalysts with unique properties from extremophilic microorganisms. Curr. Opin. Biotechnol. 29: 116–123.

Escobar-Zepeda, A., A. Vera-Ponce de León and A. Sanchez-Flores. 2015. The road to metagenomics: From microbiology to DNA sequencing technologies and bioinformatics. Front. Genet. 6: 348.

Esteves, A.M., S.K. Chandrayan, P.M. McTernan, N. Borges, M.W. Adams and H. Santos. 2014. Mannosylglycerate and di-myo-inositol phosphate have interchangeable roles during adaptation of *Pyrococcus furiosus* to heat stress. Appl. Environ. Microbiol. 80: 4226–4233.

Faria, C., C.D. Jorge, N. Borges, S. Tenreiro, T.F. Outeiro and H. Santos. 2013. Inhibition of formation of α-synuclein inclusions by mannosylglycerate in a yeast model of Parkinson's disease. Biochim. Biophys. Acta. 1830: 4065–4072.

Ferrer, M., A. Beloqui, K.N. Timmis and P.N. Golyshin. 2009. Metagenomics for mining new genetic resources of microbial communities. J. Mol. Microbiol. Biotechnol. 16: 109–123.

Frias-Lopez, J., Y. Shi, G.W. Tyson, M.L. Coleman, S.C. Schuster, S.W. Chisholm and E.F. DeLong. 2008. Microbial community gene expression in ocean surface waters. Proc. Natl. Acad. Sci. U.S.A. 105: 3805–3810.

Furubayashi, M., M. Ikezumi, J. Kajiwara, M. Iwasaki, A. Fujii, L. Li, K. Saito and D. Umeno. 2014. A high-throughput colorimetric screening assay for terpene synthase activity based on substrate consumption. PLoS One 9: e93317.

Gabani, P. and O.V. Singh. 2013. Radiation-resistant extremophiles and their potential in biotechnology and therapeutics. Appl. Microbiol. Biotechnol. 97: 993–1004.

Garcia-Ruiz, E., D. Gonzalez-Perez, F.J. Ruiz-Dueñas, A.T. Martínez and M. Alcalde. 2012. Directed evolution of a temperature-, peroxide- and alkaline pH-tolerant versatile peroxidase. Biochem. J. 441: 487–498.

Gatti-Lafranconi, P., A. Natalello, S. Rehm, S.M. Doglia, J. Pleiss and M. Lotti. 2010. Evolution of stability in a coldactive enzyme elicits specificity relaxation and highlights substrate-related effects on temperature adaptation. J. Mol. Biol. 395: 155–166.

Giger, L., S. Caner, R. Obexer, P. Kast, D. Baker, N. Ban and D. Hilvert. 2013. Evolution of a designed retro-aldolase leads to complete active site remodeling. Nat. Chem. Biol. 8: 494–498.

Gilbert, J.A., D. Field, Y. Huang, R. Edwards, W. Li, P. Gilna and I. Joint. 2008. Detection of large numbers of novel sequences in the metatranscriptomes of complex marine microbial communities. PLoS One 3: e3042.

Gtari, M., I. Essoussi, R. Maaoui, H. Sghaier, R. Boujmil, J. Gury, P. Pujic, L. Brusetti, B. Chouaia, E. Crotti, D. Daffonchio, A. Boudabous and P. Normand. 2012. Contrasted resistance of stone-dwelling *Geodermatophilaceae* species to stresses known to give rise to reactive oxygen species. FEMS Microbiol. Ecol. 80: 566–577.

Guo, N., I. Puhlev, D.R. Brown, J. Mansbridge and F. Levine. 2000. Trehalose expression confers desiccation tolerance on human cells. Nat. Biotechnol. 18: 168–71.

Han, M.J., S.Y. Lee, S.T. Koh, S.G. Noh and W.H. Han. 2010. Biotechnological applications of microbial proteomes. J. Biotechnol. 145: 341–349.

Han, T., F. Zeng, Z. Li, L. Lui, M. Wei, Q. Guan, X. Liang, Z. Peng, M. Liu, J. Qin, S. Zhang and B. Jia. 2013. Biochemical characterization of a recombinant pullulanase from *Thermococcus kodakarensis* KOD1. Lett. Appl. Microbiol. 57: 336–343.

Handelsman, J. 2004. Metagenomics: Application of genomics to uncultured microorganisms. Microbiol. Mol. Biol. Rev. 68: 669–685.

Hellinga, H.W. and F.M. Richards. 1991. Construction of new ligand binding sites in proteins of known structure. I. Computer-aided modelling of sites with predefined geometry. J. Mol. Biol. 222: 763–785.

Hess, M., A. Sczyrba, R. Egan, T.W. Kim, H. Chokhawala, G. Schroth, S. Luo, D.S. Clark, F. Chen, T. Zhang, R.I. Mackie, L.A. Pennacchio, S.G. Tringe, A. Visel, T. Woyke, Z. Wang and E.M. Rubin. 2011. Metagenomic discovery of biomass-degrading genes and genomes from cow rumen. Science 331: 463–467.

Horikoshi, K. 1999. Alkaliphiles: Some applications of their products for biotechnology. Microbiol. Mol. Biol. Rev. 63: 735–750.

Hough, D.W. and M.J. Danson. 1999. Extremozymes. Curr. Opin. Chem. Biol. 3: 39–46.

Huang, P., S.E. Boyken and D. Baker. 2016. The coming of age of *de novo* protein design. Nature 537: 320–327.

Huang, S.H., M.H. Liao and D.H. Chen. 2003. Direct binding and characterization of lipase onto magnetic nanoparticles. Biotechnol. Prog. 19: 1095–1100.

Ibrahim, A.S.S., A.M. El-Toni, A.A. Al-Salamah, K.S. Almaary, M.A. El-Tayeb, Y.B. Elbadawi and G. Antranikian. 2016. Development of novel robust nanobiocatalyst for detergents formulations and the other applications of alkaline protease. Bioproc. Biosyst. Eng. 39: 793–805.

Inokuma, K., T. Hasunuma and A. Kondo. 2014. Efficient yeast cell-surface display of exo- and endo-cellulase using the SED1 anchoring region and its original promoter. Biotechnol. Biofuels 7: 8.

Ishii, J., F. Okazaki, A.C. Djohan, K.Y. Hara, N. Asai-Nakashima, H. Teramura, A. Andriani, M. Tominaga, S. Wakai, P. Kahar, Yopi, B. Prasetya, C. Ogino and A. Kondo. 2016. From mannan to bioethanol: Cell surface co-display of β-mannanase and β-mannosidase on yeast *Saccharomyces cerevisiae*. Biotechnol. Biofuels 9: 188.

Ishino, S. and Y. Ishino. 2014. DNA polymerases as useful reagents for biotechnology—the history of developmental research in the field. Front. Microbiol. 5: 465.

Jayakumar, R., S. Jayashree, B. Annapurna and S. Seshadri. 2012. Characterization of thermostable serine alkaline protease from an alkaliphilic strain *Bacillus pumilus* MCAS8 and its applications. Appl. Biochem. Biotechnol. 168: 1849–1866.

Jo, H.J., J.W. Lee, J.S. Noh and K.H. Kong. 2012. Site-directed mutagenesis of cysteine residues in Phi-class glutathione S-transferase F3 from *Oryza sativa*. Bull. Korean Chem. Soc. 33: 4169–4172.

Jochens, H., D. Aerts and U.T. Bornscheuer. 2010. Thermo-stabilization of an esterase by alignment-guided focused directed evolution. Protein Eng. Des. Sel. 23: 903–909.

Johnson, D.B., T. Kanao and S. Hedrich. 2012. Redox Transformations of Iron at Extremely Low pH: Fundamental and Applied Aspects. Front. Microbiol. 3: 96.

Johnson, D.B. 2014. Biomining-biotechnologies for extracting and recovering metals from ores and waste materials. Curr. Opin. Biotechnol. 30: 24–31.

Johnson, P.A., H.J. Park and A.J. Driscoll. 2011. Enzyme nanoparticle fabrication: Magnetic nanoparticle synthesis and enzyme immobilisation. Methods Mol. Biol. 679: 183–191.

Joseph, B., P.W. Ramteke and G. Thomas. 2008. Cold active microbial lipases: Some hot issues and recent developments. Biotechnol. Adv. 26: 457–470.

Joshi, S. and T. Satyanarayana. 2013. Biotechnology of cold-active proteases. Biology 2: 755–83.

Kakirde, K.S., L.C. Parsley and M.R. Liles. 2010. Size does matter: Application-driven approaches for soil metagenomics. Soil. Biol. Biochem. 42: 1911–1923.

Kato, C., L. Li, Y. Nogi, Y. Nakamura, J. Tamaoka and K. Horikoshi. 1998. Extremely barophilic bacteria isolated from the Mariana Trench, Challenger Deep, at a depth of 11,000 meters. Appl. Environ. Microbiol. 64: 1510–1513.

Kelly, R.M., H. Leemhuis, H.J. Rozeboom, N. Van Oosterwijk, B.W. Dijkstra and L. Dijkhuizen. 2008. Elimination of competing hydrolysis and coupling side reactions of a cyclodextrin glucanotransferase by directed evolution. Biochem. J. 413: 517–525.

Kiss, G., S.A. Johnson, G. Nosrati, N. Çelebi-Ölçüm, S. Kim, R. Paton and N.H. Kendal. 2011. Computational design of new protein catalysts. *In*: Comba, P. (ed.). Modeling of Molecular Properties. Wiley-VCH Verlag GmbH & Co. KGaA, Weinheim, Germany.

Kiss, G., N. Celebi-Olcum, R. Moretti, D. Baker and K.N. Houk. 2013. Computational enzyme design. Angew. Chem. Int. Ed. 52: 5700–5725.

Klein, J., T. Schwarz and G. Lentzen. 2007. Ectoine as a natural component of food: detection in red smear cheeses. J. Dairy. Res. 74: 446–451.

Labrou, N.E. 2010. Random mutagenesis methods for *in vitro* directed enzyme evolution. Curr. Protein Pept. Sci. 11: 91–100.

Lam, M.K., K.T. Lee and A.R. Mohamed. 2010. Homogeneous, heterogeneous and enzymatic catalysis for transesterification of high free fatty acid oil (waste cooking oil) to biodiesel. Biotechnol. Adv. 28: 500–518.

Latiffi, A.A., A.B. Salleh, R.N. Rahman, S.N. Oslan and M. Basri. 2013. Secretory expression of thermostable alkaline protease from *Bacillus stearothermophilus* FI by using native signal peptide and alpha-factor secretion signal in *Pichia pastoris*. Genes. Genet. Syst. 88: 85–91.

Lee, H.K., J.K. Lee, M.J. Kim and C.J. Lee. 2010. Immobilisation of lipase on single walled carbon nanotubes in ionic liquid. Bull. Kor. Chem. Soc. 31: 650–665.

Lee, S.Y., J.H. Choi and Z. Xu. 2003. Microbial cell-surface display. Trends Biotechnol. 21: 45–52.

Lee, S.Y., J.H. Park, S.H. Jang, L.K. Nielsen, J. Kim and K.S. Jung. 2008. Fermentative butanol production by *Clostridia*. Biotechnol. Bioeng. 101: 209–228.

Lentzen, G. and T. Schwarz. 2006. Extremolytes: natural compounds from extremophiles for versatile applications. Appl. Microbiol. Biotechnol. 72: 623–634.

Lipscomb, G.L., G.J. Schut, M.P. Thorgersen, W.J. Nixon, R.M. Kelly and M.W. Adams. 2014. Engineering hydrogen gas production from formate in a hyperthermophile by heterologous production of an 18-subunit membrane-bound complex. J. Biol. Chem. 289: 2873–2879.

Liu, Z., K. Inokuma, T. Hasunuma and A. Kondo. 2015. Combined cell-surface display and secretion-based strategies for production of cellulosic ethanol with *Saccharomyces cerevisiae*. Biotechnol. Biofuels 8: 162.

López-López, O., M.E. Cerdán and M.I. Gonzalez-Siso. 2002. New extremophilic lipases and esterases from metagenomics. Curr. Protein Pept. Sci. 15: 445–455.

Luca, P.D., A. Musacchio and R. Taddei. 1981. Acidophilic algae from the fumaroles of Mount Lawu (Java, locus classicus of *Cyanidium caldarium* Geitler. Plant Biosyst. 115: 1–9.

Lugani, Y., R. Rai, A.A. Prabhu, P. Maan, M. Hans, V. Kumar, S. Kumar, A.K. Chandel and R.S. Sengar. 2020. Recent advances in bioethanol production from lignocelluloses: A Comprehensive review with a Focus on Enzyme Engineering and Designer Biocatalysts.

Ma, S.K., J. Gruber, C. Davis, L. Newman, D. Gray, A. Wang, J. Grate, G.W. Huisman and R.A. Sheldon. 2010. A green-by-design biocatalytic process for atorvastatin intermediate. Green Chem. 12: 81–86.

Machielsen, R., N.G. Leferink, A. Hendriks, S.J. Brouns, H.G. Hennemann, T. Daussmann and J. Van Der Oost. 2008. Laboratory evolution of *Pyrococcus furiosus* alcohol dehydrogenase to improve the production of (2S,5S)-hexanediol at moderate temperatures. Extremophiles 12: 587–594.

Maciver, B., R.H. McHale, D.J. Saul and P.L. Bergquist. 1994. Cloning and sequencing of a serine proteinase gene from a thermophilic *Bacillus* species and its expression in *Escherichia coli*. Appl. Environ. Microbiol. 60: 3981–3988.

Mak, W.S. and J.B. Siegel. 2014. Computational enzyme design: Transitioning from catalytic proteins to enzymes. Curr. Opin. Struct. Biol. 27: 87–94.

Mariani, V., F. Kiefer, T. Schmidt, J. Haas and T. Schwede. 2011. Assessment of template based protein structure predictions in CASP9. Proteins 10: 37–58.

Martinez, X., M. Pozuelo, V. Pascal, D. Campos, I. Gut, M. Gut, F. Azpiroz, F. Guarner and C. Manichanh. 2016. MetaTrans: An open-source pipeline for metatranscriptomics. Sci. Rep. 6: 26447.

Marti-Renom, M.A., A.C. Stuart, A. Fiser, R. Sanchez and F. Melo. 2000. Comparative protein structure modeling of genes and genomes. Annu. Rev. Biophys. Biomol. Struct. 29: 291–325.

Maruthamuthu, M., D.J. Jimenez, P. Stevens and J.D. Van Elsas. 2016. A multi-substrate approach for functional metagenomics-based screening for (hemi)cellulases in two wheat straw-degrading microbial consortia unveils novel thermoalkaliphilic enzymes. BMC Genomics 17: 86.

Messia, M.C., T. Candigliota and E. Marconi. 2007. Assessment of quality and technological characterization of lactose-hydrolyzed milk. Food Chem. 104: 910–917.

Miettinen-Oinonen, A. and P. Suominen. 2002. Enhanced production of *Trichoderma reesei* endoglucanases and use of the new cellulase preparations in producing the stonewashed effect on denim fabric. Appl. Environ. Microbiol. 68: 3956–3964.

Morana, A., M. Moracci, A. Ottombrino, M. Ciaramella, M. Rossi and M. De Rosa. 1995. Industrial-scale production and rapid purification of an archaeal beta-glycosidase expressed in *Saccharomyces cerevisiae*. Biotechnol. Appl. Biochem. 22: 261–268.

Moran-Reyna, A. and J.A. Coker 2014. The effects of extremes of pH on the growth and transcriptomic profiles of three haloarchaea. F1000Res. 3: 168.

Mukhopadhyay, A., A.K. Dasgupta and K. Chakrabarti. 2015. Enhanced functionality and stabilization of a cold active laccase using nanotechnology based activation immobilization. Bioresour. Technol. 179: 573–584.

Nair, S., J. Kim, B. Crawford and S.H. Kim. 2007. Improving biocatalytic activity of enzyme-loaded nanofibers by dispersing entangled nanofiber structure. Biomacromolecules 8: 1266–1270.

Nguyen, H., G. Mathiesen, E.M. Stelzer, M.L. Pham, K. Kuczkowska, A. Mackenzie, J.W. Agger, V.G.H Eijsink, M. Yamabhai, C.K. Peterbauer, D. Haltrich and T.H. Nguyen. 2016. Display of a β-mannanase and a chitosanase on the cell surface of *Lactobacillus plantarum* towards the development of whole-cell biocatalysts. Microb. Cell Fact. 15: 169.

Niehaus, F., E. Gabor, S. Wieland, P. Siegert, K.H. Maurer and J. Eck. 2011. Enzymes for the laundry industries: Tapping the vast metagenomic pool of alkaline proteases. Microbiol. Biotechnol. 4: 767–776.

Nishimura, H. and Y. Sako. 2009. Purification and characterization of the oxygenthermostable hydrogenase from the aerobic hyperthermophilic archaeon *Aeropyrum camini*. J. Biosci. Bioeng. 108: 299–303.

Nyyssonen, M., H.M. Tran, U. Karaoz, C. Weihe, M.Z. Hadi, J.B. Martiny and E.L. Brodie. 2013. Coupled high-throughput functional screening and next generation sequencing for identification of plant polymer decomposing enzymes in metagenomic libraries. Front. Microbiol. 4: 282.

Orell, A., F. Remonsellez, R. Arancibia and C.A. Jerez. 2013. Molecular characterization of copper and cadmium resistance determinants in the biomining thermoacidophilic archaeon *Sulfolobus metallicus*. Archaea 2013: 289236.

Oren, A. 2010. Industrial and environmental applications of halophilic microorganisms. Environ. Technol. 31: 825–834.

Oren, A. 2013. Life at high salt concentrations, intracellular KCl concentrations, and acidic proteomes. Front. Microbiol. 4: 315.

Pace, N.R. 1997. A molecular view of microbial diversity and the biosphere. Science 276: 734–740.

Packer, M.S. and D.R. Liu. 2015. Methods for the directed evolution of proteins. Nat. Rev. Genet. 16: 379–394.

Park, C.W., I.C. Kang and Y. Choi. 2012. Activity-based screening system for the discovery of neuraminidase inhibitors using protein chip technology. Biochip. J. 6: 133–138.

Patel, S.N., M. Sharma, K. Lata, U. Singh, V. Kumar, R.S. Sangwan and S.P. Singh. 2016. Improved operational stability of d-psicose 3-epimerase by a novel protein engineering strategy, and d-psicose production from fruit and vegetable residues. Bioresour. Technol. 216: 121–127.

Pearce, D.A., K.K. Newsham, M.A. Thorne, L. Calvo-Bado, M. Krsek, P. Laskaris, A. Hodson and E.M. Wellington. 2012. Metagenomic analysis of a southern maritime antarctic soil. Front. Microbiol. 3: 403.

Perner, M., N. Ilmberger, H.U. Köhler, J. Chow and W.R. Streit. 2011. Emerging fields in functional metagenomics and its industrial relevance: Overcoming limitations and redirecting the search for novel biocatalysts. pp. 483–498. *In*: de Bruijn, F.J. (ed.). Handbook of Molecular Microbial Ecology II: Metagenomics in Different Habitats. John Wiley & Sons Inc, Hoboken, NJ.

Prosser, J.I. 2015. Dispersing misconceptions and identifying opportunities for the use of 'omics' in soil microbial ecology. Nat. Rev. Microbiol. 13: 439–446.

Pulz, O. and W. Gross. 2004. Valuable products from biotechnology of microalgae. Appl. Microbiol. Biotechnol. 65: 635–648.

Quin, M.B. and C. Schmidt-Dannert. 2011. Engineering of biocatalysts: From evolution to creation. ACS Catal. 1: 1017–1021.

Raddadi, N., A. Cherif, D. Daffonchio and F. Fava. 2013. Halo-alkalitolerant and thermostable cellulases with improved tolerance to ionic liquids and organic solvents from *Paenibacillus tarimensis* isolated from the Chott El Fejej, Sahara desert, Tunisia. Bioresour. Technol. 150: 121–128.

Raheja, Y., B. Kaur, M. Falco, A. Tsang and B. Chadha. 2020. Secretome analysis of *Talaromyces emersonii* reveals distinct CAZymes profile and enhanced cellulase production through response surface methodology. Ind. Crop. Prod. 152: 112554.

Rai, R., B. Kaur and B.S. Chadha. 2016a. A method for rapid purification and evaluation of catalytically distinct lignocellulolytic glycosyl hydrolases from thermotolerant fungus *Acrophialophora* sp. Renewable Energy 98: 254–263.

Rai, R., B. Kaur, S. Singh, M. Di Falco, A. Tsang and B.S. Chadha. 2016b. Evaluation of secretome of highly efficient lignocellulolytic *Penicillium* sp. Dal 5 isolated from rhizosphere of conifers. Bioresource Technology 216: 958–967.

Rai, R., M. Bibra, B.S. Chadha and R.K. Sani. 2019. Enhanced hydrolysis of lignocellulosic biomass with doping of a highly thermostable recombinant laccase. International Journal of Biological Macromolecules 137: 232–237.

Rai, R., N. Basotra, B. Kaur, M. Di Falco, A. Tsang and B.S. Chadha. 2020. Exoproteome profile reveals thermophilic fungus Crassicarpon thermophilum (strain 6GKB; syn. Corynascus thermophilus) as a rich source of cellobiose dehydrogenase for enhanced saccharification of bagasse. Biomass and Bioenergy 132: 105438.

Rastogi, R.P. and A. Incharoensakdi. 2014. Characterization of UV-screening compounds, mycosporine-like amino acids, and scytonemin in the cyanobacterium *Lyngbya* sp. CU2555. FEMS. Microbiol. Ecol. 87: 244–256.

Rege, K., N.R. Raravikar, D.Y. Kim, L.S. Schadler, P.M. Ajayan and J.S. Dordick. 2003. Enzyme-polymer-single walled carbon nanotube composites as biocatalytic films. Nano Lett. 3: 829–832.

Ren, N., G. Cao, A. Wang, D.J. Lee, W. Guo and Y. Zhu. 2008. Dark fermentation of xylose and glucose mix using isolated *Thermoanaerobacterium thermosaccharolyticum* W16. Int. J. Hydrog. Energy 33: 6124–6132.

Resch, V., J.H. Schrittwieser, E. Siirola and W. Kroutil. 2011. Novel carboncarbon bond formations for biocatalysis. Curr. Opin. Biotechnol. 22: 793–799.

Roberts, M.F. 2005. Organic compatible solutes of halotolerant and halophilic microorganisms. Saline Syst. 1: 5.

Röske, I., W. Sabra, H. Nacke, R. Daniel, A.P. Zeng, G. Antranikian and K. Sahm. 2014. Microbial community composition and dynamics in high-temperature biogas reactors using industrial bioethanol waste as substrate. Appl. Microbiol. Biotechnol. 98: 9095–9106.

Rubio-Infante, N. and L. Moreno-Fierros. 2016. An overview of the safety and biological effects of *Bacillus thuringiensis* Cry toxins in mammals. J. Appl. Toxicol. 36: 630–48.

Ruller, R., J. Alponti, L.A. Deliberto, L.M. Zanphorlin, C.B. Machado and R.J. Ward. 2014. Concommitant adaptation of a GH11 xylanase by directed: Evolution to create an alkali-tolerant/thermophilic enzyme. Protein Eng. Des. Sel. 27: 255–262.

Sangwan, N., F. Xia and J.A. Gilbert. 2016. Recovering complete and draft population genomes from metagenome datasets. Microbiome 4: 8.

Santiago, M., C.A. Ramírez-Sarmiento, R.A. Zamora and L.P. Parra. 2016. Discovery, molecular mechanisms, and industrial applications of cold-active enzymes. Front. Microbiol. 7: 1408.

Sarethy, I.P., Y. Saxena, A. Kapoor, M. Sharma, S.K. Sharma, V. Gupta and S. Gupta. 2011. Alkaliphilic bacteria: Applications in industrial biotechnology. J. Ind. Microbiol. Biotechnol. 38: 769–790.

Schloss, P.D. and J. Handelsman 2003. Biotechnological prospects from metagenomics. Curr. Opin. Biotechnol. 14: 303–310.

Schröder, C., S. Elleuche, S. Blank and G. Antranikian. 2014. Characterization of a heat-active archaeal beta-glucosidase from a hydrothermal spring metagenome. Enzyme. Microb. Technol. 57: 48–54.

Serour, E. and G. Antranikian. 2002. Novel thermoactive glucoamylases from the thermoacidophilic Archaea *Thermoplasma acidophilum*, *Picrophilus torridus* and *Picrophilus oshimae*. Antonie Van Leeuwenhoek 81: 73–83.

Shao, H., L. Xu and Y. Yan. 2013. Thermostable lipases from extremely radioresistant bacterium *Deinococcus radiodurans*: Cloning, expression, and biochemical characterization. J. Basic. Microbiol. 54: 984–995.

Sharma, A. and T. Satyanarayana. 2012. Cloning and expression of acidstable, high maltose-forming, Ca2+-independent α-amylase from an acidophile *Bacillus acidicola* and its applicability in starch hydrolysis. Extremophiles 16: 515–522.

Sharma, D. and R. Rai. 2020. Biological pretreatment strategies for enhanced saccharification of lignocellulosic biomass in 2G ethanol biorefineries. European Journal of Molecular & Clinical Medicine 7(7): 4073–4073.

Sharma, V.K., N. Kumar, T. Prakash and T.D. Taylor. 2010. MetaBioME: A database to explore commercially useful enzymes in metagenomic datasets. Nucleic. Acids. Res. 38: D468–472.

Shim, J.H., Y.W. Kim, T.J. Kim, H.Y. Chae, J.H. Park, H. Cha, J.W. Kim, Y.R. Kim, T. Schaefer, T. Spendler, T.W. Moon and K.H. Park. 2004. Improvement of cyclodextrin glucanotransferase as an antistaling enzyme by error prone PCR. Protein Eng. Des. Sel. 17: 205–211.

Siddiqui, K.S., D.M. Parkin, P.M.G. Curmi, D.D. Francisci, A. Poljak and K. Barrow, M.H. Noble, J. Trewhella and R. Cavicchioli. 2009. A novel approach for enhancing the catalytic efficiency of a protease at low temperature: Reduction in substrate inhibition by chemical modification. Biotechnol. Bioeng. 103: 676–686.

Singh, V., K. Rakshit, S. Rathee, S. Angmo, S. Kaushal and P. Garg, J.H. Chung, R. Sandhir, R.S. Sangwan and N. Singhal. 2016. Metallic/bimetallic magnetic nanoparticle functionalization for immobilization of a-amylase for enhanced reusability in bio-catalytic processes. Bioresour. Technol. 214: 528–533.

Sivaramakrishnan, R. and K. Muthukumar. 2012. Isolation of thermo-stable and solvent tolerant *Bacillus* sp. lipase for the production of biodiesel. Appl. Biochem. Biotechnol. 166: 1095–1111.

Smith, M.D., A. Zanghellini and D.G. Röthlisberger. 2014. Computational design of novel enzymes without cofactors. Methods Mol. Biol. 1216: 197–210.

Sogin, M.L., G. Hilary, J.A.H. Morrison, D.M. Welch, S.M. Huse, P.R. Neal, J.M. Arrieta and G.J. Herndl. 2006. Microbial diversity in the deep sea and the underexplored rare biosphere. Proc. Natl. Acad. Sci. U.S.A. 103: 12115–12120.

Soule, T., F. Garcia-Pichel and V. Stout. 2009. Gene expression patterns associated with the biosynthesis of the sunscreen scytonemin in *Nostoc punctiforme* ATCC 29133 in response to UVA radiation. J. Bacteriol. 191: 4639–4646.

Stickney, Z., J. Losacco, S. McDevitt, Z. Zhang and B. Lu. 2016. Development of exosome surface display technology in living human cells. Biochem. Biophys. Res. Commun. 472: 53–59.

Sun, H., L. Wang, T. Wang, J. Zhang, Q. Liu, P. Chen, Z. Chen, F. Wang, H. Li, Y. Xiao and X. Zhao. 2014. Display of *Eimeriatenella* EtMic2 protein on the surface of *Saccharomyces cerevisiae* as a potential oral vaccine against chicken coccidiosis. Vaccine 32: 1869–1876.

Takai, K., K. Nakamura, T. Toki, U. Tsunogai, M. Miyazaki and J. Miyazaki. 2008. Cell proliferation at 122 degrees C and isotopically heavy CH4 production by a hyperthermophilic methanogen under high-pressure cultivation. Proc. Natl. Acad. Sci. U.S.A. 105: 10949–10954.

Tan, J.L., N. Ueda, D. Heath, A.A. Mercer and S.B. Fleminga. 2012. Development of virus as a bifunctional recombinant vaccine: Surface display of *Echinococcus granulosus* antigen EG95 by fusion to membrane structural proteins. Vaccine 30: 398–406.

Taylor, M.P., K.L. Eley, S. Martin, M.I. Tuffin, S.G. Burton and D.A. Cowan. 2009. Thermophilic ethanolgenesis: Future prospects for second-generation bioethanol production. Trends Biotechnol. 27: 398–405.

Tchigvintsev, A., O. Reva, J. Polaina, R. Bargiela, A. Canet, F. Valero, Eugenio Rico Eguizabal, F. Alexander, M.M. Martínez, M. Alcaide and M. Ferrer. 2013. Biochemical diversity of carboxyl esterases and lipases from Lake Arreo (Spain): A metagenomic approach. Appl. Environ. Microbiol. 79: 3553–3562.

Torsvik, V., L. Overas and T.F. Thingstad. 2002. Prokaryotic diversity–magnitude, dynamics, and controlling factors. Science 296: 1064–1066.

Uchiyama, T., T. Abe, T. Ikemura and K. Watanabe. 2005. Substrate induced gene expression screening of environmental metagenome libraries for isolation of catabolic genes. Nat. Biotechnol. 23: 88–93.

Urich, T., A. Lanzen, R. Stokke, R.B. Pedersen, C. Bayer, I.H. Thorseth, C. Schleper, I.H. Steen and L. Overas. 2014. Microbial community structure and functioning in marine sediments associated with diffuse hydrothermal venting assessed by integrated meta-omics. Environ. Microbiol. 16: 2699–2710.

Varadarajan, N., J. Gam, M.J. Olsen, G. Georgiou and B.L. Iverson. 2005. Engineering of protease variants exhibiting high catalytic activity and exquisite substrate selectivity. Proc. Natl. Acad. Sci. U.S.A. 102: 6855–6860.

Venter, J.C., K. Remington, J.F. Heidelberg, A.L. Halpern, D. Rusch, J.A. Eisen, D. Wu, I. Paulsen, K.E. Nelson, W. Nelson, D.E. Fouts, S. Levy, A.H. Knap, M.W. Lomas, K. Nealson, O. White and J. Peterson. 2004. Environmental genome shotgun sequencing of the Sargasso Sea. Science 304: 66–74.

Ventosa, A., R.R. de la Haba, C. Sanchez-Porro and R.T. Papke. 2015. Microbial diversity of hypersaline environments: A metagenomic approach. Curr. Opin. Microbiol. 25: 80–87

Vera, M., A. Schippers and W. Sand. 2013. Progress in bioleaching: Fundamentals and mechanisms of bacterial metal sulfide oxidation—part A. Appl. Microbiol. Biotechnol. 97: 7529–7541.

Vester, J.K., M.A. Glaring and P. Stougaard. 2015. An exceptionally cold-adapted alpha-amylase from a metagenomic library of a cold and alkaline environment. Appl. Microbiol. Biotechnol. 99: 717–727.

Wang, G., Q. Wang, X. Lin, T.B. Ng, R. Yan, J. Lin, T.B. Ng, R. Yan, J. Lin and X. Ye. 2016. A novel cold-adapted and highly salt-tolerant esterase from *Alkalibacterium* sp. SL3 from the sediment of a soda lake. Sci. Rep. 6: 19494.

Wang, M., T. Si and H. Zhao. 2012. Biocatalyst development by directed evolution. Bioresour. Technol. 115: 117–125.

Wang, P. 2006. Nanoscale biocatalyst systems. Curr. Opin. Biotechnol. 17: 574–579.

Warnecke, F. and M. Hess. 2009. A perspective: Metatranscriptomics as a tool for the discovery of novel biocatalysts. J. Biotechnol. 142: 91–95.

Warnecke, F., P. Luginbühl, N. Ivanova, M. Ghassemian, T.H. Richardson, J.T. Stege, M. Cayouette, A.C. McHardy, G. Djordjevic, N. Aboushadi, R. Sorek, S.G. Tringe, M. Podar, H.G. Martin, V. Kunin, D. Dalevi, J. Madejska, E. Kirton, D. Platt, E. Szeto, A. Salamov, K. Barry, N. Mikhailova, N.C. Kyrpides, E.G. Matson, E.A. Ottesen, X. Zhang, M. Hernández, C. Murillo, L.G. Acosta, I. Rigoutsos, G. Tamayo, B.D. Green, C. Chang, E.M. Rubin, E.J. Mathur, D.E. Robertson, P. Hugenholtz and J.R. Leadbetter. 2007. Metagenomic and functional analysis of hindgut microbiota of a wood-feeding higher termite. Nature 450: 560–565.

Wilmes, P. and P.L. Bond. 2006. Metaproteomics: Studying functional gene expression in microbial ecosystems. Trends. Microbiol. 14: 92–97.

Wilmes, P., A. Heintz-Buschart and P.L. Bond. 2015. A decade of metaproteomics: Where we stand and what the future holds. Proteomics 15: 3409–3417.

Winkler, J.D. and K.C. Kao. 2014. Recent advances in the evolutionary engineering of industrial biocatalysts. Genomics 104: 406–411.

Yang, C., Y. Xia, H. Qu, A. Li, R. Liu, Y. Wang and T. Zhang. 2016. Discovery of new cellulases from the metagenome by a metagenomics-guided strategy. Biotechnol. Biofuels 9: 138.

Yang, C.H., K.I. Lin, G.H. Chen, Y.F. Chen, C.Y. Chen, W.L. Chen and Y.C. Huang. 2010. Constitutive expression of *Thermobifida fusca* thermostable acetylxylan esterase gene in Pichia pastoris. Int. J. Mol. Sci. 11: 5143–5151.

Yang, H., L. Liu, H.D. Shin, R.R. Chen, J. Li and G. Du. 2013. Structure- based engineering of histidine residues in the catalytic domain of alpha-amylase From *Bacillus subtilis* for improved protein stability and catalytic efficiency under acidic conditions. J. Biotechnol. 164: 59–66.

Yang, S.J., I. Kataeva, S.D. Hamilton-Brehm, N.L. Engle, T.J. Tschaplinski, C. Doeppke, M. Davis, J. Westpheling and M.W.W. Adams. 2009. Efficient degradation of lignocellulosic plant biomass, without pretreatment, by the thermophilic anaerobe *Anaerocellum thermophilum* DSM 6725. Appl. Environ. Microbiol. 75: 4762–4769.

Ye, P., Z.K. Xu, J. Wu, C. Innocent and P. Seta. 2006. Nanofibrous membranes containing reactive groups: Electrospinning from poly(acrylonitrile-co-maleic acid) for lipase immobilisation. Macromolecules 39: 1041–1045.

Yildiz, S.Y., N. Radchenkova, K.Y. Arga, M. Kambourova and E.T. Oner. 2015. Genomic analysis of *Brevibacillus thermoruber* 423 reveals its biotechnological and industrial potential. Appl. Microbiol. Biotechnol. 99: 2277–2289.

Zambare, V., A. Bhalla, K. Muthukumarappan, R.K. Sani and L. Christopher. 2011. Bioprocessing of agricultural waste to ethanol utilizing a cellulolytic extremophile. Extremophiles 15: 611–618.

Zanghellini, A., L. Jiang, A.M. Wollacott, G. Cheng and J. Meiler. 2006. New algorithms and an *in silico* benchmark for computational enzyme design. Protein Sci. 15: 2785–2794.

Zhang, H., F. Zhang and Z. Li. 2009. Gene analysis, optimized production and property of marine lipase from *Bacillus pumilus* B106 associated with South China Sea sponge *Halichondria rugosa*. World J. Microbiol. Biotechnol. 25: 1267–1274.

Zhang, Y., X. Li, D.H. Bartlett and X. Xiao. 2015. Current developments in marine microbiology: High-pressure biotechnology and the genetic engineering of piezophiles. Curr. Opin. Biotechnol. 33: 157–164.

Zhang, Z., Y. Shimizu and Y. Kawarabayasi. 2015. Characterization of the amino acid residues mediating the unique amino-sugar-1-phosphate acetyltransferase activity of the archaeal ST0452 protein. Extremophiles 19: 417–427.

Zheng, F., H. Huang, X. Wang, T. Tu, Q. Liu, K. Meng, Y. Wang, X. Su, X. Xie and H. Luo. 2016. Improvement of the catalytic performance of a *Bispora antennata* cellulase by replacing the N-terminal semi-barrel structure. Bioresour. Technol. 218: 279–85.

Zheng, J., C. Liu, L. Liu and Q. Jin. 2013. Characterisation of a thermo-alkali-stable lipase from oil-contaminated soil using a metagenomic approach. Syst. Appl. Microbiol. 36: 197–204.

Zheng, M., X. Xiang, S. Wang, J. Shi, Q. Deng, F. Huang and R. Cong. 2017. Lipase immobilized in ordered mesoporous silica: A powerful biocatalyst for ultrafast kinetic resolution of racemic secondary alcohols. Process Biochem. 53: 102–108.

Zhou, C., Y. Xue and Y. Ma. 2015. Evaluation and directed evolution for thermostability improvement of a GH 13 thermostable a-glucosidase from *Thermus thermophilus* TC11. BMC Biotechnol. 15: 97.

Zhu, S., D. Song, C. Gong, P. Tang, X. Li, J. Wang and G. Zheng. 2013. Biosynthesis of nucleoside analogues via thermostable nucleoside phosphorylase. Appl. Microbiol. Biotechnol. 97: 6769–6778.

6

Exploration of Extremophiles for Value-Added Products
A Step Towards Sustainable Bio-refinery

Surojit Bera,[1,*] *Trinetra Mukherjee,*[2] *Subhabrata Das,*[3] *Sandip Mondal,*[4]
Suprabhat Mukherjee[2] and *Sagnik Chakraborty*[5]

1. Extremophiles in Bio-refinery Concept

Extremophiles (from Latin extremus meaning "extreme" and Greek philiā meaning "love") are microorganisms who can endure extreme environmental conditions like intense extreme temperature, bold acidic environments, tremendous pressure and tremendous cold. Allegedly, these conditions do not support life in normal conditions. Scientists believe that organisms may have evolved by wide genomic complexity and phylogenetic diversity to accommodate and thrive in initial harsh conditions on Earth at the time of beginning of life (Demirjian et al. 2001).

In order to support development under such harsh environment, extremophiles adapted themselves in several ways; however, little information is available regarding this till now. Mostly, they do protect themselves by altering membrane components and by secreting defensive molecules, e.g., compatible solutes, extracellular enzymes, etc. It has been anticipated that charged exterior side and steady ion connections might be responsible for thermostability of extremozymes. Additionally, constancy regarding raised temperatures can be explained through augmented quantity of acidic and basic amino acids which may help to develop a compactly packed enzyme core. These resultant excellent extremozymes permit for an extensive range of reactions such as biocatalysis in non-aqueous liquid, in elevated solvent concentrations, at high pressures or pH or temperatures up to 140°C (Cavicchioli et al. 2011).

Thriving in insensitive surroundings make extremophiles suitable candidate for the exploitation of bioprocesses applications. Recognising the vital genes accountable for the pursuit of the microbes, the constancy of chemical possessions, and their adaptableness under tough environmental situations may have variety of industrial and therapeutic applications. Bio based production of industrially important chemicals may act as a justifiable and dependable alternative to petrochemicals because

---
[1] Department of Microbiology, School of Bioengineering and Biosciences, Lovely Professional University, Punjab, India-144411.
[2] Department of Animal Science, Kazi Nazrul University, West Bengal-Asansol, West Bengal-713340.
[3] Department of Chemical & Biomolecular Engineering, National University of Singapore, Singapore-117585.
[4] Shri Indra Ganesan Institute of Medical Science - College of Pharmacy, Manikandam, Tiruchirapalli, Tamil Nadu, India 620012.
[5] School of Environment, Jiangsu University, Zhenjiang, Jiangsu, China-212013.
* Corresponding author: suromic@gmail.com

they are renewable, provincially obtainable, and decreases the carbon footprint of chemicals and fuels. It may also contribute to a further steady and cost-effective agricultural economy. Additionally, these kind of alternative source for bioactive chemicals from extremophiles may sound economically stimulating as it has an enormously huge added value compared to energy fabrication, whereas resources requirement is too low. In a rough estimation, only 3% of the fossil assets are presently being exploited for the manufacturing of chemicals, against 97% for energy expenditure, though the auxiliary worth of this 3% chemicals is $375 billion, against $520 billion for the remaining 97% directed at energy usage (Cavicchioli et al. 2011).

By the definition of biorefinery, tiny microbes ferment biomass which has already undergone pre-treatment to produce chemical compounds and fuels. These molecules then separated out followed by purification through downstream processing. However, the use of edible biomass like sugar cane, beet or corn, etc., for first generation bioprocess is already facing food versus fuel conundrum. The cause for this is that these kind of primary staple crops are very good source of fermentable sugars like glucose and sucrose (Demirjian et al. 2001). To avoid these interference of the substrates with food chain, researchers are escalating their interest towards non-food biomass like sugar cane bagasse, wheat straw or different grasses, e.g., switchgrass or miscanthus. These type of substrates are cheap, easy to procure or grow, thus making the biorefinery concept more economical and sustainable. These non-food substrates comprise mostly of lignocellulose, a complex of cellulose, hemicellulose and lignin. Cellulose contains glucose chain, whereas hemicellulose comprises of a combination of xylose, glucose, arabinose, galactose and mannose. Lignin contains of polyphenolic rings which can be utilized by most bacteria to generate low molecular weight bio-molecule. Pre-treatment of lignocellulose is mainly done to make them more bioavailable towards saccharolytic enzymes. Extremophiles make a promising source for these enzymes nowadays. The problem is, the phenolic compounds produced from lignin through pre-treatment are hydrophobic in nature and tend to increase membrane fluidity and permeability, triggering cell drip. Extremophiles get some advantage in such case as they have more resistance to hard membrane compared to mesophilic organisms. Other noxious compounds like aldehydes, organic acids and ketones may be released during pre-treatment. The model organism for biorefinery must have the resistant power to those compounds in order not to hinder the fermentation process (Robb et al. 2008). In extremophiles, these mechanisms remained an unsolved mystery, though there are suggestions towards degradation routes that aim phenolic compounds and furans. During bioconversion, polymers are converted into monomers by enzymatic action of microbes. Consolidated bioprocessing (CBP) is one suitable cohesive variant where microbes produce all enzymes essential to hydrolyse the pre-treated biomass with the help of no additional enzymes. Another variant is simultaneous saccharification and fermentation (SSF), in which saccharification and fermentation both are executed in single bioreactor; however, enzymes are still added to generate monomers which are then instantaneously fermented within the same reactor by the organisms. The most cost-inefficient route is separated hydrolysis and fermentation (SHF) where distinct stages are necessary for accomplishment of enzymes and microbes, since enzymes and organisms may have dissimilar optimum temperatures or pH. Commercial hydrolytic enzymes generally work efficiently at 45–50°C and pH 4–6, whereas to achieve those kind of parameters extremophiles are the only option. Extremophiles also deal with the cost effectiveness by reducing the cooling costs after pre-treatment of the feedstock which is required during the use of mesophiles. Furthermore, contamination risks are lowered since most common contaminants are mesophiles. Extremophiles also reduce the chance of contamination because growth parameters for them are extreme and most common contaminants are unable to grow in this situation (Robb 2008). Additional benefit of extremophiles is that they are extremely tough and commonly proficient in surviving fluctuating conditions which may happen in large scale industrial sector. Though more than 30 nations are in pursuit of philosophies, plans or policy agendas towards a sustainable bioeconomy worldwide, bioeconomy is still in its beginning. Investigation in basic and engineering science, however, is indispensable to allow the application of a sustainable bio-refinery; of course, this development needs to be monitored, leading to new challenges related to assessing the bioeconomy efficiency.

The present chapter describes the increasing use of extremophiles for different value added products like enzymes, biofuel, pharma products, etc. Additionally, this chapter sheds some light onto emerging bioremediation technologies using extremophiles.

2. Extremophiles and Enzymes

2.1 Enzyme Producing Extremophiles

Enzymes are often considered as the catalyst of choice in industrial processes due to several advantages like high substrate selectivity, stereo-selectivity, function selectivity, minimal by-product formation, environmental-friendly nature and high catalytic efficiency (Rozzell 1999, Van den Burg 2003, Mukherjee et al. 2013). Many of the available industrial enzymes are mesophilic and cannot withstand the harsh reaction conditions of the industrial processes (Rozzell 1999, Mukherjee et al. 2013). Extremophile organisms can grow in a wide range of extreme environments like high temperature, like thermophiles (optimum growth temperature 45°C–80°C) and hyperthermophiles (optimum growth temperature above 80°C), low temperature (psychrophiles), high salt concentrations (halophiles), acidic environment (acidophiles), alkaline environments (alkaliphiles), high pressure environment (piezophiles), highly radioactive environment (radiophiles) and high metal concentration (metalophiles) (Rozzell 1999, Demirjian et al. 2001, Mukherjee et al. 2013). They have several survival mechanisms that help them adapt and grow under unusual conditions and their enzymes are also optimized for structure maintenance and proper functioning under those extreme conditions. Enzymes from various extremophilic organisms have recently gained high interest in the industrial market as they are capable of functioning under those conditions where normal enzymes are unable to function. These enzymes are either purified from their original organisms or their enzyme coding genes are cloned and expressed in suitable host cells. Nowadays, metagenomics and metaproteomics are also employed in the search for novel extremozymes (Escuder-Rodríguez et al. 2018). Sometimes, a pool of enzymes from an extremophile is necessary for carrying out a particular function and in such cases, the whole cell is used as biocatalyst (Sharma et al. 2016).

2.1.1 Thermophiles

Thermophiles and hyperthermophiles are organisms that can grow at temperatures well above the normal mesophilic growth temperature limit of 45°C. Some of them can even reproduce at 121°C (Strain 121 *Geogemma barossii*) (Kashefi and Lovley 2003). Though thermophiles have been isolated from many habitats, they are predominantly found in hot springs, hydrothermal vents, volcanic deposits, composts and geothermal water. Most of the thermophiles belong to archaeal and bacterial domain. Some thermophiles can also grow at extremely high or low pH (thermoacidophiles and thermoalkaliphile) (Robb 2008). The temperature optima of enzymes from thermophiles generally range between 50°C–105°C. Several strains of thermophilic bacteria and archaea produce enzymes which can degrade complex carbohydrate polymers like cellulose, starch, xylan, pectin and chitin, proteins and lipids. Some of the important producers of cellulase, amylase, xylanase, laccase, pectinase, chitinase, protease, lipase and esterase are *Acidothermus cellulolytica, Acidobacterium capsulatum, Pyrococcus furiosus, Pyrococcus horikoshii, Sulfolobus solfaricus, Clostridium thermocellum, Thermotoga maritima, Thermus thermophilus., Thermus caldophilus, Thermomonospora fusca, Clostridium cellulovorans, Thermoanaerobacte*r sp., *Thermoanerobacterium* sp., *Dictyoglomus thermophilum, Rhodothermus marinus, Thermococcus chitinophagus, Thermococcus kodakaraensis, Methanocaldococcus jannaschii, Bacillus* sp., *Picrophilus oshimae, Picrophilus torridus, Aeropyrum pernix, Aquifex aelolicus, Desulfurococcus mucosus, Geobacillus caldoproteolyticus, Staphylothermus marinus, Caldicoprobacter algeriensis, Geobacillus* sp. and *Thermosachharolyticum* sp. (Basit et al. 2018, Escuder-Rodriguez et al. 2018, Kristjansson 1989, Robb 2008, Singh et al. 2019). The thermal stability of thermophilic fungal cellulases is generally not high, though cellulases from few thermophilic fungi like *Chaetomium thermophilum, Humicola grisesea, H. insolens* and *Thermoascus aurantiacus* have been cloned and expressed in host cells (Li and Papageorgiou 2019). Several thermophilic fungi like *Malbranchea pulchella,* and *Myceliophthora thermophila* and *Humicola* sp. are excellent producers of xylanase, the optimum activity ranging between 60°C–95°C, respectively, with good thermostability (Basit et al. 2018). Other industrially useful enzymes produced by thermophiles are alcohol dehydrogenase *(Pyrococcus furiosus, Thermococcus hydrothermalis, Aeropyrum pernix, Thermoanaerobacter brockii, Thermoanaerobacter ethanolicus, Thermomicrobium roseum* and *Thermococcus litoralis),* aldolase *(Methanocaldococcus jannaschii, P. furiosus, S. solfataricus, T. aquaticus* and *Thermoproteus tenax),* amidase *(Pseudonocardia thermophila* and *S. solfataricus),* aminoacyclase *(P. furiosus, P. horikoshii* and *T. litoralis),* arabinose isomerase *(Alicyclobacillus acidocaldarius, Geobacillus stearothermophilus, Thermoanaerobacter mathranii* and *T. maritima),* catalase *(Thermus brockianus),* cysteine synthase *(Aeropyrum pernix),* galactosidase *(T. maritima, Thermus* sp.), β-glucuronidase *(T. maritima),* glucose isomerase *(Thermus* sp. and *Thermoanaerobacterium* sp.), N-methyltransferase *(P. horikoshii)* and nitrilase *(P. abyssi)* (Robb 2008). Several thermophiles produce thermophilic DNA polymerases like Pfu pol *(Pyrococcus furiosus),* Taq pol *(Thermus aquaticus),* Deep Vent pol like *(Pyrococcus abyssi)* which are used in thermal cyclers during PCR amplification. Other DNA modifying enzymes produced by thermophiles include DNA ligase *(A. pyrophilus, P. furiosus, Rhodothermus marinus, Thermus* sp. and *Thermococcus* sp.), alkaline phosphatase *(Thermotoga* sp., *Thermus* sp. and *Pyrococcus* sp.), and ssDNA binding proteins *(Archaeglobus fulgidus, Methanocaldococcus* sp. and *Methanothermobacter* sp.). Many of these thermozymes are also produced by cloning and expression of the respective genes in suitable host cells (Robb 2008).

2.1.2 Psychrophiles

Psychrophilic organisms are found in cold regions like poles and deep sea (Feller et al. 1996). Organisms which can grow at low temperatures between –20°C and 10°C are known as psychrophiles. They are unable to grow above 15°C. There are some organisms which are psychrotolerant organisms and have a high growth rate at temperatures below 0°C but show optimum growth at 21–25°C (Santiago et al. 2016). Since they are adapted to living at cold temperatures, the temperature optima of enzymes from psychrophiles/psychrotrophs range between 10°C–30°C (in most cases retaining a significant amount of activity at 0°C–5°C). Some psychrophilic bacteria and fungus which can produce cellulase, xylanase, lipase, esterase, galactosidase, amylase, chitinase and protease belong to psychrophilic strains of *Flavobacterium* sp., *Pseudomonas* sp., *Paracoccus* sp., *Arthrobacter* sp., *Paenibacillus xylanilyticus,* other *Paenibacillus* sp., *Geomyces pannorum, Psychrobacter* sp., *Psychrobacter cryohalolentis, Photobacterium* sp., *Pseudoaltermomonas* sp., *Shewanella arctica, Bacillus* sp., *Lactococcus lactis, Psychrobacter pacificensis, Zunongwangia profunda, Micrococcus antarcticus, Colwellia psychrerythraea* and *Moritella marina* (Santiago et al. 2016). Among them, several of the enzymes are produced by cloning and expression of their genes in suitable host cells. Several DNA modifying enzymes from psychrophiles are used in molecular biology. One of the most important psychrophilic DNA modifying enzymes is alkaline phosphatase produced by psychrophilic strains of *Shewanella* sp., unclassified bacterium TAB5, *Cobetia marina* and *Arthrobacter* sp. Psychrophilic yeasts like *Cystofilobasidium capitatum, Mrakia frigida* and *Cryptococcus cylindricus* are highly preferred sources of pectinases (Nakagawa et al. 2004). Other industrially important enzymes produced by psychrophilic strains are catalase *(Bacillus* sp. N2a) (Wang et al. 2008) and oxidases *(Pseudoalteromonas haloplanktis)* (Pulicherla et al. 2011).

2.1.3 Halophiles

Halophiles are those organisms that grow best in high salt concentrations ranging from 0.2 M to 5.2 M and are frequently found in hypersaline environments like marine water and salterns. Halophiles enzymes have a high adaptation to high salt concentrations and low water potential;

some are also thermotolerant. However, since they are adapted to functioning under high salt concentrations, the purification and handling of these enzymes are somewhat difficult. Thus, only a few have biotechnological applications (Ventosa and Nieto 1995). Some important halophilic, moderately halophilic and halotolerant organisms producing enzymes are *Micrococcus halobius* (amylase), *Micrococcus varians* subsp. *halophilus* (nuclease H, 5'nucleotidase, amylase), *Bacillus* sp. (protease and nuclease), *Halobacterium halobium* and *H. sodomense* (amylase and endonuclease), *Natronomonas pharaonis* and *Natrialba magadii* (protease), *Halobacterium* sp., *Natronococcus* sp., *Salinivibrio* sp., *Haloarcula marismortui, Pelagibacterium halotolerans, Marinobacter lipolyticus SM19* and *Chromohalobacter salexigens* (protease, lipases and esterases), *Nesterenkonia halobia* (amylase), *Virgibacillus* spp. and *Halobacterium salinarum* (chitinase) (Kamekura 1986).

2.1.4 Acidophiles

There are some areas where huge amounts of sulphur or ferrous iron in pyrite are exposed to oxygen with the resultant conversion of sulphur to sulphuric acid and ferrous ion to ferric iron. These areas are generally present around sulphur and coal mines and have an acidic environment with a pH of 3.0 or lower. These acidic environments support the growth of acidophilic organisms which grow optimally at a pH of 3.0 or lower. Some of them are also thermophilic in nature. Some of the industrially important acid-stable amylases, proteases, cellulases, pectinases and xylanases are produced by *Alicyclobacillus acidocaldarius, Bacillus acidicola, Thermoplasma acidophilum, Picrophilus torridus, P. oshimae, Xanthomonas* sp., *Pseudomonas* sp., *Sulfolobus acidocaldarius, Thermoplasma volcanium, Aspergillus niger CH4, A. foetidus, A. awamori, Penicillium frequentans, Sclerotium rolfsii, Rhizoctonia solani* and *Mucor pusillus.* The acidophiles *Ferroplasma acidiphilum, Sulfolobus solfataricus* and *S. tokodaii* are producers of the DNA modifying enzymes DNA ligase, endonuclease and acid phosphatase, respectively (Sharma et al. 2012). Whole cells of acidophiles are used as biocatalysts in fuel cells (*Acidiphilium cryptum, Acidithiobacillus* and *Ferroplasma*), removal of contaminants like heavy metals like aluminium, copper, lead, zinc, cadmium, nickel and arsenic, cyanides, hydrocarbons from soil and treatment of acidic effluents (*Acidiphilium rubrum, Acidisphaera, Acidithiobacillus* and *S. solfataricus*) (Sharma et al. 2012, 2016).

2.1.5 Alkaliphiles

Alkaliphiles are organisms that cannot grow properly at neutral pH and grow at a pH value above 9. They are found in large numbers in alkaline environment and soda lakes, though they have been isolated from non-alkaliphilic places too. Some alkaliphiles may also be halophilic, thermophilic or psychrophilic. Similar to other groups of extremophiles, many enzymes from alkaliphilic organisms are alkaliphilic (Horikoshi 1996). The first report of an alkaliphilic enzyme was an alkaline protease from *Bacillus* sp. Later, other alkaline proteases were also reported from *Thermoactinomyces* sp., *Streptomyces pactum, Stenotrophomonas maltophilia, Paenibacillus tezpurensis,* and *Natronococcus occultus.* Amylase, cellulase, protease, pectinase, chitinase and lipase are produced by several alkaliphilic strains of *Bacillus* sp., *Pseudomonas* sp., *Natronococcus* sp., *Micrococcus* sp., *Nocardiopsis albus* and *Streptomyces* sp. (Horikoshi 1999, Sarethy et al. 2011).

2.1.6 Other Groups of Extremophiles

There are some other groups of extremophiles which produce biotechnologically useful enzymes. These groups have not been studied as much as the other extremophilic groups and application studies of these organisms are still at a nascent stage. Piezophiles are extremophiles which grow under extremely high hydrostatic pressure and show optimal growth above 40 Mpa of pressure; some are piezotolerant and show optimal growth at pressures less than 40 MPa and can also grow at normal atmospheric pressure. The piezophile *Methanococcus jannaschii* produces piezophilic protease (Horikoshi 1998). *Deinococcus radiodurans* is a radiophile which can survive under high radiation conditions and its whole cell can be used as a biocatalyst for bioremediation of radionucleotide contaminated sites (Demirjian et al. 2001). Lipases from this organism showed thermostability

(Shao et al. 2014). Metalophiles have a general high tolerance to different heavy metals and many of their enzymes require certain amounts of metals for their optimal activity. Lipases have been reported from the metalophile *R. solanacearum* where Na^{2+}, Fe^{2+} and Ca^{2+} stimulate the activity of lipases (Moayad et al. 2018). Whole cells of metalophiles like *R. solanacearum*, *Acidithiobacillus ferrooxidans*, *Cupriavidus metallidurans*, *Geobacter metallireducens* and *Metallosphaera sedula* can be used for bioremediation and bioconversion of metals like copper, arsenic, uranium, iron, gold and silver (Marques 2018).

2.2 Biochemical Attributes of the Enzymes of Extremophiles

2.2.1 Thermophiles

Thermophilic enzymes have several biochemical properties which make them thermostable and functionally active at high temperatures particularly close to the growth temperature of the host organism. When they were compared with their mesophilic counterparts, it was found that a few critical amino acid substitutions confer thermostability to these thermozymes. Substitutions of glycine to alanine and lysine to arginine are frequently seen. However, among thermophiles and hyperthermophiles, variations of amino acid concentrations are seen. The core of thermophilic enzymes is hydrophobic, making them more resistant to unfolding. Generally, they have a small surface-to-volume ratio and reduced glycine content which improves stability. Extensive ionic interactions, hydrogen bonds and disulfide bonds are also seen. Charged amino acids like lysine, arginine, glutamate and aspartate are frequently found on the surface of these enzymes. At mesophilic temperatures, thermozymes have low activity and high stability, providing a high energetic barrier for unfolding and catalysis. Several heat shock proteins and chaperones also aid in thermostability of the enzymes (Madigan and Oren 1999, Robb 2008, Vieille and Zeikus 2001).

2.2.2 Psychrophiles

Psychrophilic enzymes have many properties which enable them to maintain structural integrity and functionality at low temperatures. It has been observed that the catalytic activity of psychrozymes at low temperatures is like mesophilic enzymes at warm temperature. The high catalytic efficiency with a high turn-over number acts as a compensation for the reduction in reaction rate at low temperatures. The activation energy is lowered due to easy accommodation of the substrate (Feller et al. 1996). Amino acid substitutions at various regions of the enzymes makes them more flexible at lower temperatures compared to their mesophilic counterparts (Santiago et al. 2016). The properties of these enzymes are almost opposite to thermozymes. They have decreased hydrophobicity in their cores, and their surface is more hydrophobic. They have a low arginine to lysine ratio and more glycine. They have low interactions (hydrogen bonds, ionic interactions, aromatic interactions and disulphide bridges) between their subunits and domains. An overall decrease in bonding renders the enzymes more flexible (Cavicchioli et al. 2011).

2.2.3 Halophiles

Halophiles, which pump out excess salt ions, don't have any particular biochemical attributes for adaptation; however, organisms which follow "salt-in-strategy" have several features which help them adapt to high salt concentrations. These enzymes have more acidic amino acid residues on their surfaces and less hydrophobic residues. The hydrophobic residues are of short chain length. Thus, they can retain in harsh low water content areas and maintain stability when organic solvents are used. A large number of electrostatic and ionic interactions are found in halophilic enzymes. Isoleucine is frequently substituted by valine. They need high salt concentrations to maintain their structure and the enzyme activity of these enzymes is highly dependent on the optimal salt concentration. The enzymes are highly soluble in high salt concentrations (Madern et al. 2000). High concentrations of chloride ions stimulate transcription and translation of these enzymes (Delgado-García et al. 2012, Schreck and Grunden 2014). The hydrogen bonds formed between the negative side chains of the enzymes and water are very important for maintaining the stable hydration shells. Halophilic enzymes contain numerous disulfide bonds as a stabilizing factor (DasSarma and DasSarma 2015).

2.2.4 Acidophiles

Acidophilic enzymes possess many features which enable them to catalyse reactions at low pH. A large number of acidic amino acid residues of glutamate and aspartate are present on the surface of the enzymes. Thus, a high negative charge is present on the surfaces of those proteins (Sharma et al. 2016).

2.2.5 Alkaliphiles

Alkaliphilic enzymes also have many mechanisms by which they function and remain stable at high pH. They show a decrease in the number of negatively charged amino-acids and lysine and increase in arginine, glutamine, histidine and asparagine (Horikoshi 1999). The extra arginine residues account for the increase in hydrogen bonds and ionic pairs. Formation of arginine-aspartic acid ion pairs increases the stability of alkaliphilic enzymes (Fujinami and Fujisawa 2010).

2.2.6 Other group of extremophiles

Piezophiles are predominantly present in deep sea regions, where temperatures are either very cold or very hot (hot around hydrothermal vents). As a result, many of their enzymes possess characteristic attributes of thermozymes or psychrozymes and halozymes. They generally have a compact dense hydrophobic core with a large number of amino-acids which form small hydrogen bonds and multimers (Horikoshi 1998). Barophiles have many adaptations in their enzymes which confer thermostability and they have a lot of similarity to thermophilic enzymes (McDonald 2001). The precise mechanisms of metalophilicity and the properties of metalophilic enzymes are still being investigated.

2.3 Biotechnological and Environmental Applications of the Enzymes

The enzymes from extremophiles have a wide variety of biotechnological and environmental applications. A detailed list of different extremophilic enzymes with their biotechnological applications are given in Table 1. In some of the applications, extremophilic enzymes are used exclusively. For example, in PCR reactions, only thermophilic enzymes like Taq pol, Pfu pol can be used. Polymerase chain reaction comprises of a series of steps where temperatures as high as 95°C are required and this role can't be fulfilled by any mesophilic enzyme. In case of alkaline phosphatase, psychrophilic enzymes are preferred. Other DNA modifying enzymes and their applications are given in Table 1 (Robb 2008). Several extremozymes like cellulase, amylase, protease, xylanase, chitinase, pectinase, protease, laccase and lipase have applications in biomass utilization, biofuel production, paper and pulp industry, detergents, textile industry, juice processing, animal feed and food industry (Atalah et al. 2019).

Thermoacidophilic enzymes are particularly useful in the starch processing and baking industry. Acidophilic proteases are valuable for pharmaceuticals, food and beverage industry as many of the processes occur at low pH conditions (Sharma et al. 2016).

These enzymes are also environment friendly and play role in bioremediation and biodegradation. Utilization of biomass plays a crucial role in the production of renewable biofuel. The oligosaccharides produced from the action of these enzymes have several pharmaceutical and medical uses such as immunomodulation, probiotics, antioxidants, regulators of blood glucose and lipids, drug delivery, agriculture, cosmetics feed, dietary fibre and sweetener (Patel and Goyal 2011). Some enzymes like alcohol dehydrogenase, aldolase, amidase, nitrilase and glucose isomerase are useful in the synthesis and transformation of various chemicals, production of oligosaccharides and other pharmaceuticals (Demirjian et al. 2001).

Table 1. Different enzymes from extremophiles and their applications.

Enzymes from extremophiles	Applications
Cellulases (Endoglucanase EC 3.2.1.4, β-Glucosidase EC 3.2.1.21, Cellobiohydrolase EC 3.2.1.91) (From thermophiles, psychrophiles, halophiles, acidophiles and alkaliphiles)	1. Conversion of plant tissues into bioethanol (biofuel). 2. Used in a number of industries like food and breweries, textile industry, laundry, pulp and paper industry. 3. Degradation of polymers, biopolishing of cotton, brightening of colours and saccharification of agricultural wastes. 4. Used in animal feed, juice extraction. 5. Synthesis of sugars and optically pure oligosaccharides (Robb 2008).
Xylanase (Endo-1,4 β-Xylanase EC 3.2.1.8, Acetyl xylan esterase EC 3.1.1.6, 1,4-β-Xylosidase EC 3.2.1.37) (From thermophiles, psychrophiles, halophiles, acidophiles and alkaliphiles)	1. Used in animal feed processing. 2. Used in paper and pulp industry. 3. Used in fruit juice processing and bakery (Basit et al. 2018).
Chitinase (Endochitinase EC 3.2.1.14, Exochitinase EC 3.2.1.52, Chitobiase EC 3.2.1.30) (From thermophiles, psychrophiles, halophiles and alkaliphiles)	1. Marine biomass utilization. 2. Chito-oligosaccharide production for various pharmaceutical, industrial and agricultural uses including antifungal and antibacterial properties. 3. Used as food quality enhancer (Swiontek Brzezinska 4. et al. 2014).
Pectinase EC 3.1.1.11 (From thermophiles, psychrophiles, acidophiles and alkaliphiles)	1. Fruit juice industry for extraction and clarification. 2. Liquefaction and saccharification of biomass. 3. Processing of fibre crops. 4. Used in paper industry (Kashyap et al. 2001).
Amylase (α-Amylase EC 3.2.1.1, β-Amylase EC 3.2.1.2, Glucoamylase EC 3.2.1.3, α-Glucosidase EC 3.2.1.20, Pullulanase EC 3.2.1.41, Cyclodextrin glucosyl transferase EC 2.4.1.19, Amylomaltase EC 2.4.1.25, Cyclomaltodextrinase EC 3.2.1.54, Branching enzyme EC 2.4.1.18) (From thermophiles, psychrophiles, halophiles, acidophiles and alkaliphiles)	1. Bread and baking industry. 2. Production of bioethanol. 3. Starch liquefaction and saccharification. 4. Industries related to glucose production, paper and pulp, textile designing, oligosaccharide synthesis and food industry (Robb 2008). 5. Used as detergents in laundry.
Proteolytic enzymes EC 3.4.21. (serine protease, acidic protease, thiol protease, metalloprotease) (From thermophiles, psychrophiles, halophiles, acidophiles, alkaliphiles and piezophiles)	1. Used as detergents. 2. Used in bakery, brewing and production of important amino acids. 3. Animal feed (acidophilic). 4. Hide-dehairing (alkaliphilic). 5. Recovery of silver from X-ray films (alkaliphilic). 6. Cheese production. 7. Meat processing. 8. Silk degumming (Robb 2008, Sarethy et al. 2011).
Alcohol dehydrogenase EC 1.1.1.1 (From thermophiles)	Transformation of ketones to alcohols (Robb 2008).
Aldolase EC 4.1.2.13 (From thermophiles)	Used for synthesising carbohydrate (Robb 2008).
Amidase EC 3.5.1.4 (From thermophiles)	Synthesis of fine chemicals (Robb 2008).
Aminoacylase EC 3.5.1.14 (From thermophiles)	Used in pharmaceutical industry (Robb 2008).
Catalase EC 1.11.1.6 (From thermophiles, psychrophiles and alkaliphile)	Used as bio-bleaching agent (Robb 2008).
Lysozyme EC 3.2.1.17, Oxidase (From psychrophiles)	Preservation of food (Pulicherla et al. 2011).
Arabinose isomerase EC 5.3.13 (From thermophiles)	Used for preparing sweeteners (Robb 2008).

Table 1 contd. ...

...Table 1 contd.

Enzymes from extremophiles	Applications
Cysteine synthase EC 4.2.1.22 (From thermophiles)	Organo-sulphur compound synthesis (Robb 2008).
Esterase EC 3.1.1.1. and Lipase EC 3.1.1.3 (From thermophiles, psychrophiles, halophiles, alkaliphiles, metalophile and radiophile)	1. Used in food industry (dairy products, cheese), pharmaceutical industry (production of chiral compounds, in esterification of vegetable oil), paper industrym biomass transformation, textile and cosmetics industry. 2. Used as detergent in different industry (Houde et al. 2004).
α-Galactosidase EC 3.2.1.22, β-Galactosidase EC 3.2.1.23, β-Glucuronidase EC 3.2.1.31 (From thermophiles and psychrophiles)	Oligosaccharide synthesis for dietary milk products (Robb 2008).
Glucose isomerase EC 5.3.1.5 (From thermophiles)	Sweeteners (Robb 2008).
N-methyl transferase EC 2.1.1.17 (From thermophiles)	Food and medicine (phosphatidyl choline synthesis) (Robb 2008).
Nitrilase EC 3.5.5.1 (From thermophiles)	Mononitrile production (Robb 2008).
DNA polymerase EC 2.7.7.7 (e.g., Pfu pol, Deep Vent pol, Taq), DNA ligase EC 6.5.1.1., Alkaline phosphatase EC 3.1.3.1, DNA topoisomerase, ssDNA binding proteins, Acid phosphatase (From thermophiles, psychrophiles and acidophiles)	DNA modifying enzymes useful in various molecular biology and genetic engineering processing (DNA amplification, sequencing, cloning and expression studies) (Cavicchioli et al. 2011, Robb 2008).
Nuclease, 5'nucleotidase (psychrophiles, halophiles and acidophiles)	Useful in various molecular biology processes.
Whole cell of extremophiles as a biocatalyst (acidophiles, radiophiles and metalophiles)	1. Acidophilic microbe as microbial fuel cell for generating electricity. 2. Bioremediation of toxic organic contaminants, hydrocarbons, heavy metals from extremely contaminated environments. 3. Extraction of metals. 4. Treatment of acidic effluents by acidophilic bacteria. 5. Desulfurization of coal and sulphur metabolism. 6. Bioremediation of radionucleotide contaminated sites by radiophiles. 7. Bioremediation of heavy metals, leaching of ores and biomineralization (Demirjian et al. 2001, Sharma et al. 2016).

In food industry, low temperatures are essential for maintaining the freshness and quality of food. Use of psychrophilic pectinase, cellulase, protease and lysozyme are an excellent way to maintain the product quality. Similarly, use of halophilic lipases can be an effective way to split oil spills in ocean and in industrial processes with low water activity (Delgado-García et al. 2012).

Whole cells of acidophiles are frequently used for bioremediation of toxic contaminants, fuel cells and extraction of metals, radiophiles for bioremediation of radioisotopes and metalophiles for ore-leaching and biomineralization (Demirjian et al. 2001, Sharma et al. 2016). Major advantages and notable limitations are presented in Fig. 1 and Table 2.

3. Extremophiles and Biopolymers

Synthetic polymers are widely used in packaging and other industries; however, they are not environment friendly and also lead to depletion of fossil carbon. Over the last few years, a number of biopolymers have gained importance as an eco-friendly substitute of synthetic polymers.

Detergent Food Textile & Tannery Biorefinery Dairy Diagnostics Pharmaceutics

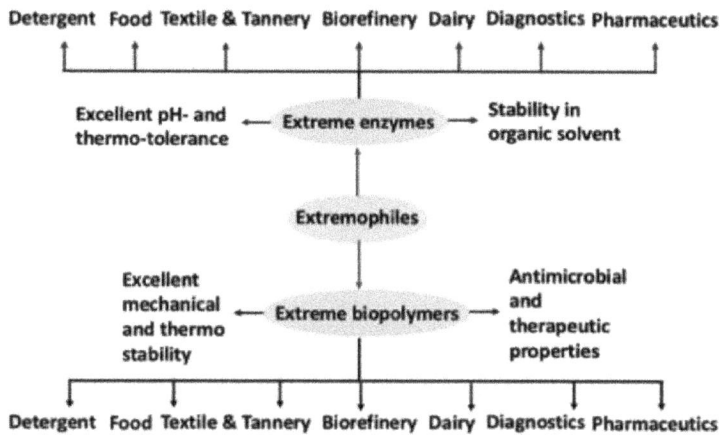

Fig. 1. Different advantages of enzymes from extremophiles and their applications.

Biopolymers from normal living organisms cannot withstand the harsh conditions of extremes of temperature, pH or salinity, so biopolymers produced by extremophiles are of high commercial value as they are expected to tolerate and function at harsh conditions. The extracellular polymeric substances (EPS) produced by extremophiles are considered to have high biotechnological value (Mitra et al. 2020). Most of the EPS from extremophiles may consist of homo and heteropolymers of carbohydrates, though some may produce polyesters or polyamides (Kazak et al. 2010). Extracellular glycosyl transferases catalyze the polymerization of polysaccharides. Modifications like acetylation, pyruvylation and phosphorylation frequently occur (Nicolaus et al. 2010). The exact composition of the biopolymers produced depends largely on the carbon source used to grow the organism (Mohammadipanah et al. 2015).

3.1 Extremophiles Producing EPS

Extremophiles producing EPS have been isolated from deep sea extremophiles growing under extremely high temperature, pressure and salts, shallow vents, shallow hydrothermal vents, halophilic, acidic and alkaliphilic environments.

3.1.1 Thermophiles and Psychrophiles

Some of the important EPS producing thermophiles are *Alteromonas infernos, Anoxybacillus amylolyticus, Thermotoga maritime, Bacillus thermoantarcticus, Alteromonas* sp. *Methanococcus jannaschii, Thermococcus* sp. and *Streptococcus thermophilus. Geobacillus tepidamans* produces high molecular weight galacto-glucans which are highly thermostable. The EPS produced by *Geobacillus thermodenitrificans* has high antiviral and immunomodulatory activity. The EPS produced by *Bacillus licheniformis* are tetrasaccharide repeating units with mannosides, while that produced by *B. thermoantarcticus* are repeating units of mannose and glucose. *Geobacillus* produces several types of EPS, some of which are polymers of glucose, galactose and mannose, and some are polymers of galactose, mannose, glucosamine and arabinose or mannose, glucose, galactose and mannosamine (Kazak et al. 2010, Nicolaus et al. 2010). Most of the monosaccharides are linked by α-1,2 and α- 1,6 linkages or 1,4-β, 1,3-β linkages to form polysachharides. The secondary structures consist of mainly helices and random coils. Most of the thermophilic EPS have a high molecular weight of several hundred kilodalton (Kambourova et al. 2016). Cyanophycin and related polyamides are produced by the thermophilic cyanobacteria *Synechococcus* sp. (Hai et al. 2002). Biopolyesters of polyhydroxybutyrate are accumulated by the thermophilic strain *Geobacillis* sp. strain AY 946034 (Giedraitytė and Kalėdienė 2015). Some of the psychrophiles producing EPS are *Pseudoalteromonas* sp. *Colwellia psychrerythraea, Olleya marilimosa* (Kazak et al. 2010, Nicolaus

Table 2. Advantages and drawbacks of enzymes from extremophiles.

Enzyme	Advantages	Drawbacks
Thermophilic enzymes	• Long lasting and can tolerate high temperature, high and low pH, of industrial processes. • Can tolerate organic solvents and harsh purification steps with better yield. • Reaction rates are enhanced, with increased solubility of reaction components and decreased viscosity of reaction media, mass transfer rate is increased. • Mesophilic growth is prevented minimizing contamination. • The biological activity is minimized in raw materials (Kristjansson 1989).	• Damage of heat labile raw material, co-factors and other chemicals. • Decreased content of gases like oxygen in reaction media. • Since they are thermostable, it is difficult to inactivate them. • Equipment encounter thermal stress. • Due to high optimum temperature, not suitable for some applications (Kristjansson 1989).
Psychrophilic enzymes	• Several industrial processes require low temperatures; therefore, psychrophilic enzymes are well suited. • High specific activity at low temperature. • Less ion bonds. • Energy saving. • Protection of labile and volatile compounds. • Low risk of contamination. • Easy inactivation of enzymes. • Minimize production cost. • Psychrophilic enzymes are specifically useful in food industry as they keep the fragrance and freshness and thus maintain food quality (Pulicherla et al. 2011).	• The enzyme itself is heat labile and loses activity at mesophilic temperatures. • Due to low optimum temperature, not suitable for some applications.
Halophilic enzymes	• Easy to maintain and grow halophile. • Low risk of contamination. • May be useful in harsh industrial processes where concentrated salt concentrations can inhibit non-halophilic enzymes and under low water activity regions. • Lipases from halophilic enzyme are useful for bioremediation of oil spills in marine environment • High organic solvent stability (Schreck and Grunden 2014).	• Difficult to purify and handle the enzymes due to the high concentrations of salt required for functionality. • Limited demand for salt tolerant enzymes industrially.
Acidophilic enzymes	• Useful in those industrial processes where pH adjustment is required such as starch processing. • Many of them are also thermostable. • Dough pH is acidic and need acid stable xylanases for use in bakery industry. • Many fruit and vegetable juices are acidic and need acidophilic enzymes for processing. • Highly useful for treatment of acidic environments; not possible by other neutral or basophilic enzymes.	• Overexpression in mesophilic hosts is troublesome (Sharma et al. 2012). • Extremely acidophilic enzymes are not useful in those processes which require higher pH.
Alkaliphilic enzymes	• Useful in industrial processes where an alkaline pH is required, especially in dehairing of hide. • Stability in detergents where pH is generally high.	• Extremely alkaliphilic enzymes are not useful in those processes which require lower pH.
Piezophiles	• May be capable of carrying out high pressure reactions.	• Application studies still in nascent stage. • Difficulty in collection of samples for piezophile culture (Ichiye 2018).
Radiophiles	• Only radiophiles can be used in bioremediation of radionucleotide contaminated places.	• Application studies still in nascent stage.
Metalophiles	• Highly suited for heavy metal infested regions.	• Studies still in nascent stage.

et al. 2010). *Pseudoalteromonas* sp. produces neutral sugars and uronic acids with sulphates. The psychrophilic yeast *Cryptococcus laurentii* produces exopolysachharides containing arabinose, mannose, glucose, galactose and rhamnose (Pavlova et al. 2011).

3.1.2 Halophiles

Among halophiles, species of *Halomonas* include *H. Maura, H. anticariensis, H. cerina, H. euihalina, Cyanothece* sp. *Halobacterium* sp. *Idiomarina* sp., *Palleronia marisminoris* and *Salipiger mucosus*. Some Archaea like *Haloarcula japonica, Haloferax* sp., *Haloferax gibbonsii, Haloferax denitrificans, Haloarcula* and *Haloferax mediterranei* and the halophilic algae *Dunaliella salina* and yeast *Aureobasidium pullulans* also produce EPS. *Halomonas* sp. produce various types of EPS like levans with repeating units of fructose, polymers of glucose, mannose and galactose or glucose and mauran which consist of repeating units of mannose, galactose and glucose. Mauran is very important due to its pseudoplastic, rheological and thixotrophic characteristics. Some of the EPS have high contents of uronic acid and sulphur (Kazak et al. 2010, Nicolaus et al. 2010). *Haloarcula* produces EPS of polymers of glucose acetate, mannose and galactose. *Haloferax* produces polymers of glucose acetate, mannose and galactose which sometimes contain amino sugars, uronic acid and mannose as major components (Nicolaus et al. 2010). *Haloferax mediterranei* produces polymers of poly β-hydroxybutyrate internally and sulphated polysacchardies externally (Chen et al. 2019, 2020). Poly3-hydroxybutyrate-co-hydroxyvalerate is produced by the halophilic archaea *Natrinema* sp. (Danis et al. 2015). The extremely halophilic *Natrialba* sp. produces exopolymers of poly (γ-D-glutamic acid) (Margesin and Schinner 2001).

3.1.3 Alkaliphiles and Acidophiles

Alkaliphilic strains of *Halomonas* sp., *Bacillus* sp., *Vagococcus carniphilus* and *Salinivibrio* sp. are very good EPS producers, which contain mainly sugars like galactose and mannose and proteins (Joshi and Kanekar 2011, Nicolaus et al. 2010). Several acidophiles help in bioleaching with the help of EPS. These include *Leptospirillum ferrooxidans, Acidithiobacillus ferrooxidans* and *Acidithiobacillus thiooxidans* (Nicolaus et al. 2010). EPS from acidophiles generally consist of varying quantities of carbohydrates like galactose, glucose, heptose, rhamnose, glucuronic acid, fucose, xylose and N-acetyl glucosamine, metals like Fe, Al, Mg and Zn, lipopolysachharides like 3-deoxy-2-manno-2-octulosonic acid and very small amounts of DNA and protein (Jiao et al. 2010, Sand and Gehrke 1999).

3.2 Applications of Extremophilic Biopolymers

Extremophiles produce EPS as a protection mechanism against various stress conditions like high acidity, predation, UV radiation and high salinity. EPS can also act as an energy resource (Kambourova et al. 2016). They also help in cell adhesion, concentration of nutrients and water retention and cell signalling and biofilm formation. Lack of nutrients enhance their production. Polysaccharides including dextran, levan, fructan and mutan are also important ecologically as they stabilize and improve soil quality and aid in bioremediation. Biotechnologically, they have applications in a number of industries like textile, oil recovery (halophilic), adhesives, detergents, brewery, water treatment, food (viscosifier and gelling agent), cosmetics and pharmaceuticals. It is predicted that biopolymers extracted from extremophiles growing at extremes of temperature, pH, salt concentration and metal concentration will have those unique properties which will help the polymers resist those extreme conditions. Besides, the risk of contamination is also low while using extremophiles. The most important microbial biopolymer is xanthan gum which is useful in food and other industries. Mauran has a high capacity to viscosity similar to xanthan with a high stability and also pseudoplastic properties. EPS from the halophiles *Halobacterium* sp. and *Haloferax* sp. have the capacity to emulsify hydrocarbons. EPS from *Pseudoalteromonas* sp. (psychrophilic) and haloakaliphilic *Bacillus* sp. I-471 can function as bioflocculants in fermentation and water treatment

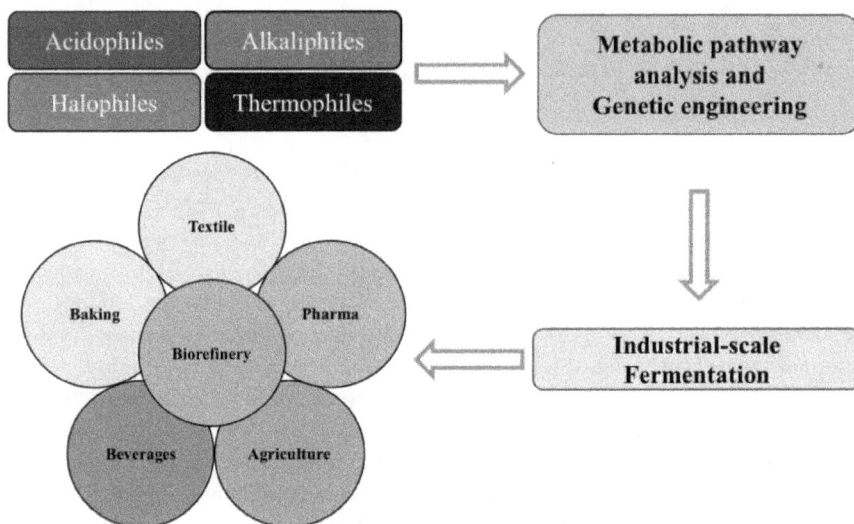

Fig. 2. Process flow for extremozymes production and their application.

processes. EPS from the halophilic bacteria *Halomonas anticariensis* and *Halomonas ventosae* are capable of adsorption of cations and sulphate. They are highly useful in sewage treatment facilities as they remove toxic metal ions and synthetic dyes and also degrade several types of hydrocarbons. Dextrans from thermophiles may be used for producing sephadex which is a very useful adsorbent. EPS like pullulans also have various roles in coating of medicines, drug delivery, vaccine production, cancer therapy and treatment of diseases (Kazak et al. 2010, Nicolaus et al. 2010, 2004). Thermostable cyanophycin and related polyamides may also be used as stable biopolymers (Hai et al. 2002). The polyβ hydroxybutyrate polymers produced by several extremophiles can act as an excellent source for bioplastic production, drug delivery, medical implants, and tissue engineering (Giedraitytė and Kalėdienė 2015, Mohammadipanah et al. 2015, Rodriguez-Valera et al. 1991). Polymers of γ-D-glumate play role in the food and pharmaceutical industry as a biodegradable humectant, thickener and drug carrier (Margesin and Schinner 2001). The EPS of *Halomonas xianhensis* has been reported to have antioxidant properties (Biswas and Paul 2017). Thus, biopolymers from extremophiles have huge biotechnological potential. A generalized overview on the properties and applications of the biopolymers obtained from the extremophiles is schematically summarized as Fig. 2.

4. Extremophiles and Biofuel

Biofuels, by definition, indicate the class of fuel products derived from biomass (e.g., sugarcane, soybean, palm, lignocellulosic substrates and Jatrpha) and biodegradable components of industrial and municipal wastes (Dufey 2006). Though the cost of fossil fuels remained cheaper compared to biofuels, there has been a significant increase in the research and development of biofuel products. The world after COVID-19 pandemic outbreak will move more towards personal transportation options than public transport. This will increase vehicle sales, which will result in increased consumption of global liquid fossil fuel by 2030 (Barnard et al. 2010). Increasing crude oil price volatility, reduction in fuel supply and increasing awareness regarding global warming issues have resulted in growing interest for development of biofuels and alternative fuel sources (Luque et al. 2008). Interest for biofuel is also reflected in the increasing global biofuel production and consumption figures. Total generation of biofuel in the EU27 countries increased from 5639 million tonnes of oil equivalent (Mtoe) to 8165 Mtoe while total biofuel consumption too increased from 5625 kilotonnes of oil equivalent (Ktoe) to 10064 Ktoe for the period between 2006 and 2008 (Barnard et al. 2010).

Biorefinery is the complete facility that enables the production of various bio-based products and biofuel with the use of biomass substrates (Zhu et al. 2020). Lignocellulosic biomass is the best candidate for biofuel production because of its low cost, widespread availability and lack of competition with food provisions (Ábrego et al. 2017). However, presence of recalcitrant lignin is one of the biggest hurdles for successful biofuel production from lignocellulosic biomass. Pre-treatment of biomass can result in effective delignification, which can result in higher performance of biorefinery units. Therefore, it is necessary to develop catalytic molecules with higher stability, activity, pH and temperature tolerance to increase effectiveness of biorefinery process for biofuels.

Extremophiles are a genre of microorganisms that grow under harsh environmental conditions such as extreme temperature, pH, salinity and pressure. Extremophiles can be subdivided in different categories depending on their preferred growing conditions: acidophiles (thrives at low pH conditions), alkaliphiles (grows optimally in alkaline pH conditions), halophiles (can grow in high salt concentration environment), thermophiles (thrives at high temperatures), psychrophiles (capable of growing in low temperatures), and xerophilies (thrives in dry environment). Extremophiles have gained a lot of interest recently, due to their ability to catalyse reactions under harsh operating conditions (Chen and Jiang 2018). Market for industrial enzymes to be used for biorefinery framework is expected to cross 5 billion USD by 2021 (Zhu et al. 2020). Thus, research and development are now being focused on the genetic engineering of the extremozymes using biotechnological tools to increase activity and specificity (Fig. 2). The chapter gives an overview on the use of extremophiles in biorefinery model for the production of biofuels from biomass.

4.1 Types of Extremophiles and Extremozymes

Various kinds of extremophiles and their enzymes have been isolated from their natural habitat and have been used in different processes of biofuel production. Table 3 summarizes some of the isolated extremophile and their enzymes.

Table 3. Popular extremophiles and extremozymes used in industry.

Extremophiles	Habitat	Extremozymes	Operating temperature (°C)	Application	Reference
Bacillus mojavensis SO-10	Hot spring	α-amylase	70	Biorefinery, food, detergent	(Ozdemir et al. 2018)
Alteromonas sp. ML117	Marine	β-galactosidase	10	Food	(Yao et al. 2019)
Bacillus subtilis Lucky9	Soil	Alkali tolerant xylanase	60	Biofuel	(Chang et al. 2017)
Bacillus cereus FT1	Soil	Alkaline protease	35	Pharma, leather	(Chang et al. 2017)
Thermophilic *Anoxybacillus* sp. HBB16	Hot spring	Alkaline lipase	50	Biodiesel	(Zhu et al. 2020)
Paenicibacillus barengoltzii	Marine	Chitinase	55	Bioethanol	(Yang et al. 2016)

4.1.1 Thermophiles

Microorganisms that can grow at extremely high temperature (40–122°C) with optimal growth ranging between 60 to 108°C are defined as thermophiles. Thermophiles are most popular for industrial use because of their ability to produce thermally stable enzymes (Singh et al. 2011). Thermostable enzymes either have a high unfolding temperature (Tm) or long half-life at a high operating temperature, allowing them to catalyze high-temperature chemical processes. Taq polymerase is by far the most popular thermophilic extremozyme produced from *Thermus aquaticus*. Thermally stable enzymes obtained from thermophiles have a huge potential in biofuel production and anaerobic fermentation. Few popular and recently isolated thermophiles are as follows: *Geobacillus* sp., *Thermoanaerobacterium* sp., *Pyrococcus* sp. (Zhu et al. 2020). However, the lack of scientific understanding and insufficient metabolic engineering tools for genetic engineering are the major hurdles for the integration of thermostable enzymes into industrial scenario.

4.1.2 Alkaliphiles and Acidophiles

Extremophiles that can thrive in alkaline environments with pH ranging from 8.5–11 are defined as alkaliphiles. They are natural habitants of soda lakes, vents, deep sea sediments and carbonate-rich soils (Zhu et al. 2020). On the other hand, acidophiles can grow in acidic environment (pH 1–5). Acidophiles are largely present in sulphuric lakes, coal mines, and sulfur mines.

Alkali stable enzymes can be used for the volarization of lignin to increase the robustness and efficiency of biofuel production process from lignocellulosic biomass. Bioleaching process of kraft and soda pulps have employed thermally stable alkaliphilic xylanase. Alkaliphilic enzymes were able to catalyze the reaction without adjusting the pH, thus improving process efficiency and cost (Zhu et al. 2020). Xylanase derived from acidophile *Penicillium oxalicum* GZ-2 has shown tremendous promise in biofuels, animal food, and food supplement industries. Biodegradation of polyaromatic hydrocarbons has also been reported using acidophilic *Stenotrophomonas maltophilia*. Acidophiles can also be employed for membrane potential reversal, to induce membrane impermeability, and other key applications in biological industry.

4.1.3 Halophiles

Microorganisms that have the ability to grow sustainably in saline conditions (deep-sea sediments, saline lakes, sea water and saline soils) are defined as halophiles. Enzymes such as cellulase, xynalase, etc., obtained from halophiles are more stable than enzymes of terrestrial origin (Zhu et al. 2020). Halophile *Nesterenkonia* sp. F is able to tolerate organic solvent and can also produce α-amylase. The isolated wild strain halophile *Nesterenkonia* sp. F is the only bacteria known to have the ability to produce both butanol and ethanol except *Clostridia* (Amiri et al. 2016). Recent studies have been focused to improve the performance of halophiles for industrial application. Researchers are working hard on the development of genetic tools using metabolic engineering to increase cell growth and enzyme yield. A recent study reported an increase of 14.6 times α-amylase yield by a mutated halophile compared to that of enzyme released under salt-free conditions (Pan et al. 2020).

4.2 Biofuel Production using Extremophiles

The robustness of enzymes produced by the extremophiles is high and is able to withstand variation in operating conditions such as pH, salinity and temperatures. Thermophiles are more popular in the biorefinery processes for the production of biofuel from biomass due to their ability to tolerate temperature fluctuations. Thermophiles can ferment sugars obtained from biomass for biofuels production (Sommer et al. 2004). As they have lower energy demands due to less cooling requirements for growth, use of thermophiles for biofuels also makes the process economically viable. In the following section, application of various extremophiles for the production of different biofuel has been discussed.

4.2.1 Bioethanol

Bioethanol is a renewable biofuel alternative for current hydrocarbon fuels, and it is produced by fermentation of sugars. Some of the major challenges of bioethanol productions are low yield of ethanol, by-products, substrate inhibitions and ethanol toxicity. *Saccharomyces cerevisiae* has the ability to convert sugars obtained from sugarcane into bioethanol with a yield of just 20% (v/v) (Das et al. 2015). However, to realize an economically viable bioethanol production operation facility, the substrate has to be a low-cost lignocellulosic biomass. Lignocellulosic biomass contains cellulose and hemicellulose. Hydrolysis of hemicellulose generates pentose sugars that cannot be utilized by *S. cerevisiae* and *Z. mobilis* for bioethanol production (Kumar et al. 2008). However, many thermophiles have the ability to utilize both hexose and pentose sugars and ferment them into bioethanol, giving them a huge advantage such as high yield, metabolic and growth rates over *S. cerevisiae* and *Z. mobilis*.

Thermophilic clostridia have the ability to grow between temperatures 60 to 65°C (Lamed and Zeikus 1980). They also can utilize lignocellulosic biomass for biofuel production due to the production of multiple extremozymes (Demain et al. 2005) from their cellulosomes. Cellulosomes are found on the outer surface of the cell membrane and it plays a critical role in the enzymatic degradation of recalcitrant lignocellulosic biomass. The enzymes produced in cellulosomes of clostridia are as follows: endo-β-glucanases, exoglucanases, cellodextrin, galacosidases and mannosidases (Demain et al. 2005). Using the enzyme cocktail released from cellulosomes, clostridia can successfully hydrolyze lignocellulosic biomass into simple sugars that can be fermented in bioethanol. Thus, *Clostridium themocellum* can be a good choice for ethanol production from lignocellulosic biomass. To mitigate the challenges of ethanol toxicity and by-product generation, researchers have resorted to genetic engineering of thermophiles. Genetically modified *C. thermocellum* with knockouts of acetate kinase and lactate hydrogenase have been reported to have increased ethanol tolerance up to 60 g/L. Although *C. thermocellum* has the ability to hydrolyze cellulose and hemicellulose into hexose and pentose sugars, it can only ferment hexose sugars into ethanol. To solve this problem, *C. thermocellum* has been used along with other thermophiles such as *Thermoanaerobacterium thermosaccharolyticum*, *Thermoanaerobacter ethanolicus*, and *Thermoanaerobacterium saccharolyticum* for complete utilization of lignocellulosic biomass (Barnard et al. 2010).

Thermoanaerobacterium is a type of hemicellulolytic thermophilic anaerobe. It has the ability to ferment pentose sugars to ethanol and organic acids. Researches have been able to reduce organic acid by-products and increase bioethanol production by metabolic engineering and creating knockout mutants of lactate dehydrogenase in *Thermoanaerobacterium* sp. (Desai et al. 2004). *T. saccharolyticum* ALK2 has the ability to ferment xylose, glucose, mannose and galactose, generating 33 g/L ethanol in continuous culture and 37 g/L in fed-batch cultures (Barnard et al. 2010). *Thermoanaerobacter* species are extremophiles similar to clostridia with primary products lactic acid and ethanol (Lamed and Zeikus 1980). *Geobacillus* thermophiles have high catabolic flexibility, enabling them to be good candidates for metabolic engineering. *G. stearothermophilus* is capable of generating ethanol at operating temperature of 70°C at yields similar to that of *S. cerevisiae*.

These microbial strains are tolerant to microbial contamination due to their extreme operating conditions. To improve biofuel process efficiency, more development is required to find genetically engineered strains with higher ethanol tolerance and yield. Such model extremophiles would help us to generate new engineering solutions for biofuel production.

4.2.2 Biodiesel

Biodiesel is a fatty acid ester compound generated by transesterification of triglycerides and fatty acids from vegetable oils or microalgae. It has higher energy content compared to ethanol and butanol. Biodiesel production is a developed technology that is already in use for compression-ignition engines. To make biodiesel economically viable, more research and development is required to find alternative efficient source of oil, such as microalgae or extremophiles.

Microalgae are photosynthetic microorganisms that can utilize sunlight, water and carbon di-oxide to produce algal biomass. Under optimal growth environment, these algae can synthesise fatty acids for transesterification into glycerol based membrane lipids which can amount to 5–20% of total dry cell weight (Hu et al. 2008). Under extreme growth conditions, some microalgae can generate long chain hydrocarbons similar to petroleum, with amount exceeding 80% of their dry cell weight (Metzger and Largeau 2005). There are large number of extremophilic microalgae, such as *Cynidium caldarium* and *Galdieria sulphuraria*, which can withstand extreme environment (temperature and pH) (Barnard et al. 2010). The major advantage of selecting an extremophile microalgae would be its ability to avoid contamination within photobioreactors. *Spirulina, Dunaliella* and *Chlorella* sp. have shown sufficient growth rates in halophilic nutrient rich aerobic environment. Though we obtain higher lipid content under stressful environment, it also reduces growth rates of the microalgae in such harsh conditions. This is the biggest hurdle for the incorporation of microalgae in biodiesel production. Genetic engineering to unlock the high growth rates and high lipid yield would make the process economically viable.

In summary, extremophiles can play a significant role in the progress of biodiesel production. Extremophilic sources of algal oils are important for open aerobic photobioreactor operations to prevent possible microbial contaminations.

4.2.3 Biobutanol

Biobutanol was first produced from *Clostridium* sp. It has higher energy density compared to ethanol, and can be gasoline and diesel without the requirement of engine modification (Baruah et al. 2018). Butanol also has the potential to be used as an aviation jet fuel due to its high energy content and low volatility. However, low butanol tolerance (2–3% v/v) of *S. cerevisiae* is the major bottleneck for the production of biobutanol on an industrial scale. This leads to higher downstream processing cost and requirement for larger production vessels. Extremophile of the genera *Bacillus* and *Pseudomonas* sp. have been reported to be able to tolerate butanol concentrations in the range between 6–7%. Another hurdle for biobutanol production using *Clostridium* sp. is the generation of mixed solvent system known as ABE fermentation (30% acetone, 60% butanol and 10% ethanol). This poses a serious downstream butanol separation problem along with growth inhibition of the microorganism in these mixed solvent systems. Ongoing research are focused on increasing the biobutanol yield by metabolic engineering of *C. acetobytylicum* and other clostridial strains (Barnard et al. 2010). Tetravitae Biosciences has reported improved efficiency and selectivity towards butanol production using ABE fermentation process by employing mutated *C. beijerinckii*. Other drawbacks of ABE fermentation that needs to fixed to make the process economically viable is the low tolerance of butanol of the microbes and the ability to utilize low cost lignocellulosic biomass for butanol production. This can be achieved by designing solvent-tolerant *Clostridium* with overexpression of chaperones (Tomas et al. 2003). Another method to increase biobutanol yield is expressing the genes responsible for butanol production and solvent tolerance present in extremophiles in a host organism using recombinant DNA technology. *Escherichia coli* is often the preferred host organism due to its well-studied genome sequence and availability of tools required for performing the recombination. Butanol production in *E. coli* was made possible for the first time by modifying the native keto-acid pathway in *E. coli* (Baruah et al. 2018). Later on, CimA enzyme from *Methanococcus jannaschii* was expressed in *E. coli* to generate 1-butanol and propanol (Barnard et al. 2010).

In summary, butanol has huge potential as a biofuel due to its energy density and compatibility with current fossil fuels. Low butanol yield in a mixed solvent ABE fermentation system with butanol intolerance is the major bottleneck for realizing industrial scale production of economically viable biobutanol. Genetic modification of microbial strains to increase productivity, robustness and tolerance towards butanol can make its production attractive in comparison to other biofuels.

4.2.4 Biohydrogen and Biogas

Compared to other fossil fuels and biofuels, hydrogen can be converted into electrical energy in fuel cells. Thus, it has huge potential to be a clean energy carrier with a high energy density (143 MJ/kg) (Barnard et al. 2010). Currently, most of the hydrogen demands are met through steam reforming by using fossil fuels. There is a huge opportunity for the generation of biohydrogen by using thermophiles. Reports suggest that thermophilic hydrogen production surpasses by almost two times to that of the hydrogen yield from mesophiles, because of the suppression of methanogens and sulphur-reducing bacteria (Lin et al. 2008). *Caldiecellulosiruptor saccharolyticus* is able to produce hydrogen by fermenting pentose sugars, while *Clostridium themocellum* is able to use hexose sugars, and mutant *Thermoanaerobacterium thermosaccharolyticum* W16 is able to utilize a biomass hydrolysate containing both glucose and xylose for biohydrogen production. However, mixed culture has been reported to produce higher hydrogen yield due to the ability to utilize complete mix of available sugar sources from the hydrolysate (Liu et al. 2008). Biohydrogen production from waste water and agricultural waste has also been studied in detail due to their rich carbohydrate content (Gavala et al. 2006, Kästner and Mahro 1996, Luo et al. 2010). Hydrogenases metalloenzymes can catalyze the reversible reduction reaction of proton to hydrogen using the energy received from organic matter or light (Baker et al. 2009). However, the sensitivity towards oxygen is one of the major challenges of hydrogenases. Hydrogenases derived from thermophiles have high thermal stability and oxygen tolerance. Hydrogenases from *Aeropyrum camini, Thermotoga maritima, Pyrococcus furiosus* have shown great hydrogen production capability.

Utilization of industrial and municipal wastes for the generation of biofuels such as methane by anaerobic digestion can be a profitable value-added product with numerous environmental benefits. For optimal production of biomethane, mixed bacterial communities can be utilized. Extremophiles that have methane production capabilities are *Methanosarcina lacustri, Methabolobus psychrophilus, Methanothermococcus okinawensis,* and *Methanosaeta* sp. (Barnard et al. 2010). Research is ongoing to optimize the process parameters to increase biogas production from anaerobic digestion using extremophiles.

5. Extremophiles and Pharmaceutical Products

Extremophiles have different cellular machinery that helps them fit and survive in extreme environmental conditions where mammalian life cannot survive. The harsh environmental conditions include extreme temperature, pH, salinity, hydrokinetic pressure, ionizing radiation, low-level oxygen hostility, dehydration, and heavy metals' presence. For many years, extremophiles were exploited by pharmaceutical biotechnology companies to use novel enzymes and other biologic materials for medicinal drug synthesis as well as for health diagnosis. The variety of commercially critical possible applications for extremophile products means the focus of research and development is on biotechnology companies around the world.

5.1 Extremophiles in Drug Discovery

A significant extremophile contribution to the medical field comes from an innovative method for distributing vaccines. Several microorganisms develop internal vesicles of gas, small gaseous protein-filled structures, the best known from the halophilic archaea. There are no literature findings that include archaea in animal disease, regardless of temperature dependence. However, the immune system in mammals responds to membrane lipid components in archaea, which is of particular pharmaceutical interest because these are being investigated as possible vehicles for drug delivery. The peculiar lipid properties found in Archaea have resulted in the creation of archaeal liposomes, or archaeosomes. Their properties allow for the development of liposomes at or below any temperature within the physiological range, enabling the encapsulation of thermally labile compounds and serving as novel drug delivery agents. They also target mononuclear phagocytes, and these cells can take on more readily than liposomes made from ester lipids. This property makes them ideal candidates for antigen transmission, as carriers, or as directly stimulating adjuvants to the immune system (Jacquemet et al. 2009). Fragments of the cell membrane from the hyperthermophilic archaeon *Sulfolobus solfataricus* culminated in a comparable immune reaction to that seen with lipopolysaccharide and other Gram-negative and Gram-positive bacteria (Abrevaya 2012).

5.2 Extremophiles in Cancer Treatment

Thermophiles require high temperatures to survive, and they have many medical applications since their cell membranes have evolved to withstand heat. As a consequence, these extremophiles are in demand to treat forms of heat-inducing cancers.

Extremophiles such as halophiles respond to increases in osmotic pressure in a variety of ways. The extremely halophilic Halobacteriaceae accumulate K+ in the cytosol to counterbalance the concentration of salt in the atmosphere, but other organisms react by generating high concentrations of compatible solutes, such as glycine, betaine, sugars, polyols, and ectoins (Oren 2008). Such 'extremolytes' are a new area of manipulation, and ectoine and its hydroxyl derivative, hydroxyectoine, are of particular interest as protein stabilizers and membrane protectors from desiccation. Most extremolytes come from marine animals, particularly mannosylglycerate (firoin) and mannoslyglyceramide (firoin A). Such extremolytes are present abundantly in *Rhodothermus Marinus* thermophilic bacteria and are used against some cancers and associated diseases (Kumar and Singh 2012).

Many developments have culminated in the ability of compatible solutes to stabilize proteins and nucleic acids. Hydroxyectoin can stabilize *in vitro* immunotoxins and have been found to decrease the side effects of immunotoxins in an animal model when both were co-administered. Therefore, it may have the potential for cancer therapy as an excipient. The use of compatible solutes in cancer therapies that protect against vascular leak syndrome, a cytotoxic side effect of cancer treatment, was also patented by Bitop AG (Irwin and Baird 2004).

5.3 Extremophiles in Mental Health Treatment

Significant neurodegenerative diseases like Huntington's and Parkinson's are defined by protein folding disorders and the presence of structures within the brain neurons (Jorge et al. 2011). Thermophiles may have extremozymes with abilities such as inhibiting structural changes in native or mutant proteins that can be useful in developing neurodegenerative disease cures. Jorge et al. (2011) proposed that new organic solutes known as thermolytes could preserve the native structures of proteins. Thermolytes were discovered to influence the growth curves of HEK293 cells, which were cultivated to assist in Parkinson's disease treatment. This work's findings showed decreased cell density in the thermolyte-treated cells relative to untreated control cells. The research was conducted to test specific effects of thermolytes in Huntington's disease with Htt-103QEGFP transfected HEK293 cells and revealed a significant effect of Mg- and DGP on the emergence of Huntington aggregates (Ventura et al. 2011).

Neurodegenerative disorders such as Machado-Joseph disease are regarded as a central event in pathogenesis via the misfolding of mass proteins. Chaperone protein overexpression accepts misfolding as a frequent target for successful therapy. Furusho et al. (2005) investigated molecules that could theoretically affect protein folding from the extremophiles. Ectoine, initially discovered in halophiles, is an organic low-molecular-mass molecule and acts as an osmoprotective agent; it is ideal for maintaining enzymatic activity against freezing protein stabilization treatments. Ectoine was found to decrease large cytoplasmic inclusions and increase the rate of nuclear inclusions, while the nuclei's integrity continued to remain. Furthermore, ectoine has been shown to protect cells from polyglutamine (Furusho et al. 2005).

Ectoines are also useful in the diagnosis of Alzheimer's disease. They play a crucial role in preventing amyloid formation, a protein aggregation factor involved in misfolding proteins (Kumar and Singh 2012).

5.4 Extremophiles in Skin Disease Treatment

A specific category of microbial species is geophilic or "soil-loving" microorganisms; they live in soil and cannot be replicated in lab conditions. Geophilous is known for developing skin injury therapies. Factors participating in geophile pathways are acid proteinases, keratinases that help supply keratin, and antibiotic cells to minimize injury (Summerbell 1995). Arena et al. (2009) proposed that the *Geobacillus thermodenitificans* biofilm formation serves as an adjuvant in regulating the immune response in viral infections (Arena et al. 2009).

Today, dermatophytes are the primary source of fungal infections (Achterman and White 2012). Managing these fungal infections can be significant; geophiles can provide a permanent diagnosis and cure for these kinds of fungal infections. Knowing the pathogenicity of the fungi concerned is necessary in order to arrive at the appropriate form of treatment, as it is where geophiles are administered (Achterman and White 2012). Geophiles now have tremendous potential for medical work to be underway.

There are several current and future uses for compatible solutes produced by extremophiles. One use of these compounds is for cosmetics, and ectoine is used in products for skincare. It is used in animal skin treatments or protection against harm caused by ultraviolet light. Patents have been filed by the German firm bitop AG for the use of these and similar compounds as free-radical scavengers in medicine, cosmetics, dermatology, and nutrition (Schwarz 2003). The company bitop has developed Med Ectoin Syxyl, a dermatologic product for treating neurodermatitis and psoriasis. Products of the bitop AG were tested on animal models and optimized for human use. The RonaCareTM Ectoin, developed by Merck KgaA, Darmstadt, Germany, is also used as a moisturizer in skincare products (Roberts 2005).

5.5 Extremophiles in Diagnostic and Genome Sequencing based on PCR

Extremophiles are rich sources of biomolecules with diverse applications in biotechnology, nutrition, veterinary medicine, and human medicine. The best-studied and well-known are enzymes, particularly thermophile enzymes and, increasingly, psychrophiles, which are used in applications of all kinds in molecular biology. DNA polymerase from thermophiles, a cornerstone of PCR-based diagnostics for a wide variety of animal pathogens, is the most commonly recognized use of an extremophile drug in pharmaceutics. Thermophilic eubacterial microorganism *Thermus aquaticus* produces a thermostable DNA polymerase enzyme called Taq polymerase. These enzymes can withstand at 50–80°C, thus allowing the development of a polymerase chain reaction (PCR) in which double-stranded DNA is denatured at temperatures > 90°C and copied using DNA polymerase (Ishino and Ishino 2014). The creation of various forms of PCR and the availability of a variety of extremophile DNA polymerases has revolutionized human diagnostics and has allowed researchers to reveal comprehensive genomic information on many domestic and wild animal species. Real-time PCR has improved and simplified laboratory methods and has allowed researchers and clinicians to obtain more knowledge from laboratory-submitted specimens. Many significant animal viruses are detected using PCR (RT–PCR) reverse transcription technologies, including foot-and-mouth, swine fever, bluetongue, and the avian influenza virus and Newcastle disease virus (Hoffmann et al. 2009). PCR-based approaches are also used to identify associated animal viruses, equine parasites, new aquatic animal diagnostics, and parasite detection.

5.6 Extremophiles in Antimicrobial Drugs

Extremophiles are known to produce antimicrobial peptides and diketopiperazines. Antimicrobial peptides (halocins) were found to contain halobacteriaceae (phylogenetic family representing all halophilic archaea) and thermoacidophile (genus Sulfolobus). Growing halocine has a broad activity spectrum and some function in the wider variety of microorganisms than others. Halocins are successful in breaking down archaeal cells by targeting the cell membranes to create pores or inhibit specific enzymes. However, no evidence suggests that halocins destroy pathogenic microorganisms in humans. Ironically, evidence exists that they are helping dogs heal from surgery.

Diketopiperazines (also known as cyclic dipeptides) have been shown to affect blood coagulation functions and have antimicrobial, antifungal, antiviral, and antitumour properties. They are present in halophiles such as *Naloterrigena hispanica*, and *Natronococcus occultus*, and they have been shown to activate and inhibit pathways of quorum sensing (Martins and Carvalho 2007). Such mechanisms are essential in pathogens like *Pseudomonas aeruginosa*, one of the pneumonia-inducing agents, and a common infection seen in cystic fibrosis patients (Kalia 2013). It may also be a potential treatment for tens of thousands of drug-resistant infections of *Pseudomonas aeruginosa* that occur each year (Coker 2016).

Many psychrophiles and psychrotolerant microorganisms also develop compounds similar to microcin, with activity against Gram-negative and Gram-positive bacteria (Sánchez et al. 2009). In the case of hyperthermophiles, cytoplasmic disulphide bond machinery helps protein folding which can offer a clue to identify antiviral drug development and large-scale therapeutic production (Saaranen and Ruddock 2013).

5.7 Extremophiles in Blood Disorder Treatment

Many compounds present in halophilic archaea, such as siderophores, include iron-chelating agents that may be used to treat iron deficiency diseases or increase antibiotic activity against bacteria (Abrevaya 2012).

Such microorganisms have tremendous promise for the medical industry because they may have information about how cells deal with cold conditions and successfully defend and control the cell's membrane fluidity. Whenever it falls to psychrophiles, the cell membrane is by far the most significant element because it governs and regulates the cell's homeostasis.

5.8 Extremophiles in Digestive System and Ulcer Treatment

The industrial application of predecessor amylases was first used in 1984, as a medicinal aid to treat intestinal disorders. Amylases are amongst the most critical hydrolytic enzymes for all starch-based industries. Microbial amylases have entirely replaced chemical hydrolysis in starch production (John 2017).

For those susceptible to infection, acidophiles are often used in the mitigation of gastric cancers and stomach ulcers (Foster 2004). They were also proven helpful for iron cycling and performing *in situ* experiments, however. The acidophilus species are commonly classified as probiotic and intestinal residents. Some of those residents inside are more beneficial than others. Acidophilus improves the function of the digestive tract and eliminates the presence of harmful organisms. Because of that, probiotic use can help prevent infectious diarrhoea. *Lactobacillus acidophilus* was used to avoid certain infections with diarrhoea, such as traveler's diarrhoea, viral diarrhea, and antibiotic diarrhoea (Ouwehand et al. 2014). Crohn's disease and symptoms of ulcerative colitis are considered inflammatory bowel diseases. Chronic diarrhoea is a typical symptom of both conditions. The Helicobacter pylori microorganism is known to cause stomach and duodenum ulcers.

Studies were performed that used probiotics to inhibit *H. pylori* growth (Gotteland et al. 2008). A comprehensive study on microorganisms, *Lactobacillus* sp., *S. boulardii*, and other probiotics by Elmer (2001) showed a detailed understanding of the usage of acidophiles for different types of diarrhoea and identified them as beneficial in controlled Crohn's disease for mild diarrhoea (Elmer 2001). Some acidophiles have been researched regarding acid stability by studying their crystal structures. Acidophilus produces many substances that can be used in novel drug therapies for those vulnerable to gastric cancer infections and stomach ulcers (de Saro et al. 2013). These organisms inject hydrogen ions into the membrane to avoid fluctuations in the cellular membrane, which controls the cell's pH (Irwin and Baird 2004).

5.9 Extremophiles in Enzyme Biosensors

Studies are underway to develop robust blood-glucose sensors, based on fluorescent assay technology. Glucose oxidase from *Aspergillus niger* is unstable in glucose sensors, so the use of thermophile glucose oxidase has been proposed as a solution to the problem of long-term protein stability and to allow the design of an implantable glucose sensor which does not deplete flavin adenine dinucleotide over time. Thermostable glucokinase from the thermophile *Bacillus stearothermophilus* has been studied as a component of a glucose sensor (Staiano et al. 2017).

5.10 Extremophiles in Other Potential Pharmaceutical Applications

The emerging environmental problem arising from higher temperatures is water scarcity because the heat evaporates water in reservoirs. This problem has become a growing challenge for the world's water-deprived communities. Conditions that enable mesophilic species to live at high temperatures and biologically deficient water need to be better studied to produce outcomes that could be useful in making animals and humans less reliant on water.

Proteases are one significant example of products that allow mesophiles to live. Proteases are proteins that are used every day in therapeutic fields (Fornbacke and Clarsund 2013), and that can quickly respond to shifts in cold temperatures (Singh et al. 2009). Even they may be used in conjunction with other extremophiles, such as psychrophiles, geophiles, etc.

Barophiles are essential in medicine since they can flourish at high pressure and are not impaired by repeated pressure level shifts. Barophilic cells are most susceptible to pressure changes after-effects, as stable pressure is necessary for biological cells. Changes in pressure may have adverse effects resulting in cellular consequences. Barophilic microorganisms utilize mechanotransduction, a system that detects the pressure changes and converts them into a signal that the cell can use effectively. Undesirable stress may cause this to happen due to mediating stresses from various environments (Tan et al. 2006). Pressure in the treatment of diseases has been established for some time; pressure affects all living organisms, and the correct pressure range is essential for cell stability. Barophiles live under high-pressure ranges and can respond to pressure changes as well as biochemical process changes (Kato and Bartlett 1997). Because of unique characteristics such as an abundance of pressure sensing mechanisms on their cell membranes, osmotolerance, and pressure-controlled operations, barophiles may play an essential role in therapy (Marquis and Keller 1975). In order to understand how eye cells respond to pressure-induced glaucoma, Tan et al. (2006) investigated how human and prokaryotic cells respond to mechanic forces (Tan et al. 2006). The capacity of barophiles to survive under intense pressure can guide the development of therapies for pressure-induced injuries such as concussions and associated athletic injury.

Cold-adapted proteases are used in broad ranges, including molecular biology, cosmetics, and pharmaceutical products (Craik et al. 2011). Cold-adapted proteases are especially valuable in low water conditions and high structural rigidity (Karan et al. 2012). Psychrophiles are possible sources of enzymes appropriate for the production of polyunsaturated fatty acids for the pharmaceutical industry, while halophiles are investigated as sources of pharmaceutical surfactants (Rothschild and Mancinelli 2001). Ectoine has respiratory medicine applications. It has been shown that ectoine in a rat model can prevent inflammation of the lung induced by nanoparticles and inhibit signal transduction and IL-8 induction in a human bronchial epithelial cell line (Sydlik et al. 2009). Bitop AG and another company, Activaero, have marketed an aerosolized inhalable Ectoin ® solution for the treatment of asthma.

6. Extremophiles and Bioremediation

The process of employing living organisms, primarily microorganisms, to degrade and reduce the concentration of hazardous materials in contaminated environmental sites is known as bioremediation (Boopathy 2000). Bioremediation has gained much public acceptance compared to other traditional methods such as incineration and land filling (Vidali 2001). In this process, microorganisms often transform the harmful contaminants into compounds that are less toxic to the environment. Generally, these microbes are naturally occurring in the contaminated sites or are imported from other locations. Several microbes have the ability to utilize these contaminants as their substrate. Microbes are bestowed with several active and passive transport system, enzymatic process and metabolic pathways that help them to utilize these contaminants (Bera et al. 2016). The ability of the microbes for bioremediation depends on its physiology. Other environmental factors including temperature, pH, and oxygen demand also directly affect their efficacy. Compared to the physical and chemical remediation processes, bioremediation is less expensive, environmental friendly, renewable, generates less secondary pollutants and chemical byproducts (Akar et al. 2013). However, one of the major limitations of bioremediation is the inability to eliminate some specific hazardous elements, for example heavy metals and radionuclides (Boopathy et al. 2000). Another drawback is that the normal microbes fail to function at extreme conditions such as high or low temperature, high salinity, and extreme acidic or alkaline conditions. Every microorganism requires an optimum culture condition for their growth and metabolic activity. Unfavourable and extreme environmental conditions abolish their growth and activity. Extremophiles are a viable solution to this problem because of their ability to thrive under extreme environmental conditions. Extremophiles are attractive biocatalysts for bioremediation of heavy metals and other environmental wastes. Owing to their molecular and physiological properties, extremophiles, including thermophiles, psychrophiles, acidophiles, alkaliphiles, xerotolerant, and radiotolerant, have immense potential in waste treatment and remediation.

Thermophiles and hyperthermophiles have the ability to grow at temperatures higher than 44°C and 80°C, respectively. The adaptation of thermophilic enzymes to high temperature might be helpful in waste remediation at high temperature conditions. Antarctica is known as the coldest continent on Earth but it harbours several geothermal sites that inhabit thermophiles. *Pyrococcus* sp. M24D13, a hyperthermophilic archaeon, was isolated from such geothermal sites and its nitrilase enzyme was purified. The enzyme showed remarkable activity and stability at 85°C. It also exhibited cyanidase activity in presence of KCN as sole substrate (Dennett and Blamey 2016). This implied nitrilase enzyme from *Pyrococcus* sp. M24D13 can potentially remediate cyanide contaminated environments. *Geobacillus thermoleovorans* T80, a thermophile, was capable of degrading 90% of hexadecane at 60°C (Perfumo et al. 2007). *Mastigocladus* sp. strain CHP1, a thermophilic cyanobacterium, isolated from Porcelana hot spring (Chile) was capable of fixing nitrogen at 60°C and displayed high affinity for nitrate (Alcamán et al. 2017). Thus, this strain might have potential in treating nitrate-contaminated waste stream at high temperatures. Thermophilic bacterium, *Nocardia otitidiscaviarum* TSH1, degraded different hydrocarbons, such as polycyclic aromatic hydrocarbons (phenanthrene and pyrene), *n*-alkanes and phenol at 50°C. An obligate thermophilic methanogen, *Methanothermobacter thermautotrophicus*, reduced hexavalent chromium ion to less toxic trivalent chromium ion, demonstrating its applicability in metal bioremediation (Singh et al. 2015). Acidophiles have bioremediation ability in acidic environments. Thermoacidophilic archaeon *Sulfolobus solfataricus Rhodothermus Marinus* 98/2 efficiently degraded phenol (Christen et al. 2011). Similarly, moderately thermoacidophilic *Sulfobacillus acidophilus* TPY was capable of degrading phenol (Zhou et al. 2016). Acidophilic *Stenotrophomonas maltophilia* strain AJH1was able to degrade low and high molecular weight polycyclic aromatic hydrocarbons such as anthracene, phenanthrene, naphthalene, fluorene, pyrene, benzo(e)pyrene and benzo(k)fluoranthene (Arulazhagan et al. 2017). Additionally, the strain potentially treated petroleum wastewater under acidic conditions in a lab scale continuous stirred tank reactor. Two acidophilic algal species, *Cyanidium caldarium* and *Euglena mutabilis,* were isolated from Agrio River-Lake Caviahue system, Argentina, and served as a potential bioindicator of polycyclic aromatic hydrocarbon contamination in soil (Diaz et al. 2015). Thermophilic acidophilus *Sulfobacillus thermosulfdooxidans* acted as a bioabsorbent for removal of heavy metal ions such as (Cd^{2+}, Cu^{2+}, Zn^{2+} and Ni^{2+}) (Huang et al. 2020). The heavy metal removal by the strain proceeded through monolayer adsorption process and the carboxyl, phosphoryl and amino functional groups played a key role. Acidophilic bacterium *Thiobacillus ferrooxidans* removed 80% of cadmium, 78% of lead, 87% of zinc and 69.9% of copper from urban sewage sludge (Azhdarpoor et al. 2019). Psychrotolerant *Pseudomonas putida* SB32 and alkalophilic *Pseudomonas monteilli* SB35, isolated from soil of Semera Mines, India, was found to be cadmium-resistant. When these strains were used as bioinoculant in cadmium polluted soil, it reduced the cadmium accumulation in the root and shoot of the plant grown on the soil (Jain and Bhatt 2014). Moreover, the strains also reduced the cadmium accumulation in the soil. Thus, these strains had high bioremediation potential in cadmium contaminated agricultural fields. Halophiles have the ability to tolerate high levels of heavy metals and hydrocarbons, and are essential candidates for bioremediation in saline environment. [Co(III)–EDTA]⁻ is generated during nuclear waste management process. Due to its high solubility and stability, it is difficult to remove from the radioactive wastes (Paraneeiswaran et al. 2016). Halophilic solar-salt-pan isolate *Pseudomonas aeruginosa* SPB-1 was found to be Co(III)–EDTA⁻ resistant and showed 80.4% removal by adsorption process (Paraneeiswaran et al. 2014). This was a promising study which showed the use of halophile for bioremediation of nuclear wastes. Halophilic bacterium *Halobacillus* EG1HP4QL is an exopolysaccharide (EPS) producing microbe (Ibrahim et al. 2020). The hydroxyl, amino, and carbonyl groups present in EPS complexed with heavy metals such as copper, nickel, lead, cadmium, and zinc and conferred resistance against them. It could also utilize crude oil such as paraffins, naphthenes, mono- and bicyclic aromatic hydrocarbons, polycyclic aromatic hydrocarbons, and alcohol–benzene resins as sole carbon source and achieved removal of 34.8%, 49.6%, 51.2%, 43.5%, and 25.5%, respectively (Ibrahim et al. 2020). Thus, *Halobacillus* EG1HP4QL was a potential strain for treating oil-polluted environment. An alkaliphilic halotolerant bacterium, *Pseudochrobactrum saccharolyticum* LY10, isolated from the chromium contaminated soil, accumulated chromium both intracellularly and extracellularly (Long et al. 2013). The strain was characterized by its Cr(VI)-reducing ability. A gram positive halophilic bacterium *Salinicoccus iranensis* showed high resistance to tellurite. The cell extract of this halophile reduced tellurite which indicated the presence of tellurite reducing enzyme (Alavi et al. 2014). The strain exhibited tellurite removing ability and thus, had promising applications in bioremediating sites contaminated with oxyanions. In another study, alkaliphilic halotolerant bacterium *Nesterenkonia lacusekhoensis* EMLA3 efficiently decolourized methyl red dye, an azo dye present in textile effluent (Bhattacharya et al. 2017). Thus, bioremediation using extremophiles is an attractive approach to treat environmental pollutants. Their remarkable capacity in bioremediation of heavy metals, poly aromatic hydrocarbons, phenolic compounds, nuclear wastes, and synthetic dyes presents a bright future in sustainable waste management.

7. Conclusion and Future Outlook

In the past decade, many countries have increased their interest in the development of sustainable biorefinery system by putting efforts into production of value-added products and alternative energy sources to meet the United Nation Global Sustainable Goals. Reduction of fossil fuel reserves has been another drive for the search of alternative biofuel source, yet most of these researches are still in laboratory and pilot-plant scale. Most of the bio refineries face major hurdles in the form of cost of production, maintenance, research cost, and the lack of commercial competitiveness. While low cost lignocellulosic biomass solves the problem of substrate cost, pre-treatment of lignocellulosic biomass incurs about 40% of the total biomass processing cost. Lack of proper delignification technology is one of the major bottleneck in the green revolution for value added products and precursors for biofuel production by extremophiles. Extremophiles and their enzymes show a glimpse of hope for the biorefinery industry due to their specificity, robustness, high tolerance and ability to operate in extreme conditions. This helps to produce value added product as well as biofuels in an effective way with more competitive chassis cells and catalysts. The cost of extremozymes is still higher compared to that of conventional enzymes. Additionally, special production vessels are also required for the growth of the extremophiles due to the operation of harsh environments. This adds to the overall cost of operation, and maintenance of the biorefinery process. Metabolic engineering for screening and modification of extremophiles to increase yield and tolerance of the microbes can improve process profitability. Recent developments in artificial intelligence and nanotechnology can also bring different innovation possibilities to improve the competitiveness of biorefinery plants for bioactive compounds' production like enzymes, pharmaceutical products, polymers, biofuel, etc.

Acknowledgment

Authors express their thanks to Lovely Professional University (Punjab, India) for necessary support during the work.

References

Ábrego, U., Z. Chen and C. Wan. 2017. Consolidated bioprocessing systems for cellulosic biofuel production. pp. 143–182. *In*: Advances in Bioenergy, Vol. 2, Elsevier.

Abrevaya, X.C. 2012. Features and Applications of Halophilic Archaea. John Wiley & Sons, New Jersey.

Achterman, R.R. and T.C. White. 2012. Dermatophyte virulence factors: Identifying and analyzing genes that may contribute to chronic or acute skin infections. Int. J. Microbiol. 358305.

Akar, T., S. Arslan and S.T. Akar. 2013. Utilization of *Thamnidium elegans* fungal culture in environmental cleanup: A reactive dye biosorption study. Ecol. Eng. 58: 363–370.

Alavi, S., M.A. Amoozegar and K. Khajeh. 2014. Enzyme (s) responsible for tellurite reducing activity in a moderately halophilic bacterium, *Salinicoccus iranensis* strain QW6. Extremophiles 18(6): 953–961.

Alcamán, M.E., J. Alcorta, B. Bergman, M. Vásquez, M. Polz and B. Díez. 2017. Physiological and gene expression responses to nitrogen regimes and temperatures in *Mastigocladus* sp. strain CHP1, a predominant thermotolerant cyanobacterium of hot springs. Syst. Appl. Microbiol. 40(2): 102–113.

Amiri, H., R. Azarbaijani, L.P. Yeganeh, A.S. Fazeli, M. Tabatabaei, G.H. Salekdeh and K. Karimi. 2016. *Nesterenkonia* sp. strain F, a halophilic bacterium producing acetone, butanol and ethanol under aerobic conditions. Sci. Rep. 6(1): 1–10.

Arena, A., C. Gugliandolo, G. Stassi, B. Pavone, D. Iannello, G. Bisignano and T.L. Maugeri. 2009. An exopolysaccharide produced by *Geobacillus thermodenitrificans* strain B3-72: Antiviral activity on immunocompetent cells. Immunol. Lett. 123: 132–137.

Arulazhagan, P., K. Al-Shekri, Q. Huda, J.J. Godon, J.M. Basahi and D. Jeyakumar. 2017. Biodegradation of polycyclic aromatic hydrocarbons by an acidophilic *Stenotrophomonas maltophilia* strain AJH1 isolated from a mineral mining site in Saudi Arabia. Extremophiles 21(1): 163–174.

Atalah, J., P. Cáceres-Moreno, G. Espina and J.M. Blamey. 2019. Thermophiles and the applications of their enzymes as new biocatalysts. Bioresour. Technol. 280: 478–488.

Azhdarpoor, A., R. Hoseini and M. Dehghani. 2019. Removal of heavy metals from urban sewage sludge using acidophilic *Thiobacillus ferrooxidans*. Journal of Health 10(2): 169–178.

Baker, S.E., R.C. Hopkins, C.D. Blanchette, V.L. Walsworth, R. Sumbad, N.O. Fischer, E.A. Kuhn, M. Coleman, B.A. Chromy and S.E. Létant. 2009. Hydrogen production by a hyperthermophilic membrane-bound hydrogenase in water-soluble nanolipoprotein particles. J. Am. Chem. Soc. 131(22): 7508–7509.

Barnard, D., A. Casanueva, M. Tuffin and D. Cowan. 2010. Extremophiles in biofuel synthesis. Environ. Tech. 31(8-9): 871–888.

Baruah, J., B.K. Nath, R. Sharma, S. Kumar, R.C. Deka, D.C. Baruah and E. Kalita. 2018. Recent trends in the pretreatment of lignocellulosic biomass for value-added products. Front. Energy Res. 6. 141.

Basit, A., J. Liu, K. Rahim, W. Jiang and H. Lou. 2018. Thermophilic xylanases: From bench to bottle. Crit. Rev. Biotechnol. 38: 989–1002.

Bera, S., V.P. Sharma, S. Dutta and D. Dutta. 2016. Biological decolorization and detoxification of malachite green from aqueous solution by *Dietzia maris* NIT-D. J. Taiwan Inst. Chem. Eng. 67: 271–284.

Bhattacharya, A., N. Goyal and A. Gupta. 2017. Degradation of azo dye methyl red by alkaliphilic, halotolerant *Nesterenkonia lacusekhoensis* EMLA3: Application in alkaline and salt-rich dyeing effluent treatment. Extremophiles 21(3): 479–490.

Biswas, J. and A.K. Paul. 2017. Diversity and production of extracellular polysaccharide by halophilic microorganisms. Biodivers. Int. J. 1: 1–9.

Boopathy, R. 2000. Factors limiting bioremediation technologies. Bioresour. Technol. 74(1): 63–67.

Cavicchioli, R., T. Charlton, H. Ertan, S.M. Omar, K.S. Siddiqui and T.J. Williams. 2011. Biotechnological uses of enzymes from psychrophiles. Microb. Biotechnol. 4: 449–460.

Chang, S., Y. Guo, B. Wu and B. He. 2017. Extracellular expression of alkali tolerant xylanase from *Bacillus subtilis* Lucky9 in *E. coli* and application for xylooligosaccharides production from agro-industrial waste. Int. J. Biol. Macromol. 96: 249–256.

Chen, G.-Q. and X.R. Jiang. 2018. Next generation industrial biotechnology based on extremophilic bacteria. Curr. Opin. Biotechnol. 50: 94–100.

Chen, J., R. Mitra, S. Zhang, Z. Zuo, L. Lin, D. Zhao, H. Xiang and J. Han. 2019. Unusual phosphoenolpyruvate (PEP) synthetase-like protein crucial to enhancement of polyhydroxyalkanoate accumulation in *Haloferax mediterranei* revealed by dissection of PEP-pyruvate interconversion mechanism. Appl. Environ. Microbiol. 85(19): e00984–19.

Chen, J., R. Mitra, H. Xiang and J. Han. 2020. Deletion of the pps-like gene activates the cryptic phaC genes in *Haloferax mediterranei*. Appl. Environ. Microbiol. 104(22): 9759–9771.

Christen, P., S. Davidson, Y. Combet-Blanc and R. Auria. 2011. Phenol biodegradation by the thermoacidophilic archaeon *Sulfolobus solfataricus* 98/2 in a fed-batch bioreactor. *Biodegradation* 22(3): 475–484.

Coker, J.A. 2016. Extremophiles and biotechnology: Current uses and prospects. F1000Res 5.

Craik, C.S., M.J. Page and E.L. Madison. 2011. Proteases as therapeutics. Biochem. J. 435: 1–16.

Danis, O., A. Ogan, P. Tatlican, A. Attar, E. Cakmakci, B. Mertoglu and Birbir, M.. 2015. Preparation of poly (3-hydroxybutyrate-co-hydroxyvalerate) films from halophilic archaea and their potential use in drug delivery. Extremophiles 19: 515–524.

Das, S., A. Bhattacharya, S. Haldar, A. Ganguly, S. Gu, Y. Ting and P. Chatterjee. 2015. Optimization of enzymatic saccharification of water hyacinth biomass for bio-ethanol: Comparison between artificial neural network and response surface methodology. SM & T. 3: 17–28.

DasSarma, S. and P. DasSarma. 2015. Halophiles and their enzymes: Negativity put to good use. Curr. Opin. Microbiol. 25: 120–126.

de Saro, F.J.L., M.J. Gómez, E. González-Tortuero and V. Parro. 2013. The dynamic genomes of acidophiles. Joseph Seckbach Aharon Oren Helga Stan-Lotter. Polyextremophiles.Springer, Dordrecht, pp. 81–97.

Delgado-García, M., B. Valdivia-Urdiales, C.N. Aguilar-González, J.C. Contreras-Esquivel and R. Rodríguez-Herrera. 2012. Halophilic hydrolases as a new tool for the biotechnological industries. J. Sci. Food Agric. 92: 2575–2580.

Demain, A.L., M. Newcomb and J.D. Wu. 2005. Cellulase, clostridia, and ethanol. Microbiol. Mol. Biol. R. 69(1): 124–154.

Demirjian, D.C., F. Morís-Varas and C.S. Cassidy. 2001. Enzymes from extremophiles. Curr. Opin. Chem. Biol. 5: 144–151.

Dennett, G.V. and J.M. Blamey. 2016. A new thermophilic nitrilase from an Antarctic hyperthermophilic microorganism. Front Bioeng. Biotechnol. 4: 5.

Desai, S., M. Guerinot and L. Lynd. 2004. Cloning of L-lactate dehydrogenase and elimination of lactic acid production via gene knockout in *Thermoanaerobacterium saccharolyticum* JW/SL-YS485. App. Microbiol. Biotechnol. 65(5): 600–605.

Diaz, M., V. Mora, F. Pedrozo, D. Nichela and G. Baffico. 2015. Evaluation of native acidophilic algae species as potential indicators of polycyclic aromatic hydrocarbon (PAH) soil contamination. J. Appl. Phycol. 27(1): 321–325.

Dufey, A. 2006. Biofuels Production, Trade and Sustainable Development: Emerging Issues. Iied.

Elmer, G.W. 2001. Probiotics: "living drugs". Am. J. Health-Syst. Ph. 58: 1101–1109.

Escuder-Rodríguez, J.-J., M.-E. DeCastro, M.-E. Cerdán, E. Rodríguez-Belmonte, M. Becerra and M.-I. González-Siso. 2018. Cellulases from thermophiles found by metagenomics. Microorganisms 6: 1–26.

Feller, G., E. Narinx, J.L. Arpigny, M. Aittaleb, E. Baise, S. Genicot and Gerday, C. 1996. Enzymes from psychrophilic organisms. FEMS Microbiol. Rev. 18: 189–202.

Fornbacke, M. and M. Clarsund. 2013. Cold-adapted proteases as an emerging class of therapeutics. Infec. Dis. T. 2: 15–26.

Foster, J.W. 2004. *Escherichia coli* acid resistance: Tales of an amateur acidophile. Nat. Rev. Microbiol. 2: 898–907.

Fujinami, S. and M. Fujisawa. 2010. Industrial applications of alkaliphiles and their enzymes—past, present and future. Environ. Technol. 31: 845–856.

Furusho, K., T. Yoshizawa and S. Shoji. 2005. Ectoine alters subcellular localization of inclusions and reduces apoptotic cell death induced by the truncated Machado–Joseph disease gene product with an expanded polyglutamine stretch. Neurobiol. Dis. 20: 170–178.

Gavala, H.N., I.V. Skiadas, B.K. Ahring and G. Lyberatos. 2006. Thermophilic anaerobic fermentation of olive pulp for hydrogen and methane production: Modelling of the anaerobic digestion process. Water Sci. Technol. 53(8): 271–279.

Giedraitytė, G. and L. Kalėdienė. 2015. Purification and characterization of polyhydroxybutyrate produced from thermophilic *Geobacillus* sp. AY 946034 strain. Chemija. 26: 38–45.

Gotteland, M., M. Andrews, M. Toledo, L. Muñoz, P. Caceres, A. Anziani, E. Wittig, H. Speisky and G. Salazar. 2008. Modulation of Helicobacter pylori colonization with cranberry juice and *Lactobacillus johnsonii* La1 in children. Nutrition 24: 421–426.

Hai, T., F.B. Oppermann-Sanio and A. Steinbüchel. 2002. Molecular characterization of a thermostable cyanophycin synthetase from the thermophilic cyanobacterium *Synechococcus* sp. strain MA19 and *in vitro* synthesis of cyanophycin and related polyamides. Appl. Environ. Microbiol. 68: 93–101.

Hoffmann, B., M. Beer, S.M. Reid, P. Mertens, C.A.L. Oura, P.A. Van Rijn, M.J. Slomka, J. Banks, I.H. Brown and D.J. Alexander. 2009. A review of RT-PCR technologies used in veterinary virology and disease control: Sensitive and specific diagnosis of five livestock diseases notifiable to the World Organization for Animal Health. Vet. Microbiol. 139: 1–23.

Horikoshi, K. 1996. Alkaliphiles—from an industrial point of view. FEMS Microbiol. Rev. 18: 259–270.

Horikoshi, K. 1998. Barophiles: Deep-sea microorganisms adapted to an extreme environment. Curr. Opin. Microbiol. 1: 291–295.

Horikoshi, K. 1999. Alkaliphiles: Some applications of their products for biotechnology. Microbiol. Mol. Biol. Rev. 63: 735–750.

Houde, A., A. Kademi and D. Leblanc. 2004. Lipases and their industrial applications: An overview. Appl. Biochem. Biotechnol. 118: 155–170.

Hu, Q., M. Sommerfeld, E. Jarvis, M. Ghirardi, M. Posewitz, M. Seibert and A. Darzins. 2008. Microalgal triacylglycerols as feedstocks for biofuel production: Perspectives and advances. The Plant J. 54(4): 621–639.

Huang, Y., M. Li, Y. Yang, Q. Zeng, P. Loganathan, L. Hu, H. Zhong and Z. He. 2020. *Sulfobacillus thermosulfidooxidans*: An acidophile isolated from acid hot spring for the biosorption of heavy metal ions. Int. J. Environ. Sci. Technol. 17: 2655–2665.

Ibrahim, I.M., S.A. Konnova, E.N. Sigida, E.V. Lyubun, A.Y. Muratova, Y.P. Fedonenko and K. Elbanna. 2020. Bioremediation potential of a halophilic *Halobacillus* sp. strain, EG1HP4QL: Exopolysaccharide production, crude oil degradation, and heavy metal tolerance. Extremophiles 24(1): 157–166.

Ichiye, T. 2018. Enzymes from piezophiles. Semin. Cell Dev. Biol. 84: 138–146.

Irwin, J.A. and A.W. Baird. 2004. Extremophiles and their application to veterinary medicine. Ir. Vet. J. 57: 348.

Ishino, S. and Y. Ishino. 2014. DNA polymerases as useful reagents for biotechnology—the history of developmental research in the field. Front. Microbiol. 5: 465–465.

Jacquemet, A., J. Barbeau, L. Lemiègre and T. Benvegnu. 2009. Archaeal tetraether bipolar lipids: Structures, functions and applications. Biochimie. 91: 711–717.

Jain, S. and A. Bhatt. 2014. Molecular and in situ characterization of cadmium-resistant diversified extremophilic strains of *Pseudomonas* for their bioremediation potential. 3 Biotech. 4(3): 297–304.

Jiao, Y., G.D. Cody, A.K. Harding, P. Wilmes, M. Schrenk, K.E. Wheeler, K.E., Banfield, J.F. and Thelen, M.P. 2010. Characterization of extracellular polymeric substances from acidophilic microbial biofilms. Appl. Environ. Microbiol. 76: 2916–2922.

John, J. 2017. Amylases-bioprocess and potential applications: A review. Int. J. Bioinf. Biol. Sci. 5: 41–50.

Jorge, C.D., R. Ventura, C. Maycock, T.F. Outeiro, H. Santos and J. Costa. 2011. Assessment of the efficacy of solutes from extremophiles on protein aggregation in cell models of Huntington's and Parkinson's diseases. Neurochem. Res. 36: 1005–1011.

Joshi, A.A. and P.P. Kanekar. 2011. Production of exopolysaccharide by *Vagococcus carniphilus* MCM B-1018 isolated from alkaline Lonar Lake, India. Ann. Microbiol. 61: 733–740.

Kalia, V.C. 2013. Quorum sensing inhibitors: An overview. Biotechnol. Adv. 31: 224–245.

Kambourova, M., N. Radchenkova, I. Tomova and I. Bojadjieva. 2016. Thermophiles as a promising source of exopolysaccharides with interesting properties. pp. 117–139. *In:* Rampelotto, R. (ed.). Biotechnology of Extremophiles. Springer, Cham, Switzerland.

Kamekura, M. 1986. Production and function of enzymes of eubacterial halophiles. FEMS Microbiol. Lett. 39: 145–150.

Karan, R., M.D. Capes and S. DasSarma. 2012. Function and biotechnology of extremophilic enzymes in low water activity. Aquat. Biosyst. 8: 4.

Kashefi, K. and D.R. Lovley. 2003. Extending the upper temperature limit for life. Science 301: 934.

Kashyap, D.R., P.K. Vohra, S. Chopra and R. Tewari. 2001. Applications of pectinases in the commercial sector: A review. Bioresour. Technol. 77: 215–227.

Kästner, M. and B. Mahro. 1996. Microbial degradation of polycyclic aromatic hydrocarbons in soils affected by the organic matrix of compost. Appl. Microbiol. Biotechnol. 44(5): 668–675.

Kato, C. and D.H. Bartlett. 1997. The molecular biology of barophilic bacteria. Extremophiles 1: 111–116.

Kazak, H., E.T. Oner and R.F.H. Dekker. 2010. Extremophiles as sources of exopolysaccharides. pp. 605–619. *In:* R. Ito and Y. Matsuo (eds.). Handbook on Carbohydrate Polymers: Development, Properties and Applications. Nova Science Publishers, Inc. NY (USA).

Kristjansson, J.K. 1989. Thermophilic organisms as sources of thermostable enzymes. Trends Biotechnol. 7: 349–353.

Kumar, R., S. Singh and O.V. Singh. 2008. Bioconversion of lignocellulosic biomass: Biochemical and molecular perspectives. Journal of Industrial Microbiology & Biotechnology 35(5): 377–391.

Kumar, R. and A. Singh. 2012. Smart therapeutics from extremophiles: Unexplored applications and technological challenges. Extremophiles: Sustainable Resources and Biotechnological Implications, pp. 389–401.

Lamed, R. and J. Zeikus. 1980. Glucose fermentation pathway of *Thermoanaerobium brockii*. Journal of Bacteriology 141(3): 1251–1257.

Li, D.-C. and A.C. Papageorgiou. 2019. Cellulases from thermophilic fungi: Recent insights and biotechnological potential. pp. 395–417. *In:* Tiquia-Arashiro, S.M. and M. Grube (eds.). Fungi in Extreme Environments: Ecological Role and Biotechnological Significance. Springer International Publishing, Switzerland.

Lin, C.-Y., C.-C. Wu and C.H. Hung. 2008. Temperature effects on fermentative hydrogen production from xylose using mixed anaerobic cultures. Int. J. Hydrogen. Energ. 33(1): 43–50.

Liu, Y., P. Yu, X. Song and Y. Qu. 2008. Hydrogen production from cellulose by co-culture of *Clostridium thermocellum* JN4 and *Thermoanaerobacterium thermosaccharolyticum* GD17. Int. J. Hydrogen. Energ. 33(12): 2927–2933.

Long, D., X. Tang, K. Cai, G. Chen, L. Chen, D. Duan, J. Zhu and Y. Chen. 2013. Cr (VI) reduction by a potent novel alkaliphilic halotolerant strain *Pseudochrobactrum saccharolyticum* LY10. J. Hazard Mater. 256: 24–32.

Luo, G., L. Xie, Z. Zou, W. Wang and Q. Zhou. 2010. Evaluation of pretreatment methods on mixed inoculum for both batch and continuous thermophilic biohydrogen production from cassava stillage. Bioresour. Technol. 101(3): 959–964.

Luque, R., L. Herrero-Davila, J.M. Campelo, J.H. Clark, J.M. Hidalgo, D. Luna, J.M. Marinas and A.A. Romero. 2008. Biofuels: A technological perspective. Energy & Environmental Science 1(5): 542–564.

Madern, D., C. Ebel and G. Zaccai. 2000. Halophilic adaptation of enzymes. Extremophiles 4: 91–98.

Madigan, M.T. and A. Oren. 1999. Thermophilic and halophilic extremophiles. Curr. Opin. Microbiol. 2: 265–269.

Margesin, R. and F. Schinner. 2001. Potential of halotolerant and halophilic microorganisms for biotechnology. Extremophiles 5: 73–83.

Marques, C.R. 2018. Extremophilic microfactories: Applications in metal and radionuclide bioremediation. Front. Microbiol. 9: 1–10.

Marquis, R.E. and D.M. Keller. 1975. Enzymatic adaptation by bacteria under pressure. J. Bacteriol. 122: 575–584.

Martins, M.B. and I. Carvalho. 2007. Diketopiperazines: Biological activity and synthesis. Tetrahedron. 63: 9923–9932.

McDonald, J.H. 2001. Patterns of temperature adaptation in proteins from the bacteria *Deinococcus radiodurans* and *Thermus thermophilus*. Mol. Biol. Evol. 18: 741–749.

Metzger, P. and C. Largeau. 2005. *Botryococcus braunii*: A rich source for hydrocarbons and related ether lipids. Appl. Microbiol. Biotechnol. 66(5): 486–496.

Mitra, R., T. Xu, H. Xiang and J. Han. 2020. Current developments on polyhydroxyalkanoates synthesis by using halophiles as a promising cell factory. Microb. Cell Fact. 19: 1–30.

Moayad, W., G. Zha and Y. Yan. 2018. Metalophilic lipase from *Ralstonia solanacearum*: Gene cloning, expression, and biochemical characterization. Biocatal. Agric. Biotechnol. 13: 31–37.

Mohammadipanah, F., J. Hamedi and M. Dehhaghi. 2015. Halophilic bacteria: Potentials and applications in biotechnology. pp. 277–321. *In:* Maheshwari, D.K. and S. Meenu (eds.). Halophiles: Biodiversity and Sustainable Exploitation. Springer International Publishing, Switzerland.

Mukherjee, S., B. Basak, B. Bhunia, A. Dey and B. Mondal. 2013. Potential use of polyphenol oxidases (PPO) in the bioremediation of phenolic contaminants containing industrial wastewater. Rev. Environ. Sci. Bio/Technol. 12: 61–73.

Nakagawa, T., T. Nagaoka, S. Taniguchi, T. Miyaji and N. Tomizuka. 2004. Isolation and characterization of psychrophilic yeasts producing cold-adapted pectinolytic enzymes. Lett. Appl. Microbiol. 38: 383–387.

Nicolaus, B., V. Schiano Moriello, L. Lama, A. Poli and A. Gambacorta. 2004. Polysaccharides from extremophilic microorganisms. Orig. Life Evol. Biosph. 34: 159–169.

Nicolaus, Barbara, M. Kambourova and E.T. Oner. 2010. Exopolysaccharides from extremophiles: From fundamentals to biotechnology. Environ. Technol. 31: 1145–1158.

Oren, A. 2008. Microbial life at high salt concentrations: Phylogenetic and metabolic diversity. Sal. Syst. 4: 2.

Ouwehand, A.C., S.J.M. Ten Bruggencate, A.J. Schonewille, E. Alhoniemi, S.D. Forssten and I.M.J. Bovee-Oudenhoven. 2014. *Lactobacillus acidophilus* supplementation in human subjects and their resistance to enterotoxigenic *Escherichia coli* infection. Br. J. Nutr. 111: 465–473.

Ozdemir, S., S.A. Fincan, A. Karakaya and B. Enez. 2018. A novel raw starch hydrolyzing thermostable α-amylase produced by newly isolated *Bacillus mojavensis* SO-10: Purification, characterization and usage in starch industries. Braz. Arch. Biol. Technol. 61.

Pan, S., T. Yao, L. Du and Y. Wei. 2020. Site-saturation mutagenesis at amino acid 329 of *Klebsiella pneumoniae* halophilic α-amylase affects enzymatic properties. J. Biosci Bioeng. 129(2): 155–159.

Paraneeiswaran, A., S.K. Shukla, T.S. Rao and K. Prashanth. 2014. Removal of toxic Co-EDTA complex by a halophilic solar-salt-pan isolate *Pseudomonas aeruginosa* SPB-1. Chemosphere 95: 503–510.

Paraneeiswaran, A., S.K. Shukla, R. Kumar and T.S. Rao. 2016. Reduction of [Co (iii)–EDTA]—complex by a novel process using phototrophic granules: A step towards sustainable bioremediation. RSC Advances 6(49): 43656–43662.

Patel, S. and A. Goyal. 2011. Functional oligosaccharides: Production, properties and applications. World J. Microbiol. Biotechnol. 27: 1119–1128.

Pavlova, K., S. Rusinova-Videva, M. Kuncheva, M. Kratchanova, M. Gocheva and S. Dimitrova. 2011. Synthesis and characterization of an exopolysaccharide by Antarctic yeast strain *Cryptococcus laurentii* AL100. Appl. Biochem. Biotechnol. 163: 1038–1052.

Perfumo, A., I.M. Banat, R. Marchant and L. Vezzulli. 2007. Thermally enhanced approaches for bioremediation of hydrocarbon-contaminated soils. Chemosphere 66(1): 179–184.

Pulicherla, K.K., M. Ghosh, P.S. Kumar and K.R.S.S. Rao. 2011. Psychrozymes—the next generation industrial enzymes. J. Mar. Sci. Res. Dev. 1: 1–7.

Robb, F., G. Antranikian, D. Grogan and A. Driessen. 2008. Thermophiles: Biology and technology at high temperatures. CRC Press, Roca Baton.

Roberts, M.F. 2005. Organic compatible solutes of halotolerant and halophilic microorganisms. Sal. Syst. 1: 5.

Rothschild, L.J. and R.L. Mancinelli. 2001. Life in extreme environments. Nature 409: 1092–1101.

Rozzell, J.D. 1999. Commercial scale biocatalysis: Myths and realities. Bioorganic Med. Chem. 7: 2253–2261.

Saaranen, M.J. and L.W. Ruddock. 2013. Disulfide bond formation in the cytoplasm. Antioxid Redox Signaling 19: 46–53.

Sánchez, L.A., F.F. Gómez and O.D. Delgado. 2009. Cold-adapted microorganisms as a source of new antimicrobials. Extremophiles 13: 111–120.

Sand, W. and T. Gehrke. 1999. Analysis and function of the eps from the strong acidophile *Thiobacillus ferrooxidans*. pp. 127–141. *In:* Wingender, J., T.R. Neu and H. Flemming (eds.). Microbial Extracellular Polymeric Substances. Springer Berlin, Heidelberg.

Santiago, M., C.A. Ramírez-Sarmiento, R.A. Zamora and L.P. Parra. 2016. Discovery, molecular mechanisms, and industrial applications of cold-active enzymes. Front. Microbiol. 7: 1408–1440.

Sarethy, I.P., Y. Saxena, A. Kapoor, M. Sharma, S.K. Sharma, V. Gupta and Gupta, S. 2011. Alkaliphilic bacteria: Applications in industrial biotechnology. J. Ind. Microbiol. Biotechnol. 38: 769–790.

Schreck, S.D. and A.M. Grunden. 2014. Biotechnological applications of halophilic lipases and thioesterases. Appl. Microbiol. Biotechnol. 98: 1011–1021.

Schwarz, T. 2003. Use of Compatible Solutes as Substances Having Free Radical Scavenging Properties, Google Patents.

Shao, H., L. Xu and Y. Yan. 2014. Thermostable lipases from extremely radioresistant bacterium *Deinococcus radiodurans* : Cloning, expression, and biochemical characterization. J. Basic Microbiol. 54: 984–995.

Sharma, A., Y. Kawarabayasi and T. Satyanarayana. 2012. Acidophilic bacteria and archaea: Acid stable biocatalysts and their potential applications. Extremophiles 16: 1–19.

Sharma, A., D. Parashar and T. Satyanarayana. 2016. Acidophilic microbes: Biology and applications. pp. 215–241. *In:* Rampelotto, P.H. (ed.). Biotechnology of Extremophiles: Advances and Challenges. Springer, Switzerland.

Singh, A.K., P.K. Pindi, S. Dube, V.R. Sundareswaran and S. Shivaji. 2009. Importance of trmE for growth of the psychrophile *Pseudomonas syringae* at low temperatures. Appl. Environ. Microbiol. 75: 4419–4426.

Singh, R., H. Dong, D. Liu, L. Zhao, A.R. Marts, E. Farquhar, D.L. Tierney, C.B. Almquist and B.R. Briggs. 2015. Reduction of hexavalent chromium by the thermophilic methanogen *Methanothermobacter thermautotrophicus*. Geochimica et cosmochimica acta 148: 442–456.

Singh, S., L. Shukla, L. Nain and S. Khare. 2011. Detection and characterization of new thermostable endoglucanase from *Aspergillus awamori* strain F 18. J. Mycol. Plant Pathol. 41(1): 97–103.

Singh, S., D.K. Jaiswal, N. Sivakumar and J.P. Verma. 2019. Developing efficient thermophilic cellulose degrading consortium for glucose production from different agro-residues. Front. Energy Res. 7: 1–13.

Sommer, P., T. Georgieva and B.K. Ahring. 2004. Potential for using thermophilic anaerobic bacteria for bioethanol production from hemicellulose, Portland Press Ltd.

Staiano, M., A. Pennacchio, A. Varriale, A. Capo, A. Majoli, C. Capacchione and S. D'Auria 2017. Enzymes as sensors. 115–131. Melvin Simon John Abelson, Gregory Verdine, Anna Pyle. Methods in enzymology. Academic Press.

Summerbell, I.W.R.C. 1995. The dermatophytes. Clin. Microbiol. Rev. 8: 240–259.

Swiontek Brzezinska, M., U. Jankiewicz, A. Burkowska and M. Walczak. 2014. Chitinolytic microorganisms and their possible application in environmental protection. Curr. Microbiol. 68: 71–81.

Sydlik, U., I. Gallitz, C. Albrecht, J. Abel, J. Krutmann and K. Unfried. 2009. The compatible solute ectoine protects against nanoparticle-induced neutrophilic lung inflammation. Am. J. Respir. Crit. Care Med. 180: 29–35.

Tan, J.C.H., F.B. Kalapesi and M.T. Coroneo. 2006. Mechanosensitivity and the eye: Cells coping with the pressure. Brit. J. Ophthalmol. 90: 383–388.

Tomas, C.A., N.E. Welker and E.T. Papoutsakis. 2003. Overexpression of groESL in *Clostridium acetobutylicum* results in increased solvent production and tolerance, prolonged metabolism, and changes in the cell's transcriptional program. Appl. Environ. Microbiol. 69(8): 4951–4965.

Van den Burg, B. 2003. Extremophiles as a source for novel enzymes. Curr. Opin. Microbiol. 6: 213–218.

Ventosa, A. and J.J. Nieto. 1995. Biotechnological applications and potentialities of halophilic microorganisms. World J. Microbiol. Biotechnol. 11: 85–94.

Vidali, M. 2001. Bioremediation. An overview. Pure Appl. Chem. 73(7): 163–1172.

Vieille, C. and G.J. Zeikus. 2001. Hyperthermophilic enzymes: Sources, uses, and molecular mechanisms for thermostability. Microbiol. Mol. Biol. Rev. 65: 1–43.

Wang, W., M. Sun, W. Liu and B. Zhang. 2008. Purification and characterization of a psychrophilic catalase from Antarctic *Bacillus*. Can. J. Microbiol. 54: 823–828.

Yang, S., X. Fu, Q. Yan, Y. Guo, Z. Liu and Z. Jiang. 2016. Cloning, expression, purification and application of a novel chitinase from a thermophilic marine bacterium *Paenibacillus barengoltzii*. Food Chem. 192: 1041–1048.

Yao, C., J. Sun, W. Wang, Z. Zhuang, J. Liu and J. Hao. 2019. A novel cold-adapted β-galactosidase from *Alteromonas* sp. ML117 cleaves milk lactose effectively at low temperature. Process Biochem. 82: 94–101.

Zhou, W., W. Guo, H. Zhou and X. Chen. 2016. Phenol degradation by *Sulfobacillus acidophilus* TPY via the meta-pathway. Microbiol. Res. 190: 37–45.

Zhu, D., W.A. Adebisi, F. Ahmad, S. Sethupathy, B. Danso and J. Sun. 2020. Recent development of extremophilic bacteria and their application in bio-refinery. Front. Bioeng. Biotechnol. 8: 483.

7

Advancements in Extremozymes and their Potential Applications in Biorefinery

Nivedita Sharma* and *Nisha Sharma*

1. Introduction

Today, the biorefinery has become a basic idea to use in the development of strategies as well as in environmental and economic concerns. A biorefinery is a combined system of renewable raw materials source, industrial intermediary processes and final products obtained by using different technologies (Alvarado et al. 2017). The aim is to produce both type of products, i.e., high value with low volume products and low value with high volume products. The substrate/biomass should be cheap and easily available for the product formation during fermentation processes. Different type of raw materials have been suggested which include corn, wheat, sugar cane, cotton, cassava, lignocellulose and algae (Bohme 2019). The simplest biorefinery systems used only one type of feedstock, i.e., grains, which have in principal fixed processing to one main product, while the most flexible ones use a mixture of feedstocks to produce value added products. Different types of biomass feedstock can be used as whole crop (e.g., cereals and corn), or lignocellulose feedstock (e.g., biomass from wood or waste) (Yuan et al. 2018).

In order to achieve efficient conversion of the raw material, a mixture of mechanical, biocatalytic and chemical treatments are expected to be combined.

The biorefining technology for biofuels and chemicals from lignocellulosic biomass has made great progress in the world (Chen et al. 2018). However, mobilization of laboratory research toward industrial setup needs to meet a series of criteria, including the selection of appropriate pretreatment technology, breakthrough in enzyme screening, pathway optimization, and production technology, etc. (Ebaid et al. 2019). Extremophiles play an important role in biorefinery by providing novel metabolic pathways and catalytically stable/robust enzymes that are able to act as biocatalysts under harsh industrial conditions on their own (Counts et al. 2017). In this chapter, the potential application of thermophilic, psychrophilic alkaliphilic, acidophilic, and halophilic bacteria and extremozymes in the pretreatment, saccharification, fermentation, and lignin valorization process has been discussed (Crawford et al. 2016). Besides, the latest studies on the engineering bacteria of extremophiles using metabolic engineering and synthetic biology technologies for high-efficiency biofuel production are also introduced. Furthermore, the comprehensive application potential of extremophiles and extremozymes in biorefinery, which is partly due to their specificity and efficiency, and points out the necessity of accelerating the commercialization of extremozymes, have also been included here (Zhu et al. 2020).

Microbiology Research Laboratory, Dr. Y S Parmar University of Horticulture & Forestry, Nauni-Solan (H.P.)-173230, India.
* Corresponding author: niveditashaarma@yahoo.com

2. Background of Enzymes

Enzymes are natural catalysts which are used to increase the rate and specificity of chemical reactions by reducing the required activation energy. Enzymes have been used as biotechnological tools since ancient times for the production of food, alcoholic beverages and other industrial applications but only recently have significant knowledge and understanding of enzymes been cultivated for their vast commercial exploitation (Ebaid et al. 2019).

In 1914, Otto Rohm prepared the first enzyme for a commercial application. He isolated trypsin from animal pancreases and added it to washing detergents to degrade proteins. Since the 1960s, scientists have known that most enzymes have a range of functionality under different conditions. Enzymes were sought so as to be used in harsh industrial chemical processes due to their unique properties that allow catalytic reactions to occur in a more efficient way in the interest of profits and environmental protection. After sometimes the demand called for higher product output, the use of the chemical processes continued to increase in order to keep up with demand which led to the need for enzymes that could perform in conditions where their predecessors could not. In the 1980s, scientists from the University of Regensburg, Germany, found enzyme that could withstand abnormal conditions. Karl Stetter and his colleagues discovered organisms that grew optimally at the boiling point of water (100°C) or greater in geothermal sediments and the heated waters of the Italian Volcano Island. After this, Karl Stetter went on to discover more than 20 genera of microbes that grew in nearly the same conditions, two of which are *Thermotoga* and *Aquifex* bacteria, while the others were archaea (Zhu et al. 2020).

Need of Extremozymes: Shifting from Enzymes to Extremozymes

Among the different relevant sources of enzymes, microorganisms are the most preferred organism for industrial enzymes production due to their clear cut advantages viz. rapid multiplication, easy cultivation, manipulation of physical and chemical growth parameters to enhance yield, easy harvesting and genetic modification, etc. (Zhu et al. 2020).

Recent enzyme discoveries and developments in genetics and protein engineering have also increased the reach of enzymes in industry. Indeed, industrial catalysis is increasingly dependent on enzymes. Enzymes have become important tools for diverse industrial markets such as biofuels food and beverage, animal feed, detergents, and technical enzymes, including biofuels, leather, pulp and paper and textile industries and also growing in markets, such as diagnostics, pharmaceuticals, and research and development.

Moreover, the diversity and unique properties of microbial enzymes, e.g., consistency, reproducibility, high yields, and economic feasibility among others, have elevated their biotechnological interest and application to different industrial areas (Ebaid et al. 2019). However, the vast majority of current industrial processes are performed under harsh conditions, including extremely high and low temperatures, acidic or basic pH, and elevated salinity. Standard enzymes have specific requirements for maximal function, performing optimally in narrow ranges of physical and chemical conditions. These requirements are quite different from industrial processing settings, where standard enzymes are easily denatured. In many cases, traditional chemical solutions are still the only viable option under such harsh conditions. There is a clear need for more sustainable and environmentally friendly methods to replace the current potentially harmful chemical processes.

Extremophiles and Extremozymes

Microorganisms have the ability to live in all extreme conditions of the environments. The optimum preferred range of environmental conditions for the growth of microorganisms are temperature (20–40°C), pH (6.5–7.0), pressure (1 atm), adequate amount of oxygen, water supply, nutrients and salts concentration. But there are microorganisms which survive in extremely harsh conditions; these organisms are known as "extremophiles". The term "extremophiles" was coined by R D

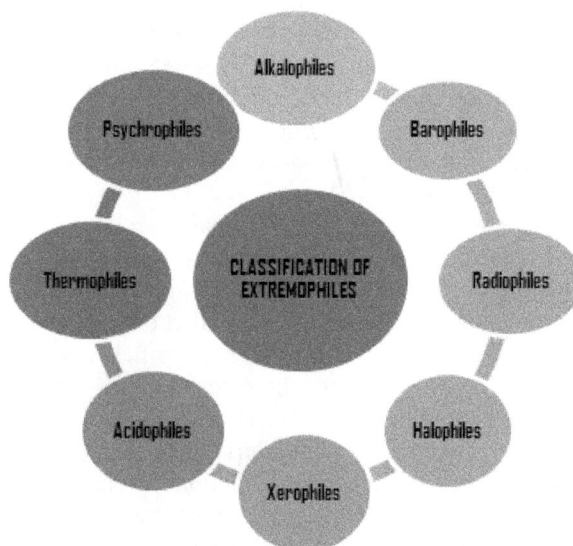

Fig. 1. Classification of extremophiles.

Mac Elroy in 1971 and the first extremophile discovered was *Thermus aquaticus*. On the basis of different parameters like pH, temperature, salt concentration, pressure and water availability, etc., extremophiles are divided into the categories like alkalophile, acidophiles, thermophiles, psychrophiles, barophiles, xerophiles, radiophiles and halophiles, etc. (Fig. 1).

What are Extremozymes?

Extremozymes are enzymes which are known as extremophiles which function under extreme environments. Examples of such are those in highly acidic/basic conditions, high/low temperatures, high salinity, or other factors, that would otherwise denature typical enzymes (e.g., catalase, rubisco, carbonic anhydrase). This feature makes these enzymes of interest in a variety of biotechnical applications in the energy, pharmaceutical, agricultural, environmental, food, health, and textile industries (Ebaid et al. 2019).

The biocatalysts produced by extremophilic microorganisms, so-called extremozymes, are proteins with outstanding stability to temperature, pH, and organic solvents, thus becoming excellent candidates to improve industrial bio-transformations. The enzymes which degrade polymers are extracted from hyperthermophiles, psychrophiles, and acidophiles and they play an important role in food, detergent, and pulp and paper industry (e.g., amylases, pullulanases, xylanases, proteases). Cellulases, proteases, pectinases, keratinases, lipases, esterases, catalases, peroxidases, and phytases are the few examples of extremozymes (Zhu et al. 2020). Extremophilic microorganisms are a rich source of natural enzymes, which are more superior over their mesophilic counterparts for applications at extreme conditions. Out of these, lignocellulolytic, amylolytic and other biomass processing extremozymes with unique properties are widely distributed in thermophilic prokaryotes and are of high potential for versatile industrial processes.

Properties of Extremophiles to Survive and Work under Extreme Conditions

To survive under the harsh environment, extremophiles have developed a variety of strategies to struggle with the adverse conditions. The physiological high temperature readily destroys the intracellular bonds present in the amino acids and lead to the unfolding as well as denaturation of proteins which is harmful to the normal microorganisms.

On the other hand, thermophilic microorganisms are able to restore their protein structure and function by producing chaperones or thermosomes under extreme condition to resist the destruction of protein by high temperature. In order to resist the protein unfolding caused by high temperature, thermophilic bacteria have developed special hydrogen bonds that can interact with hydrophobicity (Zhu et al. 2017).

Psychrophiles are able to live in extreme cold conditions and this is mainly due to their cellular cold-adaptability mechanisms: the regulation of cold-shock proteins, small RNA-binding proteins, and extracellular polymeric substances to protect the cells against mechanical disruption to the cell membrane caused by low temperature. The genome of psychrophiles contain higher G+C-rich regions encoding tRNAs, elongation factors, and RNA polymerases, and the presence of plasmids, transposable/mobile genetic elements related to the biosynthesis of unsaturated fatty acids improves their cold adaptability (Rastogi et al. 2011).

Acidophiles use various homeostatic pH mechanisms that involve restricting/passive proton entry into the cytoplasmic membrane and purging off protons. They also have a highly impermeable cell membrane which help in restricting the proton influx into the cytoplasm by active proton pumping. In order to survive in the high-saline environment, halophiles have developed various strategies to keep the osmotic balance between intracellular membranes with the environment to prevent water loss. Out of these, one such strategy is the intracellular synthesis or accumulation of compatible solutes/osmolytes, such as ectoine, trehalose, proline, dimethylsulfoniopropionate K-glutamate, betaine, and carnitine. Ectoine, a halophile derived solute,is used in bio-industry for biofuel production. Supplementation of ectoine in the growth medium of *Zymomonas mobilis* has been shown to improve the ethanol production (Sheng et al. 2017).

(1) ***Thermophiles and Thermozymes:*** The microorganism having optimal growth temperatures ranging from 45°C to 60°C. Thermophiles grows at high temperatures and are more stable at higher temperature due to their thermostable activity (Fig. 2). Therefore, it is the most explored class of extremophiles due to the production of thermostable enzymes. They are further classified into three sub classes:

(i) Simple thermophile, which thrive in the temperature range of 45°C to 60°C. Examples are: *Lactobacillus, Streptococcus thermophilus* and *Bifidobacteria.*

(ii) Extreme thermophiles, capable of surviving in the temperature of 60°C to 80°C. Examples are: *Caldicellulosiruptor, Sulfolobus, Thermotoga* and *Pyrococcus.*

(iii) Hyperthermophiles, in which temperature adaptive range is from 80°C to 120°C. For example: *Pyrococcus furiosus, Pyrolobus fumarii.* These enzymes are of great industrial and biotechnological interest. Most of the industrial processes are conducted at high temperature due to the increased solubility of many polymeric substrates, increased bioavailability, faster reaction rate, and the decreased risk of microbial contamination.

(2) ***Halophiles and Halozymes:*** Halophiles are the microorganisms which can grow in high salt concentration (> 0.3 M) of NaCl. Halophiles are divided into sub-classes based on their survival in different salt concentration.

(i) Extreme halophiles, which can survive at a salt concentration of 1.5 to 5.2 M. For example: Fungus *Wallemia ichthyophaga*, green alga *Dunaliella salina, Hologeometricum, Halococcus, Halorubrum, Haloarcula, Halobacterium, Natronococcus*, etc.

(ii) Moderate halophiles can grow at 0.2 to 0.5 M salt concentration. For example: *Halomonadaceae* spp.

(iii) Non-halophiles show best growth at a lower concentration (< 0.2 M NaCl). For example, *Staphylococcus aureus, E. coli.*

Halophiles have different mechanism to survive in different saline environments. It is found in all three domains: archaea, bacteria and eukaryotes. Halophiles play an important role in biotech industries like detergent, textile and pulp and paper industries. Halophillic enzymes produced by halophiles completely depend on salt to fold properly and to get properly fold they adapt themselves

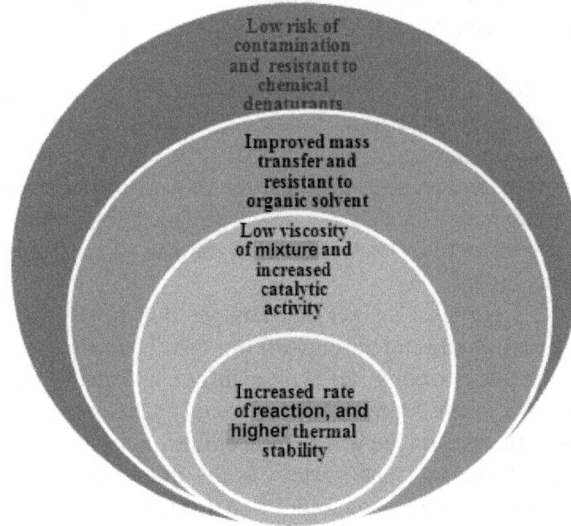

Fig. 2. Properties of thermozymes.

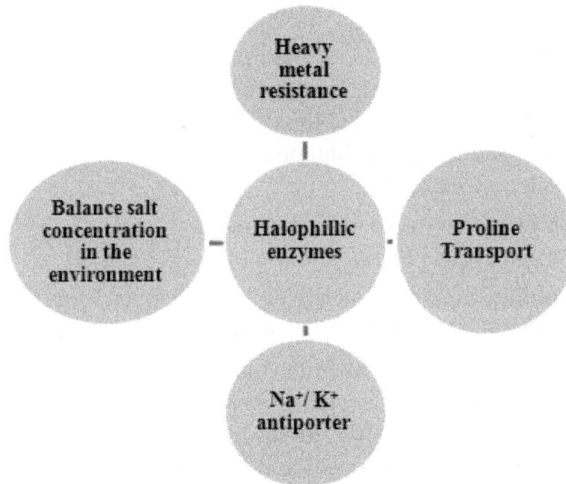

Fig. 3. Properties of halophiles.

for some properties like large number of acidic residues on surface, a smaller number of hydrophobic residues and many amino acids which enhance flexibility of an enzyme (Fig. 3) (Zhang et al. 2018). These inherent characteristics of halophilic enzymes are therefore one of the best choice for biofuel production and other industrial processes (Singh et al. 2019). Halophillic enzymes like as cellulase, xylanase, laccase, etc., are relatively more stable than those of a terrestrial origin (Li and Yu 2013).

(3) *Alkalophiles and Acidophiles:* The microbes grow at neutral pH but there are many microorganisms which can grow at low or high pH (Fig. 4). They are classified as acidophiles, which grow at low pH (below 3.0 pH) and alkalophiles, which grow at high pH (at pH 8.0 or more) depending on their survival at acidic and basic pH conditions (Zhu et al. 2013). Examples of alkalophiles are: *Bacillus halodurans, Bacillus firmus, Halorhodospira halochloris, Natromonas pharaonis Thiohalospira alkaliphila,* etc. Acidiphiles include: *F. acidiphilum, T. acidophilum, P. torridus, S. acidocaldarius,* and *Acidithiobacillus ferrooxidans, Sulfolobus acidocaldarius,* and *Acetobacter aceti.*

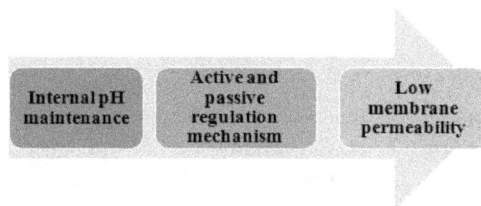

Fig. 4. Properties of alkalophillic enzymes.

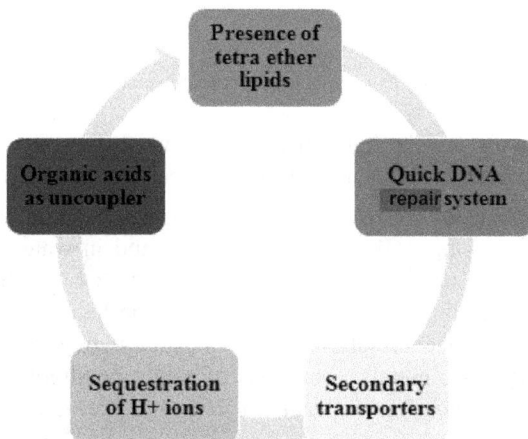

Fig. 5. Basic properties of acidophiles.

Enzymes produced from alkalophiles are highly thermostable, active in alkaline conditions and substrate specific (Annamalai et al. 2016). An alkaliphilic *Bacillus ligniniphilus* L1 has been isolated from the South China Sea. It grows well at 30°C and a pH of 9.0 and it has a potential application value in the lignin valorizations (Zhu et al. 2017).

Acididophillic enzymes are produced by the microorganisms which survive in highly acidic conditions. There are many adaptation strategies to sustain in such environments such as impermeable plasma membrane which resists any change in pH. It is mainly due to fixed nature, large isoprenoid core, membrane channel's internal buffers and ether linkages (Patel et al. 2005). The unique structures and functions of acidophilic bacteria, such as membrane potential reversal, high membrane impermeability, and the presence of secondary transporters, make them have a broad application prospect in the biofuel industry (Fig. 5).

(4) *Psychrophiles and Psychrozymes:* Psychrophiles are those microorganisms which can grow at temperature of 15°C and below. Common species of psychrophiles are *Bacillus, Arthrobacter, Pseudomonas, Pseudoalteromonas, Vibrio, Penicillium Halorubrum, Methanogenium, Cladosporium Crystococcus* and *Candida*, etc. These are "cold-adaptive" enzymes and are called psychrozymes. Their cold-activity property gives them many advantages in biotechnology industry for the production of biofuels, energy and other bioproducts (Martin and McMinn 2017). To survive in low temperature environments, psychrophilic bacteria have evolved all of their cellular components, including their membranes, energy-generating systems, protein synthesis machinery, enzymes and the components help in nutrient uptake (Fig. 6). These enzymes can treat and digest lignocellulose at low temperatures, reduce the energy input required to heat the bioreactor, and avoid chemical side-effects that can occur at higher temperatures and the generation of harmful byproducts (Hamid et al. 2014). A novel psychrotolerant *Sanguibacter gelidistatuariae* has been isolated from an ice sculpture in Antarctica which grows at 3°C (Pikuta et al. 2017). A psychrotolerant novel species *Carnobacterium antarcticum* CP1 has

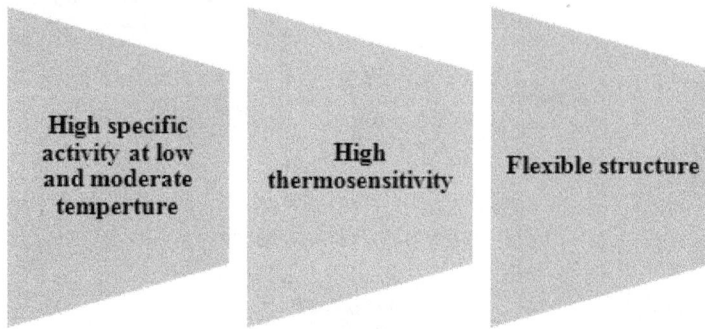

Fig. 6. Properties pscychrophillic enzymes.

been isolated from sandy soil in Antarctica (Zhu et al. 2018). The enzymes extracted from psychrophiles are also used in detergents, textiles, processing of fruit juices and food industries.

(5) *Metallophiles:* Metallophiles are the organism that grow in the presence of high metal ion concentration. The increasing pollution level on land and in water produce immense threat to every living being. In order to remove heavy metal toxicity from soil and water, we need to exploit more metallophiles. Metallophiles can be helpful in different processes like biomining, bioleaching, bioaccumulation, biomineralization, bioremediation, etc., to reduce the concentration of heavy toxic metals. Examples of some metallophiles are as nickel resistant *Proteobacteria, Viola calaminaria, Armeria maritime* and *Festuca.*

(6) *Barophiles:* Barophiles grow at very high-pressure conditions found in deep oceans. These organisms produce enzymes that are stable at very high pressure and temperature and commonly found species are *Pyrococcus, Moritella, Methanococcus* and *Shewanella.* Barophiles have main applications in food industries. Enzymes from barophiles adapt at high pressures which can be used in sterilization process for packed food products.

(7) *Xerophiles:* Xerophiles grow in dry environmental conditions where water availability is very less. The organisms which adapt in desiccation develop anhydrobiosis, i.e., ability to survive from very less intercellular water and they are metabolically inactive. The enzyme from these organisms is used in agriculture industry to improve the water management in desert plants. Xerophillic microorganisms include: *Trichosporonoides nigrecens* and *Cacti.*

(8) *Radiophiles:* Radiophiles are those microbes which can tolerate high dose of radiations. They have ability to survive in high radiations, stress and even with regular DNA damage. These organisms have great potential in nuclear waste remediation. *Deinococcus radiodurans* is the potent microorganism which has potential to survive at high radiations, high temperature, high pressure and other extreme conditions.

(9) *Microaerophiles:* These microorganisms grow at oxygen concentration of approximately 2 to 10%. They do not require more oxygen for their growth.

Applications of Extremozymes in Biorefinery

There is a consistent focus of scientists and researchers on the screening of novel enzymes from various sources to obtain the necessary characteristics amenable for industrial and biotechnological applications as well as genetic engineering of existing enzymes to increase their activity (Fig. 7). Many of the common commercial enzymes cannot meet the industrial requirements, like being able to withstand industrial requirements with high reproducibility at different pH, temperature, and aerification conditions. Therefore, extremozymes have received increased attention as a strategy of industry process and biorefining. Industrial enzymes, such as those used in biorefinery, are considered as technical enzymes. There are many research groups and companies which are committed to engineer microorganisms genetically with desirable industrial characteristics suitable

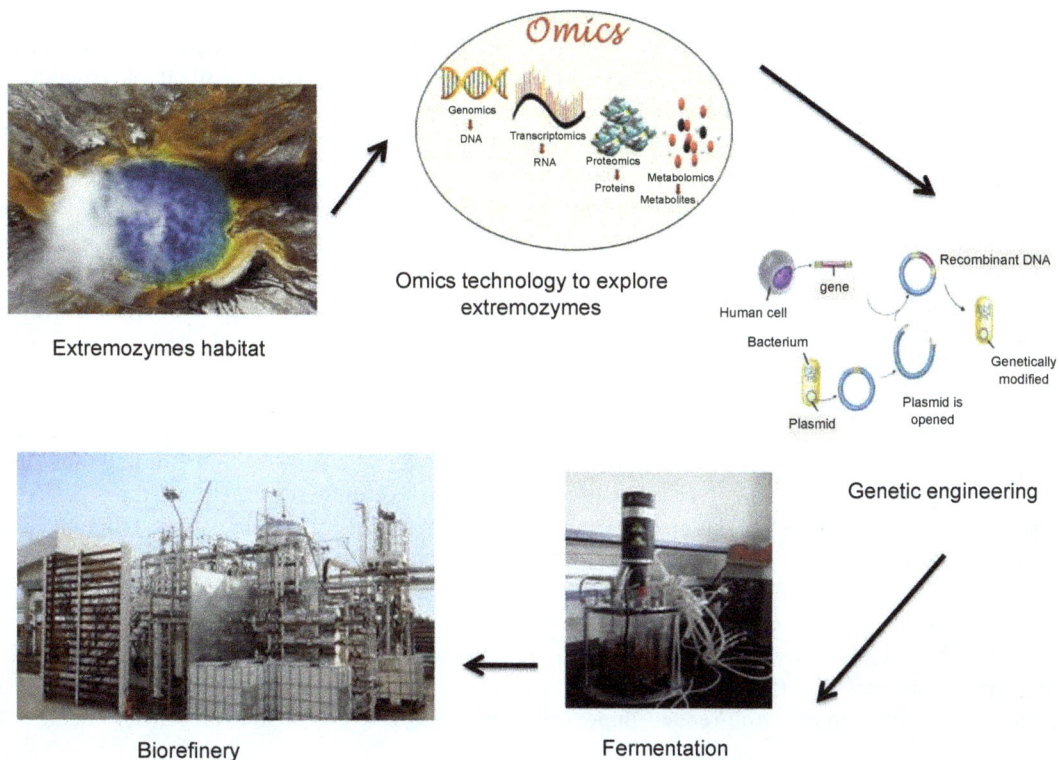

Fig. 7. Application of extremozymes in bio-energy production.

for their industrial operations. This opens a new window for researchers to meet the ever-growing global market demand, and extremozymes are the best candidates for consideration (Zhu et al. 2020).

Lignocellulose: A Recalcitrant but Valuable Biomass Resource

The term bioenergy refers to solid, liquid or gaseous substances that have to be used as an energy source which may be bioethanol, butanol, biodiesel and biomethane (Moreno and Olsson 2017). A biorefinery is a widely used concept which integrates full use of biomass generated into a spectrum of bio-based products and bioenergy (Cherubini 2010). Lignocellulose is the main structural component of plants that contains cellulose, hemicellulose and lignin as its constituents. Lignocelluloses are the best candidate feedstock for bioenergy production because of their high mass availability, relatively low price, and lack of competition with food provision. However, the recalcitrance of lignocelluloses caused by lignin is the key hindrance in the utilization of this valuable resource (Geng et al. 2018). Therefore, the effective delignification of the biomass will play an important role in the economic feasibility of biofuel processing (Arevalo-Gallegos et al. 2017).

First generation biofuels are sourced from starch and sugar cane (sucrose). The sensitivity over the use of food grains for fuel has stimulated interest for second generation fuels derived from abundant lignocellulosic materials such as agricultural and forest wastes. Plant cell walls are the source of lignocellulose, which is comprised of cellulose (35–50% of plant dry weight), hemicellulose (20–35%) and lignin (5–30%) (Sharma et al. 2019a) (Fig. 8).

Second generation biofuel production is gaining huge attention and is recognized as a green and sustainable fuel alternative. The main bottle-neck in a 2nd generation process is efficient conversion of recalcitrant lignocellulosic biomass into fermentable sugars (Tanwar et al. 2019). This may be achieved by combining pretreatment strategies; e.g., steam explosion or dilute acid exposure, with enzyme hydrolysis and downstream fermentation. Physical, chemical, physico-chemical, and biological processes have been used for the pretreatment of lignocellulosic materials, a process that may employ extreme conditions of temperature, pH, pressure or biotoxic chemicals. Pretreatment partially solubilizes hemicellulose and cellulose fractions, while removing lignin components (Sharma et al. 2019a). When coupled to effective enzyme hydrolysis, harsh pretreatment conditions may be reduced. Harsh process conditions have been a driving force behind the development of extremophilic whole cell 10 biocatalysts and the industrial use of extremozymes, the latter being discussed in the remainder of the chapter. Industry requires enzymes with high specific activity, low sensitivity to end-product inhibition, and reasonable stability and flexibility when placed in fluctuating process conditions (Kuhad et al. 2011). Enzymes isolated from thermophilic and acidophilic sources are gaining interest due to "up-front" delivery of desirable characteristics. For in-depth information, the reader is referred to excellent and recent reviews (Yeoman et al. 2010). A wide range of glycosyl hydrolases (GHs), including cellulases and hemicellulases, are involved in the deconstruction of lignocellulose to monosaccharides (D-xylose, L-arabinose and D-glucose). GHs are largely characterized by similarity in catalytic domains and carbohydrate binding modules. Based on the arrangement and positioning of active sites, GHs can be separated into three catalytic domain architecture types: (a) tunnel, specialized for processive hydrolysis where reducing and non-reducing ends of the substrate are fed into the active site of the enzyme (b) cleft, specialized for non-processive internal hydrolysis, and (c) crater/pocket, for end-on-attack hydrolysis, releasing monosaccharides (Rani et al. 2019).

The huge technological gap between producing enzymes in laboratory conditions and obtaining the final commercial product is a challenge when developing extremozymes. It can be seen that a large number of extremozyme-related papers are published every year, but they rarely achieved industrialization (Varrella et al. 2020).

Algae are currently being promoted as an ideal third generation biofuel feedstock because of their rapid growth rate, CO_2 fixation ability and high production capacity of lipids; they also do not compete with food or feed crops, and can be produced on non-arable land (Vyas et al. 2019).

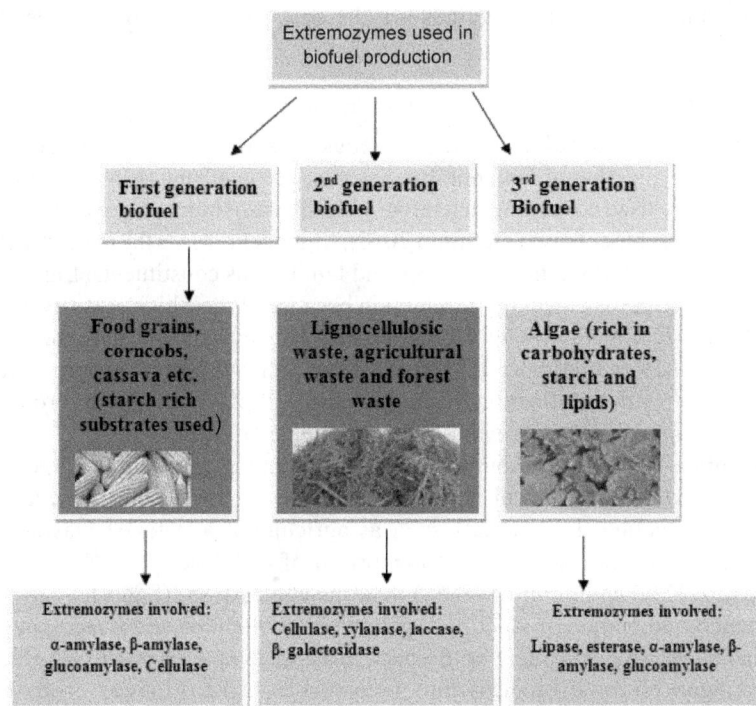

Fig. 8. Extremozymes involved in biofuel production from different substrates.

It is clear that several species of microalgae can have oil contents up to 80% of their dry body weight. Some microalgae can double their biomasses within 24 hours and the shortest doubling time during their growth is around 3.5 hours which makes microalgae an ideal renewable source for biofuel production. There are different ways algae can be cultivated. However, two widely used cultivation systems are (a) suspended cultures, including open ponds and closed reactors, and (b) immobilized cultures, including matrix-immobilized systems and biofilms. The most common large scale production systems in practice are high rate algal ponds or raceway ponds (Sharma et al. 2019b) (Fig. 8).

Mechanism of Action of Extremozymes on Cellulosic Biomass

The conversion of lignocellulosic biomass to fermentable sugars through biocatalyst cellulase and xylanase derived from cellulolytic and hemicellulolytic organisms has been suggested as a feasible process and offers potential to reduce use of fossil fuels and reduce environmental pollution. Cellulose accessibility to hydrolytic enzymes is believed to be the most important substrate characteristic limiting enzymatic hydrolysis. Cellulose solvents effectively break linkages among cellulose, hemicellulose and lignin, and also dissolve highly-ordered hydrogen bonds in cellulose fibers accompanied with great increases in substrate accessibility (Hamid et al. 2014). The inherent properties of native lignocellulosic materials make them resistant to normal enzymatic attack. To achieve instant and fast enzymatic degradation in production of ethanol by enzymatic hydrolysis, extremozymes play important role (Rani et al. 2009) (Fig. 9).

Cellulose deconstruction: Cellulose is a glucan homopolysaccharide comprised of β-D-glucopyranose units linked together by β-1,4-glycosidic bonds. Cellulases hydrolyse glucosidic bonds of amorphous and crystalline cellulose and depending on the type of cellulase, different regions and cellulose chains are targeted. Cleft-shaped catalytic domains of endoglucanases (1,4-β-D-glucan 4-glucanhydrolases) hydrolyse internal glycosidic bonds within soluble amorphous cellulose, releasing cellulo-oligosaccharides.

Exoglucanases (1,4-β-D- glucan cellobiohydrolase) display tunnel-shaped catalytic modules that progressively release cellobiose residues from the reducing and non-reducing ends of cellulose chains (Gilad et al. 2003).

Cellulases may therefore be grouped according to preferred substrates, with exoglucanases being more specific for crystalline cellulose and endoglucanases for soluble amorphous cellulose. Exoglucanase and endoglucanase end-products are either hydrolysed extracellularly by β-glucosidases or transported across the cell membrane and metabolized intracellularly.

Thermostable endoglucanases have been described for various archaeal and bacterial genera including *Pyrococcus*, *Sulfolobus*, *Thermotoga* and *Geobacillus* (Rastogi et al. 2011). Numerous thermophilic exocellulases and glucosidases have been reported including examples from *Pyrococcus* and *Thermus* spp. Protein engineering has yielded enzymes with significantly improved

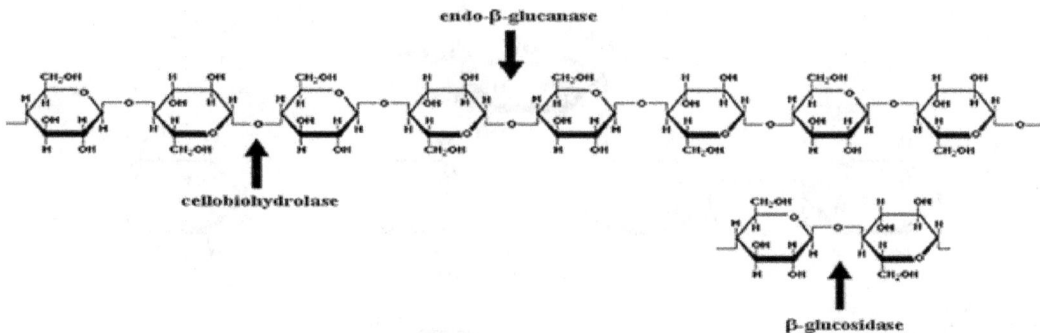

Fig. 9. Polymeric chemical structure of cellulose and site of action of endoglucanase, cellobiohydrolase and β-glucosidase (Kumar et al. 2008).

specificity and activity. Large quantities of free enzymes are required to synergistically degrade complex substrate such as lignocelluloses (Lynd et al. 2002). Therefore, currently, most industrial cellulases are produced by engineered strains of aerobic fungi, specifically *Trichoderma reesei* (syn. *Hypocrea jecorina*) and *Humicola insolens* due to very high yields and high specific activities (Karlsson et al. 2002).

Hemicellulose deconstruction: Hemicellulose consists of a combination of complex branched heteropolymers (Sharma et al. 2019b). Xylan and glucomannan constitute the largest components in hardwood and softwood, respectively. Other components include xyloglucan, glucomannan, galactoglucomannan and arabinogalactan. Hemicelluloses are hydrolysed by a combination of enzymes releasing valuable fermentable sugars such as pentoses and hexoses as well as sugar acids. The xylan β-1,4-linked backbone and respective side chains are hydrolysed by a combination of endoxylanases, βxylosidases, α-glucuronidases, α-L-arabinofuranosidases and acetylxylan esterases. Glucomannans in turn are hydrolyzed by β-mannanase and β-mannosidase (Subramaniyan and Prema 2000) (Fig. 10).

The growing importance of xylanolytic enzymes in industry lies in the assistance xylanases impart in disassociating lignin from the cellulose and hemicellulose fractions. Xylanases are, however, not commonly sourced from extremophiles. A hyperthermophilic *Pyrodictium abyssi* xylanase shows optimal activity at 110°C and thermophilic examples from the genera *Pyrococcus*, *Thermococcus* and *Sulfolobus* have been reported. Xylanolytic thermophiles belonging to the genera *Clostridium*, *Rhodothermus* and *Thermotoga* have also been reported (Rani et al. 2019).

Lignin deconstruction: Lignin has a highly recalcitrant complex aromatic structure synthesized by the oxidative coupling of three aromatic alcohol precursors: coniferyl alcohol, sinapyl and p-coumaryl. Precursors form, respectively, guiacyl, syringyl and hydroxyphenyl phenylpropanoid subunits. While of little use as a source of fermentable carbon, lignin needs to be removed for efficient biomass processing as it effectively binds the cellulose and hemicellulose into complex matrices. Numerous microorganisms have been implicated with the ability to degrade lignin by enzymatic and oxidative non-enzymatic mechanisms. Fungi may hydrolyse lignin with the help of lignin peroxidases, manganese peroxidases, versatile peroxidases and phenol oxidases (laccases). Several accessory enzymes such as cellobiose dehydrogenase, glyoxal oxidase, aryl alcohol oxidase and cellobiose/quinone oxidoreductase are involved in oxidative degradation of lignin (Maki et al. 2009). These enzymes are capable of producing H_2O_2 that release hydroxyl (•OH) free radicals which in turn degrade the lignin polymer non-specifically.

Lignin is highly resistant to chemical and enzymatic degradation. Biological degradation is attained chiefly by fungi, mostly by brown rot fungi and white-rot basidiomycetes. The three enzymes involved in degradation of lignin are lignin peroxidase, manganese peroxidase and versatile peroxidase.

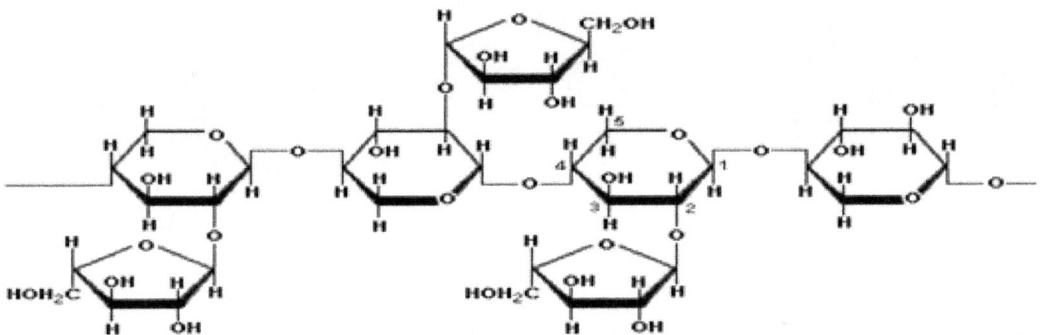

Fig. 10. Structure of xylan (Roubroeks et al. 2001).

(a) Lignin peroxidase (LP): Lignin peroxidase is a glycosylated enzyme containing heme protein with an iron protoporphyrin prosthetic group that requires H_2O_2 to catalyze the oxidation by electron transfer, non-catalytic cleavages of various bonds and aromatic ring opening of non-phenolic lignin aromatic compounds. Catalytic cycle of lignin peroxidase consists of one oxidation and two reduction steps as follows:

Step 1 Two-electron oxidation of the native ferric enzyme ([LiP]-Fe(III)) by H2O2 to form Compound I oxo-ferryl intermediate [Fe(IV)].

Step 2 The non-phenolic aromatic reducing substrate (A) gaining one electron to form Compound II.

Step 3 The oxidation cycle ends with a gain of one more electron from the reducing substrate when Compound II is returned to the resting ferric state (Singh et al. 2019).

(b) Manganese Peroxidase: Manganese (Mn) is crucial for the formation of manganese peroxidase (MnP). The enzyme MnP plays a critical part in the early stages of lignin degradation. As compared to laccase, MnP causes greater degradation of phenolic lignin due to its higher redox potential with the eventual release of carbon dioxide (Singh et al. 2019). MnP is mainly produced by white-rot basidiomycetes such as *Phanerochaete chrysosporium.*

The catalytic cycle of MnP is similar to that of LiP. Like LiPs, MnPs are also heme-containing glycoproteins which require H_2O_2 as an oxidant. Manganese acts as a mediator during MnP enzymatic activity. To begin with, MnP oxidizes Mn^{2+} to Mn^{3+}. The enzymatically generated Mn^{3+} oxidant is freely diffusible and participates in the oxidation reaction as are doxcouple (Bhuiyan et al. 2009).

(c) Versatile Peroxidase: Versatile peroxidase has catalytic properties for both MnP and LiP. VP was first purified from the genera of fungi *Bjerkandera* (Moreno et al. 2015) and was found to transform lignin even without an external mediator. The VP enzyme possesses a hybrid molecular architecture with several binding sites including Mn^{2+} and is able to oxidize Mn^{2+} like MnP and LiP. However, unlike MnP, VP has the dual ability to oxidize Mn^{2+} in the independent oxidation of simple amines and phenolic monomers. VP can also oxidize a variety of substrates (with high and low redox potentials) including Mn^{2+}, phenolic and non-phenolic lignin dimers, and aromatic alcohols (Martinez et al. 2005).

(d) Laccase: Lac (EC 1.10.3.2, *p*-diphenol oxidase) belonging to oxidoreductase group is a copper-containing enzyme which oxidizes a wide variety of organic and inorganic substances (Fig. 11).

Complex Synergism

The cellulosome bacterial cellulase production processes are mainly dominated by anaerobic clostridia including *Clostridium thermocellum, C. cellulolyticum* and *C. cellulovorans. Clostridia* are able to degrade natural crystalline cellulose through complexed synergism, implying that the cellulolytic process as a whole is more efficient than the sum of the individual cellulase enzymes (Demain et al. 2005). Synergism derives from cellulolytic machinery complexed within large, regulated, multienzyme structures called cellulosomes (Xu et al. 2010). A cellulosome consists

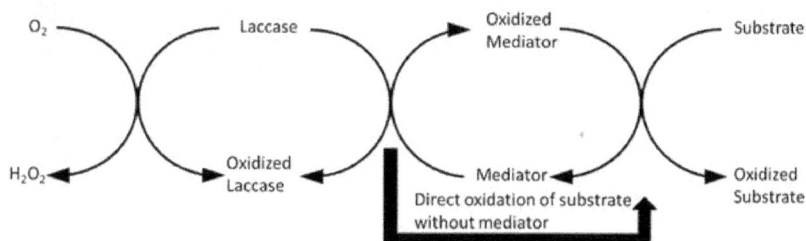

Fig. 11. The catalytic cycle of laccase (Wells et al. 2006).

of a multi-domain scaffoldin that carries at least one Carbohydrate Binding Module (CBM) and several cohesion modules. Functionally different GHs share a highly conserved dockerin module which binds to cohesion modules displayed on the scaffoldin, resulting in a large multifunctional enzyme complex. Multiple cellulosomes are displayed on the cell surface and ensure proximity between synergistic compounds. Catalytic 13 enzymes that bear dockerin modules are able bind to complementary cohesin modules through highly specific "plug and socket" cohesion-dockerin interactions. Although primarily produced by clostridia, cellulosomes have been reported for Acetivibrio cellulolyticus and Ruminococcus flavefaciens, both of which are able to simultaneously hydrolyze and ferment lignocellulosic derived sugar residues into ethanol at high temperatures (Jindou et al. 2006). The use of cellulosomes in lignocellulose degradation has several advantages over free enzyme systems. Synergistic interactions between free enzymes, on the other hand, are diluted by large substrate volumes. These complexes have inspired the formulation of designer cellulosomes with enhanced substrate binding, proximity between enzymes and functional diversity, e.g., structures with cellulytic and xylanlytic activity (Maki et al. 2009).

Extremozymes in Pretreatment of Lignocellulosic Biomass

Lignocellulosic raw materials come from a wide range of natural sources and are considered an abundant renewable resource for the production of biofuels and value-added chemicals. In recent years, the preparation of fuel ethanol from lignocellulose has attracted much attention due to its eco-friendly nature (Zabed et al. 2016). Lignocellulose is decomposed into three different polymers, such as lignin, hemicellulose, and cellulose by pretreatment, and cellulose is then converted into a monosaccharide by cellulase. Biofuel (bioethanol) is then produced by fermentation. The development of lignocellulosic energy technology has reduced the costs and protects the environment, which is conducive for the sustainable development of biorefinery based economy (Birch 2019). Enzymes from microbial origin play a leading role in the biofuel production. When compared with chemical methods, the application of enzymes in industrial bioprocesses reduces the risk of pollution; hence, they are considered to be better substitutes for lignocellulose pretreatment (Ummalyma et al. 2019). Owing to the stability and robust catalytic activity of extremozymes, further development of methods involving extreme pH and temperature conditions is the much needed research input to accelerate the industrial processes. The acid- and alkali-tolerant nature of extremozymes are helpful for pretreatment as well as the complete hydrolysis of cellulose/hemicellulose at high temperature. Lignocellulose hydrolase is limited by many factors, including crystallinity and polymerization degree, water expansion, water content, surface area, and lignin content (Karimi and Taherzadeh 2016). The combination of different pretreatment technologies is currently being employed, for example, a hydrothermal assisted enzyme degradation process is one of the effective and widely used pretreatment methods (Kirsch et al. 2011). The advantages of this combination of multiple pretreatment processes include energy saving by omitting the cooling step and better operability at high temperatures, such as improving the accessibility of the substrate while reducing viscosity (Bhalla et al. 2013). Biological pretreatment method involving laccase, along with other pretreatment methods, have been reported to significantly improve the delignification of lignocellulose. During removal of lignin from biomass using chemical process, hemicellulose and cellulose can be partially degraded. However, biological pretreatment of biomass with extremophiles and lignin decomposition enzymes reduces the amount of the degradation products as well as time required for the pretreatment. In addition, the combination of chemical and biological pretreatment processes of eucalyptus, wheat straw, pine, and corn straw have increased the delignification efficiency (Moreno et al. 2015) (Fig. 12).

Extremozymes in Biofuel Production

Among all the biofuels, bioethanol is often regarded as the most promising alternative and additive for gasoline. The production of bioethanol from lignocellulosic biomass includes four main steps:

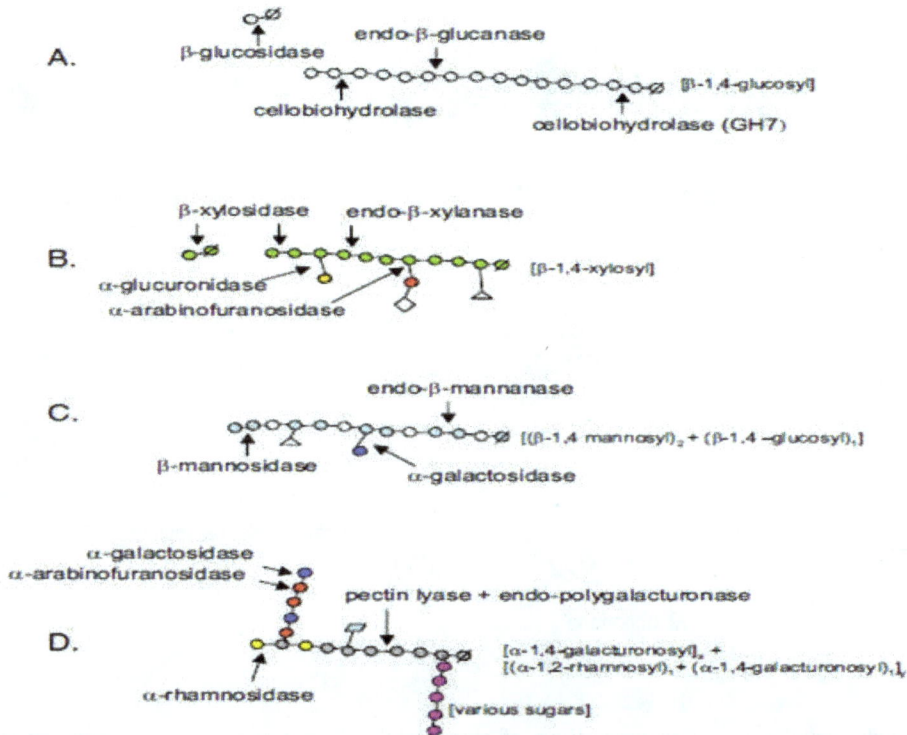

Fig. 12. Simplified structures and sites of enzymatic attack on polymers from lignocellulose. http://www. microbialcellfactories.com/content/6/1/9.

biomass pretreatment, enzymatic hydrolysis, fermentation and distillation (Indira et al. 2018). The production of bioethanol from lignocellulosic raw materials is an ecofriendly approach for sustainable development but the second-generation technology still have problems of high cost, and there are several areas of production technology that still require improvement to cut down the cost (Fig. 13). Because of the unique characteristics, extremophiles are resistant to the adverse conditions involved in bioethanol production and they harbor more advantages than terrestrial microorganisms. In particular, thermophiles and their enzymes have great potential for the bioconversion of lignocellulose into bioethanol. Certain thermophilic bacteria are known to produce both cellulase and xylanase, which can completely hydrolyze biomass at high temperature. For example, treatment of biomass using a thermostable cellulase produced by thermophilic *Geobililus* sp. R7 has been shown to yield a hydrolysate that was readily fermented by *Saccharomyces cerevisiae* ATCC 24860T to produce 0.45–0.50 g ethanol/g glucose with a 99% utilization rate of glucose (Zambare et al. 2011). Thermophilic *Caldicellulosiruptor bescii* and *Clostridium thermocellum* have been reported for its potential to use cellulose, hemp, as well as pretreated lignocellulosic biomass as a substrate to yield bioethanol. In addition, thermophile *Thermoanaerobacterium thermosaccharolyticum* M18 is able to directly utilize cellulose and xylan for the production of bioethanol (Ábrego et al. 2017). Although thermophilic bacteria have many advantages and utilize a broad spectrum of degradable carbohydrates and the fermentation of hexose and pentose and offer a low risk of pollution, the problems associated with the low G+C content, the formation of endospores, and the low permeability of plasma membrane increase the difficulty in the genetic engineering of thermophilic bacteria (Jiang et al. 2017).

In recent years, the metabolic engineering tools have already been in use for biobutanol production. To conquer the intractable gene operation of *Clostridium*, a series of methods has been developed to improve the efficiency of electroporation, increasing the dissolution of cell membrane by adding solvent, weakening the cell wall by glycine or lysozyme and optimizing the operation

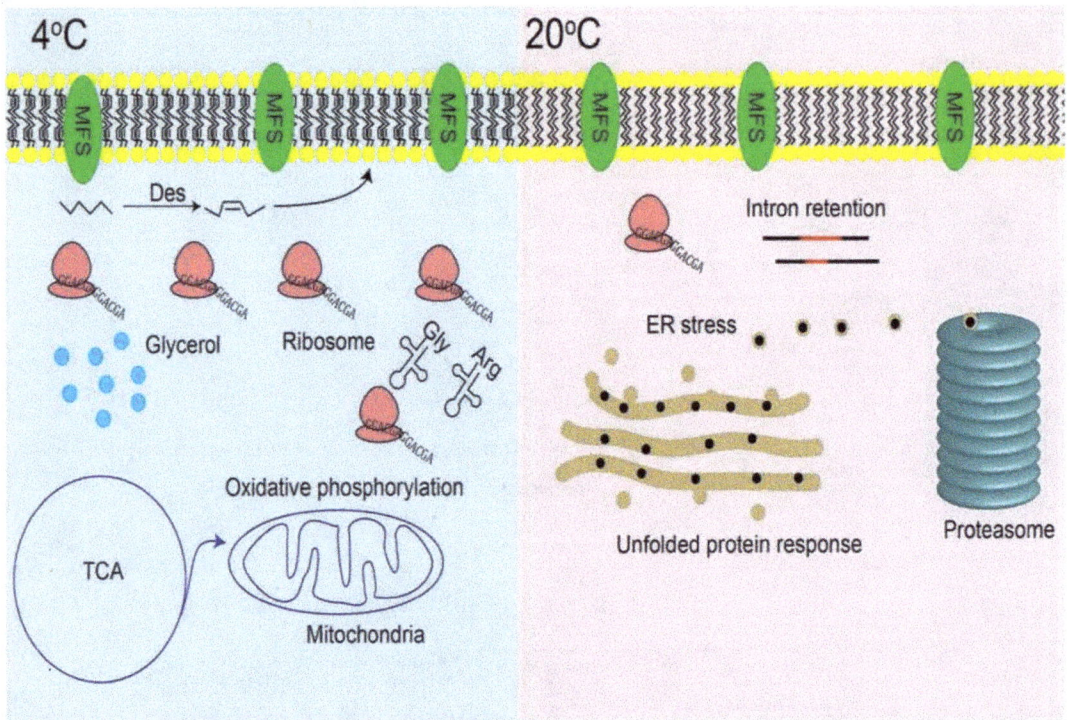

Fig. 13. Extremophilic adaptation mechanisms of extremophiles in response to environmental stress during fermentation process. https://www.frontiersin.org/articles/10.3389/fbioe.2020.00483/full.

parameters of electroporation (Pyne et al. 2014). CRISPR-Cas tools have also been successfully applied to the genetic transformation of *Clostridium*. For example, the application of *Streptococcus pyogenes* II CRISPR-Cas9 system for the genome editing in *Clostridium acetobutylicum* DSM792T has promoted it to utilize both glucose and xylose (Bruder et al. 2016).

Valorization Process of Lignin by Extremozymes

Lignin is the most abundant natural aromatic compound in nature, composed of about 15 to 40% of the dry weight of plants. However, lignin cannot be fully utilized in the process of biofuel production due to the presence of aromatic compounds. Some microorganisms have evolved metabolic pathways, which can convert these aromatic substances through the process of biological funneling. "Biological funneling" provides a direct biological conversion for the high-value added utilization of lignin to overcome the heterogeneity problem in the lignin appreciation of modern biological refineries (Galkin and Samec 2016). Extreme microorganisms and their enzymes are the best solution for this problem. Several extremophiles have shown their potential for high-value utilization of lignin, such as the halotolerant and alkalophilic bacterium *Bacillus ligniniphilus* L1 which is able to significantly degrade lignin at the optimal pH 9 and produces aromatic compounds like vanillic acid, vanillin, etc. (Zhu et al. 2017). A thermophilic strain *Bacillus* sp. B1 has been reported to degrade cinnamic acid, ferulic acid and coumaric acid into catechol, protocatechuic acid and gentisic acid. In addition, a few thermo and halotolerant laccase have been obtained from *Bacillus* sp. SS4, *Thermobifida fusca* and *Trametes trogii* and these laccase with their laccase-mediator systems (LMS) have potential applications for lignin valorization due to LMS being able to depolymerize lignin into low-molecular weight phenolics and aromatics (Yang et al. 2020) (Fig. 14).

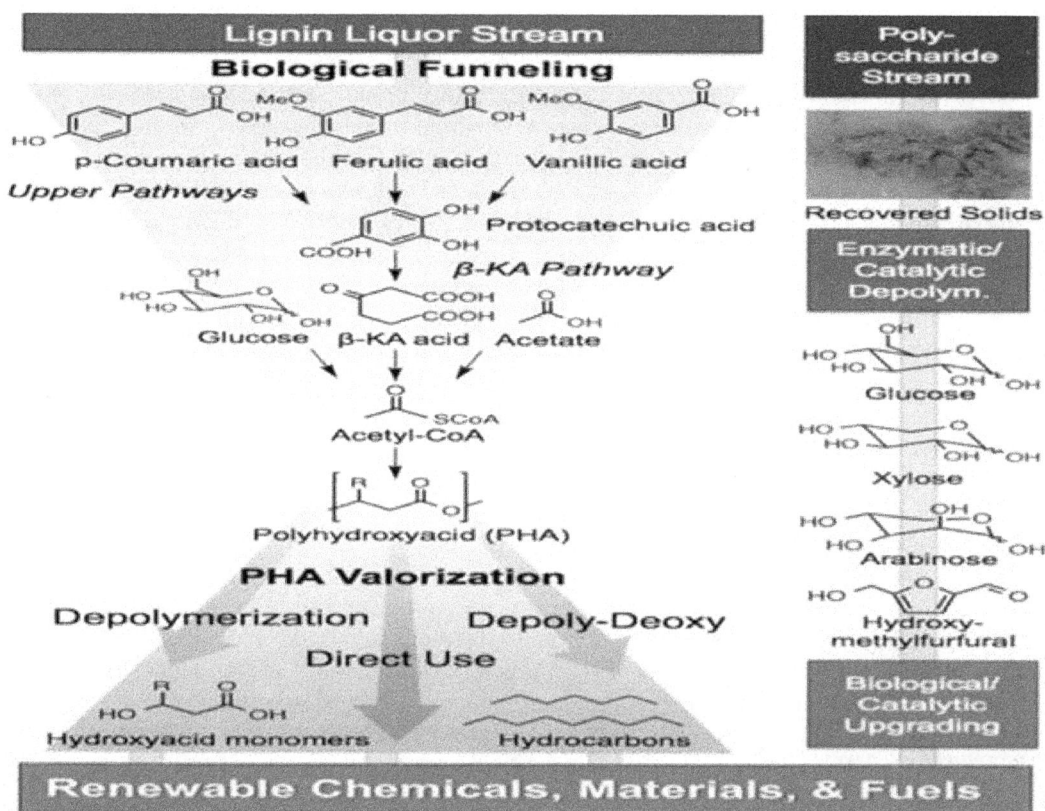

Fig. 14. Lignin valorization through biological funnelling. https://www.pnas.org/content/111/33/12013.

Extremozymes for Starch Hydrolysis

Starch is basically composed of two polymers, i.e., amylose (15–25%) and amylopectin (75–85%) and both are high molecular weight components. Both these components are composed of alpha 1,4-glycosidic and alpha 1,6-glycosidic linkages, respectively. Because of this complex structure of starch, starch processing is performed at high temperature which requires a combination of enzymes. Starch hydrolysis can result in a variety of products, principally glucose, fructose, dextrins, maltose and trehalose. Thermophilic (extremophile) enzymes may be applied to act in synergy with a high temperature starch solubilization process, while psychrophilic enzymes may assist in the cold starch hydrolysis process. The enzyme α-amylase hydrolyses starch polymers to a variety of oligomeric chains. Thermophilic α-amylases have been mostly reported for hyperthermophilic archaea of the genera *Methanocaldococcus, Pyrococcus, Sulfolobus* and *Thermococcus* (Van et al. 2007). Optimal activity typically resides around 90°C and the enzymes display impressive thermostability with activity being retained after 4 hours of autoclaving (120°C). Notable examples include amylases from *Methanocaldococcusjannaschii* (Topt 120°C and T1/2 of 50 hours at 100°C) (Antranikian and Egorova 2007).

A number of halophilic enzymes tolerant of high salt and solvent concentrations have been reported. Amylases from haloarchaea have been reported to display activity in various solvents like toluene, benzene and chloroform and to tolerate up to 4.5M NaCl and pH values as high as 10. Glucoamylase hydrolyses terminal residues on α-1,4-linked D-glucose chains, liberating D-glucose. Thermostable and acidophilic glucoamylases from the archaeal genera *Sulfolobus* (Kim et al. 2004) and thermoplasma (Topt 75–90°C and T1/2 of 24–40 hours at 60–90°C, pHopt of 2–5) are particularly noteworthy.

Transesterification of Oils by Extremozymes

The most commonly used method for biodiesel production is the transesterification of oils or fats involving lipases as a catalyst, producing high purity product and easy separation from the glycerol byproduct (Bisen et al. 2010). Lipases differ from esterases primarily in their preference for substrate chain length. Lipases prefer longer chain true lipids while esterases are more active on short chain esters. A huge variety of lipases and esterases has been reported and many are derived from extremophiles. Thermophilic examples include enzymes described for *Thermoanaerobacter* spp. and the archaeal genus *Sulfolobus* (Rao et al. 2010).

Commercial application of psychrophilic lipases has been moderate, save for a few well known examples such as the *Candida antarctica* lipases available from Sigma Aldrich and Novozymes (Cavicchioli et al. 2010). *C. antarctica* lipases have been applied for the production of biodiesel. The cost of the enzyme, however, remains the major barrier for industrial implementation of biodiesel. Several strategies including enzyme immobilization and protein engineering are being explored to develop this as a commercially viable source of biodiesel.

Extremozymes Used in Alcohol Production: Alcohol Dehydrogenase

Alcohol dehydrogenases (ADH) and monomeric sugar isomerases have application in the production of ethanol, butanol and secondary alcohols. *Thermoanaerobacter* and *Geobacillus* are well known ADH producers (Peng et al. 2008). Sugar isomerases play a significant role in tailoring carbohydrate streams (glucose, xylose or arabinose) to respective monomeric forms. Glucose isomerase is currently widely used in the production of High Fructose Corn Syrup in the food industry (Lim and Saville 2007), while arabinose isomerase is used in the production of sweeteners.

Enhancement in Enzyme Efficiency

Despite the potential advantages offered by extremozymes for application in biofuels, enzymes often require modification to achieve optimal activity in a particular process so as to improve process economics. Engineering of extremophilic enzymes is a routine approach taken by companies such as Novozymes and Verenium. The properties of individual enzyme components may be improved by rational design or directed evolution. Rational design relies on detailed knowledge of protein structure and structure–function relationships, ideally from high resolution crystallographic studies (Zhang and Fang 2006). Even with detailed knowledge of the enzyme structure, targets for rational design are difficult to predict. Directed evolution does not require detailed knowledge of enzyme structure or interactions between enzyme and substrate, but rather employs selective pressures, such as pH, thermostability and catalytic activity to 'evolve' enzymes according to desired characteristics. Directed evolution relies on the creation of large mutant libraries with DNA mutation techniques such as error-prone PCR (epPCR), DNA shuffling or staggered extension process (StEP) PCR (Zhao and Zha 2006). Rational and irrational designs have resulted in cellulases and hemicellulases with increased catalytic activity, enzyme stability, recombinant expression and tolerance to hydrolysis product inhibition (Maki et al. 2009). Specific modules targeted within enzymes include CBMs, catalytic sites, and surface structures. SCHEMA structure-guided recombination uses a modelling approach to generate novel enzymes by randomly shuffling "blocks" of amino acids between structurally closely related parent proteins. This approach was designed to reduce the number of inactive clones by limiting the extent of conformational disruption of the tertiary structure (Volkers et al. 2009). General trends and potential strategies for increasing internal stabilization have been reviewed and include the increase of ion-pair networks, disulphide- and salt-bridging, hydrogen bonding, hydrophobic and aromatic interactions and stabilization of surface exposed amino acids (Li et al. 2005).

Immobilization of Extremozymes

As compared to free enzymes in solution, immobilized enzymes are more robust and more resistant to environmental changes. More importantly, the heterogeneity of the immobilized enzyme systems

Table 1. Supporting material used for the immobilization of enzymes.

Sr. No.	Supporting materials	Example of supporting materials used	Reference
1.	Natural polymers	Cellulose, chitosan, agarose, starch, chitin, etc.	Monier 2010
2.	Synthetic polymers	Hydrophobic polypeptides, nylon fibers, glycidyl methacrylate, etc.	Sakai 2010
3.	Nano-particles	Cellulose coated nanoparticle, silver nanoparticles, TiO_2 particles, Nickle, etc.	Ahmed et al. 2013 Petkova et al. 2014

allows an easy recovery of both enzymes and products, multiple re-use of enzymes, continuous operation of enzymatic processes, rapid termination of reactions and greater variety of bioreactor designs.

Recently, various enzymes immobilized on carrier materials such as modified glass or polymethyl methacrylate have been developed for industrial use (Honda et al. 2017). Enzyme immobilization is usually achieved by chemical covalent linking, affinity labeling, physical adsorption, or entrapment (Table 1). These immobilization methods have been described in recent reviews (Elleuche et al. 2014). Adsorption is a simple and low-cost method. Enzyme immobilization by this method can provide high retention of enzymatic activity. However, adsorbed enzymes are likely to leak gradually from the carrier materials; therefore, the operational stability is lower than that provided by covalent linking. For covalent or noncovalent linking methods, surface modification of the carrier materials is generally necessary to improve the operational stability. Noncovalent binding is often achieved by affinity interactions such as that between a histidine-tagged enzyme and metal-ion-chelated beads (Wang et al. 2009). Although the interaction is specific, it requires the use of recombinant tagged enzymes. This method is therefore unsuitable for routine immobilization of all other enzymes. In contrast, covalent linking methods involve chemical reactions between the side-chains in enzymes and surface-modified carrier materials. These methods do not require any modification of the target enzymes. However, it requires expensive surface-modified carrier materials. Furthermore, multiple reaction points often alter the three-dimensional structure of the enzyme, leading to loss of activity and operational stability (Yiu et al. 2005). In addition, the reported procedures for preparing immobilized enzymes are usually multistep procedures. Facile techniques for routine immobilization of enzymes are therefore needed. Enzyme cross-linking, in which enzymes are covalently connected to each other by cross-linkers, can be partly categorized as covalent linking. This method enables the formation of miniature structures that consist of polymerized enzymes. Cross-linked enzyme aggregates (CLEAs) have recently been prepared for use in carrier-free enzyme-immobilization methods (Majumder et al. 2010).

They are easily prepared, and carrier costs are circumvented. The prepared CLEAs showed good retention of catalytic activity, high operational stability, and reusability. Improvement of the biophysical properties can significantly reduce the cost of enzymes and this makes industrial applications economically feasible. In this review, we specifically introduce recent advances in CLEA techniques for industrial applications (Franssen et al. 2013). Typical and modified methods for CLEA preparation are summarized. The immobilization of two or more enzymes is also covered. Finally, the use of CLEA reactors for chemical synthesis and environmental applications is introduced (Bilal et al. 2018).

Nanoparticles act as very efficient and attractive supporting material to bind up the enzymes. Nanoparticles are used for immobilization of extremozymes due to their unique properties like the biocatalysts' efficiency of enzyme, mass transfer resistance as well as specific surface area and easy separation method (Kalkan et al. 2014) (Table 2).

Production and Synthesis of Extremozymes

Extremozymes are produced commonly under two main processes of fermentations, i.e., submerged fermentation (SmF) and solid state fermentation (SSF) method traditionally (Fig. 15).

Table 2. Immobilized enzymes and their applications.

Sr. No.	Immobilized enzymes	Nanoparticle used	Application	Reference
1.	β-glucosidae	Iron oxide	Biofuel production	Verma et al. 2013
2.	Laccase	Chitosan magnetic nanoparticle	Treatment of lignocellulosic biomass	Kalkan et al. 2014
3.	Alpha-amylase	Cellulose coated nanoparticle, silver nanoparticles, TiO$_2$ particles	Degradation of starch	Ahmed et al. 2013
4.	Diastase	Nickle nanoparticles	Hydrolysis of starch	Prakasham et al. 2007
5.	Cellulase	TiO$_2$ particles	Hydrolysis of cellulose subunit, carboxy methyl cellulose	Ahmed et al. 2014
6.	Alcohol dehydrogenase	Gold and silver nanoparticles	Alcohol synthesis	Petkova et al. 2014

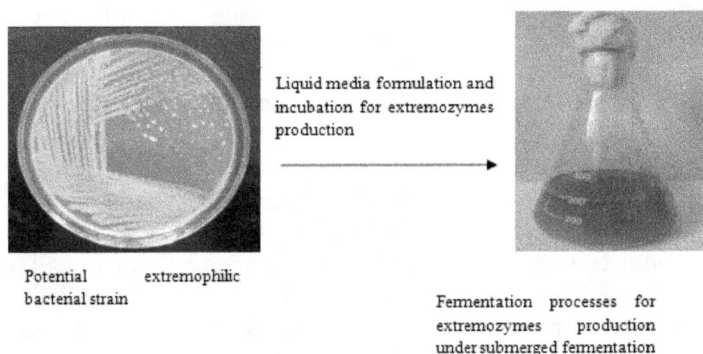

Liquid media formulation and incubation for extremozymes production

Potential extremophilic bacterial strain

Fermentation processes for extremozymes production under submerged fermentation

Fig. 15. Extremozymes' production under submerged fermentation (SmF).

Cellulolytic and xylanolytic enzymes are synthesized by a number of microorganisms. Among the microbes, bacteria are more reliable due to their adaptability to different environmental conditions, relatively fast growth than any other microbes and have the capacity to produce highly stable enzymes complement which serves as highly potent sources of individually important enzymes. Some of the examples are as *Bacillus* spp., *Micrococcus* spp., *Pseudomonas* spp. (Bischoff et al. 2006).

Submerged fermentation technology is used at the industrial level due to short period, low cost and high yield and purification of products is easier. In liquid culture, the control of the fermentation is simpler and consequently significant reductions in fermentation time can be achieved (Sanchez et al. 2009).

Analternative to this traditional SmF method is the solid state fermentation (SSF) method (Fig. 16), which involves the growth of microorganisms on solid materials in the absence of free liquids, which is best suited for fungal enzyme producers. SSF has three advantages viz. (i) lower consumption of water and energy (ii) reduced was test ream and (iii) more highly concentrated product. Some of the examples of potential fungal strains are *Trichoderma* spp., *Fusarium oxysporum*, *Aspergillu*s spp., *Rhizopus* spp., etc. (Jeffrey 2008).

Recent Advancements in Extremozymes

In the current time, the techniques like meta-genomics, meta-proteomics, meta-transcriptomics or metabolomics playan important role in the discovery of new potential enzymes for biorefinery with their improved activity and yield. In addition, bioinformatics and algorithms play an irreplaceable role in designing the *in situ* mutagenesis and gene shuffling to improve the stability of protein for

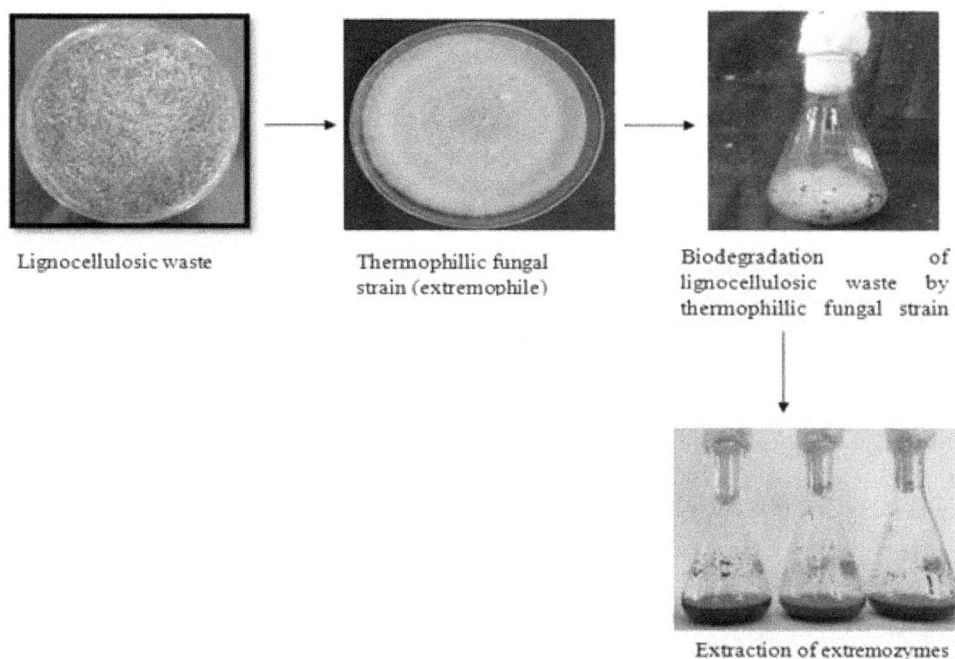

Lignocellulosic waste	Thermophillic fungal strain (extremophile)	Biodegradation of lignocellulosic waste by thermophillic fungal strain

Extraction of extremozymes

Fig. 16. Extremozymes' production under solid state fermentation (SSF).

potential industrial biorefinery purposes (Annamalai et al. 2016). These techniques have proved to be very useful for the development of extremozymes for biotechnology and provided a powerful tool for the discovery of new enzymes from nature (Juerges and Hansjusrgens 2018).

Whole Genome Sequencing

In the last decades, whole genome sequencing technology has been of great interest to understand the survival strategies of extremophiles under the extreme conditions. It provides the knowledge about the application of the metabolic pathways, substrate biotransformation, transport mechanism and enzymatic mechanism of extremophiles. Genomics research helps us to better understand the mechanism of robust enzymes and convey information about the three-dimensional structure of extreme enzymes. Based on the annotation and analysis of the genomic data of *Comamonas* SP 35 and combined with the metabolic analysis using GC-MS, the lignin degradation pathways have been elucidated (Zhu et al. 2018). Bioinformatics analysis of the whole genome sequence of psychrotolerant extremophile *Pseudomonas* sp. MPC6 revealed the identification of metabolism mechanism of toxic aromatic compounds and the synthetase system of natural PHAs and this *Pseudomonas* sp. MPC6 can be exploited as a biopolymer factory (Orellana-Saez et al. 2019). In recent years, the combination of computational and structural analysis with evolutionary driven methods has been significantly enhanced by identifying novel extreme enzymes with high industrial application potential (Ibrahim et al. 2016).

With the establishment of a large public database of genomicinformation, sequence-based methods, such as metagenomics, meta-transcriptomics, and meta-proteomics, have greatly increased the discovery of new biological systems (Caspi et al. 2019). Genomic data for more than 120 thermophilic bacteria are available in public database and the genomes of *Pyrococcus, Anaerobranca, Thermotoga, Thermoplasma,* and *Thermus* genera have been studied in detail and provide enzymes appropriate for industrial application (Counts et al. 2017). The availability of whole genome sequences, the construction of marker-free in-frame deletion mutants, and the homologs expression of proteins via ectopic integration of foreign genes in *Sulfolobus acidocaldarius* and *Sulfolobus islandicus* allow the modulation of the regulatory mechanisms and rechanneling of metabolic pathways to improve the production (Wagner et al. 2014).

Transcriptomics

Transcriptomics approaches help to study the organism's total content of ribonucleic acid transcripts in a cell including coding and non-coding RNAs, which can provide genome wide data and information on gene functions to reveal molecular mechanisms related to specific biological processes. Transcriptome analysis of extremophiles is helpful to reveal the dynamic changes of gene expression in harsh environment and a more comprehensive understanding of the functional and regulating networks of microorganisms adapting to the living environments (Jorquera et al. 2019).

A Multi-Omics Analysis

This approach revealed the thermal adaptation strategies of *Thermus filiformis* including oxidative stress induced by high temperature, which lead to the inhibition of genes involved in glycolysis and tricarboxylic acid cycle; glucosemetabolism is achieved mainly through the pentose phosphate pathway or the glycolysis pathway and the accumulation of oxaloacetic acid, a-ketoglutarate, and antioxidant enzymes related to free radical scavenging (Mandelli et al. 2017). Based on the transcriptome analysis of the digestive system of termites, it has been revealed that more than 14 kinds of auxiliary oxidoreductases and glycosyl hydrolase genes may be involved in the decomposition of lignin components and their redox networks during the process of biomass pretreatment (Geng et al. 2018).

Transcriptional Engineering

It has been shown to be a powerful tool to improve recombinant bacteria. It can change the pH value, ion demand and the product specificity and increases enzyme activity of the strains (Harman-Ware et al. 2017). A mutant alpha-amylase has been observed from the halophilic thermotolerant *Bacillus strains* cu-48 which is widely used for better industrial applications (Bibra et al. 2017).

Proteomics Analysis

The proteomic analysis of extremophiles has paid great attention to the revelation of its special adaptation to severe climate and environmental conditions. Proteome technology provides the knowledge for exploring the survival mechanism of extremophiles and promoted the further application of extremophiles in the field of bioenergy (Blachowicz et al. 2019). The proteome of *Bacillus ligniniphilus* L1 with lignin as a substrate revealed that there are more than 30 kinds of upregulated enzymes involved in lignin degradation like peroxiredoxin, cytochromeoxidase, oxidoreductase, ferredoxin, etc. Many environmental responses viz. repressor LexA, the DNA integrity scanning protein, the catabolite repression HPr-like protein, the central glycolytic genes regulator and the transcriptional regulator, which positively regulates lignin as a substrate, have been found (Zhu et al. 2017). The application of proteomics and gene-recombinant protein modification technologies helps to improve the characteristics of enzymes, such as thermal stability, higher activity, specificity, pH, solvent tolerance, etc. (Tesei et al. 2019). The information provided by genomics, proteomics, and transcriptome technologies can be used to identify new targets and metabolic engineering for the production of strains in the biorefinery industry. In this way, the metabolic regulations and the pathways can be extensively studied by multi-omics technologies to further improve the production and the performance of strains. Metabolic versatility and stability of *Thermoacidophilic sulfolobus* species has attracted researchers to use it as one of the promising platforms for synthetic biology and metabolic engineering (Schocke et al. 2019).

LC–MS/MS Shotgun Proteomics Analysis

It reveals that the obligate hydrocarbon-degrading psychrophile Oleispira Antarctica RB-8 expressed n-alkane oxidation pathway, which includes two alkane monooxygenases, two alcohol dehydrogenases, two aldehyde dehydrogenases, a fatty-acid-CoA ligase and a fatty acid desaturase. When grown on tetradecane (n-C14), the synthesis of these proteins increased 3- to 21-fold compared with the control group (Gregson et al. 2020).

The *Saccharomyces cerevisiae* INVSc1 strain, equipped with a synthetic genetic circuit, containing heat shock protein and superoxide dismutase from *Thermus thermophiles* HB8 and *Thermoanaerobacter tengcongensis* MB4, can grow well at 42°C and produces significantly more ethanol than its wild type. It is noteworthy to mention that alcohol dehydrogenases from extremophiles have been proved to be an excellent catalyst to produce butanol using cell-free systems. Metabolically engineered *P. putida* KT2440 MA-9 has been designed to produce cis and cis-muconic acid by using hydrothermally depolymerized lignin aromatics as a source, which is hydrogenated to form adipic acid and finally polymerized into nylon (Kohlstedt et al. 2018).

Deletion of vanillin dehydrogenase from the industrially important strains has been shown to enhance the production of vanillin from lignocellulose biomass (Linger et al. 2014). Alternatively, a thermo regulated-genetic system, i.e., the heterologous expression of two key enzymes, such as feruloyl-CoA synthetase (Fcs) and enoyl-CoA hydratase/aldolase (Ech) of thermophilic actinomycete Amycolatopsis thermoflava N1165in *E. coli* can also be used. This system allows *E. coli* to produce vanillyl alcohol using ferulic acid as a source at 30°C and subsequent conversion of vanillyl alcohol into vanillin at 50°C by the enzymatic activities of Fcs and Ech (Ni et al. 2018).

Utilization of renewable sources for the production of biofuels and bioenergy is a much-needed industrial sector to cope with global warming and draining of fossil fuels. Recent advancements in multi-omics technologies, the discovery of highly efficient lignin-degrading enzymes, the integration of physical, chemical,and biological methods in lignin depolymerization and the need for economically viable and eco-friendly production of biofuel and bioenergy necessitate the engineering of industrially important microorganisms through a synthetic biology approach.

3. Conclusion and Future Perspectives

At present, one of the biggest problems in biofuels industry is that it has not yet achieved the full utilization of lignin, cellulose, and hemicellulose. One of the key bottlenecks in the green revolution is difficulties associated with biodegradation of lignin. These problems have to be addressed in a scientific, eco-friendly, and cost-effective way, which in turn will boost the industrial processes for sustainable production of biofuel. Extremophiles and extremozymes bring the dawn of success for biorefinery owing to their specificity, robustness in action, and high tolerance to the adverse conditions of the biorefinery process. It is expected that more extreme bacteria and their enzymes/proteins will provide more competitive chassis cells and catalysts for the production of biofuels in an efficient way. Of course, the realization of all these goals necessitates the assistance of available traditional and modern cutting edge molecular biological tools. However, the production cost of extreme enzyme is still much higher than that of conventional enzyme preparation.

Finally, as the structure function rules that impart stability to enzymes under various extreme conditions become better understood, it will be possible to tailor specific extremophilic traits into any protein of interest by protein engineering or directed evolution possibly improving upon nature.

Acknowledgement

Authors gratefully acknowledge the financial support given by National Mission on Himalayan studies (NMHS), Ministry of Environment, Forest and Climate Change (MoEF&CC), Govt. of India, New Delhi and G.B. Pant National Institute of Himalayan Environment and Sustainable Development (GBPNIHESD), Kosi-Kataramal, Almora, Uttarakhand.

References

Ábrego, U., Z. Chen and C. Wan. 2017. Consolidated bioprocessing systems for cellulosic biofuel production. pp. 143–182. *In*: Li, Y. and X. Ge (eds.). Advances in Bioenergy. Amsterdam: Elsevier.

Annamalai, N., M.V. Rajeswari and T. Balasubramanian. 2016. Thermostable and alkaline cellulases from marine sources. pp. 91–98. *In*: Gupta, V.K. (ed.). New and Future Developments in Microbial Biotechnology and Bioengineering. Amsterdam: Elsevier.

Antranikian, G. and K. Egorova. 2007. Extremophiles, a unique resource of biocatalysts for industrial biotechnology. Physiology and biochemistry of extremophiles. C. Gerday and N. Glansdorff. Washington DC, ASM Press: 361–406.

Arevalo-Gallegos, A., Z. Ahmad, M. Asgher, R. Parra-Saldivar and H.M.N. Iqbal. 2017. Lignocellulose: A sustainable material to produce value-added products with a zero waste approach—A review. Int. J. Biol. Macromol. 99: 308–318.

Arnoldas Kaunietis, A.B., J. Donaldas, Citaviˇcius and P. Oscar. 2019. Heterologous biosynthesis and characterization of a glycocin from a thermophilic bacterium. Nat. Commun. 10: 1115.

Asha, B. and M. Palaniswamy. 2018. Optimization of alkaline protease production by *Bacillus cereus* FT 1 isolated from soil. J. Appl. Pharm. Sci. 8: 119–127.

Bhalla, A., N. Bansal, S. Kumar, K.M. Bischoff and R.K. Sani. 2013. Improved lignocellulose conversion to biofuels with thermophilic bacteria and thermostable enzymes. Bioresour. Technol. 128: 751–759.

Bhuiyan, N.H., G. Selvaraj, Y. Wei and J. King. 2009. Role of lignification in plant defense. Plant Signal Behav. 4: 158–159.

Bibra, M., R. Navanietha Krishnaraj and R.K. Sani. 2017. An overview on extremophilic chitinases. pp. 225–247. *In*: Sani, R.K. and R.N. Krishnaraj (eds.). Extremophilic Enzymatic Processing of Lignocellulosic Feedstocks to Bioenergy. Cham: Springer International Publishing.

Bilal, M., H.M.N. Iqbal, S. Guo, H. Hu, W. Wang and X. Zhang. 2018. State-of-the-art protein engineering approaches using biological macromolecules: A review from immobilization to implementation view point. Int. J. Biol. Macromol. 108: 893–901.

Birch, K. 2019. Background to emerging bio-economies in: Neoliberal Bio-Economies. New York, NY: Springer. 45–77.

Bischoff, K.M., A.P. Rooney, X.L. Li, S. Liu and S.R. Hughes. 2006. Purification and characterization of a family 5 endoglucanase from a moderately thermophilic strain of *Bacillus licheniformis*. Biotechnol. Lett. 28: 1761–1765.

Bisen, P.S., B.S. Sanodiya, G.S. Thakur, R.K. Baghel and G.B.K.S. Prasad. 2010. Biodiesel production with special emphasis on lipase-catalysed transesterification. Biotechnol. Lett. 32: 1019–1030.

Blachowicz, A., A.J. Chiang, A. Elsaesser, M. Kalkum, P. Ehrenfreund, J.E. Stajich, T. Torok, C.C.C. Wang and K. Venkateswaran. 2019. Proteomic and metabolomic characteristics of extremophilic fungi under simulated mars conditions. Front. Microbiol. 10: 1013.

Bohme, B., B. Moritz, J. Wendler, T. Hertel, C. Ihling, W. Brandt and M. Pietzsch. 2019. Enzymatic activity and thermoresistance of improved microbial transglutaminase variants. Amino Acids 52: 313–26.

Bruder, M.R., M.E. Pyne, M. Moo-Young, D.A. Chung and C.P. Chou. 2016. Extending CRISPR-Cas9 technology from genome editing to transcriptional engineering in the genus Clostridium. Appl. Environ. Microbiol. 82: 6109–6119.

Caspi, R., R. Billington, C.A. Fulcher, I.M. Keseler, A. Kothari, M. Krummenacker, P.E. Midford, W.K. Ong, S. Paley, P. Subhraveti and P.D. Karp. 2019. BioCyc: A genomic and metabolic web portal with multiple omics analytical tools. FASEB J. 33: 473.

Cavicchioli, R., T. Charlton, H. Ertan, S.M. Omar, K.S. Siddiqui and T.J. Williams. 2010. Biotechnological uses of enzymes from psychrophiles. Microb. Biotechnol. 4: 449–460.

Chen, G.Q. and X.R. Jiang. 2018. Next generation industrial biotechnology based on extremophilic bacteria. Curr. Opin. Biotechnol. 50: 94–100.

Cherubini, F. 2010. The biorefinery concept: Using biomass instead of oil for producing energy and chemicals. Energy Convers. Manage 51: 1412–1421.

Counts, J.A., B.M. Zeldes, L.L. Lee, C.T. Straub, M.W. Adams and R.M. Kelly. 2017. Physiological, metabolic and biotechnological features of extremely thermophilic microorganisms. Wiley Interdisc. Rev. 9: 1377.

Crawford, J.T., C.W. Shan, E. Budsberg, H. Morgan, R. Bura and R. Gustafson. 2016. Hydrocarbon bio-jet fuel from bioconversion of poplar biomass: Techno economic assessment. Biotechnol. Biofuels. 9: 141.

Demain, A.L., M. Newcomb and J.H. Wu. 2005. Cellulase, clostridia, and ethanol. Microbiol. Mol. Biol. Rev. 69: 124–154.

Ebaid, R., H. Wang, C. Sha, A.E.F. Abomohra and W. Shao. 2019. Recent trends in hyperthermophilic enzymes production and future perspectives for biofuel industry: A critical review. J. Cleaner Prod. 117925.

Elleuche, S., C. Schröder, K. Sahm and G. Antranikian. 2014. Extremozymes—Biocatalysts with unique properties from extremophilic microorganisms. Curr. Opin. Biotechnol. 29: 116–123.

Franssen, M.C., P. Steunenberg, E.L. Scott, H. Zuilhof and J.P. Sanders. 2013. Immobilised enzymes in biorenewables production. Chem. Soc. Rev. 42: 6491–6533.

Galkin, M.V. and J.S. Samec. 2016. Lignin valorization through catalytic lignocellulose fractionation: A fundamental platform for the future biorefinery. Chem. Susb. Chem. 9: 1544–1558.

Geng, A., Y. Cheng, Y. Wang, D. Zhu, Y. Le, J. Wu, R. Xie, J.S. Yuan and J. Sun. 2018. Transcriptome analysis of the digestive system of a wood-feeding termite (Coptotermes formosanus) revealed a unique mechanism for effective biomass degradation. Biotechnol. Biofuels. 11: 24.

Gilad, R., L. Rabinovich, S. Yaron, E.A. Bayer, R. Lamed, H.J. Gilbert and Y. Shoham. 2003. CelI, a noncellulosomal family 9 enzyme from *Clostridium thermocellum*, is a processive endoglucanase that degrades crystalline cellulose. J. Bacteriol. 185: 391–398.

Gregson, B.H., G. Metodieva, M.V. Metodiev, P.N. Golyshin and B.A. Mckew. 2020. Protein expression in the obligate hydrocarbon-degrading psychrophile Oleispira antarctica RB-8 during alkane degradation and cold tolerance. Environ. Microbiol. 22: 1870–1883.

Hamid, B., R.S. Rana, D. Chauhan, P. Singh, F.A. Mohiddin, S. Sahay and I. Abidi. 2014. Psychrophilic yeasts and their biotechnological applications-A review. Afr. J. Biotechnol. 13: 13644.

Harman-Ware, A.E., R.M. Happs, B.H. Davison and M.F. Davis. 2017. The effect of coumaryl alcohol incorporation on the structure and composition of lignin dehydrogenation polymers. Biotechnol. Biofuels 10: 281.

Honda, T., H. Yamaguchi and M. Miyazaki. 2017. Development of enzymatic reactions in miniaturized reactors. pp. 99–166. *In*: Innovations and Future Directions Applied Bioengineering, Wiley-VCH: Hoboken, NJ, USA.

Ibrahim, A.S., A.M. El-Toni, A.A. Al-Salamah, K.S. Almaary, M.A. El-Tayeb, Y.B. Elbadawi and G. Antranikian. 2016. Development of novel robust nanobiocatalyst for detergents formulations and the other applications of alkaline protease. Bioprocess Biosyst. Eng. 39: 793–805.

Indira, D., B. Das, P. Balasubramanian and R. Jayabalan. 2018. Sea water as a reaction medium for bioethanol production. pp. 171–192. *In*: Microbial Biotechnology, New York, NY: Springer.

Jeffrey, L.S.H. 2008. Isolation, characterization and identification of actinomycetes from agriculture soils at Semongok, Sarawak. Afr. J. Biotechnol. 7: 3697–3702.

Jiang, Y., F. Xin, J. Lu, W. Dong, W. Zhang, M. Zhang, H. Wu, J. Ma and M. Jiang. 2017. State of the art review of biofuels production from lignocellulose by thermophilic bacteria. Bioresour. Technol. 245: 1498–1506.

Jindou, S., I. Borovok, M.T. Rincon, H.J. Flint, D.A. Antonopoulos, M.E. Berg, B.A. White, E.A. Bayer and R. Lamed. 2006. Conservation and divergence in cellulosome architecture between two strains of *Ruminococcus flavefaciens*. J. Bacteriol. 188: 22, 7971–7976.

Jorquera, M.A., S.P. Graether and F. Maruyama. 2019. Bioprospecting and biotechnology of extremophiles. Front. Bioeng. Biotechnol. 7: 204.

Juerges, N. and B. Hansjürgens. 2018. Soil governance in the transition towards a sustainable bioeconomy–A review. J. Cleaner. Prod. 170: 1628–1639.

Kalkan, N.A., S. Aksoy, E.A. Aksoy and N. Hasirci. 2012. Preparation of chitosan coated magnetic nanoparticles and applications for immobilization of laccase. J. Appl. Polym. Sci. 123: 707–716.

Karimi, K. and M.J. Taherzadeh. 2016. A critical review of analytical methods in pretreatment of lignocelluloses: Composition, imaging, and crystallinity. Bioresour. Technol. 200: 1008–1018.

Karlsson, J., D. Momcilovic, B. Wittgren, M. Schulein, F. Tjerneld and G. Brinkmalm. 2002. Enzymatic degradation of carboxymethyl cellulose hydrolyzed by the endoglucanases Cel5A, Cel7B, and Cel45A from Humicola insolens and Cel7B, Cel12A and Cel45Acore from *Trichoderma reesei*. Biopolymers 63: 32–40.

Kim, M.S., J.T. Park, Y.W. Kim, H.S. Lee, R. Nyawira, H.S. Shin, C.S. Park, S.H. Yoo, Y.R. Kim, T.W. Moon and K.H. Park. 2004. Properties of a novel thermostable glucoamylase from the hyperthermophilic archaeon *Sulfolobus solfataricus* in relation to starch processing. Appl. Environ. Microbiol. 70: 3933–3940.

Kirsch, C., C. Zetzl and I. Smirnova. 2011. Development of an integrated thermal and enzymatic hydrolysis for lignocellulosic biomass in fixed-bed reactors. Holzforschung 65: 483–489.

Kohlstedt, M., S. Starck, N. Barton, J. Stolzenberger, M. Selzer, K. Mehlmann, R. Schneider, D. Pleissner, J. Rinkel, J.S. Dickschat, J. Venus, J.B.J.H.V. Duuren and C. Wittmann. 2018. From lignin to nylon: Cascaded chemical and biochemical conversion using metabolically engineered *Pseudomonas putida*. Metab. Eng. 47: 279–293.

Kuhad, R.C., R. Gupta and A. Singh. 2011. Microbial cellulases and their industrial applications. Enzyme Res. 280696.

Li, W.F., X.X. Zhou and P. Lu. 2005. Structural features of thermozymes. Biotechnol. Adv. 23: 271–281.

Li, X. and H.Y. Yu. 2013. Halostable cellulase with organic solvent tolerance from *Haloarcula* sp. LLSG7 and its application in bioethanol fermentation using agricultural wastes. J. Indust. Microbiol. Biotechnol. 40: 1357–1365.

Lim, L.H. and B.A. Saville. 2007. Thermoinactivation mechanism of glucose isomerase. Appl. Biochem. Biotechnol. 140: 115–130.

Linger, J.G., D.R. Vardon, M.T. Guarnieri, E.M. Karp, G.B. Hunsinger, M.A. Franden, C.W. Johnson, G. Chupka, T.J. Strathmann, P.T. Pienkes and G.T. Beckham.. 2014. Lignin valorization through integrated biological funneling and chemical catalysis. Proc. Natl. Acad. Sci. 111: 12013–12018.

Lynd, L.R., P.J. Weimer, W.H. van Zyl and I.S. Pretorius. 2002. Microbial cellulose utilization: Fundamentals and biotechnology. Microbiol. Mol. Biol. Rev. 66: 506–577.

Majumder, A.B. and M.N. Gupta. 2010. Stabilization of *Candida rugosa* lipase during transacetylation with vinyl acetate. Bioresour. Technol. 101: 2877–2879.

Maki, M., K.T. Leung and W. Qin. 2009. The prospects of cellulase-producing bacteria for the bioconversion of lignocellulosic biomass. Int. J. Biol. Sci. 5: 500–516.

Mandelli, F., M. Couger, D. Paixão, C. Machado, C. Carnielli, J. Aricetti, I. Polikarpov, R. Prado, C. Caldana, A.F.P. Leme, A.Z. Mercadante, D.M.R. Pachon and F.M. Squina. 2017. Thermal adaptation strategies of the extremophile bacterium *Thermus filiformis* based on multi-omics analysis. Extremophiles 21: 775–788.

Martin, A. and A. McMinn. 2017. Sea ice, extremophiles and life on extra-terrestrial ocean worlds. Int. J. Astrobiol. 17: 1–16.

Martinez, A.T., M. Speranza, F.J. Ruiz-Duenas, P. Ferreira, S. Camarero, F. Guillen, M.J. Martinez, A. Gutierrez and J.C. del Rio. 2005. Biodegradation of lignocellulosics: Microbial, chemical, and enzymatic aspects of the fungal attack of lignin. Int. Microbiol. 8: 195–204.

Miyazaki, M., M. Portia Nagata, T. Honda and H. Yamaguchi. 2013. Bioorganic and biocatalytic reactions. In Microreactors in Organic Synthesis and Catalysis, 2nd ed.; Wiley-VCH: Hoboken, NJ, USA.

Moreno, A.D., D. Ibarra, P. Alvira, E. Tomás-Pejó and M. Ballesteros. 2015. A review of biological delignification and detoxification methods for lignocellulosic bioethanol production. Critic. Rev. Biotechnol. 35: 342–354.

Moreno, A.D. and L. Olsson. 2017. Pretreatment of lignocellulosic feedstocks. pp. 31–52. *In*: Sani, R. and R. Krishnaraj (eds.). Extremophillic Enzymatic Processing Pf Lignocellulosic Feedstocks to Bioenergy. Springer, Cham.

Ni, J., Y.Y. Gao, F. Tao, H. Y. Liu and P. Xu. 2018. Temperature-directed biocatalysis for the sustainable production of aromatic aldehydes or alcohols. Angew. Chem. Int. Ed. 57: 1214–1217.

Orellana-Saez, M., N. Pacheco, J.I. Costa, K.N. Mendez, M.J. Miossec, C. Meneses, E.C. Nallar, A.E. Marcoleta and I.P. Castro. 2019. In-depth genomic and phenotypic characterization of the antarctic psychrotolerant strain *Pseudomonas* sp. MPC6 reveals unique metabolic features, plasticity, and biotechnological potential. Front. Microbiol. 10: 1154.

Patel, M.A., M.S. Ou, L.O. Ingram and K. Shanmugam. 2005. Simultaneous saccharification and co-fermentation of crystalline cellulose and sugar cane bagasse hemicellulose hydrolysate to lactate by a thermotolerant acidophilic *Bacillus* sp. Biotechnol. Prog. 21: 1453–1460.

Peng, H., Y. Gao and Y. Xiao. 2008. The high ethanol tolerance in a thermophilic bacterium *Anoxybacillus* sp. WP06. Sheng Wu Gong Cheng Xue Bao. 24: 1117–1120.

Pikuta, E.V., Z. Lyu, M.D. Williams, N.B. Patel, Y. Liu, R.B. Hoover, H.J. Busse, P.A. Lawson and W.B. Whitman. 2017. *Sanguibacter gelidistatuariae* sp. nov., a novel psychrotolerant anaerobe from an ice sculpture in Antarctica, and emendation of descriptions of the family Sanguibacteraceae, the genus Sanguibacter and species *S. antarcticus*, *S. inulinus*, *S. kedieii*, *S. marinus*, *S. soli* and *S. suarezii*. Int. J. Syst. Evol. Microbiol. 67: 1442–1450.

Pyne, M.E., M. Bruder, M. Moo-Young, D.A. Chung and C.P. Chou. 2014. Technical guide for genetic advancement of underdeveloped and intractable Clostridium. Biotechnol. Adv. 32: 623–641.

Rani, V., P. Sharma and K. Dev. 2019. Characterization of thermally stable b galactosidase from *Anoxybacillus flavithermus* and *Bacillus licheniformis* isolated from tattapani hotspring of North Western Himalayas, India. Int. J. Curr. Microbiol. App. Sci. 8: 2517–2542.

Rao, L., Y. Xue, C. Zhou, J. Tao, G. Li, J. R. Lu and Y. Ma. 2010. A thermostable esterase from *Thermoanaerobacter tengcongensis* opening up a new family of bacterial lipolytic enzymes. Biochim. Biophys. Acta. 1814(12): 1695–1702.

Rastogi, G., A. Bhalla, A. Adhikari, K.M. Bischoff, S.R. Hughes, L.P. Christopher and R.K. Sani. 2011. Characterization of thermostable cellulases produced by *Bacillus* and *Geobacillus* strains. Bioresour. Technol. 101(22): 8798–8806.

Sánchez, C. 2009. Lignocellulosic residues: Biodegradation and bioconversion by fungi. Biotechnol. Adv. 27: 185–194.

Schocke, L., C. Bräsen and B. Siebers. 2019. *Thermoacidophilic Sulfolobus* species as source for extremozymes and as novel archaeal platform organisms. Curr. Opin. Biotechnol. 59: 71–77.

Sharma, P., N. Sharma and N. Sharma. 2019a. Exploration of *Rhizoclonium* sp. algae potential under different ethanol production strategies with SEM analysis of biomass and detoxification of hydrolysate. Life Sci. J. 16(6): 73–83.

Sharma, N., N. Sharma and D. Tanwar. 2019b. An evaluation study of different white rot fungi for degradation of pine needles under solid state fermentation. Int. J. Curr. Microbiol. Appl. Sci. 8(6): 588–601.

Sheng, L., K. Kovács, K. Winzer, Y. Zhang and N.P. Minton. 2017. Development and implementation of rapid metabolic engineering tools for chemical and fuel production in *Geobacillus thermoglucosidasius* NCIMB 11955. Biotechnol. Biofuels 10: 5.

Singh, G., S. Singh, K. Kaur, S.K. Arya and P. Sharma. 2019. Thermo and halo tolerant laccase from *Bacillus* sp. SS4: Evaluation for its industrial usefulness. J. Gen. Appl. Microbiol. 65: 26–33.

Subramaniyan, S. and P. Prema. 2000. Cellulase-free xylanases from Bacillus and other microorganisms. FEMS Microbiol. Lett. 183: 1–7.

Tanwar, D., N. Sharma and N. Sharma. 2019. Evaluation of different process parameters for enhanced enzymes production using pine needles as substrate by *Trichoderma guizhouense* S5 [Accession No. MN170570] isolated from rotten wood under solid state fermentation. J. Chem. Biol. Phys. Sci. 9(4): 617–636.

Tesei, D., K. Sterflinger and G. Marzban. 2019. Global proteomics of extremophilic fungi: Mission accomplished. pp. 205–249. *In*: Fungi in Extreme Environments: Ecological Role and Biotechnological Significance. New York, NY: Springer.

Ummalyma, S.B., R.D. Supriya, R. Sindhu, P. Binod, R.B. Nair, A. Pandey, H.J. Busse, P.A. Lawson and W.B. Whitman. 2019. Biological pretreatment of lignocellulosic biomass—Current trends and future perspectives. pp. 197–212. In Second and Third Generation of Feedstocks. Amsterdam: Elsevier.

Van, T.T., S.I. Ryu, K.J. Lee, E.J. Kim and S.B. Lee. 2007. Cloning and characterization of glycogen-debranching enzyme from hyperthermophilic archaeon *Sulfolobus shibatae*. J. Microbiol. Biotechnol. 17: 792–799.

Varrella, S., M. Tangherlini and C. Corinaldesi. 2020. Deep hypersaline anoxic basins as untapped reservoir of polyextremophilic prokaryotes of biotechnological interest. Mar. Drugs. 18: 91.

Volkers, R.J., H. Ballerstedt, H. Ruijssenaars, J.A. de Bont, J.H. de Winde and J. Wery. 2009. TrgI, toluene repressed gene I, a novel gene involved in toluene-tolerance in *Pseudomonas putida* S12. Extremophiles 13: 283–297.

Vyas, G., N. Sharma, N. Sharma and P. Sharma. 2019. Scale up of bioethanol production under Separate hydrolysis and fermentation as well as simultaneous saccharification and fermentation from an Indigenous Algae-*Hydrodictyon* species of Indowestern Himalayas and its kinetic modeling. Trends Carbohyd. Res. 11(2): 68–79.

Wagner, M., A. Wagner, X. Ma, J.C. Kort, A. Ghosh, B. Rauch, B. Siebers and S. V. Albers. 2014. Investigation of the malE promoter and MalR, a positive regulator of the maltose regulon, for an improved expression system in *Sulfolobus acidocaldarius*. Appl. Environ. Microbiol. 80: 1072–1081.

Wang, L.S., F. Khan and J. Micklefield. 2009. Selective covalent protein immobilization: Strategies and applications. Chem. Rev. 109: 4025–4053.

Xu, C., Y. Qin, Y. Li, Y. Ji, J. Huang, H. Song and J. Xu. 2010. Factors influencing cellulosome activity in consolidated bioprocessing of cellulosic ethanol. Bioresour. Technol. 101: 24 9560–9569.

Yang, X., Y. Wu, Y. Zhang, E. Yang, Y. Qu, H. Xu, Y. Chen, C. Irbis and J. Yan. 2020. A thermo-active laccase isoenzyme from Trametes trogii and its potential for dye decolorization at high temperature. Front. Microbiol. 11: 241.

Yeoman, C.J., Y. Han, D. Dodd, C.M. Schroeder, R.I. Mackie and I.K. Cann. 2010. Thermostable enzymes as biocatalysts in the biofuel industry. Adv. Appl. Microbiol. 70: 1–55.

Yiu, H.H.P. and P.A. Wright. 2005. Enzymes supported on ordered mesoporous solids: A special case of an inorganic–organic hybrid. J. Mater. Chem. 15: 3690–3700. ·

Zabed, H., J. Sahu, A.N. Boyce and G. Faruq. 2016. Fuel ethanol production from lignocellulosic biomass: An overview on feedstocks and technological approaches. Renew. Sustain. Energ. Rev. 66: 751–774.

Zambare, V.P., A. Bhalla, K. Muthukumarappan, R.K. Sani and L. Christopher. 2011. Bioprocessing of agricultural residues to ethanol utilizing a cellulolytic extremophile. Extremophiles 15: 611.

Zhang, G. and B. Fang. 2006. Support vector machine for discrimination of thermophilic and mesophilic proteins based on amino acid composition. Protein Pept Lett. 13: 965–970.

Zhang, X., Y. Lin and G.Q. Chen. 2018. Halophiles as chassis for bioproduction. Adv. Biosyst. 2: 1800.

Zhao, H. and W. Zha. 2006. *In vitro* 'sexual' evolution through the PCR-based staggered extension process (StEP). Nat. Protoc. 1: 1865–1871.

Zhu, D., P. Li, S.H. Tanabe and J. Sun. 2013. Genome sequence of the alkaliphilic bacterial strain *Bacillus ligninesis* L1, a novel degrader of lignin. Genome Announce. 1: e0004213: 42–13.

Zhu, D., P. Zhang, C. Xie, W. Zhang, J. Sun, W.J. Qian and B. Yang. 2017. Biodegradation of alkaline lignin by *Bacillus ligniniphilus* L1. Biotechnol. Biofuels. 10: 44.

Zhu, D., W.A. Adebisi, F. Ahmed, B.D. Sethapathy and J. Sun. 2020. Recent development of extremophillic bacteria and their applications in Biorefinery. Front. Bioeng. Biotechnol. 8: 483.

Zhu, S., D. Lin, S. Xiong, X. Wang, Z. Xue, B. Dong, X. Shen, X. Ma, J. Chen and J. Yang. 2018. *Carnobacterium antarcticum* sp. nov., a psychrotolerant, alkaliphilic bacterium isolated from sandy soil in Antarctica. Int. J. Syst. Evol. Microbiol. 68: 1672–1677.

8

Psychrophilic Enzymes
Adaptations and Industrial Relevance

Shivika Sharma,[1,*] *Vikas Sharma,*[2] *Subhankar Chatterjee*[3] and *Sachin Kumar*[4]

1. Introduction

Temperature is a significant parameter that outlines the sustenance of life on this Earth. Our planet is occupied by extreme habitats that have a fluctuating temperature ranging from –89°C in Arctic and Antarctic regions to 400°C in hydro-thermal vents that lies on the deep seafloor (Siddiqui 2015). The biosphere of Earth is largely dominant by cold habitats with temperature conditions less than 5°C. This is due to the fact that approximately 80% of the Earth's shell is occupied by large oceans having temperature range 2–5°C below a depth of 1000 m. Also, the polar habitats add up for additional 15% on Earth's surface which include alpines and glaciers as well as the permafrost regions which represent approximately 20% of terrestrial soils (Feller 2012). All these cold adapted regions are occupied by a special form of organisms known as psychrophiles or can also termed as cold adapted microorganisms. Psychrophilic microorganisms thrive in permanent cold habitats such as glaciers, deep layers of ocean, alpines, shallow regions of subterranean, refrigerated devices and can sustain subzero conditions (supercooled liquid water) (Gupta et al. 2014). These cold adapted microorganisms comprise of cyanobacteria, bacteria (*Bacillus, Pseudoalteromonas, Pseudomonas, Vibrio and Arthrobacter* and *Bacillus*) (Okuda et al. 2004, Collins et al. 2002, Zeng et al. 2004), archaea (*Halorubrum* and *Methanogenium*), yeasts (*Crystococcus* and *Candida*) (Nakagawa et al. 2004), fungi (*Cladosporium* and *Penicillium*) (Sakamoto et al. 2003) and also include viruses and microalgae. The bacterial strain *Planococcus halocryophilus* Or1 isolated from permafrost Arctic region is a typical low temperature growing organism which can grow at –15°C and thus can be categorized as a typical psychrophilic microbial stain (Mykytczuk et al. 2013). Further metabolic active bacterial species have been also reported with a temperature of –30°C (Bakermans and Skidmore 2011). Thus, it can be illustrated that the psychrophilic microorganisms are the most diverse and abundant form of extremophiles with respect to diversity, distribution and biomass. This ability of psychrophiles to sustain in cold conditions is due to different adaptations at structural and functional level of all defined cellular elements (Collins and Margesin 2019). Thus, all cell

[1] Biochemical Conversion Division, Sardar Swaran Singh National Institute of Bio-Energy, Kapurthala, Punjab, India .
[2] Department of Molecular Biology and Genetic Engineering, Lovely Professional University, Jalandhar (Pb) India.
  Email: biotech_vikas@rediffmail.com
[3] Bioremediation and Metabolomics Research group, Department of Environmental Sciences, Central University of Himachal Pradesh, Temporary Academic Block-Shahpur, District-Kangra, Himachal Pradesh-176206, India.
  Email: schatt.cuhp@gmail.com
[4] Biochemical Conversion Division, Sardar Swaran Singh National Institute of Bio-Energy, Kapurthala, Punjab, India.
  Email: sachin.biotech@gmail.com
* Corresponding author: shivikasharma25@gmail.com

components of psychrophiles ranging from membranes and transport systems (which include intracellular solutes) along with proteins and nucleic acids should be properly adapted. Also, all essential cellular biological processes comprising of replication, transcription and translation should be well adapted to thrive in cold conditions (Cavicchioli et al. 2002). Psychrophilic microorganisms have to face various challenges to thrive in cold conditions. These challenges include low enzyme activity, decreased membrane fluidity, reduced thermal energy, protein cold-denaturation, modified transport systems, desiccation and ice formation, decreased diffusion, higher osmotic stress, increased viscosity of solvent, increased solubility of oxygen and also reactive oxygen species and a reduction in the solubility of nutrients and solutes (D'Amico et al. 2006). To surmount all challenges, psychrophiles are known to develop certain notable adaptations such as these cold adapted microorganisms produce large sum of unsaturated fatty acids, methyl-branched fatty acids and also shorter acyl- chain fatty-acids to amplify fluidity of membrane (Sarmiento et al. 2015, Chintalapati et al. 2004). Also, cold-adapted microorganisms have developed certain adaptive changes which include production of cold active enzymes (psychrozymes) and cold shock proteins and ice binding proteins (which restricts growth of ice crystals) (Mangiagalli et al. 2019). The cold adapted enzymes have higher specific activity at lower temperature conditions which is an important adaptation to balance the decreased rates of chemical reaction of metabolic pathways at cold conditions. The higher activity of psychrozymes is due to the loss of different non-covalent interactions which results in an increased flexibility of the conformation of enzyme. The adaptive feature of psychrophilic enzymes is genetically programmed inside the protein sequence and outcome from a long-time selection. Thus, cold adapted enzymes have to undergo different adaptation tools to maintain activity at extreme low temperature (Feller 2012). Researchers have acknowledged the importance of these cold adapted enzymes and have started to formulate different mechanistic pathways to develop these molecules. In this review, we will explore the adaptations, importance and industrial relevance of cold adapted enzymes (psychophilic enzymes).

2. Cold Adapted Enzymes: Psychrophilic Enzymes

Cold-adapted enzymes have high catalytic efficiency or high specific activity at lower temperature conditions in comparison to their mesophilic (25 to 45°C) and thermophilic (above 45°C) counterparts. Generally, these cold adapted enzymes catalyze reactions of higher magnitude at lower temperature conditions (Margesin et al. 2008). The standard temperature condition for the activity of psychrophilic enzymes (T_{opt}) falls in the range of 20–30°C (Gerday 2014). This property is dependent on the higher flexibility of certain structural regions which are very important for specific activity and thus result in the minimization of the activation energy (Santiago et al. 2016). Due to this lower activation energy and higher catalytic activity at lower temperature conditions, the cold adapted enzymes could bring reduction in energy consumption and have lower environmental impacts on metabolic reactions. The higher flexibility of these cold active enzymes balances the condition of low kinetic energy in extreme cold habitats. Due to their flexible structure, these enzymes have reduced activation energy ($\Delta H^{\#}$) and have negative entropy ($\Delta S^{\#}$) in comparison with mesophilic and thermophilic organisms. Thus, when temperature is lowered, the rate of reaction of psychrophilic enzymes also tends to decrease slowly. This equilibrium of certain thermodynamic factors is transformed into comparatively higher catalytic activity (or K_{cat}) at lower temperature conditions and also simultaneous lower structural stability of psychrozymes in comparison to enzymes of mesophilic and thermophilic microbial strains (Cavicchioli et al. 2011).

The comparative analysis of different crystal structures and protein models showed that distinguished structural and compositional factors that impart high range of flexibility to psychrophilic enzymes are usually contradictory to their stable and rigid mesophilic and thermophilic counterparts (Siddiqui and Cavicchioli 2006, Feller 2008). Cold-adapted enzymes exhibit reduction in the number and strength of various structural factors which are known to impart stabilization to protein molecules and thus have higher flexibility. The studies on psychrophilic enzymes revealed certain prominent amino acid substitutions which includes decreased proline residues in loops, lower arginine/lysine ratio, lesser

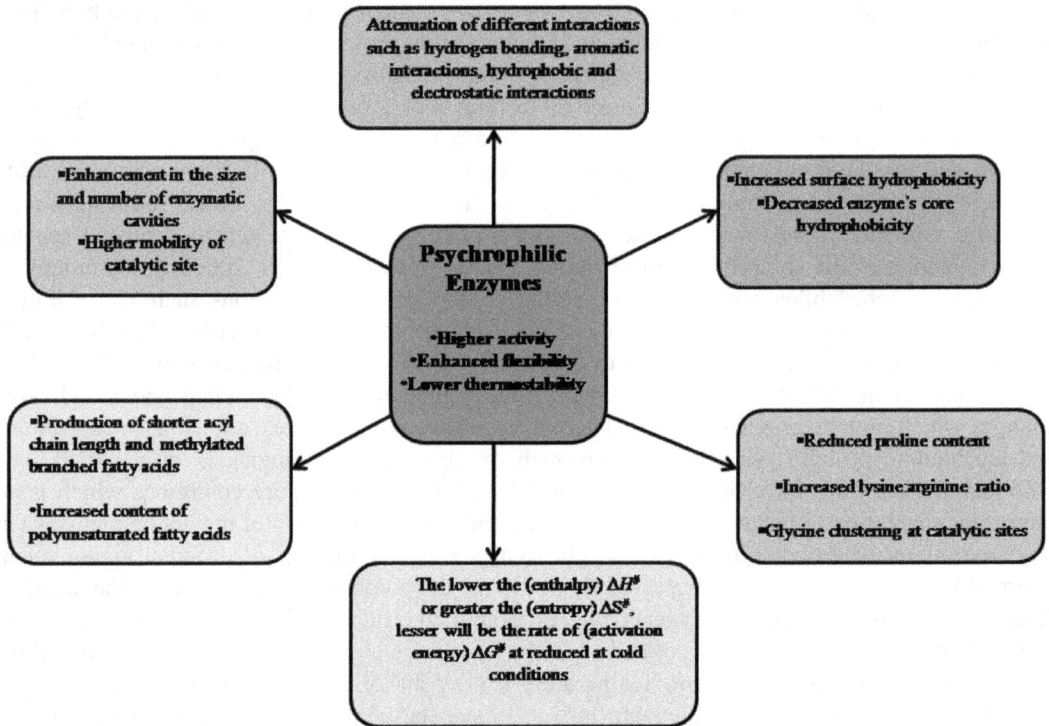

Fig. 1. Different adaptations of psychrophilic enzymes in extreme temperature conditions.

aromatic interactions, re-distribution of amino acid residues that facilitates solvent interaction, higher glycine residues, reduced hydrophobic packing, fragile inter-domain subunit interactions, reduction in disulfide bonds, lowered oligomerization, lesser electrostatic interactions (salt bridges, H-bonds, aromatic–aromatic interactions, salt-bridges and cation–pi interactions), fewer stabilizing cofactors and a higher conformational entropy of protein molecule in unfolded state (Fig. 1) (Feller and Gerday 2003). The genomic and crystallographic comparisons of psychrophilic, mesophilic and thermophilic microorganisms have showed that extremophilic adaptation of psychrophilic microorganisms comprises of diverse and significant arrangement of amino-acid composition which is a significant trademark of their adaptation in extreme cold conditions (Saunders et al. 2003). The experiments based on principles of site-directed and random mutagenesis performed on these psychrophilic enzymes showed that the catalytic activity at low temperature is not always associated with conformational volatility. Each set of cold adapted enzymes may undergo its individual significant set of amino acid modifications depending upon the extreme conditions of the habitat and the confirmation of substrate to be catalyzed. It is still not clear that the observed negotiation in relative activity and stability of these cold-adapted enzymes is due to structural alterations such as increased flexibility in specific region of enzyme's configuration to counter the effect of higher degree compactness of proteins at lower temperature conditions or due to arbitrary genetic drift and appropriately lower selective pressure required for enzymes in cold habitats (Arnold et al. 2001).

3. Adaptations in Cold Adapted Enzymes

3.1 Enzyme Kinetics

The rate of chemical reaction is defined by Arrhenius equation (Laidler 1984) which states an inversely proportional relationship between activation energy (*E*a) and the rate of reaction (*K*) and is direct proportional to the absolute range of temperature (*T*) (Dhaulaniya 2018):

Rate of reaction $(K) = Ae^{-Ea/RT}$ [I]

Generally, the rate of reaction is affected by enzyme by reducing the barrier of activation energy (*E*a) and turning higher number of substrate molecules to form enzyme-substrate intermediates and finally products (Siddiqui and Cavicchioli 2006). Thus, it is clear from Eqn (I) that *E*a is temperature dependent. In cold conditions, the pace of diffusion of solute as well as of solvents decreases, leading to significant rise in kinetic energy which is required to surmount the activation energy barrier and this enhancement in rate of kinetic energy ultimately retards the overall reaction rate (Peterson et al. 2007). The cold adapted organisms (psychrophiles) have evolved different compensatory techniques to evade the impact of cold conditions on their metabolism rate. This comprises of a higher concentration of enzyme and development of certain enzymes in which the reaction velocity would be specifically controlled by process of diffusion and would not be dependent on the temperature criteria (Georlette et al. 2004). Psychrophilic enzymes optimize their catalytic efficiency (K_{cat}/K_m) by raising the value of K_{cat} value or by lowering *Km* value (Fields and Somero 1998, Wolfenden et al. 2011).

3.2 Flexibility of Enzyme

The various factors which grant enhanced flexibility to these cold adapted enzymes are usually quite opposite to their mesophilic homologues (Violot et al. 2005). These alterations attained by these cold active enzymes comprise of decreased enzyme's core hydrophobicity and increased surface hydrophobicity. Also, the feeble subunit interactions, fewer disulphide linkages and extended loops with lower electrostatic interactions all at once enhance the flexibility of psychrozymes (Cavicchioli et al. 2011, Marx et al. 2007). The concept of flexibility in these cold active enzymes can be described by either global or local flexibility concept (Fields and Somero 1998). This can be explained by an example of zinc metalloprotease which is isolated from an Arctic-ice bacterial strain and the complete enzyme's structure appears to be evenly flexible (global flexible) due to overall reduction in hydrogen bonds in the enzyme confirmation (Xie et al. 2009). Also, in some cold adapted enzymes such as carbonic anhydrase and α-amylase, the flexibility is bound to be localized in specific regions such as regions surrounding the active site (Chiuri et al. 2009, D'Amico et al. 2003). In case of cold-active citrate synthase, the flexibility was confined to other regions of enzyme molecule instead of region surrounding active site (Bjelic et al. 2008). These research findings demonstrate that the flexibility in cold-adapted enzymes can be manifested in different specific ways depending upon classification of enzyme and conditions of prevailing environment.

3.3 Enzyme's Structural and Thermal Stability

3.3.1 Structural Stability of Enzyme

Psychrophiles have different adaptations in their enzymatic structure leading to lesser structural stability in comparison to mesophilic homologues (Lee et al. 2016, Sindhu et al. 2017). This reduction in structural stability is due to attenuation of different interactions such as hydrogen bonding, aromatic interactions, charge–dipole interactions, hydrophobic interactions and ion bonding (De Maayer et al. 2014). Also, lack of different salt bridges in these cold adapted enzymes maintains the structural stability due to the absence of arginine residues (Ramli et al. 2013). Presence of sufficient number of H-bonds imparts stability to three dimensional enzyme structure (Creighton 1991). Psychrophilic enzymes have feeble aromatic interaction in comparison to mesophilic enzymes and this is an important factor for lower structural stability of these cold adapted enzymes. Other important feature which is liable for the higher structural solidity in mesophiles is grouping of hydrophobic side chains inside the protein's core structure whereas in psychrophilic enzymes substitution of these hydrophobic side chains inside the hydrophobic center of proteins have lowered the value of hydrophobicity index (DasSarma et al. 2013).

3.3.2 Thermal Stability of Enzyme

The trend of decreased thermal solidity or stability has now been studied in maximum number of psychrozymes (Maiangwa et al. 2015). The main thermolabile constituents in these psychrophilic enzymes are active sites (Georlette et al. 2004). Studies on cold adapted enzymes (membrane-bound or unbound) have revealed that there is higher temperature sensitivity due to the deformation in the lipid bilayer as well as in its related proteins (Arcus et al. 2016). Also, the process of cooperative uncoiling occurs in cold-adapted proteins, particularly at low molecular weight proteins at higher temperature conditions (Siddiqui and Cavicchioli 2006). This can be explained in case of α-amylases isolated from psychrophilic bacterial strain *P. haloplanktis* which showed that there is appearance of cooperative uncoiling because of reduced interactions which is a key factor for maintaining the structural firmness (D'Amico et al. 2001). The other phenomenon is cold denaturation of enzyme structure that usually occurs at a temperature below T_{max}. This cold denaturation of enzyme disturbs the structure of enzyme and also hinders rate of reaction at cold conditions (Aznauryan et al. 2013). Cold denaturation is due to distortion of weak hydrophobic interactions which influences the phenomenon of protein coiling at cold conditions (Graziano 2014). Also, cold denaturation comprises of hydration of different polar as well as non-polar groups, thus suggesting that these hydrophobic interactions are of least importance in psychrophilic enzymes (Vajpai et al. 2013).

3.4 Membrane Fluidity of Enzyme

The feature of structural integrity and stability of the enzymatic membrane is reliant on a significant factor which is fluidity of enzyme membrane (Deming 2002). The physical properties of membrane are conferred by lipid composition and extreme cold surroundings have a negative effect on the composition of membrane (D'Amico 2006). At extreme low temperature conditions, membrane undergoes a change in its transition state forming a gel phase state resulting in loss of membrane's properties. Thus pyschrophiles have to undergo different adaptive measures to increase the membrane fluidity. This includes production of shorter acyl chain methylated branched fatty acids and there should be increased concentration of unsaturated fatty acids and polyunsaturated fatty acids (PUFA) (Chintalapati et al. 2004). Further higher concentration of lipid head clusters and proteins also enhance the fluidity of membrane (Chintalapati et al. 2004). In an analysis study, it was shown that increased content of isoleucine in the sequences situated outside the bilayer is also a key factor for increase in membrane fluidity. Also, decreased content of alanine in the membrane proteins was studied which results in lower helix formation leading to decreased stability (Kahlke and Thorvaldsen 2012). The transcriptomic analysis revealed that phenomenon of gene up regulation, which induces production of lipopolysaccharides, peptidoglycans and additional significant membrane proteins, has important role in counteracting the negative effects of cold conditions and increasing the fluidity of membrane (Deming 2002). Additional factors which are accountable for maintaining the fluidity factor of membrane include formation of carotenoids (polar and non-polar) and wax esters (Rodrigues et al. 2008, Chattopadhyay 2006).

4. Adaptations in Active Site Architecture of Enzyme

The various adaptations are also reported at the active site of psychrophilic enzymes. In cold adapted enzymes, a superior cavity of the catalytic cleft has been observed (Jung et al. 2008) and in case of a cold-active citrate synthase, this catalytic cleft opening has been achieved by different ways such as the bulky side chains have been replaced by smaller chain groups or by introducing minute deletions in the loops surrounding the active site and thus forming a discrete conformation of the grouped loops that outline the border of active site (Russell et al. 1998). In the study of psychrophilic protease isolated from cold adapted *Pseudomonas*, an extra calcium (Ca^{2+}) ion draws the backbone outlining the opening of site and thus enhancing its accessibility as compared to mesophilic counterpart (Aghajari et al. 2003). Due to this enhanced accessibility of active site, the psychrophilic enzymes have an increased potential to accommodate a range of substrates at lesser energy cost and thus lower down the barrier of activation energy which is required for the enzyme-substrate complex (Bjelic et al. 2008). Also, the larger catalytic site of psychrophilic enzymes would also assist in easy release of product and thus might lighten the effect of rate-limiting step on the rate of reaction (Vos et al. 2006). The major difference in electrostatic potentials around the catalytic site of psychrophilic enzymes also plays an important function in maintaining good relative activity at extreme lower temperature conditions. The occurrence of surface electrostatic potentials produced by various charged polar groups is an important parameter of catalysis pathway because as this potential expands in the medium, there is an immediate orientation and attraction of the substrate way before any type of contact between the substrate and enzyme occurs. It is very interesting to know that a range of cold active enzymes such as citrate synthase, elastase, malate dehydrogenase, trypsin and uracil-DNA glycosylase are differentiated by differences in their electrostatic potentials produced by charged and polar groups near the region surrounding the active site of psychrophilic enzymes in comparison to their moderate and thermal counterparts in which interaction with ligand improves the catalysis (Russell et al. 1998, Papaleo et al. 2007, Gorfe et al. 2000, Brandsdal et al. 2001, Raeder et al. 2010). The study of psychrophilic enzyme malate dehydrogenase illustrated the positive amplified potential at the binding site of oxaloacetate and the reduced negative potential around NADH binding region (Kim et al. 1999). Some cold adapted enzymes have additional strategies to combat the cold conditions. The crystallographic studies revealed that maximum psychrophilic enzymes exist in homo-tetrameric forms except in exclusive cold adapted enzyme $\beta\beta$-galactosidases which is a homo-hexamer and possess six catalytic sites (Skalova et al. 2005). This exclusive structure comprises of two additional active sites which would certainly help in improving the catalytic activity at extreme low temperature conditions.

The crystallographic cellulose structure displays a modular conformation comprising of a unique globular catalytic domain which is connected through linker to another cellulose binding domain. The linkers present in cold adapted cellulases were typical long linkers with above 100 amino acid residues and this length is approximately five times more than their mesophilic cellulosic counterparts (Garsoux et al. 2004). It was also studied that the catalytic confirmation of psychrophilic cellulase has appropriately 40-fold extended available surface area of substrate, which also improves the activity of psychrophilic cellulases at lower temperature conditions.

5. Enzyme Activity

The detailed studies on the extremophilic cold-adapted enzymes have exhibited different molecular adaptations to invalidate all the unfavorable consequences of cold conditions on the catalytic activity of these enzymes (Dick et al. 2016). Also, the energetic studies by Tattersall et al. (2012) showed that lower temperature conditions could reduce the reaction speed by raising the barrier of free activation energy ($\Delta G^{\#}$). Thus, to lower down this activation energy barrier and to balance the harmful effects of low temperature conditions, these psychrophilic enzymes have to undergo some survival modifications such as they have to increase the value of their *Km* to speed up the rate of reaction (Lian et al. 2015). As we know that the barrier formed between lower level and transition level is free activation energy, thus, lesser the energy barrier, higher is the reaction rate, which results in induction of higher enzymatic activity. The transition theory states that after the formation of enzyme-substrate complex, it is subjected to an energy pit, i.e., ES complex will be elevated to reach to an activated state ES$^{\#}$ (greater the enzyme affinity towards substrate, more energy is needed by ES to elevate to ES$^{\#}$). This activated ES$^{\#}$ will then finally break down into product releasing free enzyme.

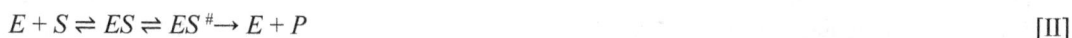

$$E + S \rightleftharpoons ES \rightleftharpoons ES^{\#} \rightarrow E + P \qquad\qquad\qquad [II]$$

In psychrophilic enzymes, the barrier of free activation energy is decreased by two main strategies. In the first strategy, it has been studied that the decreased affinity for the substrate will result in lowering the scale of energy barrier and finally increase the enzyme activity as studied by Feller (2012). The other survival strategy comprises of energetic study of cold adaptive enzymes. According to the energetic kinetics, the Gibbs–Helmholtz stated that the free activation energy ($\Delta G^{\#}$) is fully dependent on both the factors of enthalpy ($\Delta H^{\#}$) as well as entropy ($\Delta S^{\#}$).

$$\Delta G^{\#} = \Delta H^{\#} - T\Delta S^{\#} \tag{III}$$

Also, another finding of transition state theory defines that turn over number of enzyme (K_{cat}) is related to temperature and free energy of activation as proposed by Siddiqui and Cavicchioli (2006):

$$K_{cat} = (K_B T/h)e^{-\Delta G^{\#}/RT} \tag{IV}$$

Merging equations (III) and (IV) will help in considering the impact of both enthalpy $\Delta H^{\#}$ and entropy $\Delta S^{\#}$ on the *Kcat* directly by following equation (Eyring 1935):

$$K_{cat} = (K_B T/h)e^{-\{(\Delta H^{\#}/RT)+(\Delta S^{\#}/R)\}} \tag{V}$$

It is clear from the equation that lower the $\Delta H^{\#}$ or greater the $\Delta S^{\#}$, less will be the $\Delta G^{\#}$ at cold situations (Lonhienne et al. 2000). The lower $\Delta H^{\#}$ feature has been studied on almost all cold adapted enzymes (Feller 2012). The activation enthalpy is decreased by reducing enthalpy related interactions which are required to be dissociated during the formation of transition state, ultimately enhancing the flexibility feature specifically in catalytic position of the psychrozymes (Siddiqui and Cavicchioli 2006).

6. Industrial Utilization of Psychrophilic Enzymes

In current scenario, there is a lot of focus on different scientific and industrial strategies for the development of novel psychrophilic enzymes. The various internal significant features of psychrophilic enzymes such as higher activity and thermolability makes them an excellent contender for diverse biotechnological applications and can be utilized in different industries ranging from food and detergent to molecular biology, to beverage preparation and bioremediation (Table 1) (Cavicchioli et al. 2011, Feller 2012). Thus, due to their tendency to work at extreme range of temperature and pressure, these cold adapted enzymes are replacing their mesophilic homologues for various industrial purposes. As these extremozymes have higher value of activity, the expected results from physio-chemical reactions could be obtained by using minute amount of enzyme concentration under optimized conditions. Also, due to their thermostability, these psychrophilic enzymes can be inactivated in its compound mixture by raising the temperature conditions of reaction and this property in which enzyme is required for a particular time interval in a reaction could be utilized in different applications, such as tenderizing of meat by cold active proteases and also in the area of molecular biology. Since a range of psychrophilic enzymes can function at low as well as moderate temperature range, these enzymes are able to catalyze various reactions at room temperature and need no heat input into system resulting in cost-proficient process. Cold adapted microorganisms are the source of novel and interesting enzymes which have various adaptations to perform innovative reactions in different industrial applications.

6.1 Food and Pharmaceutical Industries

Due to the high catalytic activity of psychrophilic enzymes, these are very much utilized in food processing industry as use of these cold enzymes minimizes the spoiling of food and also inhibit any unwanted changes in the sensory and nutritional qualities of food. Several industrial important psychrophilic enzymes are used to increase filterability factor and enhance the aromatic flavor qualities of food products (Sarmiento et al. 2015). In the recent scenario, food industry is particularly replacing higher temperature processing activities with low temperature processing activities as low temperatures offers more environmental and economic advantages. These benefits comprise of energy saving processing, minimum spoilage and contamination, preservation of flavor compounds

Table 1. Industrial utilization of different psychrophilic enzymes and applications.

Industrial Market	Enzyme	Uses	References
Food	Pectinases	Fermentation of beer and wine, bread manufacture and fruit juice processing	Sarmiento et al. 2015
	Lipases	Slow-ripening of cheese	
	Proteases	Tenderization of meat and improvement of taste and flavors of frozen meat products	
Pharmaceutical	β-galactosidases	Improvement of digestion in patients suffering from lactose intolerance Formation of glucose-rich and galactose-rich syrups	Siddiqui 2015
	Esterases	Synthesis of chiral drugs	
Detergent	Cellulases	Wash of cotton fabrics	Margesin et al. 2008
	Lipases	Breaking down of lipid stains	
	Proteases	Breaking down of protein stains	
	Pectate lyases	Pectin-stain removal activity	
	Amylases	Breakdown of starch-based stains	
Molecular Biology	Alkaline phosphatases	Dephosphorylation of 5′ end of alinearized fragment of DNA	Mangiagalli et al. 2019
	Nucleases	Digestion of all types of DNA and RNA	
	Uracil-DNA N-glycosylases (UNGs)	Release of free uracil from uracil-containing DNA	
Textile	Cellulases	Bio-finishing combined with dyeing of cellulosic fabrics	Sarmiento et al. 2015
	Amylases	De-sizing of woven fabrics	
Biofuels	α-Amylases Glucoamylases	These enzymes perform the catalysis that help in the release of fermented sugar from the starch and is directly used by the yeast leading to process of ethanol production.	Shetty et al. 2005
	Cellulases Hemi-cellulases	Hydrolysis of lignocellulose biomass for the production of second generation biofuel.	Felby et al. 2006
Bioremediation	Flevodoxin reductase and Cytochrome P450 alkane hydrolases	Ability to hydrolyse alkane and degrade derivatives of nitrotriazine at lower temperature conditions of 5°C.	Bowman and Deming 2014

(volatile), higher degree of reaction control as these cold adapted enzymes can be deactivated at higher range of temperature and elimination of undesirable side chain reactions (Pulicherla et al. 2011). Cold-adapted lipases can reduce the manufacturing costs by acting as rennet alternatives and increasing the maturation process of slow-ripening of cheese which requires specifically lower moisture and lower temperature conditions. Psychrophilic proteases can be utilized for tenderization of meat and enhancing taste and flavors of frozen meat items (He et al. 2004). Also, these proteases are used for amputation of unwanted tissues from seafood (removal and de-scaling of skin of fish) and extraction of important carotene-proteins from marine shell-fish (Shahidi and Kamil 2001). In the fruit processing industry, cold active pectin-depolymerizing enzymes (pectate lyases and polygalacturonases) along with psychrophilic hemicellulases can be utilized for manufacture of different products, e.g., fruit juices (Nakagawa et al. 2004).

The psychrophilic β-galactosidases which can function at normal pH have the ability to improve the digestion of different dairy products and this property is very beneficial for those consumers suffering from lactose intolerance. The higher catalytic activity of cold active β-galactosidases enzymes which can operate at acidic range of pH can lower down the impact of pollution and enhance the industrial utility of whey which is an outcome of dairy (cheese) industry by forming important syrups such as glucose-rich and galactose-rich syrups. These syrups can be utilized as prominent sweeteners in a wide range of food articles (Gerday et al. 2005). In baking industry, different psychrozymes (such as amylases, lipases, proteases gluco-oxidases and xylanases) are employed for the modification of composition of starch, hemicelluloses and gluten during the preparation and processing of dough which usually occurs at moderate temperatures, i.e., below 35°C. The action of these psychrophilic enzymes might help in improving the elasticity of dough that result in a bigger quantity of loaf and also an improvement in the crumb configuration.

Psychrophilic enzymes having high specific activities at lower temperature conditions can be utilized in various pharmaceutical applications. Nowadays, there is an increased demand specifically for the enantiomer-pure drugs, which has led to the exploitation of biocatalysis in organic mediated synthesis (Cavicchioli et al. 2011). Psychrophilic enzymes have an immense potential in organic synthesis comprising of both aqueous-organic solvents and non-aqueous solvents. The flexibility of these cold adapted enzymes neutralizes the stabilization impacts of lower water activity in organic synthesis (Gerday et al. 2000, Sellek and Chaudhuri 1999). For this, psychrophilic lipases and esterases exhibited higher degree of stereo-specificity for chemical defined synthesis (Tutino et al. 2009, Joseph et al. 2008). Due to the stereo-specificity of psychrophilic enzymes, these are used for the synthesizing of chiral drugs of immense potential (Jeon et al. 2009). A psychrophilic lipase isolated from *Candida antarctica* have wide range of applications which comprises of variation of polysaccharides, resolution of amines and alcohols and de-symmetrization of complex drug intermediates (Suen et al. 2004).

6.2 Enzymes for Detergents and Personal Care

Enzymes have a major dominance in the market of detergent formulation and are very actively used in detergent industry across the globe. In the current trend, the utilization of detergents at lower temperature washing conditions is very actively employed because of decreased energy consumption, reduced cost infrastructure and for the maintenance of texture of fabrics. Among various enzymatic agents, subtilisin which is an alkaline serine protease isolated from *Bacillus* species mainly rules detergent industry (Feller 2012). Accordingly, psychrophilic subtilisins are being actively developed for washing at moderate temperature conditions. To achieve this goal, the first cold active subtilisin isolated from psychrophilic *Bacillus* species from Antarctic continent has been produced to cope up with the demand of detergent industry (Narinx et al. 1997). The cold active detergents containing engineered cold active subtilisin have various advantages such as alkaline stability and cold activity which make them ideal contender in detergent formulation.

With this trend of lower wash temperatures, different enzyme companies specifically in Europe and Japan have started investment in the formulation of detergents containing prominent cold active enzymes and ingredients which are very efficient at lower temperature conditions. The examples from Novozymes comprises of: Kannase®, a form of detergent containing protease which is competent in tough stain removal at washing temperature conditions of 10–20°C; Savinase®, a detergent containing alkaline protease and removing rough stains at room temperature conditions; Celluzyme®, a cellulase containing detergent which efficiently performs hydrolysis at a temperature below 15°C and lastly Duranyl® which is able to eliminate starch stains at ambient temperatures. A detergent formulation industry, Genencor, have developed Properase® and Purafect® Prime (proteases) which are very effective under lower temperature washing conditions (Margesin et al. 2008). There are patents that are filed for different psychrophilic proteases isolated from cold adapted microorganisms from Arctic and Antarctic continents that are used in detergent formulation industry (Asenjo et al. 2006, Brenchley et al. 2001). Studies have shown that cold active protease exhibited very high stain removing efficiency at a temperature range of 0–37°C when compared with other commercial mesophilic counterparts (Brenchley et al. 2001). Thus, utilization of psychrophilic enzymes is much more cost efficient as compared to its mesophilic counterparts as better results can be achieved by smaller enzyme concentrations. Because of these characteristics, the detergents having cold active enzymes are used for household and industrial cleaning purposes.

Also in cosmetic industry, these psychrophilic enzymes can be evaluated to increase the outcome of bio-transformations involving volatile substrates, which are prone to evaporation at higher temperature ranges (Trytek and Fiedurek 2005). A range of proteolytic cold adapted enzymes which have good activity lower and ambient temperature conditions have vast applicability in skin treatment, used in wound healing gels and in different gels for skin infection.

6.3 Psychrophilic Enzymes in Molecular Biology and Enzyme Nano-biotechnology

Cold-adapted enzymes have numerous benefits which are required for the enzymatic sequential methods used in different techniques of molecular biology. The heat lability of these enzymes is an important feature for their utilization in sequencing applications. The most common enzyme used in the field of molecular biology is alkaline phosphates which are used for removal of phosphate at the 5′ end of a linear DNA fragment to stop re-circularization while performing cloning techniques (Sarmiento et al. 2015). For this, the first cold adapted and heat-labile alkaline phosphates were separated from a bacterial strain from Antarctic (Rina et al. 2000) and was then commercialized by New England Biolabs as brand name of Antarctic phosphatase. In 2015, a novel psychrophilic alkaline phosphatse was produced from genomic library which was constructed by using oceanic-tide flat sediments procured from the Korean coast (Lee et al. 2015) and have good efficiency for commercialization purposes. Thus, psychrophilic alkaline phosphates have benefits of heat stability in comparison to its mesophilic counterparts.

Also, other psychrophilic enzymes having different significant roles are used for molecular biology applications. The first cold adapted enzyme is Uracil-DNA N-glycosylase (UNG), which catalyzes discharge of free uracil from a DNA fragment containing uracil. Also, this enzyme is used to manage the phenomenon of carry–over contagion in case of PCR and also in RT-PCR, SNP genotypic and mutagenesis applications. The prestigious enzyme company Arctic-Zymes is dealing with the manufacture of these psychrophilic UNG which can be utterly and irrevocably inactivated by higher temperature treatments, i.e., 55°C. Also, a new heat-labile cold adapted UNG enzyme was recovered by genomic analysis of *Psychrobacter* sp. HJ147; the discovered enzyme was cloned and expression was done in *E. coli*. This enzyme showed efficient performance at an optimal temperature range of 15–20°C and also has 2 minutes half-life at 45°C (Lee et al. 2009). The second important psychrozyme used in the field of molecular biology is cryonase which is a recombinant psychrophilic nuclease isolated by a cold active microorganism *Shewanella* sp. and is commercialized by Takara-Clontech, USA. It can digest various types of DNA and RNA which includes circularized, linear, single-stranded and double-stranded. The use of cryonase is very much required during the steps where the samples are placed on ice cubes and this enzyme can be fully inactivated at 70°C with an incubation period of 30 min. Also, the third example of cold adapted enzyme comprises of double strand-specific DNase which allows complete digestion of double-stranded DNA and does not have any effect on primers or probes which are single stranded. This double strand-specific DNase can be utilized for the decontamination PCR mixes. Psychrophilic heat labile double strand-specific DNase is isolated from cold active shrimp and after genetic engineering this enzyme is offered by Arctic-Zymes. Also, a recombinant DNase from psychrophilic *Pichia pastoris* is commercialized by Affymetrix, USA and this enzyme can be inactivated at 70°C for an incubation of 25–30 min. As the ligation yield increases at low temperature, so psychrophilic DNA ligases have an immense advantage in comparison to mesophilic counterparts (Georlette et al. 2000). Thus, the introduction of DNA ligases from psychrophilic strains have various advantages such as maintenance of higher specific activity at very lower temperature range and for this a new psychrophilic DNA ligase from the cold adapted strain of *Pseudoalteromonas haloplanktis* is known to get cloned, expressed and finally characterized. Also, this novel DNA ligase retains good enzymatic activity even at a lower temperature limit (4–6°C) (Georlette et al. 2000). Thus, the interest in development of expression systems regulating at lower-temperature range using various psychrophilic hosts will give rise to the requirement for different psychrophilic enzymes which can catalyze reactions at very lower temperature range (Allen et al. 2001).

In enzyme nano-biotechnology, the synthesis of different mineral or organic based chemical compounds and silica-containing material which includes molecular sieves, nano-resins, nano-fibers and other electronic matters required extreme range of temperature, pH and pressure. For this, the applicability of psychrophilic enzymes have an immense potential in the field of nanobiotechnology. A detailed study of the mechanistic pathway in the dominant species of *Tethya aurantia* (marine sponge), which was isolated from the coastal water bodies of temperate region, revealed that the biosynthesis of silica spicules consisting of catalytic active enzymes which are called silicateins (structural-directing enzymes) were actually dependable for the phenomenon of silica polymerization (Shimizu et al. 1998). Further enzymatic studies showed that the protein strands along with their subunits catalyze the synthesis of silicon and organo-silicon to yield silicones and silica at lower temperature range (Cha et al. 1999). Various novel enzymatic methods have been formulated for the production of nano-oxides at low temperature conditions as these novel enzymatic methods showed promising results for the development of new generation family polymers which comprises of efficient organic molecules. These methods include the steps of inclusion of enzymes into nano-structured matter which leads to various developments such as the formulation of design of a range of biomedical sensors, formation of specific designed sensors with effective environmental applications and also the synthesis of efficient materials which is capable of phenomenon of self-repair. Also, cold-adapted enzymes which can synthesize various efficient nano-particulate materials at lower range of temperatures are inexpensive and eco-friendly.

6.4 Cold-adapted Enzymes in Textile Industry

Many psychrozymes have important functions in the textile based industry. For this, Optisize® COOL amylase is used in Genencor-DuPont which helps in de-sizing of specifically woven fabrics at lower temperature conditions. Also, for cellulosic fabrics, Primafast® GOLD HSL cellulase can be utilized for the bio-finishing of cellulose containing materials and IndiAge® NeutraFlex cellulase which permits enzymatic stone washing at low temperatures and also at a neutral pH range. In the recent studies, gene encoding psychrotolerant enzyme catalase isolated from *Serratia* genus was cloned and then expressed in *E. coli*. This cold adapted enzyme catalase has a high activity and has antioxidant properties at lower temperature range as this enzyme has a potent role in trapping of reactive oxygen species (ROS) which is generated by condition of oxidative stress (Sarmiento et al. 2015). This enzyme catalase can be active in broad range of temperature (20 to 70°C) and retain half of its activity even at an exposure (7 h) of 50°C. These unique features of enzyme catalyze have applicability in textile, cement, cosmetic and research industries. The catalase is commercially offered by Swiss austral Company, USA.

6.5 Environmental Applications

The psychrophilic enzymes have increased specific activity at lower temperature conditions which makes them an ideal candidate for the process of bioremediation process. The inoculation of temperate region with psychrophilic enzymes in consortia has boosted the degradation of a range of hydrocarbons (Dhaulaniya 2018). For example, P450 dependent flevodoxin reductase and Cytochrome P450 alkane hydrolases have been developed from genome of psychrophilic strain and showed the ability to hydrolyze alkane and also have the potential to degrade derivatives of nitrotriazine at lower temperature conditions of 5°C and thus showed a great potential of bioremediation at cold conditions (Bowman and Deming 2014). The remote and cold habitat on the Earth surface is being polluted by different sources which include industrial pollution, tourist activities and also through migration of certain species but the ultimate source of pollution is prevalence or deposition of different harmful compounds through atmospheric transportation (Gai et al. 2014, Goutte et al. 2013). The transfer of different polycyclic aromatic hydrocarbons (PAHs) from their origin point to different parts of the ecosystem is an issue of concern as it is adversely affecting the health of environment. These PAHs are organic pollutants and due to their hydrophobic nature, these particles get bounded to the suspended particles of the environment (Sarma et al. 2016). Under the influence of different environmental factors such as hurricanes, dust storms and cyclones, these PAHs get transferred from their pin location to different layers of ecosystems including Himalayas and different extreme locating glaciers. Thus, the microbial communities of these cold adapted regions get exposed to these PAHs and have a potential of bioremediation of these PAHs. The biodegradation by psychrophilic enzymes have been studied for polychlorinated biphenyls, chlorobenzenes, chlorobenzoates and dioxins and also detailed degradation pathways have been illuminated (Bajaj and Singh 2015). For the degradation of γ-hexachlorocyclohexane at lower temperature range of 4°C, three strains of genus *Sphingobium* were reported which include *Sphingomonas japonicum* UT26, *S. indicum* B90A, and *S. francens* (Zheng et al. 2011). The cold adapted machinery of Arctic *Pseudomonas* Cam-1 revealed the biodegradation of polychlorinated biphenyls (PCBs) at extreme lower temperature of 6–7°C (Master and Mohn 1998) and also psychrophilic genus *Rhodococcus* (*Rhodococcus erythropolis* S-7 and *Rhodococcus erythropolis* MS11) showed biodegradation of 3-Chlorobenzoate at 10°C and biodegradation of chlorobenzene at 5°C, respectively (Rapp and Gabriel-Jurgens 2003).

6.6 Biofuels

In current scenario, with increase in human population, there is an unexpected pressure on the demand of fossil fuel and thus to avoid this problem, there is a tremendous interest in the production of other options for the source of energy worldwide. For this, biofuels such as bio-ethanol and bio-methanol, which are formed by the fermentation of carbohydrates (starch, sucrose), offer the mode of renewable form of energy that has important benefits such as reduced greenhouse emissions, enhanced energy security, economic benefits, and problems related with discarding of various agro-industrial wastes (Wyman 2003). The conventional process of ethanol production is significantly optimized but this process is neither energetically nor economically competent because it necessitates higher levels of heat levels and also requires specialized equipment for the process, thus restricting the biofuel production ability of biorefineries. To solve this problem, the industrial enzyme based companies are uplifting their research process for the development of low energy biofuel production which is also known as process of cold hydrolysis as it involves hydrolysis of raw starch. This cold hydrolysis process eliminates the demand of high energy and also excludes the requirement of equipment-intensive liquefaction cooking. For this, Genencor has specifically widened the STARGEN™ which is lineage of fungal based enzyme products containing glucoamylases and α-amylases which are able to perform hydrolysis of uncooked starch in saccharification and fermentation process (Solid State Fermentation) at 30–32°C (Shetty et al. 2005). These enzymes perform the catalysis that help in the release of fermented sugar from the starch (granular) and is directly used by the yeast leading to process of ethanol production.

One of the renowned enzyme industry, Novozymes, has collaborated with prestigious Broin Companies and developed Broin's BPX™ technology which is a low temperature hydrolysis procedure utilizing fungal based enzymes for the efficient conversion of starch into fermentable sugars avoiding cooking procedure. This process is energy efficient and offers various advantages such as increased ethanol production, elimination of undesirable side products, minimized emissions and ecofriendly and cost efficient. Also, the solid state fermentation enables higher enzymatic efficiencies by regular elimination of ion of maltose and glucose formed during the process, thus reducing the side product formation. The cold adapted α-amylases and glucoamylases have higher specificities and activities at lower temperature suitable for fermentation process, thus reducing the biocatalyst concentration for the reaction and is hence economically efficient as compared to mesophilic counterparts (Lin and Tanaka 2006). Also, the production of second generation biofuel which uses complex lignocelluloses matter necessitates the action of hydrolases enzymes such as cellulases and hemicellulases; researchers are developing enzymes for the hydrolysis of lignocellulose at lower range of temperatures well-suited to process of fermentation (Felby et al. 2006). Thus, use of psychrophilic hydrolases enzymes such as cellulases, hemicellulases, glucosidases and xylanases enables the conversion of lignocellulose biomass at economic rate and thus plays a vital role in encouraging the growth and production of an economically biofuel process to cope up with the demands of world's population.

7. Conclusion

Psychrophiles are a class of extremophilic microorganisms thriving at sub-zero temperatures and have different adaptations in their lifecycle which includes production of cold shock proteins and psychrophilic enzymes. These psychrophilic enzymes have various modifications which include high specific activity at lower temperature conditions and have reduced activation energy barrier of the transition state. Also, these psychrophilic enzymes are useful due to increased selectivity and higher catalytic activity rate at lower and ambient temperatures; due to their good structural lability, these cold adapted enzymes can be utilized in multistep reaction procedures which extensively require frequent and rapid inactivation actions. These psychrozymes are extensively used in different biotechnological based applications which particularly demands higher activity rates at mild temperatures and also quick heat-inactivation rate. Cold-adapted enzymes can be exposed to extreme conditions of temperature, pH and pressure as compared to mesophilic counterparts and are thus used to meet the lower energy demands and costs. Moreover, the conformational plasticity of psychrophilic enzymes is exploited in various organic synthesis functions and specifically utilized in the production of fine pharmaceutical intermediates. Current advances in technology comprising of metagenomics and genetic engineering techniques have enhanced the applicability of psychrophilic enzymes in different biotechnological industrial practices.

Ethical Standards

Authors declare no conflict of interests.

References

Aghajari, N., F. Vanpetegem, V. Villeret, J. Chessa, C. Gerday, R. Haser and J. Vanbeeumen. 2003. Crystal structures of a psychrophilic metalloprotease reveal new insights into catalysis by cold-adapted proteases. Proteins 50: 636–647.

Allen, D., A.L. Huston, L.E. Wells and J.W. Deming. 2001. Biotechnological use of psychrophiles. pp. 1–17. *In*: Bitton, G. (ed.). Encyclopedia of Environmental Microbiology. Wiley, New York.

Arcus, V.L., E.J. Prentice, J.K. Hobbs, A.J. Mulholland, M.W. Van der Kamp, C.R. Pudney, E.J. Parker and L.A. Schipper. 2016. On the temperature dependence of enzyme-catalyzed rates. Biochemistry 55: 1681–1688.

Arnold, F.H., P.L. Wintrode, K. Miyazaki and A. Gershenson. 2001. How enzymes adapt: Lessons from directed evolution. Trends Biochem. Sci. 26: 100–106.

Asenjo, J.A., B.A. Andrews, F. Reyes, M. Salamanca and L. Burzio. 2006. Protein and nucleic acid sequence encoding a krill-derived cold adapted trypsin-like activity enzyme. Patent No # WO2006022947.

Aznauryan, M., D. Nettels, A. Holla, H. Hofmann and B. Schuler. 2013. Single-molecule spectroscopy of cold denaturation and the temperature-induced collapse of unfolded proteins. J. Am. Chem. Soc. 135: 14040–14043.

Bajaj, S. and D.K. Singh. 2015. Biodegradation of persistent organic pollutants in soil, water and pristine sites by cold-adapted microorganisms: Mini review. Int. Biodeterior. Biodegradation 100: 98–105.

Bakermans, C. and M.L. Skidmore. 2011. Microbial metabolism in ice and brine at −5 degrees C. Environ. Microbiol. 13: 2269–2278.

Bjelic, S., B.O. Brandsdal and J. Aqvist. 2008. Cold adaptation of enzyme reaction rates. Biochemistry 47: 10049–10057.

Bowman, J.S. and J.W. Deming. 2014. Alkane hydroxylase genes in psychrophile genomes and the potential for cold-active catalysis. BMC Genomics 15: 1120.

Brandsdal, B.O., A.O. Smalas and J. Aqvist. 2001. Electrostatic effects play a central role in cold adaptation of trypsin. FEBS Lett. 499: 171–175.

Brenchley, J.E., J. Loveland-Curtze, K. Gutshall and V. Humphrey. 2001. Stain removing compositions containing particular isolated and pure proteolytic enzymes. U.S. Patent # 6326346.

Cavicchioli, R., K.S. Siddiqui, D. Andrews and K.R. Sowers. 2002. Low-temperature extremophiles and their applications. Curr. Opin. Biotech. 13: 253–261.

Cavicchioli, R., T. Charlton, H. Ertan, S. Mohd Omar, K.S. Siddiqui and T.J. Williams. 2011. Biotechnological uses of enzymes from psychrophiles. Microb. Biotechnol. 4: 449–60.

Cha, J.N., K. Shimizu, Y. Zhou, S.C. Christiansen, B.F. Chmelka, G.D. Stucky and D.E. Morse. 1999. Silicatein filaments and subunits from a marine sponge direct the polymerization of silica and silicones *in vitro*. Proc. Natl. Acad. Sci. USA 96: 361–365.

Chattopadhyay, M.K. 2006. Mechanism of bacterial adaptation to low temperature. J. Biosci. 31: 157–165.

Chintalapati, S., M.D. Kiran and S. Shivaji. 2004. Role of membrane lipid fatty acids in cold adaptation. Cell Mol. Biol. 50: 631–642.

Chiuri, R., G. Maiorano, A. Rizzello, L.L. del Mercato, R. Cingolani, R. Rinaldi, M. Maffia and P. P. Pompa. 2009. Exploring local flexibility/rigidity in psychrophilic and mesophilic carbonic anhydrases. Biophys. J. 96: 1586–1596.

Collins, T., M.A. Meuwis, I. Stals, M. Claeyssens, G. Feller and C. Gerday. 2002. A novel family 8 xylanase. Functional and physicochemical characterization. J. Biol. Chem. 277: 35133–35139.

Collins, T. and R. Margesin. 2019. Psychrophilic lifestyles: mechanisms of adaptation and biotechnological tools. Appl. Microbiol. Biotechnol. 103: 2857–2871.

Creighton, T.E. 1991. Stability of folded conformations: Current opinion in structural biology. Curr. Opin. Struct. Biol. 1: 5–16.

D'Amico, S., C. Gerday and G. Feller. 2001. Structural determinants of cold adaptation and stability in a large protein. J. Biol. Chem. 276: 25791–25796.

D'Amico, S., J.C. Marx, C. Gerday and G. Feller. 2003. Activity-stability relationships in extremophilic enzymes. J. Biol. Chem. 278: 7891–7896.

D'Amico, S., T. Collins, J.C. Marx, G. Feller and C. Gerday. 2006. Psychrophilic microorganisms: Challenges for life. EMBO Rep. 7: 385–389.

DasSarma, S., M.D. Capes, R. Karan and P. DasSarma. 2013. Amino acid substitutions in cold-adapted proteins from *Halorubrum lacusprofundi*, an extremely halophilic microbe from Antarctica. PLoS One 8(3): e58587.

De Maayer, P., D. Anderson, C. Cary and D.A. Cowan. 2014. Some like it cold: Understanding the survival strategies of psychrophiles. EMBO Rep. 15: 508–517.

De Vos, D., T. Collins, W. Nerinckx, S.N. Savvides, M. Claeyssens, C. Gerday, G. Feller and J. Van Beeumen. 2006. Oligosaccharide binding in family 8 glycosidases: Crystal structures of active-site mutants of the beta-1, 4-xylanase pXyl from *Pseudoaltermonas haloplanktis* TAH3a in complex with substrate and product. Biochemistry 45: 4797–4807.

Deming, J.W. 2002. Psychrophiles and polar regions. Curr. Opin. Microbiol. 5: 301–309.

Dhaulaniya, A.S., B. Balan, M. Kumar, P.K. Agrawal and D.K. Singh. 2018. Cold survival strategies for bacteria, recent advancement and potential industrial applications. Arch. Microbiol. 201: 1–16.

Dick, M., O.H. Weiergraber, T. Classen, C. Bisterfeld, J. Bramski, H. Gohlke and J. Pietruszka. 2016. Trading off stability against activity in extremophilic aldolases. Sci. Rep. 6: 17908.

Eyring, H. 1935. The activated complex and the absolute rate of chemical reactions. Chem. Rev. 17: 65–77.

Felby, C., D. Larsen, H. Joergensen and J. Vibe-Pederson. 2006. Enzymatic Hydrolysis of Biomasses having High Dry Matter (DM) Content. Patent No # WO2006056838.

Feller, G. and C. Gerday. 2003. Psychrophilic enzymes: Hot topics in cold adaptation. Nat. Rev. Microbiol. 1: 200–208.

Feller, G. 2008. Enzyme function at low temperatures in psychrophiles. In Protein Adaptation in Extremophiles. pp. 35–69. Nova Science Publishers, New York, United States.

Feller, G. 2012. Psychrophilic enzymes: from folding to function and biotechnology. Scientifica doi. org/10.1155/2013/512840.

Fields, P.A. and G.N. Somero. 1998. Hot spots in cold adaptation: localized increases in conformational flexibility in lactate dehydrogenase A4 orthologs of Antarctic notothenioid fishes. Proc. Natl. Acad. Sci. 95: 11476–11481.

Gai, N., J. Pan, H. Tang, S. Chen, D. Chen, X. Zhu, G. Lu and Y. Yang. 2014. Organochlorine pesticides and polychlorinated biphenyls in surface soils from Ruoergai high altitude prairie, east edge of Qinghai-Tibet Plateau. Sci. Total Environ. 478: 90–97.

Garsoux, G., J. Lamotte, C. Gerday and G. Feller. 2004. Kinetic and structural optimization to catalysis at low temperatures in a psychrophilic cellulase from the Antarctic bacterium *Pseudoalteromonas haloplanktis*. Biochem. J. 384: 247–253.

Georlette, D., Z.O. Jonsson, F. Van Petegem, J. Chessa, J. Van Beeumen, U. Hubscher and C. Gerday. 2000. A DNA ligase from the psychrophile *Pseudoalteromonas haloplanktis* gives insights into the adaptation of proteins to low temperatures. Eur. J. Biochem. 267: 3502–3512.

Georlette, D., V. Blaise, T. Collins, S. D'Amico, E. Gratia, A. Hoyoux, J.C. Marx, G. Sonan, G. Feller and C. Gerday. 2004. Some like it cold: Biocatalysis at low temperatures. FEMS Microbiol. Rev. 28: 25–42.

Gerday, C., M. Aittaleb, M. Bentahir, J.P. Chessa, P. Claverie, T. Collins, S. D'Amico, J. Dumont, G. Garsoux, D. Georlette and A. Hoyoux. 2000. Cold-adapted enzymes: From fundamentals to biotechnology. Trends Biotechnol. 18: 103–107.

Gerday, C., A. Hoyoux, J.M. Francois, P. Dubois, E. Baise, I. Jennes and S. Genicot. 2005. Cold-active ß-galactosidase, the process for its preparation and the use thereof. U.S. Patent # 2005196835.

Gerday, C. 2014. Fundamentals of cold-active enzymes. Cold-adapted yeasts. pp. 325–350. Springer, Berlin, Heidelberg.

Gorfe, A.A., B.O. Brandsdal, H.K.S. Leiros, R. Helland and A.O. Smalas. 2000. Electrostatics of mesophilic and psychrophilic trypsin isoenzymes: Qualitative evaluation of electrostatic differences at the substrate binding site. Proteins 40: 207–217.

Goutte, A., M. Chevreuil, F. Alliot, O. Chastel, Y. Cherel, M. Eleaume and G. Masse. 2013 Persistent organic pollutants in benthic and pelagic organisms off Adelie Land, Antarctica. Mar. Pollut. Bull. 77: 82–89.

Graziano, G. 2014. On the mechanism of cold denaturation. Phys. Chem. Chem. Phys. 16: 21755–21767.

Gupta, G.N., S. Srivastava, S.K. Khare and V. Prakash. 2014. Extremophiles: An overview of microorganism from extreme environment. Int. J. Agric. Environ. Biotechnol. 7: 371–380.

He, H., X.L. Chen, J.W. Li, Y.Z. Zhang and P.J. Gao. 2004. Taste improvement of refrigerated meat treated with cold-adapted protease. Food Chem. 84: 307–311.

Jeon, J., J.T. Kim, S. Kang, J.H. Lee and S.J. Kim. 2009. Characterization and its potential application of two esterases derived from the Arctic sediment metagenome. Mar. Biotechnol. 11: 307–316.

Joseph, B., P.W. Ramteke and G. Thomas. 2008. Cold active microbial lipases: Some hot issues and recent developments. Biotechnol. Adv. 26: 457–470.

Jung, S.K., D.G. Jeong, M.S. Lee, J.K. Lee, H.K. Kim, S.E. Ryu, B.C. Park, J.H. Kim and S.J. Kim. 2008. Structural basis for the cold adaptation of psychrophilic M37 lipase from *Photobacterium lipolyticum*. Proteins 71: 476–484.

Kahlke, T. and S. Thorvaldsen. 2012. Molecular characterization of cold adaptation of membrane proteins in the *Vibrionaceae coregenome*. PLoS One 7(12): e51761.

Kim, S.Y., K.Y. Hwang, S.H. Kim, H.C. Sung, Y.S. Han and Y. Cho. 1999. Structural basis for cold adaptation: Sequence, biochemical properties, and crystal structure of malate dehydrogenase from a psychrophile *Aquaspirillium arcticum*. J. Biol. Chem. 274: 11761–11767.

Laidler, K.J. 1984. The development of the Arrhenius equation. J. Chem. Educ. 61: 494.

Lee, M.S., G.A. Kim, M.S. Seo, J.H. Lee and S.T. Kwon. 2009. Characterization of heat-labile uracil-DNA glycosylase from *Psychrobacter* sp. HJ147 and its application to the polymerase chain reaction. Biotechnol. Appl. Biochem. 52: 167–175.

Lee, D.H., S.L. Choi, E. Rha, S. Kim, S.J. Yeom, J.H. Moon and S.G Lee 2015. A novel psychrophilic alkaline phosphatase from the metagenome of tidal flat sediments. BMC Biotechnol. 15: 1.

Lee, H.W., H.Y. Jeon, H.J. Choi, N.R. Kim, W.J. Choung, Y.S. Koo, D.S. Ko, D.S. Ko, S. You and J.H. Shim. 2016. Characterization and application of bila, a psychrophilic α-amylase from *Bifidobacterium longum*. J. Agric. Food Chem. 64: 2709–2718.

Lian, K., H.K.S. Leiros and E. Moe. 2015. MutT from the fish pathogen *Aliivibrio salmonicida* is a cold-active nucleotide-pool sanitization enzyme with unexpectedly high thermostability. FEBS Open Biol. 5: 107–116.

Lin, Y. and S. Tanaka. 2006. Ethanol fermentation from biomass resources; Current state and prospects. Appl. Microbiol. Biotechnol. 69: 627–642.

Lonhienne, T., C. Gerday and G. Feller. 2000. Psychrophilic enzymes: Revisiting the thermodynamic parameters of activation may explain local flexibility. Biochim. Biophys. Acta. (BBA) Protein Struct. Mol. Enzymol. 1543: 1–10.

Maiangwa, J., M.S.M. Ali, A.B. Salleh, R.N.Z.R.A. Rahman, F.M. Shariff and T.C. Leow. 2015. Adaptational properties and applications of cold-active lipases from psychrophilic bacteria. Extremophiles 19: 235–247.

Mangiagalli, M., S. Brocca, M. Orlando and M. Lotti. 2019. The cold revolution. Present and future applications of cold-active enzymes and ice-binding proteins. New Biotechnol. 55: 5–11.

Margesin, R., F. Schinner, J.C. Marx and C. Gerday (eds.). 2008. Psychrophiles: From Biodiversity to Biotechnology. Springer Science & Business Media, Germany.

Marx, J.C., T. Collins, S.D. Amico, G. Feller and C. Gerday. 2007.Cold-adapted enzymes from marine antarctic microorganisms. Mar. Biotechnol. 9: 293–304.

Master, E.R. and W.W. Mohn. 1998. Psychrotolerant bacteria isolated from Arctic soil that degrade polychlorinated biphenyls at low temperatures. Appl. Environ. Microbiol. 64: 4823–4829.

Mykytczuk, N.C., S.J. Foote, C.R. Omelon, G. Southam, C.W. Greer and L.G. Whyte. 2013. Bacterial growth at −15°C; molecular insights from the permafrost bacterium *Planococcus halocryophilus* Or1. ISME J. 6: 1211–1226.

Nakagawa, T., T. Nagaoka, S. Taniguchi, T. Miyaji and N. Tomizuka. 2004. Isolation and characterization of psychrophilic yeasts producing cold-adapted pectinolytic enzymes. Lett. Appl. Microbiol. 38: 383–387.

Narinx, E., E. Baise and C. Gerday. 1997. Subtilisin from psychrophilic antarctic bacteria: Characterization and site-directed mutagenesis of residues possibly involved in the adaptation to cold. Protein Eng. 10: 1271–1279.

Okuda, M., N. Sumitomo, Y. Takimura, A. Ogawa, K. Saeki, S. Kawai, T. Kobayashi and S. Ito. 2004. A new subtilisin family: Nucleotide and deduced amino acid sequences of new high-molecular-mass alkaline proteases from *Bacillus* spp. Extremophiles 8: 229–235.

Papaleo, E., M. Olufsen, L. De Gioia and B.O. Brandsdal. 2007. Optimization of electrostatics as a strategy for cold-adaptation: A case study of cold- and warm-active elastases. J. Mol. Graph. Model. 26: 93–103.

Peterson, M.E., R.M. Daniel, M.J. Danson and R. Eisenthal. 2007. The dependence of enzyme activity on temperature: Determination and validation of parameters. Biochem. J. 402: 331–337.

Pulicherla, K.K., G. Mrinmoy, S. Kumar and K.R. Sambasiva Rao. 2011. Psychrozymes—the next generation industrial enzymes. J. Mar. Sci. Res. Dev. 1: 2.

Raeder, I.L.U., E. Moe, N.P. Willassen, A.O. Smalas and I. Leiros. 2010. Structure of uracil-DNA N-glycosylase (UNG) from *Vibrio cholerae*: Mapping temperature adaptation through structural and mutational analysis. Acta Crystallogr. F. 66: 130–136.

Ramli, A.N., M.A. Azhar, M.S. Shamsir, A. Rabu, A.M. Murad, N.M. Mahadi and R.M. Illias. 2013. Sequence and structural investigation of a novel psychrophilic α-amylase from *Glaciozyma antarctica* PI12 for cold-adaptation analysis. J. Mol. Model. 19: 3369–3383.

Rapp, P. and L.H. Gabriel-Jurgens. 2003. Degradation of alkanes and highly chlorinated benzenes, and production of biosurfactants, by a psychrophilic *Rhodococcus* sp. and genetic characterization of its chlorobenzene dioxygenase. Microbiology 149: 2879–2890.

Rina, M., C. Pozidis, K. Mavromatis, M. Tzanodaskalaki, M. Kokkinidis and V. Bouriotis. 2000. Alkaline phosphatase from the Antarctic strain TAB5. Properties and psychrophilic adaptations. Eur. J. Biochem. 267: 1230–1238.

Rodrigues, D.F. and J.M. Tiedje. 2008. Coping with our cold planet. Appl. Environ. Microbiol. 74: 1677–1686.

Russell, R.J.M., U. Gerike, M.J. Danson, D.W. Hough and G.L. Taylor. 1998. Structural adaptations of the cold-active citrate synthase from an Antarctic bacterium. Structure 6: 351–362.

Sakamoto, T., H. Ihara, S. Kozakic and H. Kawasaki. 2003. A cold adapted endo-arabinanase from *Penicillium chrysogenum*. BBA-Gen Subjects 1624: 70–75.

Santiago, M., C.A. Ramirez-Sarmiento, R.A. Zamora and L.P. Parra. 2016. Discovery, Molecular Mechanisms, and Industrial Applications of Cold-Active Enzymes. Front. Microbiol. 7: 1408–1408.

Sarma, H., N.F. Islam, P. Borgohain, A. Sarma and M.N.V. Prasad. 2016. Localization of polycyclic aromatic hydrocarbons and heavy metals in surface soil of Asia's oldest oil and gas drilling site in Assam, north-east India: Implications for the bio-economy. Emerg. Contam. 2: 119–127.

Sarmiento, F., R. Peralta and J.M. Blamey. 2015 Cold and hot extremozymes: Industrial relevance and current trends. Front. Bioeng. Biotechnol. 3: 148.

Saunders, N.F.W., T. Thomas, P.M. Curmi, J.S. Mattick, E. Kuczek, R. Slade, J. Davis, P.D. Franzmann, D. Boone, K. Rusterholtz and R. Feldman. 2003. Mechanisms of thermal adaptation revealed from the genomes of the Antarctic Archaea *Methanogenium frigidum* and *Methanococcoides burtonii*. Genome Res. 13: 1580–1588.

Sellek, G.A. and J.B. Chaudhuri. 1999. Biocatalysis in organic media using enzymes from extremophiles. Enzyme Microb. Technol. 25: 471–482.

Shahidi, F. and Y.V.A.J. Kamil. 2001. Enzymes from fish and aquatic invertebrates and their application in the food industry. Trends Food Sci. Technol. 12: 435–464.

Shetty, J.K., O.J. Lantero and N. Dunn-Colemen. 2005. Technological advances in ethanol production. Int. Sugar J. 107: 605.

Shimizu, K., J. Cha, G.D. Stucky and D.E. Morse. 1998. Silicatein α: Cathespin L-like protein in sponge biosilica. Proc. Natl. Acad. Sci. USA 95: 6234–6238.

Siddiqui, K.S. and R. Cavicchioli. 2006. Cold-adapted enzymes. Annu. Rev. Biochem. 75: 403–433.

Siddiqui, K.S. 2015. Some like it hot, some like it cold: Temperature dependent biotechnological applications and improvements in extremophilic enzymes. Biotech. Adv. 33: 1912–22.

Sindhu, R., P. Binod, A. Madhavan, U.S. Beevi, A.K. Mathew, A. Abraham, A. Pandey and V. Kumar. 2017. Molecular improvements in microbial α-amylases for enhanced stability and catalytic efficiency. Bioresour. Technol. 245: 1740–1748.

Skalova, T., J. Dohnalek, V. Spiwok, P. Lipovova, E. Vondrackova, H. Petrokova, J. Duskova, H. Strnad, B. Kralova and J. Hasek. 2005. Cold-active beta-galactosidase from *Arthrobacter* sp. C2-2 forms compact 660 kDa hexamers: Crystal structure at 1.9 A resolution. J. Mol. Biol. 353: 282–294.

Suen, W.C., N. Zhang, L. Xiao, V. Madison and A. Zaks. 2004. Improved activity and thermostability of *Candida antarctica* lipase B by DNA family shuffling. Protein Eng. Des. Sel. 17: 133–140.

Tattersall, G.J., B.J. Sinclair, P.C. Withers, P.A. Fields, F. Seebacher, C.E. Cooper and S.K. Maloney. 2012. Coping with thermal challenges: Physiological adaptations to environmental temperatures. Compr. Physiol. 2: 2151–2202.

Trytek, M. and J. Fiedurek. 2005. A novel psychrotrophic fungus, *Mortierella minutissima*, for D-limonene biotransformation. Biotechnol. Lett. 27: 149–153.

Tutino, M.L., G. di Prisco, G. Marino and D. de Pascale. 2009. Cold-adapted esterases and lipases: From fundamentals to application. Protein Pept. Lett. 16: 1172–1180.

Vajpai, N., L. Nisius, M. Wiktor and S. Grzesiek. 2013. High-pressure NMR reveals close similarity between cold and alcohol protein denaturation in ubiquitin. Proc. Natl. Acad. Sci. 110: E368–E376.

Violot, S., N. Aghajari, M. Czjzek, G. Feller, G.K. Sonan, P. Gouet, C. Gerday, R. Haser and V. Receveur-Brechot. 2005. Structure of a full length psychrophilic cellulase from *Pseudoalteromonas haloplanktis* revealed by X-ray diffraction and small angle X-ray scattering. J. Mol. Biol. 348: 1211–1224.

Wolfenden, R. 2011. Benchmark reaction rates, the stability of biological molecules in water, and the evolution of catalytic power in enzymes. Annu. Rev. Biochem. 80: 645–667.

Wyman, C.E. 2003. Potential synergies and challenges in refining cellulosic biomass to fuels, chemicals, and power. Biotechnol. Prog. 19: 254–262.

Xie, B.B., F. Bian, X.L. Chen, H.L. He, J. Guo, X. Gao, Y.X. Zeng, B. Chen, B.C. Zhou and Y.Z. Zhang. 2009. Cold adaptation of zinc metalloproteases in the thermolysin family from deep sea and arctic sea ice bacteria revealed by catalytic and structural properties and molecular dynamics: new insights into relationship between conformational flexibility and hydrogen bonding. J. Biol. Chem. 284: 9257–9269.

Zeng, X., X. Xiao, P. Wang and F. Wang. 2004. Screening and Characterization of psychrotrophic lipolytic bacteria from deep sea sediments. J. Microbiol. Biotech. 14: 952–958.

Zheng, G., A. Selvam and J.W. Wong. 2011. Rapid degradation of lindane (γ-hexachlorocyclohexane) at low temperature by *Sphingobium* strains. Int. Biodeterior. Biodegrad. 65: 612–618.

9

Advancements in Hemicellulosic Extremozymes and Their Applications
A Future Perspective

Sweety Kaur,[1] *Gurmeen Rakhra*[2] and *Richa Arora*[3,]*

1. Introduction

At the end of 20th century, a revolution in the field of biotechnology established a bright vision and novelistic approach to various issues along with evolution and expansion of industrial, pharmaceutical and agricultural processing. "Extremophilic organisms" were first discovered in 1960s and have been compatible in different biotechnological techniques as a whole cell (or organism) and their derived macromolecules (such as DNA, enzymes or metabolic products). Proper implementation of such organisms can result in a massive decrease in production cost with inclination towards production rate along with equivalent expenditure and facilitate the completion of process which was not earlier feasible. Those microorganisms (mainly *Archaea* and *Bacteria* domain) that thrive under extreme environmental conditions where most of the organisms cannot survive are called as "extremophiles" (derived from Latin word 'extremus' meaning 'extreme' and Greek word 'philia' meaning 'love'), and many of them can't resist standard anthropic worldwide environments (Gomes and Steiner 2004). These extremophiles mainly grow and reproduce under extreme conditions that include high-temperature conditions (like hot springs or thermal vents), low temperature conditions (like glaciers or deep cold oceans), highly-acidic and highly alkaline pH conditions (like industrial waste effluents), high-salinity conditions (like salt lakes), high intensity of radiation, highly drought conditions (like deserts) and other extreme conditions in distinct ecological niches (Sarmiento et al. 2015).

Now the question arises that how these extremophiles strive and resist under such harsh conditions? The reason behind is that extremophiles have developed and adapted to their dwelling atmosphere by evolving peculiar mechanisms that are pretty complex, to maintain their cellular elements and metabolism in a steady and dynamic state. For instance, to withstand high-salinity conditions, these extremophiles develop higher amount of compatible solutes inside their cells; in case of high-acidic conditions, extremophiles utilize proton pumps to maintain an appropriate pH inside cell and, in case of high and low temperatures, they alter their plasma membrane composition. In most situations, extremophiles produce enzymes that have ability to adapt under extreme

[1] Nestle India Limited, Rajarhat, Kolkata-700156, India.
[2] School of Bioengineering and Biosciences, Lovely Professional University, Phagwara-144411, Punjab, India.
[3] Department of Microbiology, Punjab Agricultural University, Ludhiana-141004, India.
* Corresponding author: aroraricha@ymail.com

conditions on their own; such enzymes are known as "extremozymes" and the adaptations comply to substantial changes in sequence of amino acid that is further translated into structural modification, elasticity, charge and hydrophobic nature of the enzymes (Reed et al. 2013). Extremophiles are extremozymes that have the ability to moderate the catalysis in broad range of functional situations including extreme conditions and improved catalytic rates. Due to such uniquely fascinating properties, extremozymes assure a wide range of real-time potential applications and contribute to bioremediation, farming and cultivation, synthetics, biofuels, bioenergy, pharmaceuticals, research and other potential industrial applications. With increasing demand, these enzymes' market is getting more and more dynamic. The worldwide market of industrial enzymes was $5.5 billion in 2018, is estimated to reach its peak of $5.9 billion by the end of 2020, and is further expected to attain $8.7 billion by 2026 at a compound annual growth rate of 6.5%, in terms of value (www.marketsandmarkets.com). This market may presumably continue to increase in coming future leading to advancements in microbiology and biotechnology industry with ongoing requirements for cost-effective methods of manufacturing. Hemicellulases are diversified class of enzymes, which mainly hydrolyzes hemicelluloses, comprising of α-L-arabinofuranosidase, α-1,5-arabinanase, endo-β-1,4-xylanase, β-1,4-xylosidase, β-1,4-mannanase, β-1,4-mannosidase, β-glucosidase, α-glucoronidase, α-galactosidase, endo-galactanase, acetyl-mannan and xylan esterase, ferulic acid esterase and p-coumaric acid esterase (Beg et al. 2001). L-arabinose is a major primary sugar substituent present on the structure of the hemicellulase enzyme molecule that can be degraded by the action of α-L-arabinofuranosidase enzyme which is an accessory enzyme (Kim 2008). α-L-arabinofuranosidase enzyme hydrolyses the L-arabinose side chain residing at α-1,2-, α-1,3-, and α-1,5-positions, hence helping other enzymes to act effectively on backbone. α-L-arabinofuranosidases and α-L-arabinanases mainly cleave and hydrolyse arabinofuranosyl-containing hemicelluloses and are identified in glycoside hydrolase (GH) families-3, 43, 51, 54, and 62 (Saha 2000). They have many potential uses like bread preparation, fruit juice clarification, wine industry for its flavor enrichment, starch, coffee and plants essential oils' extraction, preparation of beneficial animal fodder and grainy feeds, paper and pulp industry, production of bioethanol and therapeutics (Thakur et al. 2019). This chapter investigates and emphasizes on the basic concept of hemicellulase and α-L-arabinofuranosidase enzymes as extremozymes including structural properties, biochemical properties (pH, temperature, etc.), microorganisms involved in their production along with their potential applications in industrial level with recent advancements.

2. Thermophiles and Thermozymes

Thermophiles (derived from two Greek terms 'thermotita' meaning heat and 'philia' meaning love) are the group of organisms that are adapted to grow and survive under extreme conditions with a wide range of growth from lowest temperature of 50°C to the highest range of 121°C. Based upon different temperature range of growth, they are classified into: (1) moderate thermophiles (40–60°C), (2) extreme thermophiles (60–85°C) and (3) hyperthermophiles (> 85°C) (Table 1) (Mehta and Satyanarayana 2013). Different thermophiles were isolated from various natural environments like geothermal regions, volcanic areas, hot water springs, deep-ocean hydrothermal vents, solar warm soil surfaces, geothermally heated oil, fossil fuel reserves and synthetic environments such as acid mine sewage, compost heaps and sewage treatment plants. Cellular organization and metabolic activities are largely affected by different parameters which also includes temperature. For any certain microorganism (more specifically thermophile) to survive and grow at high temperature conditions, its main feature should be its ability to resist heat, alongwith presence of heat resistant proteins, nucleic acids and lipids. Therefore, thermophiles have acquired different adaptations that enable them to overcome and grow properly at high temperatures (Mehta and Satyanarayana 2013).

Enzymes or biocatalysts are molecules that have high catalytic efficiency with specificity towards particular substrate or reaction and are required for various industrial processes to attain products with greater yield, easy product recovery with less environmental stress. However, their use

Table 1. Categorization of thermophiles based on different temperature range of growth.

Classification	Optimum Temperature (°C)	Genus
(1) Moderate thermophiles	40–60°C	*Tepidibacte, Clostridium, Exiguobacterium, Caminibacter, Lebetimonas, Hydrogenimonas, Nautilia, Desulfonauticus, Sulfurivirga, Caminicella, Vulcanibacillus, Marinotoga, Caldithrix, Sulfobacillus, Acidimicrobium, Hydrogenobacter, Thermoplasma, Mahella, Thermoanaerobacter, Desulfovibrio.*
(2) Extreme thermophiles	60–85°C	*Methanocaldococcus, Thermococcus, Palaeococcus, Methanotorris, Aeropyrum, Thermovibrio, Methanothermococcus, Thermosipho, Caloranaerobacter, Thermodesulfobacterium, Thermodesulfatator, Deferribacter, Thermosipho, Desulfurobacterium, Persephonella, Kosmotoga, Rhodothermus, Desulfurobacterium, Balnearium, Acidianus, Thermovibrio, Marinithermus, Oceanithermus, Petrotoga, Vulcanithermus, Carboxydobrachium, Thermaerobacter, Thermosulfidibacter, Metallosphaera.*
(3) Hyper-thermophiles	> 85°C	*Geogemma, Archaeoglobus, Methanopyrus, Pyrococcus, Sulfolobus, Thermoproteus, Methanothermus, Acidianus, Ignisphaera, Ignicoccus, Geoglobus.*

is limited because most enzymes require sustainable atmospheric conditions to work properly and usually get inactivated or denatured during industrial operations under extreme conditions, hence, there is a need to search for enzymes that can resist extreme and harsh operating environment. Microorganisms growing and surviving under high temperature conditions secrete enzymes for their metabolic activity and adaptation and are referred to as thermozymes, which is considered as an ideal enzyme to undergo catalytic activities under elevated temperatures. These thermozymes have been proven effective in diversified fields of industry such as biofuel, food and chemical manufacturing, paper and pulp, detergents, pharmaceuticals and nutraceuticals. One best example of thermozymes application is Taq polymerase and other thermophilic DNA polymerases employed in Polymerase Chain Reaction that transformed genetic engineering field. Several thermozymes have additionally proved stability under other extreme conditions like high acidity, high pressure and high salinity,which can be regarded appropriate for application under poly-extremic conditions (Kumar et al. 2019).

3. Thermophilic Hemicellulases and Thermophilic Microorganisms

Hemicellulases are the flexible proteins comprising of a catalytic domain (CD) which, depending upon the amino acid or nucleic acid composition, can be either glycosyl hydrolases (GHs) or carbohydrate esterases (CEs) and it functions as enzyme and other functional components (Juturu and Wu 2013). One of the catalytic domain among them is the carbohydrate-binding module (CBM) that directs and transports the catalytic domain (CD) towards the substrate (Oliveira et al. 2015). There are two main hemicellulases that attack the similar kind of linkages presiding in the backbone of hemicelluloses structure, endo-hemicellulase hydrolyses the internal glycosidic linkages, whereas exo-hemicellulases attack at the reducing or non-reducing terminus of the sugar, therefore releasing monomers or dimers (Mamo 2019). Table 2 summarizes the different classes of hemicellulase enzymes with their preliminary functions and their particular site of action on the hemicellulose polysaccharide (Juturu and Wu 2013).

Usually, hemicellulases are developed by the saprophytic microorganisms (that grow upon dead detritus matter), but there are other diversified sources containing microbes that grow hemicellulase enzymes such as soil and water samples from bio-reserves, national parks, forests, microbiome of bovine rumens, human gastrointestinal tract, fruit and vegetables. However, hemicellulases that are extracted from saprophytic fungi, gram negative and gram positive thermophillic bacteria are of great commercial value, because they have the ability to survive under harsh conditions like extreme

Table 2. Different hemicellulases with their functions and action site on hemicelluloses.

Hemicellulases Types	Preliminary Functions	Site of Action
i) Glycosyl hydrolases (GHs)		
Endo-xylanase	Hydrolyses β-1,4 bond of xylan backbone structure, hence releasing xylooligomers	β-1,4 xylan back bone
β-Xylosidase	Attacks on exo β-1,4 bond of xylooligomers by cleaving and releasing free xylose	β-1,4 xylooligomers
Endo-1,4-mannanase	Hydrolyses β-1,4 bond of mannan and releases mannan oligomers	β-1,4 mannan
β-Mannosidase	Attacks on exo β-1,4 bond of mannan oligomers by cleaving and releasing free mannan	β-1,4 mannan oligomers
α-L-Arabinofuranosidase	Hydrolyses arabinan at O-2 and O-3 positions on structure of xylan backbone	α-L-arabinofuranosyl oligomers
α-L-Arabinanase	Hydrolyses xylooligomers producing arabinose	α-1,5-arabinan
α-L-Glucuronidase	Hydrolyses α-1, 2 linkage between glucuronic acid side chain substitutions, liberating glucuronic acid	4-O-methyl-α-glucuronic acid
ii) Carbohydrate esterases (CEs)		
Acetyl xylan esterase	Hydrolyses acetyl side chain substitutions, releasing acetic acid	2- or 3-O-acetyl xylan
Feruloyl xylan esterase	Hydrolyses ferulic acid side chain substitutions, releasing ferulic acid	Ferulic acid substitutions

temperature and pH. Furthermore, the selected potent hemicellulase-producing microrganisms can undergo modification by doing mutagenesis of wild-strain while-cell genome. Thermophilic and hyperthermophilic hemicellulolytic enzymes have been predominantly extracted from *Thermotoga* sp. strain FjSS3-B.1 (extracted from hot-spring on Sava-Savu beach in Fiji; secreted highly thermostable xylanase and β-glucosidase) (Leuschner and Antranikian 1995), and *Pyrococcus furiosus* (secreted β-glucosidase and β-galactosidase) (Kengen et al. 1993); similarly, *Thermotoga maritima, Dictoglomus turgidus* and *Caldocellum saccharolyticum* also secrete hyperthermal hemicellulase and cellulase enzymes (Bergquist et al. 1999). Table 3 depicts the information compiled from Brenda Enzyme Database (Schomburg et al. 2002) regarding the hemicellulase-producing microorganisms isolated from their native habitats with their defined appropriative conditions for growth and survival under extreme thermophilic conditions.

4. Potential Applications of Thermophilic Hemicellulases

Hemicellulase enzymes have been employed consistently in various industrial treatments and procedures like quality advancements in animal feeds, effective conversion of lignocellulosic substrates into renewable biofuels and biochemicals, improvement of pulp-biobleaching and delignification operation proficiency in paper and pulp industry, processing of textile and cellulose fibers, preparation of detergent formulations, lowering viscosity of extraction of fruit juices, preparation of probiotics, manufacturing of bakery products for increasing bread volume, stability, dough handling and others (Juturu and Wu 2013). With such increasing interest towards "greener" potential processes and applications of high yielding enzymes, market of enzymes is growing at rapid rate both volume-wise and diversity-wise, leading to constantly increasing commercial value of industrial enzymes. Among the major classes of hydrolases, it is estimated that hemicellulases are commercially crucial hydrolases, hence accounting for total of 75% of overall market of industrial enzymes (Li et al. 2012). The annual market valuation of solely xylanase enzymes is believed to be of about US $ 200 million and compound annual growth rate of world-wide commercialization of xylanases has been inclined towards 6.6% (Parameswaran et al. 2018).

Table 3. Microorganisms (bacteria and fungi) and their substrates for producing different hemicellulases, properties and optimum conditions.

Enzyme	Microorganisms	Substrates/Source	Optimum activities and other properties	References
Feruloyl esterase	*Clostridium stercorarium*	Ethyl ferulate	Temp. −65°C, pH-8, Activity-88 U mg^{-1}	(Schomburg et al. 2002)
	Aspergillus niger	Methyl sinapinate	Temp. −55°C, pH-5 Activity-156 U mg^{-1}	(Schomburg et al. 2002)
Endo-1,4-β-xylanase	*Bacillus pumilus*	β-1,4-D-xylan	Temp. −40°C, pH-6.5 Activity-1780 U mg^{-1}	(Schomburg et al. 2002)
	Trichoderma Longibrachiatum	1,4-β-D-xylan	Temp. −45°C, pH-5 Activity-6630 U mg^{-1}	(Schomburg et al. 2002)
	Clostridium thermocellum (XynD 10)	β-1,4-D-xylan	Temp. −80°C, pH-6.4 MW-71.7 kDa, Half-life- > 30 min at 90°C	(Zverlov et al. 2005)
	Sulfolobus solfataricus	β-1,4-D-xylan	Temp. −90°C, pH-7 MW-57 kDa, Half-life- > 47 min at 100°C	(Cannio et al. 2004)
	Clostridium stercorarium	1,4-β-D-xylan	Temp. −80°C, pH-6.4 MW-71.7 kDa, Half-life- > 30 min at 90°C	(Zverlov et al. 2005)
β-1,4-xylosidase	*Thermoanaerobacter ethanolicus*	o-nitrophenyl-β-D-xylopyranoside	Temp. −93°C, pH-6 Activity-1073 U mg^{-1}	(Schomburg et al. 2002)
	Aspergillus nidulans	p-nitrophenyl-β-D-xylopyranoside	Temp. −50°C, pH-5 Activity-107 U mg^{-1}	
	Sulfolobus solfataricus	Lignocellulosic plant biomass in Solfataric hot spring, Italy	Temp. −85°C, GC content − 36% Genome size-2.99 Mb	(Morana et al. 2007)
Exo-β-1,4-mannosidase	*Aspergillus niger*	β-D-Man-(1-4)-β-D-GlcNAc-(1-4)-β-DGlcNAc-Asn-Lys	Temp. −55°C, pH-3.5 Activity-188U mg^{-1}	(Schomburg et al. 2002)
	Pyrococcus horikoshii	pNP-bglucopyranoside/ pNP-b-xylopyranoside	Temp. −90°C, pH-4.8 MW-56.5 kDa Half-time-27 min at 102°C	(Kaper et al. 2002)
	Thermobifida fusca	pNP-mannopyranoside	Temp. −53°C, pH-7.2 MW-94 kDa Half-time-10 min at 60°C	(Beki et al. 2003)
Endo-β-1,4-mannanase	*Bacillus subtilis*	Galactoglucomannan/ glucomannans	Temp. −105°C, pH-7 Activity-514 U mg^{-1}	(Schomburg et al. 2002)
	Sclerotium rolfsii	Galactoglucomannan/ mannans galactomannans/ glucomannans	Temp. −74°C, pH-3.3 Activity-380 U mg^{-1}	
	Alicyclobacillus acidocaldaius	Galactoglucomannan/ mannans galactomannans/ glucomannans	Temp. −65°C, pH-5.5	(Zhang et al. 2008)
	Caldibacillus Cellulovorans	Galactoglucomannan/ mannans galactomannans/ glucomannans	Temp. −85°C, pH-6, MW-31 kDa Half-time-15 min at 80°C	(Sunna et al. 2000)
	Rhodothermus Marinus	Galactoglucomannan/ mannans galactomannans/ glucomannans	Temp. −85°C, pH-5.4, MW-56.5 kDa Half-time-27 min at 102°C	(Politz et al. 2000)

Table 3 contd. ...

...Table 3 contd.

Enzyme	Microorganisms	Substrates/Source	Optimum activities and other properties	References
Endo-α-1,5-arabinanase	*Bacillus subtilis*	1,5-α-L-arabinan	Temp. −105°C, pH-8 Activity-429 U mg^{-1}	(Schomburg et al. 2002)
	Aspergillus niger	1,5-α-L-arabinan	Temp. −55°C, pH-5 Activity-90.2 U mg^{-1}	(Schomburg et al. 2002)
α-Glucuronidase	*Thermoanaerobacterium saccharolyticum*	4-O-methyl-glucuronosyl-xylotriose	Temp. −50°C, pH-6 Activity-9.6 U mg^{-1}	(Schomburg et al. 2002)
	Geobacillus stearothermophilus 236	4-O-methyl-glucuronosyl-xylotriose	Temp. −40°C, pH-6.5, MW-78.2 kDa Half-time-50 min at 50°C	(Choi et al. 2000)
	Thermotoga maritima	2-O-(4-O-methyla-glucopyranosyluronic acid)-xylobiose	Temp. −80°C, pH-7.8, Half-time-30 min at 80°C	(Suresh et al. 2003)
α-Galactosidase	*Escherichia coli*	Raffinose	Temp. −60°C, pH-7 Activity-27350 U mg^{-1}	(Schomburg et al. 2002)
	Mortierella vinacea	Melibiose	Temp. −60°C, pH-4 Activity-2000 U mg^{-1}	(Schomburg et al. 2002)
	Clostridium stercorarium	Raffinose/guar gum	Temp. −70°C, pH-6, MW-84.8 kDa	(Kimura et al. 2003)
	Aspergillus niger	-	Temp. −55°C, pH-3.5 Activity-6593 U mg^{-1}	(Schomburg et al. 2002)
Acetyl xylan esterase	*Schizophyllum commune*	4-methylumbelliferyl acetate/4-nitrophenyl acetate	Temp. −30°C, pH-7.7 Activity-227 U mg^{-1}	(Schomburg et al. 2002)
	Clostridium thermocellum	-	Temp. −60°C, pH-7, Activity-15 min at 70°C	(Correia et al. 2008)
β-Glucosidase	*Thermotoga* sp. FjSS3-B.1	Marine sediments; Thermotogales are capable of fermenting polymers into H$_2$, CO$_2$, acetate	Temp. −100°C, pH-5.3 MW-31 kDa	(Leuschner and Antranikian 1995)
	Thermotoga neapolitana	Marine sediments in Lucrino, Italy	Temp. −80°C, GC content-41%	(Van Ooteghem et al. 2004)

Among various potential applications of hemicellulases, the most broadly employed application is degradation of hemicellulose in different substrates enzymatically, which leads to production of value-added products; they also contribute in various processes in different industries like paper and pulp, biofuel and bioethanol, chemical products, baking, animal fodder, biorefinery, etc. Utilisation of hemicellulose from lignocellulosic substrates is crucial for the development of commercially sustainable second generation biofuels. After pretreatment of solid biomass of lignocellulose, hemicellulases are added in order to increase the hydrolysis rate of cellulose, implying that the hemicellulose residues which,though are present in small proportion, can limit the mechanism of action of cellulase enzymes. Xylanase-deprived bio-bleaching of chemical paper pulps is widely employed as a bleacher in paper and pulp industry prior to the stage of bleaching in order to enhance lignin extraction (Viikari et al. 2010). The property of specificity and moderate hydrolysis state of purified enzyme can be used in lignocellulosic substrate characterization (Oksanen et al. 2000). By implementation of progressive enzyme peeling and HPLC, localization of xylan and glucomannan biopolymers within wood pulps has been interpreted; it was also depicted that the structure of xylan and glucomannan present upon the nearby planes of fibers distinguishes greatly from the

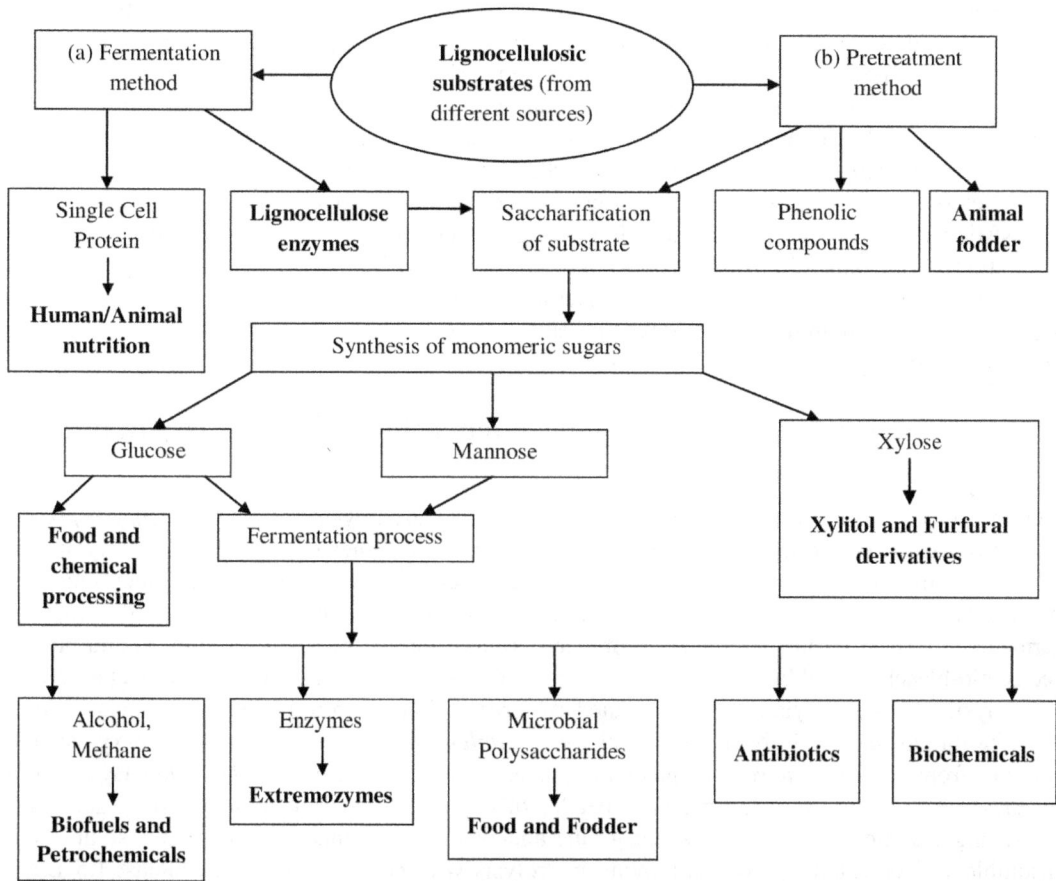

Fig. 1. Bioconversion of lignocellulosic substrates using different methods and production of different by-products with their commercial applications in industries.

components of other carbohydrates in the entire pulp of paper (Suurnakki et al. 1996). The primary application of xylanase enzymes involves bio-bleaching of paper pulp industry. Apart from this, they are also employed in other potential applications such as fruit juice and wine clarification, isolation of starch for preparation of active food components, enhancement of product quality of animal fodder and baked foods, debarking from woods of trees, de-inking of recycled paper fibers and then preparing paper pulp by dissolving using xylanase enzyme. If xylanse is incorporated into a feed mixture of maize and sorghum, where both are less consistent food, then xylanase can enhance nutrients digestion inside digestive tract, leading to better energy consumption (Polizeli et al. 2005) These xylanases, when added into rye-derived diet for broiler chickens, resulted in increment in the weight of chicks and when combined with phytases, it led to enhancement in weight of egg and albumin produced from egg-laying hens of white and brown color (Silversides et al. 2006). Some xylanses (family 11) synthesized by rumen-inhabiting bacteria of genera *Psuedobutyrivibrio* and *Butyrivibrio* had been observed cleaving the oat spelt xylan into tetramers or complex oligomers and these xylooligomers could be effective in fostering multiplication of good microflora; hence, xylanases produced by them can be utilized as a feed supplement for animals and these strains further can be employed as probiotics for animals (Van Craeyveld et al. 2008). Acetyl-xylan esterases and ferulolyl esterases are the xylan-debranching enzymes, can improve the solubilization of complex of lignin-polysaccharide by elimination bonds or linkages and substitutions within the polymers during the process of pulping of the paper fibers. Acetyl-xylan esterase can also be employed in paper-deinking process by helping in expulsion of substituent groups that obstructs main primary chain debranching enzymes (De Graaff et al. 2000). Feruloyl esterases have been described for their applications in conversion of lignocellulosic residues, ester development in organic solvents, and separation of different phenolic compounds as precursors for several value-added products (Garcia-Conesa et al. 2004). Pectinases have commercial applications in pre-treatment of softwoods that further enhances the effectiveness of treatment by increasing the permeability of wood for different chemical preservatives (De Gregorio et al. 2002). Hemicellulases have also been employed for the development of several bio-medicinal compounds like anti-inflammatory, anti-glycemic, anti-metaststic and anti-carcinogenic agents, improving product quality in dairy and poultry industries, separation of coconut oil from coconut flesh using heated hydraulic press technique yielding < 60% oil from it, pretreatment and bio-processing of textile and cellulose fibers, formulations of surfactants-based liquid detergents, development of wine, beer and vinegar by incorporating these enzymes in fermentation process and refinement of coffee beans.

5. Advancements and Patents related to Hemicellulases with their Commercial Importance

Hemicellulases, along with cellulases derived from an ascomycetous fungus, *Chrysosporium lucknowense*, were exploited to create multi-enzyme constituents, comprising of xylanases, α-L-aranbinofuranosidases, acetyl-xylan esterases, feruloyl-xylan esterases, glucuronidases, mannanases and cellulases and these hemicellulases have also been employed for releasing fermentable sugars during fermentation of dried grains, quality upgrading of chicken and poultry feeds, bio-bleaching and bio-refining of paper pulp fibers and textile fabrics (Gusakov et al. 2011). Hydrolysis of arabinoxylan can be attained by preparation of formulation comprising of α-L-aranbinofuranosidases (derived from *Meripilus giganteus* and *Humicola insolens*), β-xylosidases (derived from *Trichoderma reesei*) and endo-xylanase (derived from *Humicolinsolens*). By collating all these enzymes of about 75 mg each per kg of dry matter, hydrolysis of different substrates consisting arabinoxylan, vinasse, spent grains and wheat straw that are both water soluble and insoluble was outstanding over employing hydrolysis with commercialized Celluclast 1.5 L and Ultraflo L that liberates arabinose sugar of 33–364% (w/w) and xylose sugar of 12–109% (w/w), respectively (Soerensen et al. 2011). Preparation of purified α-L-arabinofuranosidases (40 Uml^{-1}) and xylanases (I U ml^{-1}) isolated from *Bacillus strearothermophilus* NCIMB 40221 and 40222 supernatant cultures was compiled together and utilized to improve delignification of wood pulp, thereby liberating 28% extra lignin at pH 9,65°C, in comparison to utilizing individually prepared enzymes within the same atmospheric conditions (Rosenberg et al. 1995). Thermostable enzymes α-L-arabinofuranosidases and xylanase expressed by the bacterium *Rhodothermus marinus* used together in bio-bleaching of softwood paper pulp has been observed to improve the paper brightness up to 1.8% ISO (International Standard Organization) grades in comparison to enhanced brightness (i.e., 1.9–2.1%) by using commercialized enzymes Igrazyme 40-X4 and Gamanase, either used individually or together (Gomes et al. 2000). A value-added product of palm oil extraction process from seeds, referred to as palm kernel press cake (PKC), is investigated to contain a higher proportion of hexose sugars in terms of galactomannon monomers that renders PKC as a fascinating feedstock for synthesis of bioethanol where consecutive saccharification and fermentation process exploiting *Saccharomyces cerevisae* has yielded about 200 g of ethanol kg^{-1} of PKC within just 9 days, production of biofuels and other by-products through microbial fermentation (Jorgensen et al. 2010). There are many invented patents related to the production of different types of hemicellulases including xylanases, mannanases, β-xylosidases and acetyl-xylan esterases with their commercial applications which are enlisted in Table 4 below (Soni and Kango 2013).

Table 4. Patents relevant to different hemicellulases with commercial applications.

Patent Number	Illustration	Commercial Applications
US7320741B2	Xylanase treatment method in ClO$_2$ bleaching sequence	Useful in a pulp mill for complex pulp bleaching process
US7547534B2	Method for making a composition to treat a wood, a pulp or a paper.	Provides methods of designing new xylanases
WO044818	Method for preparing kaempferol-3-O-rutinoside (K-3-R) and composition of skin external application.	K-3-R obtained by sugar removal of xylopyranose group from camelliaside B, using a selected xylanase, xylosidase
JP087401A	Alkaline Mannanase	Mannanase with optimum pH 8 to 9.5 useful in many fields
China10113343	Eosinophil beta-mannanase MAN5A and gene and application of the same	Useful for animal feeding stuff, fodder, medicament and energy industry in term of novel enzyme
China10095505	β-mannanase for feeding and its preparation	Invention belongs to the field of microbial fermentation and enzyme engineering
US0003702A1	Process for producing saccharified solution of lignocellulosic biomass	Lignocellulosic biomass saccharification is done by adding saccharifying enzyme, β-xylosidase from *Thermotoga maritime*
China Patent 101724615A	β-xylosidase and encoding gene and its application	Hydrolyze a corncob steam-exploded fluid, has mild reaction conditions and is environment friendly
WO160484	Novel GH with β-xylosidase and β-glucosidase activities and its uses	Novel glycosyl hydrolase with β-xylosidase and β-glucosidase activities
US7923234B2	Thermoacid tolerant β xylosidases, genes encoding, related organism and methods	Thermal and acid tolerant β-xylosidase, applicable in various industrial purposes
US0280541 A1	β-xylosidase for conversion of plant cell wall carbohydrates to simple sugars	Exhibits activity for hydrolyzing xylose-plant materials to xylose. Used alone or in combination with other hydrolytic or xylanolytic enzymes for treatment of lignocellulosic plant materials
US0281303A1	Polypeptides having acetyl xylan esterase activity and polynucleotides encoding same.	Relates to isolated polypeptides with acetyl xylan esterase activity and isolated polynucleotides encoding polypeptides.

5.1 Thermophilic α-L-Arabinofuranosidases and its Classification

Breakdown of polysaccharides using hemicellulase enzymes is usually restricted because of the existence of arabinose residues as side chains attached to the major skeletons of the structure. α-L-aranbinofuranosidases (AFAs : EC 3.2.1.55) are the hemicellulases that particularly catalyze the hydrolysis of α-L-(1,2)-,α-L-(1,3)-, and α-L-(1,5)-arabinosuranosyl residues residing at the non-reducing terminal of arabinoxylans (or arabinose-containing oligo- or poly-mers), thereby liberating arabinose from the substrates joined together with α-L-aranbinofuranosidic linkage, in anexolytic mode of action of polysaccharide cleavage. Particularly, these enzymes can synchronously improve the hydrolysis process of glycosidic linkages between polymers attached with other hemicellulases (Saha 2000). In hemicellulose, the arabinose substitute interrelates with other polymers like cellulose, lignin and pectin, making the structure more complex and thereby rendering endurance and physical strength to the cell wall of the plant. Substitution of arabinose and pectin residues present in the hemicellulosic structures inhibits the enzymatic activity of hemicellulolytic and pectinolytic enzymes; hence, α-L-aranbinofuranosidases can be employed in a synchronous manner for the improvement of hydrolysis of plant biomass (Shallom et al. 2002). Soluble sugars are

Table 5. Classification of α-L-aranbinofuranosidases based on two major categories.

Classification	Illustration	References
(a) Based on substrate specificity and how α-L-aranbinofuranosidases (AFAs) behaves on its polysaccharide substrate:		
Type-A α-L-arabinofuranosidases (AFAs)	Preferably cleaves α-(1,5)-L-arabino-oligosaccharides into monomers of arabinose and also able to break down synthetic p-nitrophenyl (PNP) α-L-arabinofuranosides, but inactive against arabinosyl-linkages in polymers.	(Contesini et al. 2017)
Type-B α-L-arabinofuranosidases (AFAs)	Hydrolyze branched arabino-oligomers, linear L-arabinanpolysaccharides, and artificial PNP-α-L-arabinofuranoside, therefore liberating xylose and L-arabinose monomers.	(Contesini et al. 2017)
(1→4)-β-D-arabinoxylan Arabinofuranohydrolases (AXHs)	Highly specified for arabinosidic linkages in various arabinoxylans, by liberating L-arabinose and they are not active on artificial substrates like pNP-α-L-arabinofuranoside.	(Contesini et al. 2017)
(i) AXH-m	Active on arabinose residues joined to C-2 (S2/ara) and C-3 (S3/ara) position of xylose residues with single substitution in arabinoxylan, hence referred to as AXH-m; Preferentially found in *Aspergillus awamori*, *Bifidobacterium adolescentis*, and *Bacillus subtilis*.	(Sakamoto et al. 2011)
(ii) AXH-d3	Cleaves terminal arabinose moieties linked to C-3 of the xylose residues branches of the arabinoxylan (D3/ara) with double substitutions, hence AXH-d3; Particularly found in *B. adolescentis* and *Humicola insolens*.	(Sakamoto et al. 2011)
(iii) AXH-md	Acts on both types of xylose residues (C-2 and C-3), having one and two substitutions in arabinoxylans, hence referred to as AXH-md.	(Sakamoto et al. 2011)
(b) Based on degree of amino acids sequence uniformity, α-L-aranbinofuranosidases *have been categorized into different GH families:*		
GH-2	This family of AFAs belongs to GH-A group of glycoside hydrolase family; breaks glycosidic linkages with glutamate as nucleophile and proton donor and their catalytic-site has (β/α) 3D structure.	(Shi et al. 2014)
GH-3	This family exhibits identical enzymatic action as to those of GH-2 family. Here aspartate is catalytic nucleophile and glutamate acts as a proton donor for hydrolases, while histidine acts as a proton donor for phosphorylases.	(De La Mare et al. 2013)
GH-43	Few GH-43 AFAs depictα-xylosidase and α-L-arabinofuranosidase activities, liberating arabinose from O-3 and (1→3) α-L-linked arabinofuranosyl side chains from di-substituted xylose residues. Exhibits catalytic domain with 5-fold β-propeller 3D- structure having aspartate and glutamate as nucleophile and catalytic protondonor.	(De La Mare et al. 2013)
GH-51	They can hydrolyze both mono- and di-substitutions at terminal or internal xylopyranosyl units of arabinoxylan, liberating (1→2) and (1→3)-α-L-arabinofuranosyl residues from mono-substitutedresidues and cleaves glycosidic-linkages via double displacement mechanism. Their catalytic domain has (β/α)-barrel structure with proton donating and nucleophilic two glutamates on β-4 and β-7 strands.	(Bouraoui et al. 2016)
GH-54	In this case, AFAs are located together with β-xylosidases and possess type-B substrate specificity. Their catalytic domain is identical to GH-B clan of glycoside hydrolase family, comprising of β-sandwich fold with glutamate and aspartate as nucleophilic and proton donating residues.	(De La Mare et al. 2013)
GH-62	These AFAs are single group of enzymes that belong to GH-F clan and hydrolyzesingle-substituted xylose residues residing in the arabinoxylans acting at O-2 and O-3 while liberating arabinose.	(De La Mare et al. 2013)

then liberated through enzymatic hydrolysis process, which is consumed by both prokaryotic and eukaryotic microbes for the generation of many different value-added by-products such as bio-ethanol, furfurals, animal feed, xylitol, etc. α-L-aranbinofuranosidases could not be categorized depending upon the backbone with which arabinofuranosyl residues are linked and therefore illustrates a broad range of substrate specificity (Rahman et al. 2003).

In another study, Beldman et al. (1997) classified these enzymes as: (i) exo-acting enzymes that perform in exo-pattern, releasing L-arabinose from polymers (like arabinoxylans, arabinans, arabinogalactans or arabinooligomers and from synthetic substrates including p-nitrophenyl group), which are persistently referred to as α-L-arabinofuranoside arabinofurnaohydrolases, and (ii) endo-acting enzymes that hydrolyse polysaccharides comprising of (1→5)-α-L-arabinan chain at their interior and are referred as per enzyme nomenclature as (1→5)-α-L-arabinan(1→5)-α-L-arabinohydrolases. The classification of α-L-aranbinofuranosidases is carried out on the basis of two major categories (Table 5), where the first category involves classification based on their specificity towards substrate (termed as arabinose anomeric conformation) and how α-L-aranbinofuranosidases behaves on its polysaccharide substrate (endo- or exo-acting); these enzymes are grouped with glycoside hydrolases into three major groups: α-L-arabinofuranoside hydrolases (E.C. 3.2.1.55), β-L-arabinosidases (E.C. 3.2.1.88) and endo-1,5-α-L-arabinanases (E.C. 3.2.1.99) (Fritz et al. 2008). With reference to CAZY database, α-L-aranbinofuranosidases are assembled within class glycoside hydrolase that cleaves linkage between 2 carbohydrates and carbohydrate-noncarbohydrate moieties; so the second category involves classification depending on degree of sequence uniformity of amino acids, different glycoside hydrolases have been categorized into different GH families (Thakur et al. 2019).

5.2 Microorganisms Involved in the Production of α-L-aranbinofuranosidases

There are many microorganisms that perform a crucial role in natural conversion of plant polysaccharides like cellulose, hemicelluloses, lignin and pectin by producing several hydrolytic and de-branching enzymes. Microorganisms such as bacteria, fungi and actinomycetes secrete α-L-arabinofuranosidases, in addition to some main lignocellulosic enzymes; nevertheless, vast scale α-L-arabinofuranosidases production obliges (Yang et al. 2016). In addition to microorganisms, some plants have been also recorded for the production of α-L-arabinofuranosidases (Montes et al. 2008). Being an important class of microorganisms, bacteria produce enzyme that are responsible for the breakdown of various plant polymers and comprise of bacterial genus *Alicyclobacillus*, *Bacillus*, *Bifidobacterium*, *Geobacillus*, *Paenibacillus, Pseudomonas, Thermotoga*, etc. (Poria et al. 2020). *Bifidobacterium adolescentis* has been observed to have the capability to degrade di-substituted xylose group residing in arabinoxylan, thereby liberating arabinosyl residues (Van Laere et al. 1997). A bifunctional enzyme was isolated from the bacterium *Alicyclobacillus* sp. A4 illustrating enzymatic activities for endo-xylanses and α-L-arabinofuranosidases and categorized under GH-51 family (Yang et al. 2015). Hoffmam et al. (2013) reported thatα-L-arabinofuranosidases are secreted by *Bacillus subtilis*, a mesophilic bacterium that belongs to the GH-51 family of glycoside hydrolases, produced by submerged-state fermentation. Apart from this, solid-state fermentation has also been carried out to generate 97 kDa of α-L-arabinofuranosidase enzyme from bacterium *Arthrobacter* sp., which was completely characterized after purification (Khandeparker and Jalal 2015).

Irrespective of bacterial species, actinomycetes produce most of their enzymes extracellularly, therefore decreasing the task of downstream processing. The main producer of α-L-arabinofuranosidase enzyme is genus *Streptomyces* with example of species such as *Streptomyces* sp. PC22 with specific activity of 16.34 Umg^{-1} for artificial substrate (Raweesri et al. 2008), *Streptomycescoelicolor* produces GH-62 family α-L-arabinofuranosidase enzyme (Maehara et al. 2014), and *Streptomycesdiastaticus* generates about 38 kDa and 60 kDa of α-L-arabinofuranosidase enzymes (Tajana et al. 1992).

Several fungi, which also involve ascomycetes and basidomycetes, like *Penicillium* sp., *Aspergillus* sp. and *Pleurotus* sp. were studied to secrete α-L-arabinofuranosidases. Fungus *Penicillium purpurogenum* was studied to generate GH-51 enzyme that was active for artificial substrate p-nitrophenyl (PNP) α-L-arabinofuranosides exhibiting specific activity of 1.2 Umg^{-1}

Table 6. Characterization of α-l-arabinofuranosides (AFAs) synthesized by thermophillic microorganisms.

Type	Microorganisms	Characterization of AFAs		References
		Optimum temperature and pH	Other properties	
Bacteria	*Alicyclobacillus* sp. A4	Temp. –60°C; pH-6	GH family-51; MW-56.7 kDa	(Yang et al. 2015)
	Thermobacillus xylanilyticus	Temp. –75°C; pH-5.6 to 6.2	GH family-51; MW-56 kDa	(Debeche et al. 2000)
	Bacillus pumilus	Temp. –55°C; pH-7	GH family-51; MW-60 kDa	(Degrassi et al. 2003)
	Thermotoga maritima MSB8	Temp. –90°C	GH family-51; MW-55 kDa	(Miyazaki 2005)
	Geobacillus caldoxylolyticus TK 4	Temp. –80°C; pH-6	GH family-51; MW-58 kDa	(Canakci et al. 2007)
	Anoxybacillus ketsanbolensis	Temp. –65°C; pH-5.5	GH family-51; MW-58 kDa	(Canakci et al. 2008)
	Bacillus subtilis	Temp. –60°C; pH-8	GH family-51; MW-57.5 kDa	(Inacio et al. 2008)
	Bacillus subtilis	Temp. –50°C; pH-6.5	GH family-51; MW-59.5 kDa	(Hoffmam et al. 2013)
	Bifidobacterium longum H-1	Temp. –57°C; pH-4.7	GH family-51; MW-57.4 kDa	(Lee et al. 2011)
	Caldicellulosiruptor saccharolyticus	Temp. –80°C; pH-5.5	GH family-51; MW-58 kDa	(Shin et al. 2013)
	Geobacillus vulcani GS90	Temp. –70°C; pH-5	GH family-51; MW-60 kDa	(Ilgu et al. 2018)
	Paenibacillus sp. DG-22	Temp. –60°C; pH-6	GH family-51; MW-56.5 kDa	(Lee and Lee 2014)
	Thermotoga petrophila RKU-1	Temp. –70°C; pH-6	GH family-51; MW-57 kDa	(Dos Santos et al. 2011)
	Thermotoga thermarum	Temp. –80°C; pH-5.5	GH family-2; MW-90 kDa	(Shi et al. 2014)
Actinomycetes	*Streptomyces chartreusis*	Temp. –50°C; pH-7	GH family-43; MW-37 kDa	(Matsuo et al. 2000)
	Streptomyces avermitilis NBRC 14893	Temp. –45°C; pH-6	GH family-43; MW-53 kDa	(Ichinose et al. 2008)
	Streptomyces sp. PC22	Temp. –65°C; pH-6	MW-79 kDa	(Raweesri et al. 2008)
	Streptomyces thermoviolaceus OPC520	Temp. –60°C; pH-5	GH family-62; MW-37 kDa	(Tsujibo et al. 2002)
Fungi	*Penicillium canescens*	Temp. –75°C; pH-4.7	MW-60 kDa	(Sinitsyna et al. 2003)
	Penicillium purpurogenum	Temp. –60°C; pH-5	GH family-51; MW-70 kDa	(Fritz et al. 2008)
	Aspergillus nidulans	Temp. –70°C; pH-5	GH family-51; MW-88.6 kDa	(de Lima Damásio et al. 2012)
	Penicillium janthinellum	Temp. –50°C; pH-5.5	GH family-54; MW-64 kDa	(Chadha et al. 2019)
	Phanerochaete chrysosporium	Temp. –50°C; pH-5.5	GH family-43; MW-83 kDa	(Huy et al. 2013)

(Fritz et al. 2008). Hemicellulolytic fungus *Penicillium janthinellum* was also studied to produce α-L-arabinofuranosides of about 212 Ug^{-1} of substrate for wheat bran and paddy straw dry substrate effectively. α-L-Arabinofuranosides can also be produced by utilizing inorganic nitrogen and carbon resources that comprise of substrate L-arabinose and crude beet arabinans by fungus *Aspergillus nidulans* (Poria et al. 2020).

Several thermostable α-L-arabinofuranosides were reported from different thermophilic microbes including *Geobacillus caldoxylolyticus* TK4 (Canakci et al. 2007), *Anoxybacillus kestanbolensis* AC26Sari (Canakci et al. 2008), *Caldicellulosiruptor saccharolyticus* (Lim et al. 2010), *Thermotoga thermarum* (Shi et al. 2014), *Thermotoga petrophila* (Sontos et al. 2011), and *Caldicellulosiruptor saccharolyticus* (Shin et al. 2013). The optimum range of temperature of 65–80°C and pH of 5.5–6 for α-L-arabinofuranosides has been reported through various studies. *Thermotoga maritima* MSB8 with temperature optimum of 90°C has been found to produce hyperthermophilic α-L-arabinofuranosides (Miyazaki 2005). Recently isolated, fungus *Aspergillus hortai* CRM1919, having optimum activity at 60°C and pH of 4, secretes halo-tolerant α-L-arabinofuranoside with highest enzymatic activity of 190 U L^{-1} (Terrone et al. 2020). *Geobacillus vulcani* GS90 was reported to produce a thermostable GH-51 family α-L-arabinofuranoside with its optimum temperature and pH of 70°C and 5, and showed highest enzymatic activity of 200 Umg^{-1} with pNP-α-L-arabinofuranoside and 526 Umg^{-1} with sugar beet arabinan, respectively (Ilgu et al. 2018). It was also recorded that *Massilia* sp. RBM26 produces a halo-tolerant, xylose-tolerant and alkali-tolerant, bi-functional GH-43 family α-l-arabinofuranoside and/or β-xylosidase with optimum temperature of 50°C and pH of 6.5 (Xu et al. 2019).

6. Potential Applications of Thermophilic α-L-Arabinofuranosides with Recent Advancements and Related Patents

α-l-Arabinofuranosides are very crucial hemicellulosic enzymes because they perform an important function in degradation and bioconversion of hemicellulosic substrates for the formation of sustainable and eco-friendly biofuels and chemicals. They cover a broad commercial range of applications in various industries initiating from development of beverages to processing of animal feed, altogether comprising of food-processing, production of biofuels and manufacturing of agro-based products (Binod et al. 2019). Several thermophilic arabinofuranosidases have been isolated from extremophiles and reported to acquire potential applications involving fruit juice clarification and rendering of aromatic flavors into wines by degrading monoterpenes present (Michlmayr et al. 2012), improving digestibility of animal feed (Fritz et al. 2008), improving digestibility of poultry feed (Ravn et al. 2018), enhancement of bread texture quality (Zhou et al. 2019), increase in hydrolysis rate of sugar substrates in biofuel and bioethanol industry (Denli and Ercan 2001), improves delignification of paper pulp (Numan and Bhosle 2006), production of prebiotics (Falck et al. 2018), control of disease caused by *Mycobacterium* (Shewry et al. 2010), development of less caloric sweetener (Grembecka 2015), and also plays a crucial role in cancer chemotherapeutic treatments (Chinnathambi and Lachke 1995). Companies like Megazyme and Creative Enzyme produce arabinofuranosidases on acommercial basis through bacteria and fungi. For instance, some commercialized arabinofuranosidases marketed by Megazymes are: GH-51 AFA isolated from *Aspergillus niger* (McCleary et al. 2015), GH-62 AFA isolated from *Aspergillus nidulans* (Wilkens et al. 2016) and GH-43 AFA from *Bifidobacterium adolescentis* (Sorensen et al. 2006). There are many invented patents related to the production of different thermostable arabinofuranosidases with their commercial applications which are enlisted in Table 7 below (Soni and Kango 2013).

Table 7. Patents related to thermostable arabinofuranosidases with commercial applications.

Patent Number	Commercial Applications
Japan 022301	Constitutes arabinofuranosidase that breaks down arabinan liberating arabino-oligomers.
WO06584	Effective in hydrolysis of arabinofuranosyl residues from L-arabinose linked polymers or hemicelluloses.
China80014246	Correlates to the isolated polypeptides associated with α-Larabinofuranosidase activity and isolated nucleic acid sequences that encodethis polypeptide.
US0136476	Correlates to the isolated polypeptides associated with α-Larabinofuranosidase activity and isolated nucleic acid sequences that encode this polypeptide.
AU239240	Correlates to the isolated polypeptides associated with α-Larabinofuranosidase activity and isolated nucleic acid sequences that encode this polypeptide.
US0028334	Correlates to the isolated polypeptides associated with α-Larabinofuranosidase activity and isolated nucleic acid sequences that encode this polypeptide.
US0281303A1	Correlates to the isolated polypeptides associated with acetyl xylan esterase activity and isolated polynucleotides that encode this polypeptide.

7. Conclusion and Future Perspectives

Extremozymes produced from lignocellulosic biomass with wide optimum temperature and pH tolerance makes them an outstanding prospect for many potential uses such as production of biofuel and bioethanol from lignocelluloses bioconversion, animal feed improvement for nutrient supply and digestibility, bio-remediation, textile industry for eliminating cellulose fibrils, paper and pulp and detergent industry for dirt removal and improvement of whiteness and softness. A tremendously increasing demand and concern in the area of biofuels, biochemicals, bioenergy, bioethanol, biopolymers and different natural materials creates a push to search for alternative renewable resources for various residues and effective technologies in order to transform the constituents of the biomass residues. The field of biotechnology renders a fascinating opportunity for bioprocessing of lignocelluloses and transformation of polymers, particularly when consolidated with chemical and mechanical techniques. Applications of hemicellulases derived from lignocellulosic substrates have provided a productive impact and, furthermore, diversifies biochemical structure of natural hemicelluloses, which require coordination of these enzymes which is vital in acquiring absolute hydrolysis of different hemicellulolytic structures. α-L-Arabinofuranosidases encompasses a broad commercial range of applications in various industries starting from development of beverages to processing of animal feed, altogether comprising of food-processing, production of biofuels and manufacturing of agro-based products. Inspite of many potential uses, α-L-arabinofuranosidases are yet not emphasized much at commercial level in industries because of non-availability of high productive strains. Hence, it is necessary to discover and investigate diversified niches and different metagenomes through gemone mining techniques for identification and segregation of new novelistic α-L-arabinofuranosidase-producing microbes and cloning genes with properties feasible for various industries.

Abbreviations

GH, glycoside hydrolase; CE, carbohydrate esterases; CD, catalytic domain; CBM, carbohydrate-binding module; HPLC, high performance liquid chromatography; PKC, palm kernel press cake; AFAs, α-L-aranbinofuranosidases; CAZY-database, carbohydrate-active enzymes database; p-NP, p-nitrophenyl; AXHs, arabinofuranohydrolases

References

Beg, Q., M. Kapoor, L. Mahajan and G.S. Hoondal. 2001. Microbial xylanases and their industrial applications: A review. Appl. Microbiol. Biotechnol. 56: 326–338.

Beldman, G., H.A. Schols, S.M. Pitson, M.J.F. Searle-van Leeuwen and A.G.J. Voragen. 1997. Arabinans and arabinan degrading enzymes. Adv. Macromol. Carbohydr. Res. 1: 64.

Bergquist, P.L., M.D. Gibbs, D.D. Morris, V.J. Te'o, D.J. Saul and H.W. Morgan. 1999. Molecular diversity of thermophilic cellulolytic and hemicellulolytic bacteria. FEMS Microbiol. Ecol. 28: 99–110.

Binod, P., E. Papamichael, S. Varjani and R. Sindhu. 2019. Introduction to Green Bioprocesses: Industrial Enzymes for Food Applications. Green Bio-processes: 1–8. Springer, Singapore.

Bouraoui, H., M.L. Desrousseaux, E. Ioannou, P. Alvira, M. Manaï, C. Rémond, C. Dumon, N. Fernandez-Fuentes and M.J. O'Donohue. 2016. The GH51 α-l-arabinofuranosidase from *Paenibacillus* sp. THS1 is multifunctional, hydrolyzing main-chain and side-chain glycosidic bonds in heteroxylans. Biotechnol. Biofuels 9: 140.

Canakci, S., A.O. Belduz, B.C. Saha, A. Yasar, F.A. Ayaz and N. Yayli. 2007. Purification and characterization of a highly thermostable α-L-arabinofuranosidase from *Geobacillus caldoxylolyticus* TK4. Appl. Microbiol. Biotechnol. 75: 813–820.

Canakci, S., M. Kacagan, K. Inan, A.O. Belduz and B.C. Saha. 2008. Cloning, purification, and characterization of a thermostable α-L-arabinofuranosidase from *Anoxybacillus kestanbolensis* AC26Sari. Appl. Microbiol. Biotechnol. 81: 61.

Cannio, R., N. Di Prizito, M. Rossi and A. Morana. 2004. A xylan-degrading strain of *Sulfolobus solfataricus*: Isolation and characterization of the xylanase activity. Extremophiles 8: 117–124.

Chadha, B.S., A. Monga and H.S. Oberoi. 2021. α-l-arabinofuranosidase from an efficient hemicellulolytic fungus penicillium janthinellum capable of hydrolyzing wheat and rye arabinoxylan to arabinose. J. Microbiol., Biotechnol. Food Sci., pp. 1132–1139.

Chinnathambi, S. and A.H. Lachke. 1995. An extracellular α-l-arabinofuranosidase from *Sclerotium rolfsii*. World J. Microbiol. Biotechnol. 11: 232–233.

Choi, I.D., H.Y. Kim and Y.J. Choi. 2000. Gene cloning and characterization of α-glucuronidase of *Bacillus stearothermophilus* No. 236. Biosci. Biotechnol. Biochem. 64: 2530–2537.

Contesini, F.J., M.V. Liberato, M.V. Rubio, F. Calzado, M.P. Zubieta, D.M. Riaño-Pachón, F.M. Squina, F. Bracht, M.S. Skaf and A.R. Damasio. 2017. Structural and functional characterization of a highly secreted α-L-arabinofuranosidase (GH62) from *Aspergillus nidulans* grown on sugarcane bagasse. Biochim. Biophys. Acta, Proteins Proteomics. 1865: 1758–1769.

Correia, M.A., J.A. Prates, J. Bras, C.M. Fontes, J.A. Newman, R.J. Lewis, H.J. Gilbert and J.E. Flint. 2008. Crystal structure of a cellulosomal family 3 carbohydrate esterase from *Clostridium thermocellum* provides insights into the mechanism of substrate recognition. J. Mol. Bio. 379: 64–72.

De Graaff, L.H., J. Visser, H.C. van den Broeck, F. Strozyk, F.J. Kormelink and J.C. Boonman. 2000. Gist Brocades BV. Method to alter the properties of acetylated xylan. U.S. Patent #6,010,892.

De Gregorio, A., G. Mandalari, N. Arena, F. Nucita, M.M. Tripodo and R.L. Curto. 2002. SCP and crude pectinase production by slurry-state fermentation of lemon pulps. Bioresour. Technol. 83: 89–94.

De La Mare, M., O. Guais, E. Bonnin, J. Weber and J.M. Francois. 2013. Molecular and biochemical characterization of three GH62 α-l-arabinofuranosidases from the soil deuteromycete *Penicillium funiculosum*. Enzyme Microb. Technol. 53: 351–358.

de Lima Damásio, A.R., B.C. Pessela, F. Segato, R.A. Prade, J.M. Guisan and T.M. Maria de Lourdes. 2012. Improvement of fungal arabinofuranosidase thermal stability by reversible immobilization. Process Biochem. 47: 2411–2417.

Debeche, T., N. Cummings, I. Connerton, P. Debeire and M.J. O'Donohue. 2000. Genetic and biochemical characterization of a highly thermostable α-l-Arabinofuranosidase from *Thermobacillus xylanilyticus*. Appl. Environ. Microbiol. 66: 1734–1736.

Degrassi, G., A. Vindigni and V. Venturi. 2003. A thermostable α-arabinofuranosidase from xylanolytic *Bacillus pumilus*: Purification and characterisation. J. Biotechnol. 101: 69–79.

Denli, E. and R. Ercan. 2001. Effect of added pentosans isolated from wheat and rye grain on some properties of bread. Eur. Food Res. Technol. 212: 374–376.

Dos Santos, C.R. Squina, A.M. Navarro, D.P. Oldges, A.F.P. Leme, R. Ruller, A. Mort, R. Prade and M. Murakami. 2011. Functional and biophysical characterization of a hyperthermostable GH51 α-L-arabinofuranosidase from *Thermotoga petrophila*. Biotechnol. Lett. 33: 131–137.

Falck, P., J.A. Linares-Pastén, E.N. Karlsson and P. Adlercreutz. 2018. Arabinoxylanase from glycoside hydrolase family 5 is a selective enzyme for production of specific arabinoxylooligosaccharides. Food Chem. 242: 579–584.

Fritz, M., M.C. Ravanal and C. Braet. 2008. A family 51 α-L-arabinofuranosidase from *Penicillium purpurogenum*: Purification, properties and amino acid sequence. Mycol. Res. 112: 933–942.

Garcia-Conesa, M.T., V.F. Crepin, A.J. Goldson, G. Williamson, N.J. Cummings, I.F. Connerton, C.B. Faulds and P.A. Kroon. 2004. The feruloyl esterase system of *Talaromyces stipitatus*: Production of three discrete feruloyl esterases, including a novel enzyme, TsFaeC, with a broad substrate specificity. J. Biotechnol. 108: 227–241.

Gomes, J., I. Gomes, K. Terler, N. Gubala, G. Ditzelmüller and W. Steiner. 2000. Optimisation of culture medium and conditions for α-L-arabinofuranosidase production by the extreme thermophilic eubacterium *Rhodothermus marinus*. Enzyme Microb. Technol. 27: 414–422.

Gomes, J. and W. Steiner. 2004. The biocatalytic potential of extremophiles and extremozymes. Food Technol. Biotechnol. 42: 223–225.

Grembecka, M. 2015. Sugar alcohols—their role in the modern world of sweeteners: A review. Eur. Food Res. Technol. 241: 1–14.

Gusakov, A.V., P.J. Punt, J.C. Verdoes, A.P. Sinitsyn, E. Vlasenko, S.W.A. Hinz, M. Gosink and Z. Jiang. 2011. Fungal enzymes. U.S. Patent #7,923,236.

Hoffmam, Z.B., L.C. Oliveira, J. Cota, T.M. Alvarez, J.A. Diogo, M. de Oliveira, A.P.S. Citadini, V.B. Leite, F.M. Squina, M.T. Murakami and R. Ruller. 2013. Characterization of a hexameric exo-acting GH51 α-L-arabinofuranosidase from the mesophilic *Bacillus subtilis*. Mol. Biotechnol. 55: 260–267.

https://www.marketsandmarkets.com/Market-Reports/industrial-enzymes-market-237327836.html.

Huy, N.D., P. Thayumanavan, T.H. Kwon and S.M. Park. 2013. Characterization of a recombinant bifunctional xylosidase/arabinofuranosidase from *Phanerochaete chrysosporium*. J. Biosci. Bioeng. 116: 152–159.

Ichinose, H., M. Yoshida, Z. Fujimoto and S. Kaneko. 2008. Characterization of a modular enzyme of exo-1, 5-α-L-arabinofuranosidase and arabinan binding module from *Streptomyces avermitilis* NBRC14893. Appl. Microbiol. Biot. 80: 399.

İlgü, H., Y. Sürmeli and G. Şanlı-Mohamed. 2018. A thermophilic α-l-arabinofuranosidase from *Geobacillus vulcani* GS90: Heterologous expression, biochemical characterization, and its synergistic action in fruit juice enrichment. Eur. Food Res. Technol. 244: 1627–1636.

Inacio, J.M., I.L. Correia and I. de Sa-Nogueira. 2008. Two distinct arabinofuranosidases contribute to arabino-oligosaccharide degradation in *Bacillus subtilis*. Microbiology 154: 2719–2729.

Jorgensen, H., A.R. Sanadi, C. Felby, N.E.K. Lange, M. Fischer and S. Ernst. 2010. Production of ethanol and feed by high dry matter hydrolysis and fermentation of palm kernel press cake. Appl. Biochem. Biotechnol. 161: 318–332.

Juturu, V. and J.C. Wu. 2013. Insight into microbial hemicellulases other than xylanases: A review. J. Chem. Technol. Biotechnol. 88: 353–363.

Kaper, T., H.H. van Heusden, B. van Loo, A. Vasella, J. van der Oost and W.M. de Vos. 2002. Substrate specificity engineering of β-mannosidase and β-glucosidase from Pyrococcus by exchange of unique active site residues. Biochemistry 41: 4147–4155.

Kengen, S.W., E.J. Luesink, A.J. Stams and A.J. Zehnder. 1993. Purification and characterization of an extremely thermostable β-glucosidase from the hyperthermophilic archaeon *Pyrococcus furiosus*. Eur. J. Biochem. 213: 305–312.

Khandeparker, R. and T. Jalal. 2015. Xylanolytic enzyme systems in *Arthrobacter* sp. MTCC 5214 and *Lactobacillus* sp. Appl. Biochem. Biotechnol. 62: 245–254.

Kim, T.J. 2008. Microbial exo-and endo-arabinosyl hydrolases: Structure, function, and application in L-arabinose production. In CAZymes: 229–257.

Kimura, T., K. Sakka and K. Ohmiya. 2003. Cloning, sequencing, and expression of the gene encoding the *Clostridium stercorarium* α-Galactosidase Aga36A in *Escherichia coli*. Biosci. Biotechnol. Biochem. 67: 2160–2166.

Kumar, S., A.K. Dangi, P. Shukla, D. Baishya and S.K. Khare. 2019. Thermozymes: Adaptive strategies and tools for their biotechnological applications. Bioresour. Technol. 278: 372–382.

Lee, J.H.,Y.J. Hyun and D.H. Kim. 2011. Cloning and characterization of α-L-arabinofuranosidase and bifunctional α-L-arabinopyranosidase/β-D-galactopyranosidase from *Bifidobacterium longum* H-1. J. Appl. Microbiol. 111: 1097–1107.

Lee, S.H. and Y.E. Lee. 2014. Food microbiology and biotechnology: Cloning, expression, and characterization of a thermostable GH51 α-L-arabinofuranosidase from *Paenibacillus* sp. DG-22. J. Microbiol. Biotechnol. 24: 236–244.

Leuschner, C. and G. Antranikian. 1995. Heat-stable enzymes from extremely thermophilic and hyperthermophilic microorganisms. World J. Microbiol. Biotechnol. 11: 95–114.

Li, S., X. Yang, S. Yang, M. Zhu and X. Wang. 2012. Technology prospecting on enzymes: Application, marketing and engineering. Comput. Struct. Biotechnol. J. 2: e201209017.

Lim, Y.R., R.Y. Yoon, E.S. Seo, Y.S. Kim, C.S. Park and D.K. Oh. 2010. Hydrolytic properties of a thermostable α-l-arabinofuranosidase from *Caldicellulosiruptor saccharolyticus*. J. Appl. Microbiol. 109: 1188–1197.

Maehara, T., Z. Fujimoto, H. Ichinose, M. Michikawa, K. Harazono and S. Kaneko. 2014. Crystal structure and characterization of the glycoside hydrolase family 62 α-L-arabinofuranosidase from *Streptomyces coelicolor*. J. Biol. Chem. 289: 7962–7972.

Mamo, G., 2019. Alkaline active hemicellulases. In Alkaliphiles in biotechnology. Springer, Cham. 245–291.

Matsuo, N., S. Kaneko, A. Kuno, H. Kobayashi and I. Kusakabe. 2000. Purification, characterization and gene cloning of two α-L-arabinofuranosidases from *Streptomyces chartreusis* GS901. Biochemistry J. 346: 9–15.

McCleary, B.V., V.A. McKie, A. Draga, E. Rooney, D. Mangan and J. Larkin. 2015. Hydrolysis of wheat flour arabinoxylan, acid-debranched wheat flour arabinoxylan and arabino-xylo-oligosaccharides by β-xylanase, α-L-arabinofuranosidase and β-xylosidase Carbohydr. Res. 407: 79–96.

Mehta, D. and T. Satyanarayana. 2013. Diversity of hot environments and thermophilic microbes. Thermophilic Microbes in Environmental and Industrial Biotechnology, Springer, Dordrecht.

Michlmayr, H., S. Nauer, W. Brandes, C. Schümann, K.D. Kulbe, M. Andrés and R. Eder. 2012. Release of wine monoterpenes from natural precursors by glycosidases from *Oenococcus oeni*. Food Chem. 135: 80–87.

Miyazaki, K. 2005. Hyperthermophilic α-L-arabinofuranosidase from *Thermotoga maritima* MSB8: Molecular cloning, gene expression, and characterization of the recombinant protein. Extremophiles 9: 399–406.

Montes, R.A.C., P. Ranocha, Y. Martinez, Z. Minic, L. Jouanin, M. Marquis, L. Saulnier, L.M. Fulton, C.S. Cobbett, F. Bitton and J.P. Renou. 2008. Cell wall modifications in Arabidopsis plants with altered α-l-arabinofuranosidase activity. Plant Physiol. 147: 63–77.

Morana, A., O. Paris, L. Maurelli, M. Rossi and R. Cannio. 2007. Gene cloning and expression in *Escherichia coli* of a bi-functional β-D-xylosidase/α-L-arabinosidase from *Sulfolobus solfataricus* involved in xylan degradation. Extremophiles. 11: 123–132.

Numan, M.T. and N.B. Bhosle. 2006. α-L-arabinofuranosidases: The potential applications in biotechnology. J. Ind. Microbiol. Biotechnol. 33: 247–260.

Oksanen, T., J. Pere, L. Paavilainen, J. Buchert and L. Viikari. 2000. Treatment of recycled kraft pulps with *Trichoderma reesei* hemicellulases and cellulases. J. Biotechnol. 78: 39–48.

Oliveira, C., V. Carvalho, L. Domingues and F.M. Gama. 2015. Recombinant CBM-fusion technology—Applications overview. Biotechnol. Adv. 33: 358–369.

Parameswaran, B., S. Varjani and S. Raveendran. 2018. Green Bio-processes: Enzymes in Industrial Food Processing, Springer, Singapore.

Politz, O., M. Krah, K.K. Thomsen and R. Borriss. 2000. A highly thermostable endo-(1, 4)-β-mannanase from the marine bacterium *Rhodothermus marinus*. Appl. Microbiol. Biotechnol. 53: 715–721.

Polizeli, M.L.T.M., A.C.S. Rizzatti, R. Monti, H.F. Terenzi, J.A. Jorge and D.S. Amorim. 2005. Xylanases from fungi: Properties and industrial applications. Appl. Microbiol. Biotechnol. 67: 577–591.

Poria, V., J.K. Saini, S. Singh, L. Nain and R.C. Kuhad. 2020. Arabinofuranosidases: Characteristics, microbial production, and potential in waste valorization and industrial applications. Bioresour. Technol. 304: 123019.

Rahman, A.S., K. Kato, S. Kawai and K. Takamizawa. 2003. Substrate specificity of the α-L-arabinofuranosidase from *Rhizomucor pusillus* HHT-1. Carbohydr. Res. 338: 1469–1476.

Ravn, J.L., V. Glitso, D. Pettersson, R. Ducatelle, F. Van Immerseel and N.R. Pedersen. 2018. Combined endo-β-1, 4-xylanase and α-l-arabinofuranosidase increases butyrate concentration during broiler cecal fermentation of maize glucurono-arabinoxylan. Anim. Feed Sci. Technol. 236: 159–169.

Raweesri, P., P. Riangrungrojana and P. Pinphanichakarn. 2008. α-L-arabinofuranosidase from *Streptomyces* sp. PC22: Purification, characterization and its synergistic action with xylanolytic enzymes in the degradation of xylan and agricultural residues. Bioresour. Technol. 99: 8981–8986.

Reed, C.J., H. Lewis, E. Trejo, V. Winston and C. Evilia. 2013. Protein adaptations in archaeal extremophiles. Archaea, 2013.

Rosenberg, E. and Y. Shoham. 1995. α-L-Arabinofuranosidase and xylanase from *Bacillus stearothermophilus* NCIMB 40221, NCIMB 40222 or mutant thereof for delignification. U.S. Patent #5,434,071.

Saha, B.C. 2000. α-L-Arabinofuranosidases: Biochemistry, molecular biology and application in biotechnology. Biotechnol. Adv. 18: 403–423.

Sakamoto, T., A. Ogura, M. Inui, S. Tokuda, S. Hosokawa, H. Ihara and N. Kasai. 2011. Identification of a GH62 α-L-arabinofuranosidase specific for arabinoxylan produced by *Penicillium chrysogenum*. Appl. Microbiol. Biotechnol. 90: 137–146.

Santos, C., F. Squina, A. Navarro, D. Oldiges, A. Leme, R. Ruller, A. Mort, R. Prade and M. Murakami. 2011. Functional and biophysical characterization of a hyperthermostable GH51 α-l-arabinofuranosidase from *Thermotoga petrophila*. Biotechnol. Lett. 33.

Sarmiento, F., R. Peralta and J.M. Blamey. 2015. Cold and hot extremozymes: Industrial relevance and current trends. Front. Bioeng. Biotechnol. 3: 148.

Schomburg, I., A. Chang and D. Schomburg. 2002. BRENDA, enzyme data and metabolic information. Nucleic Acids Res. 30: 47–49.

Shallom, D., V. Belakhov, D. Solomon, S. Gilead-Gropper, T. Baasov, G. Shoham and Y. Shoham. 2002. The identification of the acid–base catalyst of α-arabinofuranosidase from *Geobacillus stearothermophilus* T-6, a family 51 glycoside hydrolase. FEBS Letters 514: 163–167.

Shewry, P.R., V. Piironen, A.M. Lampi, M. Edelmann, S. Kariluoto, T. Nurmi, R. Fernandez-Orozco, A.A. Andersson, P. Aman, A. Fras and D. Boros. 2010. Effects of genotype and environment on the content and composition of phytochemicals and dietary fiber components in rye in the Health grain diversity screen. J. Agric. Food Chem. 58: 9372–9383.

Shi, H., Y. Zhang, B. Xu, M. Tu and F. Wang. 2014. Characterization of a novel GH2 family α-L-arabinofuranosidase from hyperthermophilic bacterium *Thermotoga thermarum*. Biotechnol. Letters 36: 1321–1328.

Shin, K.C., G.W. Lee and D.K. Oh. 2013. Production of ginsenoside Rd from ginsenoside Rc by α-L-arabinofuranosidase from *Caldicellulosiruptor saccharolyticus*. J. Microbiol. Biotechnol. 23: 483–8.

Silversides, F.G., T.A. Scott, D.R. Korver, M. Afsharmanesh and M. Hruby. 2006. A study on the interaction of xylanase and phytase enzymes in wheat-based diets fed to commercial white and brown egg laying hens. Poult. Sci. J. 85: 297–305.

Sinitsyna, O.A., F.E. Bukhtoyarov, A.V. Gusakov, O.N. Okunev, A.O. Bekkarevitch, Y.P. Vinetsky and A.P. Sinitsyn. 2003. Isolation and properties of major components of *Penicillium canescens* extracellular enzyme complex. Biochemistry 68: 1200–1209.

Soerensen, H.R., S. Pedersen, A. Viksoe-Nielsen, C.T. Joergensen, L.H. Christensen, C.I. Joergensen, C.H. Hansen and L.V. Kofod. 2011. Hydrolysis of arabinoxylan. U.S. Patent #7,993,890.

Soni, H. and N. Kango. 2013. Hemicellulases in lignocellulose biotechnology: Recent patents. Recent Pat. Biotechnol. 7: 207–218.

Sorensen, H.R., C.T. Jørgensen, C.H. Hansen, C.I. Jørgensen, S. Pedersen and A.S. Meyer. 2006. A novel GH43 α-L-arabinofuranosidase from *Humicola insolens*: Mode of action and synergy with GH51 α-L-arabinofuranosidases on wheat arabinoxylan. Appl. Microbiol. Biotechnol. 73: 850–861.

Sunna, A., M.D. Gibbs, C.W. Chin, P.J. Nelson and P.L. Bergquist. 2000. A gene encoding a novel multidomain β-1, 4-mannanase from *Caldibacillus cellulovorans* and action of the recombinant enzyme on kraft pulp. Appl. Environ. Microbiol. 66: 664–670.

Suresh, C., M. Kitaoka and K. Hayashi. 2003. A thermostable non-xylanolytic α-glucuronidase of *Thermotoga maritima* MSB8. Biosci. Biotechnol. Biochem. 67: 2359–2364.

Terrone, C.C., J.M. de Freitas Nascimento, C.R.F. Terrasan, M. Brienzo and E.C. Carmona. 2020. Salt-tolerant α-arabinofuranosidase from a new specie *Aspergillus hortai* CRM1919: Production in acid conditions, purification, characterization and application on xylan hydrolysis Biocatal. Agric. Biotechnol. 23: 101460.

Thakur, A., K. Sharma and A. Goyal. 2019. α-L-Arabinofuranosidase: A Potential Enzyme for the Food Industry. Green Bio-processes, Springer, Singapore.

Tsujibo, H., C. Takada, Y. Wakamatsu, M. Kosaka, A. Tsuji, K. Miyamoto and Y. Inamori. 2002. Cloning and expression of an α-L-arabinofuranosidase gene (stxIV) from *Streptomyces thermoviolaceus* OPC-520, and characterization of the enzyme. Biosc. Biotech. Bioch. 66: 434–438.

Van Craeyveld, V., K. Swennen, E. Dornez, T. Van de Wiele, M. Marzorati, W. Verstraete, Y. Delaedt, O. Onagbesan, E. Decuypere, J. Buyse and B. De Ketelaere. 2008. Structurally different wheat-derived arabinoxylooligosaccharides have different prebiotic and fermentation properties in rats. J. Nutr. 138: 2348–2355.

Van Laere, K.M.J., G. Beldman and A.G.J. Voragen. 1997. A new arabinofuranohydrolase from *Bifidobacterium adolescentis* able to remove arabinosyl residues from double-substituted xylose units in arabinoxylan. Appl. Microbiol. Biotechnol. 47: 231–235.

Van Ooteghem, S.A., A. Jones, D. Van Der Lelie, B. Dong and D. Mahajan. 2004. H 2 production and carbon utilization by *Thermotoga neapolitana* under anaerobic and microaerobic growth conditions. Biotechnol. Lett. 26: 1223–1232.

Viikari, L., S. Gronqvist, K. Kruus, J. Pere, M. Siika-aho, and A. Suurnäkki. 2010. Industrial biotechnology in the paper and pulp sector. Ind Biotechnol. Wiley-VCH Published Online: 29.

Wilkens, C., S. Andersen, B.O. Petersen, A. Li, M. Busse-Wicher, J. Birch, D. Cockburn, H. Nakai, H.E. Christensen, B.B. Kragelund and P. Dupree. 2016. An efficient arabinoxylan-debranching α-l-arabinofuranosidase of family GH62 from *Aspergillus nidulans* contains a secondary carbohydrate binding site. Appl. Microbiol. Biotechnol. 100: 6265–6277.

Xu, B., L. Dai, W. Zhang, Y. Yang, Q. Wu, J. Li, X. Tang, J. Zhou, J. Ding, N. Han and Z. Huang. 2019. Characterization of a novel salt-, xylose-and alkali-tolerant GH43 bifunctional β-xylosidase/α-l-arabinofuranosidase from the gut bacterial genome. J. Biosci. Bioeng. 128: 429–437.

Yang, W., Y. Bai, P. Yang, H. Luo, H. Huang, K. Meng, P. Shi, Y. Wang and B. Yao. 2015. A novel bifunctional GH51 exo-α-l-arabinofuranosidase/endo-xylanase from *Alicyclobacillus* sp. A4 with significant biomass-degrading capacity. Biotechnol. Biofuels 8: 197.

Yang, Y., J. Sun, J. Wu, L. Zhang, L. Du, S. Matsukawa, J. Xie and D. Wei. 2016. Characterization of a novel α-L-arabinofuranosidase from *Ruminococcus albus* 7 and rational design for its thermostability. J. Agric. Food Chem. 64: 7546–7554.

Zhang, Y., J. Ju, H. Peng, F. Gao, C. Zhou, Y. Zeng, Y. Xue, Y. Li, B. Henrissat, G.F. Gao and Y. Ma. 2008. Biochemical and structural characterization of the intracellular mannanase AaManA of *Alicyclobacillus acidocaldarius* reveals a novel glycoside hydrolase family belonging to clan GH-A. J. Biol. Chem. 283: 31551–31558.

Zhou, T., Y. Xue, Y. Zhang, Y. Dong, R. Gao and Y. Li. 2019. Improvement of the characteristics of steamed bread by supplementation of recombinant alpha-L-arabinofuranosidase containing xylan-binding domain. Food Biotechnol. 33: 34–53.

Zverlov, V.V., N. Schantz, P. Schmitt-Kopplin and W.H. Schwarz. 2005. Two new major subunits in the cellulosome of *Clostridium thermocellum*: Xyloglucanase Xgh74A and endoxylanase Xyn10D. Microbiol. 151: 3395–3401.

10

Extremophiles in Aviation Biofuel Production

S.M. Bhatt

1. Introduction

Rising prices of fuel along with high emission of CO_2 leads to great contribution in greenhouse gases and various jet-fuels are produced in chemical way. Concept of jet-biofuel production via extremophiles can be a good alternative to reduce GHG up to 1.3% or more in 2030 (Staples et al. 2018). Currently, jet-biofuel is produced via hydro-processing pathway using renewable source. In coming decades, reducing the cost of production and processing of jet biofuel is a major challenge. Application of extremozyme such as alkalophile and halophile would be highly beneficial as enzyme isolated from such organism since bio waste is available at large scale ad thus is helpful in reducing cost of bio-processing. Bio-waste such as water released from cane molasses industries, lignocellulosic waste, and algae are examples which can be used (Valdez-Ojeda et al. 2020). For aviation biofuel various substrate are available such as jatropha, algal oil and even municipal solid waste.

Biofuel production for jet started in 2011 using various substrates. Biggest thought behind jet biofuel production was increasing air pollution. A few airlines such as **Lufthansa** are now using biofuel produced from algae; see Fig. 1.

A group, Sustainable Aviation Biofuel User Group (SAFUG), was founded in 2008 as target of commercial aviation biofuel. This group was called as Algal Biomass Organisation, or ABO.

2. Classification (Based on Bioresources)

Aviation biofuel is also called as jet biofuel today and is mainly derived from hydro-processed esters and fatty acids (HEFA) (Fig. 1). There are other forms of aviation biofuel that can be obtained by using extremophiles.

Classification (based on bioresourses):

(a) Bio-waste from cane molasses
(b) Algal oil
(c) Alkane from Jatropha
(d) Soybean oil
(e) Palm oil
(f) Canola oil

Department of Agriculture, Amritsar College of Engineering and Technology, Amritsar.
Email: drsmbhatt@gmail.com

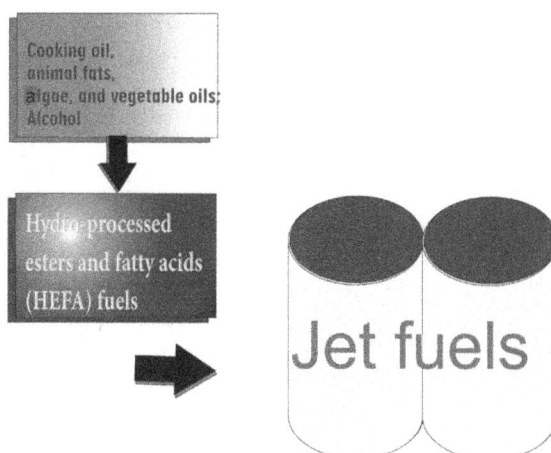

Fig. 1. Aviation biofuel obtained from HEFA.

In view of jet biofuel, long hydrocarbon chains of high molecular weight are very useful and also known as naphtha kerosene jet fuel. These are sometime called as **naphtha-A** or wide cut jet fuel or Jet-a-1, JP-5, **naphtha-B, naphtha-C**.

As compared to biodiesel, there are many advantages of using aviation bio fuel which is not going to freeze easily in cold weather or going to hydrolyze in hot summer. Recently, a very informative article was published on aviation biofuel using renewable resources (Kandaramath Hari et al. 2015), which mentioned that aviation biofuel must be produced as per ASTM standard considering following properties such as good flow properties, low freezing point so that during winter they must not freeze and good thermal stability for long storage purpose.

Currently, high cost of biofuel production is a challenge. Therefore, various efforts are underway to lower the cost by using more alternative biofuel. By 2030, IATA (International Air Transport Associations) proposed that there should be 30% increased use of such biofuel in aviation industry.

3. Alternative Sources of Jet Biofuel

Various sources such as vegetable oils (Baroutian et al. 2013) and animal fats, algal oil (Schlagermann et al. 2012), **sugar cane** (see Fig. 6) (**Energy Cane**) (Cantarella et al. 2015), and municipal solid waste (Skaggs et al. 2018) have been reported in making aviation biofuel more sustainable (Wang and Tao 2016, Yilmaz and Atmanli 2017) (Table 1 and Fig. 1).

4. Composition and Specifications of Jet Fuels as per ASTM Standard

The main composition of Jet Biofuel is linear chain of hydrocarbon which is difficult to obtained when mixture of hydrocarbon is mixed. Important properties of fuel such as high flash points, low freezing points, high energy density, sealing property are provided when only linear chain is in maximum in composition. Presence of isoalkane improves cold flow property of fuel, e.g., Jet-A fuel has low freezing property –47°C (Diaz-Pérez and Serrano-Ruiz 2020) (Table 2).

5. Problems in Aviation Biofuel Production

Main challenges reported are

(1) Production is very expensive
(2) Availability of suitable feedstock round the year
(3) Availability of suitable production technology
(4) Application of extremophiles for production of jet fuel.

Table 1. Some feedstock has been listed and their processing is mentioned.

S. No.	Bio-source of Aviation Biofuel	Technology of conversion	References
1	Oils and fats	Hydroprocessings/Decarboxylation and decarbonylation	(Vásquez et al. 2017)
2	Bio-kerosene		(Hong et al. 2013)
3	Hydrotreated vegetable oil (HVO)	Derived from oil seed plants.	(Sandquist and Güell 2012)
4	Biojet fuel	Derived from renewable sources	(Deane and Pye 2018)
5	Both Fischer-Tropsch (FT) and Hydroprocessed Esters and Fatty Acids (HEFA)	Synthetic Paraffinic Kerosine (SPK)	(Starck et al. 2016)
6.	Algae		(Chen et al. 2013) (Prabandono and Amin 2015) (Velazquez-Lucio et al. 2018) (Jiang et al. 2016) (Sirajunnisa and Surendhiran 2016) (Prabandono and Amin 2015) (Markou et al. 2012)
7	Camelina oil		(Moore et al. 2017)
8.	Municipal solid waste is food waste		(Karmee 2016)

Table 2. Bio jet fuel and some ASTM certified techniques used.

S. No.	ASTM Certified Techniques	Substrate	References
(HEFA bio-jet	Hydroprocessing	Fatty acid, oils	(Vásquez et al. 2017)
Fischer Tropsch method (FT)	Gasification	Municipal solid waste (MSW) or woody biomass	(Starck et al. 2016)
Synthesised Iso-Paraffinic fuels (SIP)	Farnesane	Sugars-to-hydrocarbon route	(Abanteriba et al. 2016)
Alcohol-to-jet based (ATJ)	On isobutanol	Alcohol	

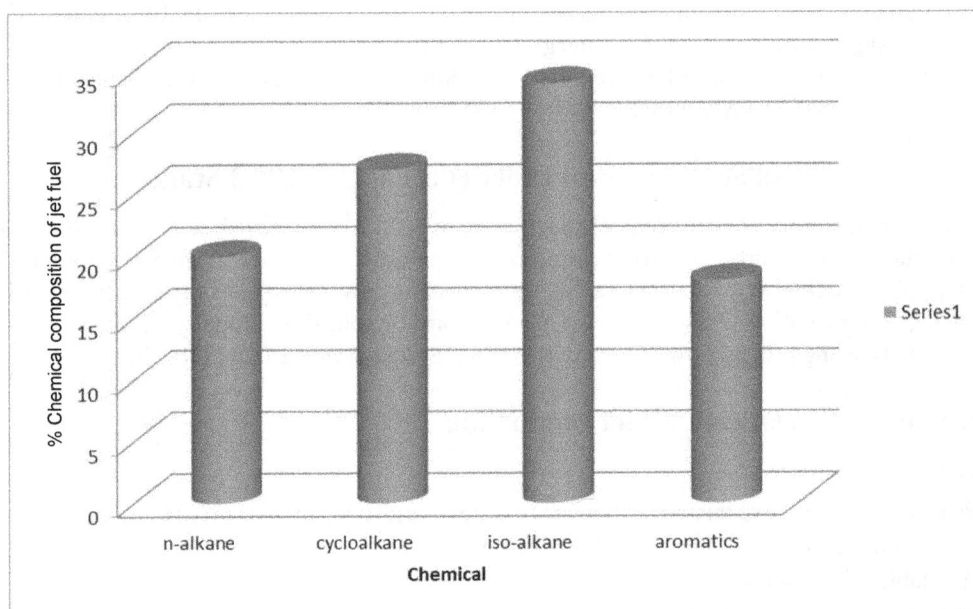

Fig. 2. Chemical composition of jet fuel.

Various alternative feedstocks currently in use for aviation biofuel are:

(1) Carbohydrates

(2) Algal oil

(3) Energy crops

(4) Municipal waste

(5) Lipid rich plants such as Jatropha

(6) Microbial based aviation biofuel (use of extremophiles)

6. Type of Aviation Biofuel Based on Technology

(a) Hydroprocessed aviation biofuel

(b) Fischer tropsch fuels (thermochemical process)

(c) Microbial based biofuel such as bioethanol (biochemical route)

(d) Biodiesel

(e) Liquid bio hydrogen and bio-methane

(f) Liquid hydrocarbon (BTL) process

(g) DHSC direct hydrocarbon process and alcohol (from biomass to sugarcane)

(a) Hydroprocessed jet fuel (HEFA)

Many feedstocks have been found to undergo hydro-deoxygenation such as vegetable oil, animal fats, and algal oil, which yields mainly water and propane. Hydroprocessed Esters and Fatty Acids (HEFA), also known as hydrotreated vegetable oil (HVO), involves hydrotreatment of oil with hydrogen under high pressure so that oxygen can be removed (Asnadi et al. 2020, Repetto et al. 2016, Rumizen 2017). In this way, long hydrocarbon chain is produced which is similar to diesel, chemically. These fuel are also called as **vegetable oil hydro treated biodiesel** which has no corrosive report free of aromatic and sulfur with high octane number. Such fuel similar to biodiesel is also called as drop-in-fuel, which means it can be used in any diesel engine without any modifications. See Fig. 3.

As per data, Sweden has produced around 160 million liters of HVO (Fig. 3) and HVO has benefits that it can reduce CO_2 emission up to 46%. Such type of HVO is also called as green diesel or Hydro processed Renewable Jet (HRJ). No corrosion has been reported in the engine of the aeroplane and can be used in all-weather without caring about pollution and also because it doesn't have any ashes after combustion. But the major issue that arises is from where we will manage such a large amount of feedstock. Many airlines such as Air France and Lufthansa are using HEFA fuels in at least 50% combinations (Figs. 2 and 4).

(b) Fischer tropsch fuels (thermochemical process)

This is chemical conversion based fuel obtained after catalytic conversion of syn gas obtained from incomplete combustion of biomasss. They are safe fuels as they are free of all major pollutants such as CO_2, hydrocarbons free of sulphur and aromatic components. It has benefit in terms of high production and can be produced around 2–3 millions barrels per day (Fig. 4).

Synthetic biology approaches are also found to be attractive options since microbes can be modified to convert sugar into alkanes. Jet biofuel produced is eitherkerosene based (Jet A, A1) or gasoline based (Jet B) but very less with microbial biofuel.

Kerosene based biofuel is mainly mixtures of hydrocarbon using alkane and linear hydrocarbon (called as paraffins), branched alkanes, cycloalkanes, olefins, and aromatic hydrocarbons which is hydro-processed to form jet biofuel. Jet B biofuel can be obtained using light hydrocarbon which can be distilled between 100°C and 250°C.

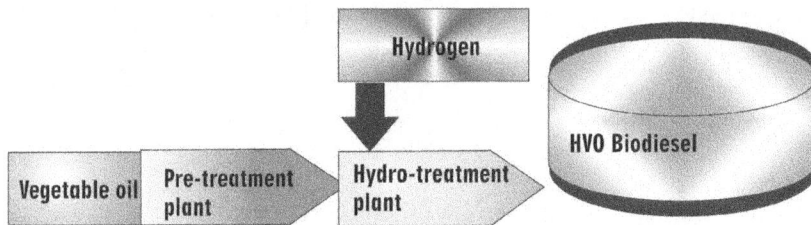

Fig. 3. Method of preparation of jet fuel (Zanata et al. 2019).

The main types of jet fuel used today are kerosene-based (Jet A and Jet A-1) and gasoline-based (Jet B). Jet B includes all light hydrocarbon oils for use in aviation, which distill between 100°C and 250°C (International Energy Agency http://www.iea.org/). Kerosene type jet fuels (Jet A and Jet A1) are mixtures of hydrocarbons, which are mainly alkanes, which include linear hydrocarbons also known as paraffins, branched alkanes, cycloalkanes, olefins and aromatic hydrocarbons (Jiménez-Diaz et al. 2017) (Fig. 5).

(c) Microbial based biofuel such as bioethanol

Bioenergy production is related to solid, liquid, and gas production in the form of CNG, biogas, ethanol, isobutanol, biodiesel, and biomethane. Using various bio-waste conversions requires various bio-catalysts in the form of lignase, cellulase, hemicellulose, pectinase, amylase and other enzymes.

In this regard, tapping some extremophile with higher alkane producing capability would have more scope in reducing the overall price with more volume productions. Its method of preparation has been shown in Fig. 3.

7. Role of Extremophiles In Jet Fuel Productions

Commercial biofuel production using enzyme isolated from mesophilic organism isn't feasible commercially because of temperature limitations in the bioreactors. Therefore, biocatalyst isolated from extremophiles become widely important in bioreactor applications, which has high temperature tolerance and enzyme has high processivity factors.

Extremophiles are known by various names such as thermophiles, psychrophiles, alkalophiles, halophiles, and acidophile.

Extremophile isolated from cold (Sánchez et al. 2009) or heat (Mandelli et al. 2017) (see Table 3 and 4) become of extreme importance in case of agri-waste based lignocellulose processing that require removal of lignin for bio-refinery applications. Other important steps include hydrolysis of cellulosic biomass at large scale which requires enzyme with extreme stability, like heat stable, pH stable and ineffective to toxic components released during processing and conversion of cellulose into glucose in the form of cellulobiose. Cyanobacteria and oligogenous algae are more attractive alternatives to produce large scale substrate for production of jet biofuels (Sánchez et al. 2009).

Table 3. List of some extremophiles.

1. *Acidithiobacillus, Arthrobacter*
2. *Bacillus, Caldicellulosiruptor*
3. *Clostridium, Coprothermobacter*
4. *Enterobacter, Geobacillus*
5. *Micrococcus, Paenibacillus*
6. *Penicillium, Picrophilus*
7. *Pseudoalteromonas*
8. *Thermobifida*

Table 4. Extremophiles' sources.

Species	Sources	Name of enzyme	Optimum temp.°C	Applications	References
Bacillus mojavensis	Soil samples	Amylase	70	Biorefinery	(Roberts et al. 1994)
Anoxybacillus thermarum	Hot spring	Amylase	60		(Poli et al. 2009)
Anoxybacillus flavithermus TWXYL3		Xylanase activity	65–85°C		
Pseudoalteromonas haloplanktis					(Garsoux et al. 2004)
halobacterium salinarum, Haloarcula marismortui, Haloquadratum walsbyi, and *the haloalkaliphile Natronomonaspharaoni*					(Falb et al. 2008)

8. Applications of Extremophiles

Various microbes living in harsh and tough conditions in environment are called as extremophiles such as alkalophiles, acidophiles, halophiles, thermohile, and hyperthermophiles. It may be animals, plants, microbes, fungi, piezophilic/barophilic (high pressure), oligotrophic (nutrient deficient environments), endolithic (within rock spaces), and xerophilic (dry area). Besides, polyextremophiles and extremotolerant microorganisms are also present in various extreme ecological niches (Dumorné et al. 2017, Gundala and Chinthala 2017, Singh et al. 2019b).

Various applications have been recorded in past for extremophile enzymes' uses in biofuel (Barnard et al. 2010). Some strain *Thermoanaerobacter* sp. x514 an anaerobic thermophilic bacterium has been found to convert various substrate into ethanol and also is ethanol tolerant (Lin et al. 2013).

(a) Bioprospecting

Various thermophilic/thermotolerant ethanologenic species have been reported such as *Clostridium thermocellum, C. thermohydrosulfuricum, C. thermosaccharolyticum, Caldicellulosiruptor* sp., *Thermotoga* sp., *Thermoanaerobiumbrockii, Thermoanaerobacter ethanolicus, T. thermohydrosulfuricus, T. mathranii,* etc. (Arora et al. 2015).

(b) In CBP

Certain other organisms such as *Clostridium thermocellum, Caldicellulosiruptor bescii,* and *Caldicellulosiruptor obsidiansis* are worked out as some good, cellulolytic, mixed-acid fermenting candidate for potential applications in consolidated bioprocessing for example Populus has been used as model feedstock which can be pretreated with dilute acid or with any harsh conditio. For efficient hydrolysis of corn stover, some thermophilic strain is more helpful as their enzyme can help in easy bioprocessing (Galanopoulou et al. 2016). Some cellulose identifying component bacterium such as *Clostridium thermocellum, Scaffoldin* or *Dockerins* from *Geobacillus* and *Caldicellulosirupt* species helps in 50% more hydrolysis of cellulosic biomass.

(c) The production of volatile industrial chemicals

Some thermophilic bacterium such as *Caldicellulosiruptor bescii* has potential for processes with simultaneous fermentation and product separation. In addition, transfer of alcohol dehydrogenase from *Clostridium thermocellum* may help in high processing of cellulose (Kahn et al. 2019).

Due to varied bioreactor condition, normal enzymes don't meet industrial expectation. Enzymes isolated from extremophiles have naturally optimized conditions to grow in such conditions. Some practical aspects have been developed already and find uses in laundry as it was isolated from cold adapted extremophiles, amylase and protease, developed by Novozyme.

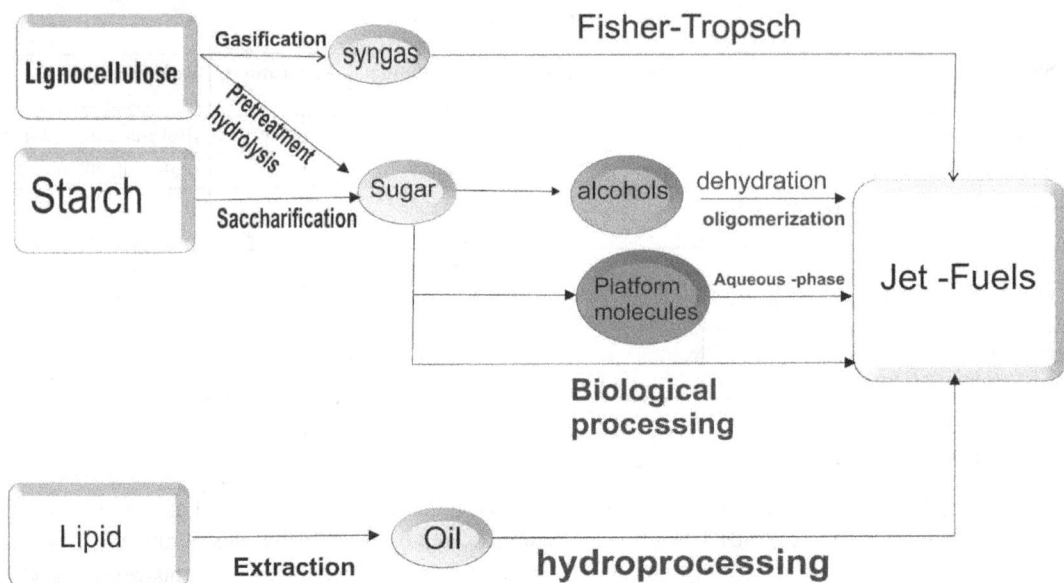

Fig. 4. Route of biomass conversion from various biofuel sources (Diaz-Pérez and Serrano-Ruiz 2020).

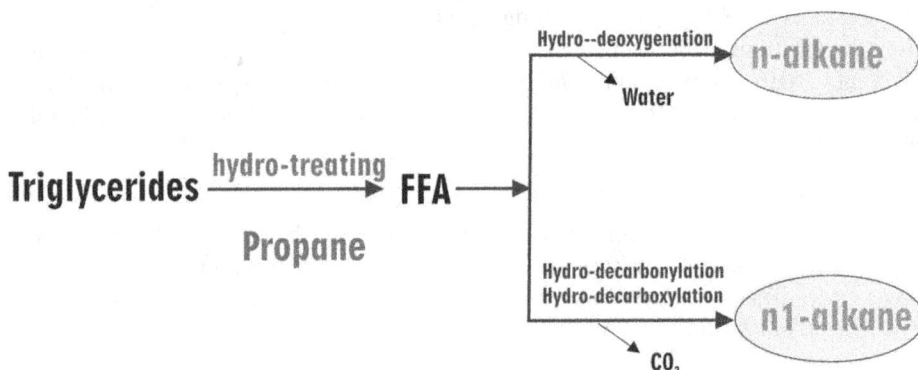

Fig. 5. Jet fuel production steps using TGA (Wang et al. 2015).

Catalytic conversion of various substrate such as lignocellulosic biomass into various forms (Zhou et al. 2011). Techniques such as liquefaction, pyrolysis,gasification, and hydrogenolysis is useful for large scale hydrogen production by using catalytic conversion.

Hydrogen can also be produced from lignin based bio-oil by technique of hydro deoxygenation. Also, in some cases, gas to jet fuels conversion has been used in the following pathway:

Gasification -→Syngas--→Fisher-tropsche

Caldicellulosiruptor grows optimally at 65 to 78°C and degrades lignocellulosic biomass without conventional pretreatment and *clostridium thermocellum* is an anaerobic, Gram positive thermophile that is capable of producing ethanol or acetone directly from cellulose at high temperature > 50°C (Straub et al. 2020).

(d) Compost-derived microbial consortia for switch grass digestion

Consortia prepared from thermophilic *Paenibacilli, Rhodothermusmarinus*, and *Thermus thermophiles* can digest switch grass directly (Gladden et al. 2011). Extracellular thermostable cellulase-free xylanase (EC 3.2.1.8) have been produced by *Bacillus licheniformis* A99 strain (Archana and Satyanarayana 1997). *Clostridium stercorarium*, a thermophilic ubiquitous soil dweller, has been isolated for hemicellulose digestions (Zverlov and Schwarz 2008). See Table 5.

Table 5. Largest biological source available on Earth, lignocellulose can be harnessed for bioethanol production.

S. No.	Biowaste Used	Process Followed/Yield	Reference
1.	Coir waste substrate	SSF/submerged (SmF) CELLULASE production Cellulase in SSF 14.6 fold > SmF A. niger were shown to be at pH 6, temperature 30°C	(Mrudula and Murugammal 2011)
2.	Wheat bran	*Trichodermareesei* NCIM 992 cellulase production by wheat bran SSF mode, cellulase production was 2.63 U ml −1	(Maurya et al. 2012)
3.	Rice straw + wheat bran	*A. niger* and *T. reseei,* respectively. Mixed-culture SSF using rice straw supplemented with wheat bran in the ratio 3:2 resulted in higher FP cellulase, β-glucosidase, endoglucanase (CMCase) and xylanase activities	(Dhillon et al. 2011)
4.	Various agro waste	Submerged fermentation (SmF)	(Srivastava et al. 2018a)
5.	Various agro waste	SSF > SmF	(Imran et al. 2016)
6.	Wheat straw	SSF mode/optimal conditions for SSF were temp 40°C, time 120 h, cellulase loading 40 FPU/g (wheat straw)	(Chen et al. 2008)
7.	Sorghum stover	Trichodermareesei RUT C-30 + solid-state fermentation (SSF)	(Idris et al. 2017)

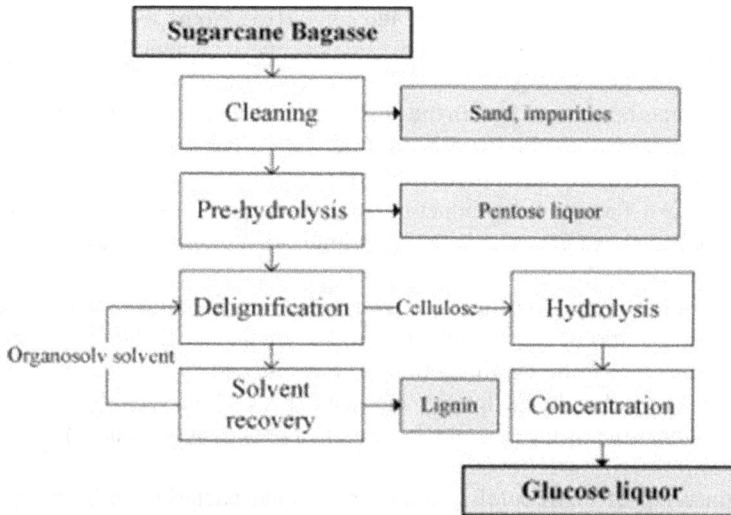

Fig. 6. Flow chart of separating sugar from bagasse (Dias et al. 2009).

Extracellular amylase has been produced by the thermophilic fungus *Thermomyces lanuginosus* for digestion of wheat bran, molasses bran, rice bran, maize meal, millet cereal, wheat flakes, barley bran, crushed maize, corncobs and crushed wheat.

DDGS digestion is quite difficult but can be digested by a protease isolated from Pseudomonas aeruginosa.

9. Extremophile Applications

Extremophilic microorganisms thrive in the harsh environments where other organisms cannot even survive. Extremophiles are taxonomically widely distributed and are a functionally diverse group (Cowan et al. 2015) that includes thermophiles, psychrophiles, acidophiles, alkalophiles, halophiles, barophiles/piezophiles, metalophiles, and radiophiles.

Extremophiles have the potential to produce biomolecules of high relevance for white, grey, and red biotechnological sectors. These microorganisms produce extremophilic enzymes (extremozymes) and protective organic biomolecules.

The variation in the biojet production cost would be highly dependent on the following parameters:

(a) Composition and cost of feedstock;
(b) Process design;
(c) Conversion efficiency or product yield;
(d) Valorization of co-products; and
(e) Energy conservation.

Therefore, decreasing the production cost of biojet fuel requires technological improvements in conversion, extraction of oil, and process energy conversion.

Extremophile applications in biofuel production is one of the hot research area; also, for large scale production, its understanding may open door to various other biorefinery options such as delignification, in pretreatment of various biomass, etc.

To produce green biofuel at large scale a very useful technique have been developed called as bio-augmentation where microbes such as *Ruminococcus* and *Clostridium* is used to form consortia. This resulted in improved yield of up to 90% from xylan to biobutanol.

High production of extremo-enzyme or isobutanol can be produced by engineered *Pichia pastoris* species. Some authors think that microalgae-based biofuel engineering can work as best "green cell factories" in future (Hong and Lee 2015). Some aspects of introducing trans-hydrogenase and NAD kinase in combination can improve isobutanol production tremendously because of enhanced production of NADH.

Many improvements have been done in the recent past in order to increase lipid and carbohydrate production, so as to improve biofuel production, and the diversion due to production of more intermediates can produce more biofuels (Gooch 2011).

Improved algal cell factory for production of oligogenous lipid using latest technology of CRISPER CAS9 tool can have great opportunity of improving intermediate component's (Banerjee et al. 2018).

For improved bio-refinery concept, actually there is requirement of whole cell enzyme for applications in harsh conditions. Biofuel ethanol production under alkaline production can be done when yeast cell surface is modified with cellulolytic enzyme and xylanolytic enzymes.

For direct conversion of cellulose into ethanol, genetically modified *Clostridium thermocellum* is useful when gene for two enzyme (acetaldehyde and alcohol dehydrogenase) is expressed (Chung et al. 2014).

Genetic engineering of thermophilic ethanol producing bacteria has been reported by many workers (Michael Scully and Orlygsson 2017).

10. Hydrolases from Extremophiles In Biofuels

The marine environment covers almost three quarters of the planet and is where evolution took its first steps. Extremophile microorganisms are found in several extreme marine environments, such as hydrothermal vents, hot springs, salty lakes and deep-sea floors. The ability of these microorganisms to support extremes of temperature, salinity and pressure demonstrates their great potential for biotechnological processes.

Hydrolases including amylases, cellulases, peptidases and lipases from hyperthermophiles, psychrophiles, halophiles and piezophiles have been investigated for these reasons. Extremozymes are adapted to work in harsh physical-chemical conditions and their use in various industrial applications such as the biofuel, pharmaceutical, fine chemicals and food industries has increased. The understanding of the specific factors that confer the ability to withstand extreme habitats on such enzymes has become a priority for their biotechnological use. The most studied marine extremophiles are prokaryotes and in this review, we present the most studied archaea and bacteria extremophiles and their hydrolases, and discuss their use for industrial applications.

Declarations

Author declares that there is no conflict of interest with any funding agencies.

Acknowledgement

Author acknowledges support of Ms. Shilpa Bhatt and authorities of AGC, Amritsar.

References

Abanteriba, S., U. Yildirim, R. Webster, D. Evans and P. Rawson. 2016. ERRATUM: derived cetane number, distillation and ignition delay properties of diesel and jet fuels containing blended synthetic paraffinic mixtures. SAE International Journal of Fuels and Lubricants. https://doi.org/10.4271/2016-01-9076.01.

Adeniyi, O.M., U. Azimov and A. Burluka. 2018. Algae biofuel: Current status and future applications. Renewable and Sustainable Energy Reviews. https://doi.org/10.1016/j.rser.2018.03.067.

Archana, A. and T. Satyanarayana. 1997. Xylanase production by thermophilic *Bacillus licheniformis* A99 in solid-state fermentation. Enzyme and Microbial Technology. https://doi.org/10.1016/S0141-0229(96)00207-4.

Arora, R., S. Behera and S. Kumar. 2015. Bioprospecting thermophilic/thermotolerant microbes for production of lignocellulosic ethanol: A future perspective. Renewable and Sustainable Energy Reviews. https://doi.org/10.1016/j.rser.2015.06.050.

Asnadi, C., S. Marno, P. Lestari, D. Islami, N. Putri and W. Rustyawan. 2020. The production and specification analysis of aviation biofuel as the alternative fuel of airplane. Ecological Engineering and Environment Protection. https://doi.org/10.32006/eeep.2020.1.5057.

Banerjee, A., C. Banerjee, S. Negi, J.S. Chang and P. Shukla. 2018. Improvements in algal lipid production: A systems biology and gene editing approach. Critical Reviews in Biotechnology. https://doi.org/10.1080/07388551.2017.1356803.

Barnard, D., A. Casanueva, M. Tuffin and D. Cowan. 2010. Extremophiles in biofuel synthesis. Environmental Technology. https://doi.org/10.1080/09593331003710236.

Baroutian, S., M.K. Aroua, A.A.A. Raman, A. Shafie, R.A. Ismail and H. Hamdan. 2013. Blended aviation biofuel from esterified *Jatropha curcas* and waste vegetable oils. Journal of the Taiwan Institute of Chemical Engineers. https://doi.org/10.1016/j.jtice.2013.02.007.

Cantarella, H., A.M. Nassar, L.A.B. Cortez and R. Baldassin. 2015. Potential feedstock for renewable aviation fuel in Brazil. Environmental Development. https://doi.org/10.1016/j.envdev.2015.05.004.

Chen, Chun-Yen, Xin-Qing Zhao, Hong-Wei Yen, Shih-Hsin Ho, Chieh-Lun Cheng, Duu-Jong Lee, Feng-Wu Bai and Jo-Shu Chang. 2013. Microalgae-based carbohydrates for biofuel production. Biochemical Engineering Journal. https://doi.org/10.1016/j.bej.2013.03.006.

Choi, I.H., J.S. Lee, C.U. Kim, T.W. Kim, K.Y. Lee and K.R. Hwang. 2018. Production of bio-jet fuel range alkanes from catalytic deoxygenation of Jatropha fatty acids on a WOx/Pt/TiO2 catalyst. Fuel. https://doi.org/10.1016/j.fuel.2017.11.094.

Chung, D., M. Cha, A.M. Guss and J. Westpheling. 2014. Direct conversion of plant biomass to ethanol by engineered *Caldicellulosiruptor bescii*. Proceedings of the National Academy of Sciences of the United States of America. https://doi.org/10.1073/pnas.1402210111.

Deane, J.P. and S. Pye. 2018. Europe's ambition for biofuels in aviation—A strategic review of challenges and opportunities. Energy Strategy Reviews. https://doi.org/10.1016/j.esr.2017.12.008.

Diaz-Pérez, M.A. and J.C. Serrano-Ruiz. 2020. Catalytic production of jet fuels from biomass. Molecules 25(4): 802.

Falb, Michaela, Kerstin Müller, Lisa Königsmaier, Tanja Oberwinkler, Patrick Horn, Susanne von Gronau, Orland Gonzalez, Friedhelm Pfeiffer, Erich Bornberg-Bauer and Dieter Oesterhelt. 2008. Metabolism of halophilic archaea. Extremophiles. https://doi.org/10.1007/s00792-008-0138-x.

Flores-Fernández, C.N., M. Cárdenas-Fernández, D. Dobrijevic, K. Jurlewicz, A.I. Zavaleta, J.M. Ward and G.J. Lye. 2019. Novel extremophilic proteases from *Pseudomonas aeruginosa* M211 and their application in the hydrolysis of dried distiller's grain with solubles. Biotechnology Progress. https://doi.org/10.1002/btpr.2728.

Fortier, M.-O.P. 2015. Integrating spatial heterogeneity into sustainability assessments of biofuels from wastewater algae. ProQuest Dissertations and Theses.

Galanopoulou, A.P., S. Moraïs, A. Georgoulis, E. Morag, E.A. Bayer and D.G. Hatzinikolaou. 2016. Insights into the functionality and stability of designer cellulosomes at elevated temperatures. Applied Microbiology and Biotechnology. https://doi.org/10.1007/s00253-016-7594-5.

Garsoux, G., J. Lamotte, C. Gerday and G. Feller. 2004. Kinetic and structural optimization to catalysis at low temperatures in a psychrophilic cellulase from the Antarctic bacterium *Pseudoalteromonas haloplanktis.* Biochemical Journal. https://doi.org/10.1042/BJ20040325.

Gladden, John M., Martin Allgaier, Christopher S. Miller, Terry C. Hazen, Jean S. VanderGheynst, Philip Hugenholtz, Blake A. Simmons and Steven W. Singer. 2011. Glycoside hydrolase activities of thermophilic bacterial consortia adapted to switchgrass. Applied and Environmental Microbiology. https://doi.org/10.1128/AEM.00032-11.

Gooch, J.W. 2011. Eukaryotic cell. In Encyclopedic Dictionary of Polymers. https://doi.org/10.1007/978-1-4419-6247-8_13707.

Hong, S.J. and C.G. Lee. 2015. Microalgal systems biology for biofuel production. In Algal Biorefineries: Volume 2: Products and Refinery Design. https://doi.org/10.1007/978-3-319-20200-6_1.

Hong, T.D., T.H. Soerawidjaja, I.K. Reksowardojo, O. Fujita, Z. Duniani and M.X. Pham. 2013. A study on developing aviation biofuel for the Tropics: Production process—Experimental and theoretical evaluation of their blends with fossil kerosene. Chemical Engineering and Processing: Process Intensification. https://doi.org/10.1016/j.cep.2013.09.013.

Inan, K., A.O. Belduz and S. Canakci. 2013. *Anoxybacillus kaynarcensis* sp. nov., a moderately thermophilic, xylanase producing bacterium. Journal of Basic Microbiology. https://doi.org/10.1002/jobm.201100638.

Jiang, R., K.N. Ingle and A. Golberg. 2016. Macroalgae (seaweed) for liquid transportation biofuel production: What is next? Algal Research. https://doi.org/10.1016/j.algal.2016.01.001.

Jiménez-Diaz, L., A. Caballero, N. Pérez-Hernández and A. Segura. 2017. Microbial alkane production for jet fuel industry: Motivation, state of the art and perspectives. Microbial Biotechnology 10(1): 103–124.

Kahn, Amaranta, Sarah Moraïs, Anastasia P. Galanopoulou, Daehwan Chung, Nicholas S. Sarai, Neal Hengge, Dimitris G. Hatzinikolaou, Michael E. Himmel, Yannick J. Bomble and Edward A. Bayer. 2019. Creation of a functional hyperthermostable designer cellulosome. Biotechnology for Biofuels. https://doi.org/10.1186/s13068-019-1386-y.

Kandaramath Hari, T., Z. Yaakob and N.N. Binitha. 2015. Aviation biofuel from renewable resources: Routes, opportunities and challenges. Renewable and Sustainable Energy Reviews. https://doi.org/10.1016/j.rser.2014.10.095.

Karmee, S.K. 2016. Liquid biofuels from food waste: Current trends, prospect and limitation. Renewable and Sustainable Energy Reviews. https://doi.org/10.1016/j.rser.2015.09.041.

Lin, L., Ji, Y., Tu, Q., Huang, R., Teng, L., Zeng, X., Song, H., Wang, K., Zhou, Q., Li, Y. and Cui, Q. 2013. Microevolution from shock to adaptation revealed strategies improving ethanol tolerance and production in Thermoanaerobacter. Biotechnology for Biofuels. https://doi.org/10.1186/1754-6834-6-103.

Mandelli, F., Couger, M.B., Paixão, D.A.A., Machado, C.B., Carnielli, C.M., Aricetti, J.A., Polikarpov, I., Prade, R., Caldana, C., Leme, A.P. and Mercadante, A.Z. 2017. Thermal adaptation strategies of the extremophile bacterium *Thermus filiformis* based on multi-omics analysis. Extremophiles. https://doi.org/10.1007/s00792-017-0942-2.

Markou, G., I. Angelidaki and D. Georgakakis. 2012. Microalgal carbohydrates: An overview of the factors influencing carbohydrates production, and of main bioconversion technologies for production of biofuels. Applied Microbiology and Biotechnology. https://doi.org/10.1007/s00253-012-4398-0.

Michael Scully, S. and J. Orlygsson. 2017. Recent advances in genetic engineering of thermophilic ethanol producing bacteria. In Engineering of Microorganisms for the Production of Chemicals and Biofuels from Renewable Resources. https://doi.org/10.1007/978-3-319-51729-2_1.

Moore, R.H., Thornhill, K.L., Weinzierl, B., Sauer, D., D'Ascoli, E., Kim, J., Lichtenstern, M., Scheibe, M., Beaton, B., Beyersdorf, A.J. and Barrick, J. 2017. Biofuel blending reduces particle emissions from aircraft engines at cruise conditions. Nature. https://doi.org/10.1038/nature21420.

Poli, A., I. Romano, P. Cordella, P. Orlando, B. Nicolaus and C.C. Berrini. 2009. *Anoxybacillus thermarum* sp. nov., a novel thermophilic bacterium isolated from thermal mud in Euganean hot springs, Abano Terme, Italy. Extremophiles. https://doi.org/10.1007/s00792-009-0274-y.

Prabandono, K. and S. Amin. 2015. Biofuel production from microalgae. In Handbook of Marine Microalgae: Biotechnology Advances. https://doi.org/10.1016/B978-0-12-800776-1.00010-8.

Repetto, S.L., J.F. Costello and D. Parmenter. 2016. Current and potential aviation additives for higher biofuel blends in Jet A-1. In Biofuels for Aviation. https://doi.org/10.1016/b978-0-12-804568-8.00011-1.

Roberts, M.S., L.K. Nakamura and F.M. Cohan. 1994. *Bacillus mojavensis* sp. nov., distinguishable from *Bacillus subtilis* by sexual isolation, divergence in DNA sequence, and differences in fatty acid composition. International Journal of Systematic Bacteriology. https://doi.org/10.1099/00207713-44-2-256.

Rumizen, M. 2017. Aviation biofuel standards and airworthiness approval. In Biokerosene: Status and Prospects. https://doi.org/10.1007/978-3-662-53065-8_24.

Sánchez, L.A., F.F. Gómez and O.D. Delgado. 2009. Cold-adapted microorganisms as a source of new antimicrobials. Extremophiles. https://doi.org/10.1007/s00792-008-0203-5.

Sandquist, J. and B.M. Güell. 2012. Overview of biofuels for aviation. In Chemical Engineering Transactions. https://doi.org/10.3303/CET1229192.

Schlagermann, P., G. Göttlicher, R. Dillschneider, R. Rosello-Sastre and C. Posten. 2012. Composition of algal oil and its potential as biofuel. Journal of Combustion. https://doi.org/10.1155/2012/285185.

Shanmugam, S., C. Sun, Z. Chen and Y.R. Wu. 2019. Enhanced bioconversion of hemicellulosic biomass by microbial consortium for biobutanol production with bioaugmentation strategy. Bioresource Technology. https://doi.org/10.1016/j.biortech.2019.01.121.

Sirajunnisa, A.R. and D. Surendhiran. 2016. Algae – A quintessential and positive resource of bioethanol production: A comprehensive review. Renewable and Sustainable Energy Reviews. https://doi.org/10.1016/j.rser.2016.07.024.

Skaggs, R.L., A.M. Coleman, T.E. Seiple and A.R. Milbrandt. 2018. Waste-to-Energy biofuel production potential for selected feedstocks in the conterminous United States. Renewable and Sustainable Energy Reviews. https://doi.org/10.1016/j.rser.2017.09.107.

Staples, M.D., R. Malina, P. Suresh, J.I. Hileman and S.R.H. Barrett. 2018. Aviation CO2 emissions reductions from the use of alternative jet fuels. Energy Policy. https://doi.org/10.1016/j.enpol.2017.12.007.

Starck, L., L. Pidol, N. Jeuland, T. Chapus, P. Bogers and J. Bauldreay. 2016. Production of Hydroprocessed Esters and Fatty Acids (HEFA)—Optimisation of process yield. Oil & Gas Science and Technology – Revue d'IFP Energies Nouvelles. https://doi.org/10.2516/ogst/2014007.

Straub, C.T., R.G. Bing, J.K. Otten, L.M. Keller, B.M. Zeldes, M.W.W. Adams and R.M. Kelly. 2020. Metabolically engineered *Caldicellulosiruptor bescii* as a platform for producing acetone and hydrogen from lignocellulose. Biotechnology and Bioengineering. https://doi.org/10.1002/bit.27529.

Thomas, D. G., Minj, N., Mohan, N., & Rao, P. H. (2016). Cultivation of microalgae in domestic wastewater for biofuel applications–an upstream approach. J. Algal Biomass Utln, 7(1): 62–70.

Toyama, T., T. Hanaoka, K. Yamada, K. Suzuki, Y. Tanaka, M. Morikawa and K. Mori. 2019. Enhanced production of biomass and lipids by *Euglena gracilis* via co-culturing with a microalga growth-promoting bacterium, *Emticicia* sp. EG3. Biotechnology for Biofuels. https://doi.org/10.1186/s13068-019-1544-2.

Valdez-Ojeda, R.A., M.G. del Rayo Serrano-Vázquez, T. Toledano-Thompson, J.C. Chavarría-Hernández and L.F. Barahona-Pérez. 2020. Effect of media composition and culture time on the lipid profile of the green microalga *Coelastrum* sp. and its suitability for biofuel production. Bioenergy Research. https://doi.org/10.1007/s12155-020-10160-5.

Vásquez, M.C., E.E. Silva and E.F. Castillo. 2017. Hydrotreatment of vegetable oils: A review of the technologies and its developments for jet biofuel production. Biomass and Bioenergy. https://doi.org/10.1016/j.biombioe.2017.07.008.

Velazquez-Lucio, Jesús, Rosa M. Rodríguez-Jasso, Luciane M. Colla, Aide Sáenz-Galindo, Daniela E. Cervantes-Cisneros, Cristóbal N. Aguilar, Bruno D. Fernandes and Héctor A. Ruiz. 2018. Microalgal biomass pretreatment for bioethanol production: A review. Biofuel Research Journal. https://doi.org/10.18331/BRJ2018.5.1.5.

Wang, Jicong, Peiyan Bi, Yajing Zhang, He Xue, Peiwen Jiang, Xiaoping Wu, Junxu Liu, Tiejun Wang and Quanxin Li. 2015. Preparation of jet fuel range hydrocarbons by catalytic transformation of bio-oil derived from fast pyrolysis of straw stalk. Energy. https://doi.org/10.1016/j.energy.2015.04.053.

Wang, W.C. and L. Tao. 2016. Bio-jet fuel conversion technologies. Renewable and Sustainable Energy Reviews. https://doi.org/10.1016/j.rser.2015.09.016.

Yilmaz, N. and A. Atmanli. 2017. Sustainable alternative fuels in aviation. Energy. https://doi.org/10.1016/j.energy.2017.07.077.

Zanata, M., S.T.W. Amelia, M.R. Mumtazy, F. Kurniawansyah and A. Roesyadi. 2019. Synthesis of bio jet fuel from crude palm oil by hefa (Hydroprocessed esters and fatty acids) using ni-mo catalyst supported by rice husk ash-based sio2. In Materials Science Forum. https://doi.org/10.4028/www.scientific.net/MSF.964.193.

Zhou, C.H., X. Xia, C.X. Lin, D.S. Tong and J. Beltramini. 2011. Catalytic conversion of lignocellulosic biomass to fine chemicals and fuels. Chemical Society Reviews. https://doi.org/10.1039/c1cs15124j.

Zverlov, V.V. and W.H. Schwarz. 2008. Bacterial cellulose hydrolysis in anaerobic environmental subsystems—*Clostridium thermocellum* and *Clostridium stercorarium*, thermophilic plant-fiber degraders. In Annals of the New York Academy of Sciences. https://doi.org/10.1196/annals.1419.008.

11

Application of Extremophiles in the Area of Bioenergy

Nibedita Sarkar

1. Introduction

In recent times, oil, coal, natural gas which are the source of fossil fuel generally control the world economy. Fossil fuel sources are utilized for liquid fuel, electricity and several valuable commodity production (Uihlein and Schbek 2009). However, a gradual increase in fossil fuel consumption, especially in metropolitan cities, has been seen in last few decades which has created a tremendous air pollution with an extensive greenhouse gas emission (Ballesteros et al. 2006). Exponential population growth and industrialization as well as modenerization has resulted in a gradual increase in universal energy consumption. Hence, it is estimated that limited fossil fuels will be running out in next two decades and thereafter, renewable energy will be the alternate source of energy like wind, water, sun, biomass, geothermal heat which are the potent substitute of fossil fuels (Campbell and Laherrere 1998, Lynd and Wang 2003).

Biorefineries are manufacturing facilities that convert biobased materials (such as agricultural residues) to various products such as food, feed, fuels, chemicals, and energy (Uihlein and Schbek 2009, Ballesteros et al. 2006). Biomass processing is analogous to petroleum refineries, which refine crude oil into several products including fuels (e.g., petrol, diesel, and kerosene) and chemical precursors like butanol for manufacturing different materials (Campbell and Laherrere 1998). In 2011, renewable energy policy network reported that fossil fuel supplies about 78% of the total energy consumed through the world whereas nuclear energy supplies 3% and renewable energy supplies remaining 19% of the total energy requirement from renewable sources such as wind, solar, geothermal, hydrothermal, and biomass (Lynd and Wang 2003). However, a considerable portion of carbon rich renewable sources are forcefully wasted by simple burning or thermochemical processing for heat and power generation. In the current scenario, approximately 85 million barrels of petroleum oil are thoroughly processed and utilized to achieve the worldwide energy demand. It is assumed that the energy requirement will be increased up to 116 million barrels by 2030 (IEA 2007). Hence, the limited source of fossil fuel reserves will be hampered and depleted by this huge energy requirement. Therefore, alternative sources for energy production should be explored which can replace fossil fuel (Lee and Lavoie 2013).

In the first decade of 21st century, biofuels were produced from food crops which led to a major controversy of food versus fuel. Several national and international agencies, such as OECD, FAO, EU reported that biofuel production from food crops resulted in a great impact on food value

Doltala, Bankura, West Bengal-722101, India.
Email: nibedta_sarkar@rediffmail.com

(OECD 2008). This worldwide controversy influenced to promote 2nd generation biofuel production, i.e., biofuel production from renewable feedstocks with no food value. Biofuels and biochemicals from nonedible feedstock have several advantages to the society like (i) exploring of renewable and sustainable source (ii) reduction of greenhouse gas emission (iii) improvement of total economy (iv) proper and worthy utilization of biomass rather than burning (v) stabilizing energy security (vi) creating job opportunities in society for chemical engineers, specialist in fermentation field, process plant engineers, and researchers (Greenwell et al. 2012).

Biorefineries have opted for several modern technologies including chemical and thermochemical processes. However, biofuel production from renewable sources using microorganisms are more economically feasible which has created a huge opportunity for the researchers to explore different types of efficient microorganisms (Luque 2008). In this aspect, extremophiles and their efficient enzymes are major concern now a days and are being explored in the industries for biofuel production as well. In research, it is explored that in extreme conditions like pH, salinity, radiation, and high temperature, extremophiles are able to thrive (Wisselink et al. 2009).

This chapter briefly explains about various types of biofuels and their production. The chapter also describes about application of extremophiles and/or their robust enzyme products for biofuel production.

2. Background

2.1 Biofuel

Since previous decade, biofuel production from renewable and nonedible sources has become prime concern due to high crude oil price as well as limited storage and tremendous environmental pollution because of high amount of greenhouse gas emission. Bioethanol and biodiesel have too much importance in biofuel market as well as in transportation sector (Antoni et al. 2007). Additionally, biobutanol, biogas and biohydrogen have been an emerging area in biofuel market.

2.1.1 Bioethanol

Ethanol has achieved highest commercial value in transport sector as it has received worldwide interest as liquid biofuel (Demirbas 2005). A gradual increase in global warming, climate change and limited source of crude oil has created a keen interest to explore a clean fuel. Bioethanol demand in transportation sector has been increasing exponentially. Bioethanol has become a potential clean fuel. However, bioethanol production from edible feedstocks is costly compared to petroleum oil synthesis from fossil fuel source. Brazil and USA are the leading countries in bioethanol production which accounts for approximately 62% of total bioethanol production throughout the world (Kim and Dale 2004). The concerned industries in Brazil are purely dependent on sugarcane for bioethanol production whereas in USA, corn is the main feedstock in commercial sector. However, high cost of the edible feedstock is the bottleneck for bioethanol production. Additionally, bioethanol production from edible feedstock has created an extreme controversy because of its food and feed value since last decades. Hence, industries have opted for 2nd generation biothanol production, i.e., bioethanol production from lignocellulosic feedstock. Lignocellulosic feedstocks are feasible source because they are abundant, renewable and low cost. Lignocellulosic biomass can lead to approximately 442 billion liters of bioethanol whereas agricultural residues and their waste products have potential to produce 491 billion liters of bioethanol per year which is around 16 times higher than 1st generation bioethanol production (Kim and Dale 2004). However, to overcome the economic barrier in commercial sector, an efficient microorganism with broad substrate utilization and high ethanol tolerance capacity has to be explored. Microorganisms having high tolerance at elevated temperature range has to be explored as well. Furthermore, one step bioethanol production through consolidated bioprocessing is more feasible than traditional production strategy. In that aspect, genetically modified or engineered microorganisms with cellulolytic activity as well as sugar utilization capacity are to be developed (Sarkar et al. 2012).

2.1.2 Biodiesel

Biodiesel is another important liquid fuel that is generally obtained from virgin vegetable oils and animal fats. Biodiesel is a very attractive and ecofriendly replacement of petro-diesel because of its low-toxicity, biodegradability, carbon-neutrality, less HC, CO and sulphur content (Babcock et al. 2014).

Babcock et al. (2014) have reported that National Renewable Energy Laboratory (NREL) carried out life-cycle analysis of biodiesel and showed approximately 78% reduction of CO_2 emission compared to petro-diesel. Fuel properties of biodiesel resemble with petro-diesel. Additionally, biodiesel is safer than petro-diesel due to high flash point. However, the calorific value of biodiesel possesses lower calorific value than petroleum diesel (w37.27 MJ L_1) (Demirbas 2003).

Approximately 90% of the energy required in transportation sector depends on crude oil. In 2016, 4% of liquid biofuel was utilized for motor vehicles. According to the reports of IEA 2017, the production of biofuel production is increasing gradually to meet the market demand and will reach 159 billion liters in 2022. In India, transportation sector used 46% of total import diesel and the demand will increase to 226 billion liters in CY 2026 from 134 billion liters during current year. The Government of India has a vision of energy security for which it has planned to use natural gas resource to half the oil import by 2030. Around 70% of biodiesel production cost is controlled by the raw materials of the process. Biodiesel synthesis from edible oil crops such as mustard oil, sunflower oil, safflower oil, rapeseed oil, coconut oil, palm oil, soybean oil, and canola oil is known as first generation biodiesel which has created a major controversy due to the food security of the nation as well as land shortage for food crops cultivation (Marousek 2014). Utilization of second-generation feedstock such as non-edible oils like Jatropha, Kachnar, jojoba and animal fats, and waste vegetable oil can nullify the issue created by first generation feedstocks. Non-edible sources are more feasible due to less water demands, cost, and maintenance for cultivation but few fuel properties of 2nd generation biodiesel need to be improved (Janaun and Ellis 2010, Marousek et al. 2013). In this context, third generation feedstock such as microalgae has been explored for efficient biodiesel production. Microalgae can be a potent source for biodiesel production due to its easy propagation, fast growth rate and high yield.

Transesterification is the most accepted method for biodiesel production in commercial sector. The process is a chemical reaction between a vegetable oil and an alcohol in the presence of a catalyst where a wide range of alcohols are used. The catalyst may be acidic, basic or enzymatic (Al-Zuhair 2007). Transesterification process solely depends on various factors such as type of catalyst employed, alcohol/oil ratios, temperature and water content. However, the traditional method faces various challenges like byproduct recovery, biodiesel purification, unwanted side reactions and necessity of pretreatment step. In this context, enzymatic transesterification may be feasible due to its low environmental impact, no pretreatment step, lower energy consumption, mild condition, easy downstream processing, etc. (Mehrasbi et al. 2017).

2.1.3 Biobutanol

Butanol is the next generation fuel which is more efficient than other existing biofuels, due to its improved properties such as high energy density, high air-fuel ratio and lower hygroscopicity. Furthermore, butanol is easy to transport through the existing gas line distribution and it does not require any engine modification (Gonz ∋lez-Ramos 2013). Butanol, or butylalcohol, is generally a type of alcohol consisting four carbon chain with chemical formula: C_4H_9OH. Butanol is distributed in nature in various isomeric forms like iso-butanol, sec-butanol, ter-butanol, n-butanol. The isomers possess different solubility, melting and boiling points. Additionally, n-butanol and iso-butanol have the physicochemical properties to be used as commercial solvent where remaining isomers are highly hydrophilic (Kataoka 2011). Apart from that, butanol has vast applications in chemical synthesis, preparation of medicine and drugs on industrial scale. Recently, butanol has become an emerging fuel in transportation sector. Butanol is an isomeric alcohol of short carbon chain. Butanol is superior than ethanol due to its structure. Butanol consists more methylene groups than ethanol which makes butanol more hydrophobic, less volatile and less soluble. Furthermore, butanol has high calorific value and improved miscibility with gasoline compared to ethanol (Liu and Qureshi 2009). Recently, biobutanol is gaining more concern in fuel sector due to its excellent fuel properties (Branduardi 2014).

Clostridium sp. is the well-known microorganism for biobutanol production through two step bioconversion process, i.e., ABE fermentation (acetone:butanol:ethanol) with 3:6:1 ratio (Mariano et al. 2008, Shapovalova and Ashkinaib 2008). During the process, in the first step, feedstock is transformed to hydrogen and butyric acid and simultaneously simplified to butanol. Generally, carbohydrate rich biomass like corn, sugar beet, sorghum is used for biobutanol production (Shapovalova and Ashkinaib 2008). However, product inhibition is the main bottleneck of biobutanol production. Two step biobutanol processes can relieve *Clostridrium* from product inhibition. However, *Clostridium* faces many more challenges in two step bioconversion process which made microbiologists to find out an alternative microorganism for advanced biobutanol production (Wisselink et al. 2009).

2.1.4 Biogas

Like other fuels, biogas also plays a great role to reduce the dependence on oil imports and to reduce environmental pollution. Additionally, biogas from waste biodegradable matters has become a great choice for waste management. At present, 80% of energy for generation of electricity comes from fossil sources to run industries and for household purposes as well (IEA 2016). However, fossil fuels are non-renewable and limited energy source. In this context, biogas generation from biodegradable matter has gained worldwide interest because it is natural, renewable carbon resource (Klass 2004). There are several processes to convert chemical compositions of the biomass into gaseous fuel and those are direct combustion, thermochemical transformation, physicochemical transformation and biochemical transformation (Nagy and Szabó 2011). Anaerobic microorganism plays a major role for biogas production where they break the complex material of the organic matter in an anaerobic digestion chamber and simultaneously produces biogas. Biogas is mainly composed of 60% methane and 40% carbon oxide (Ali et al. 2015). Generally, mesophillic or thermophilic microorganisms are involved in the bioconversion process (Torquati et al. 2014, Zhang et al. 2015). However, thermophillic microorganism has become a concerned topic for better outcome (Nagy and Szabó 2011).

2.1.5 Biohydrogen

Molecular hydrogen is the most powerful and the most efficient source of energy among all the renewable energy sources. High energy requirement can be easily fulfilled by hydrogen energy (Mona and Kaushik 2015). Hydrogen is carbon neutral energy and it possesses around 2.75 fold higher energy compared to the other existing biofuels (Singh and Mahapatra 2019). Cavendish traced molecular hydrogen synthesis for the first time while dissolving metals in dilute acids (Cavendish 1776). Hydrogen has application in the fuel cell where it generates electricity from chemical energy (Rosenbaum et al. 2005). Additionally, hydrogen energy is generally applied for the production of fertilizers (ammonia), methanol, removal of impurities in oil refineries, in pharmaceutical industries, cryogenics, rotor coolants and petroleum recovery. Furthermore, it is used in rocket engines due to its elevated energy capacity (Shaishav et al. 2013, Singh and Mahapatra 2019). Nowadays, hydrogen has become an emerging field of research due to its efficiency and high energy. Several methods are available to produce hydrogen such as gasification of coal and biomass, natural gas, and water splitting through electrolysis. However, these methods have adverse environmental effects. For the first time, Gaffron has identified biological hydrogen production in green microalgae, though some bacteria, higher plants, blue green algae are also able to produce hydrogen at a very mild condition (Gaffron 1939, Gaffron and Rubin 1942, Stevens 2001). Biological hydrogen is generally produced by photofermentation and dark fermentation. Algae produces hydrogen through photofermentation in the presence of light whereas dark fermentation occurs through anaerobic digestion of organic substrates by some anaerobic bacteria. However, thermophilic microorganisms such as *Caldicellulosiruptor saccharolyticus* and *Thermotogoelfii* are being explored for their efficiency (Antoni et al. 2005, de Vrije et al. 2002, de Vrije and Claassen 2003, Claassen et al. 2004).

2.2 *Extremophiles*

Extremophiles are the microorganisms which are known to be functionally alive in extreme environmental condition which makes them very important to explore in an efficient way. Manipulation at genetic level has modified the metabolic pathway of the microbes in such a manner so that they can exist as well as function in extreme condition. Extremophiles are known to be ancient microbes which thrived in an elevated environmental and ecological condition. Exploring these microbes has become a very interesting area as they convey the evolutionary details of Earth. The detailed study of extremophiles explored life trend at the first stage of the planet. Extremophiles are classified as per their sustainability in different environmental conditions. Types of extremophiles mainly include thermophiles, psychrophiles, acidophiles, alkalophiles and halophiles (Fig. 1). Apart from that, barophiles and xerophiles can be included in extremophiles consortium. Thermophiles can thrive in an environment with elevated temperature range such as hydrothermal vents, volcanic sites, on the mountains at high altitudes and deep inside the oceans. Barophiles can tolerate high pressure under the deep sea. Acidophiles can sustain in acidic condition at pH less than 5.0 or even near to 0.0. They are generally found in acid mine drainage sites and acidic lakes whereas alkalophiles are able to grow and function at highly basic conditions such as in sodic lakes. Halophiles are seen at very

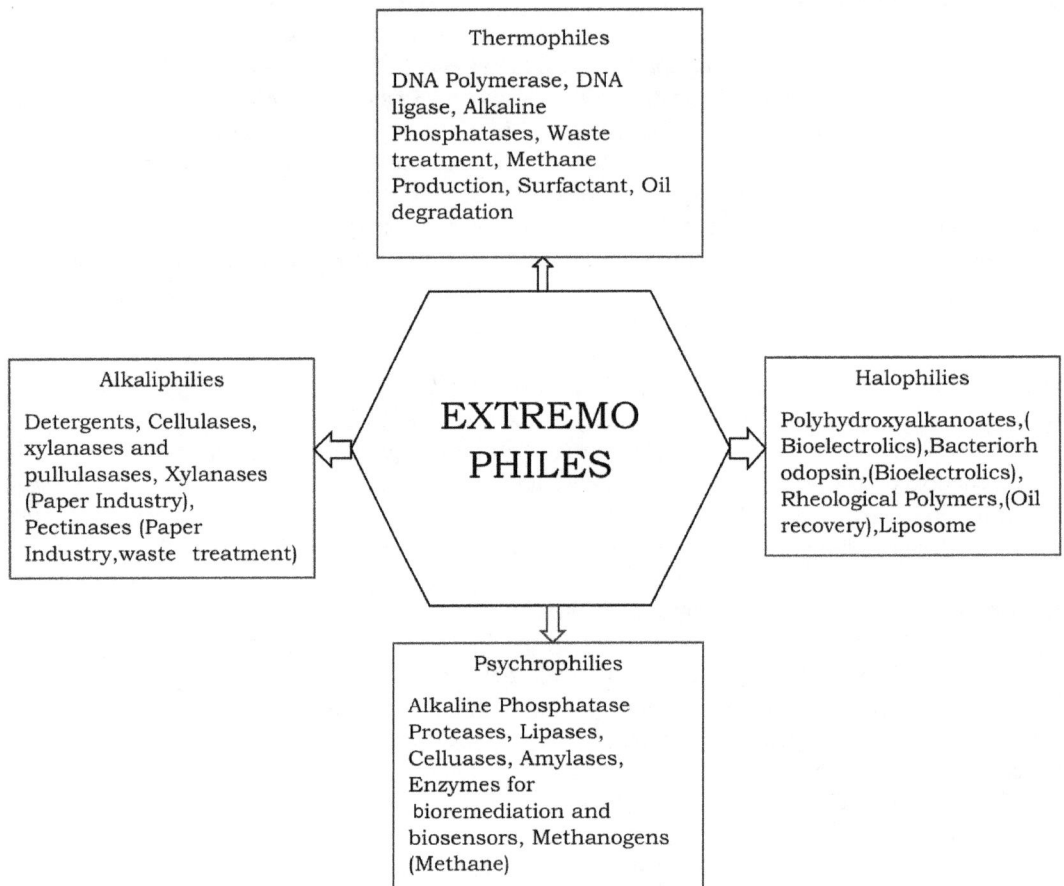

Fig. 1. Types of Extremophiles.

high salt concentrations such as in sea, and high pH saline lakes. Xerophiles can flourish in low water conditions such as in deserts. In general, extreme environmental conditions are much suitable for extremophiles. Extremophiles are pollution loving microbes. Extremophilic microorganisms are very important topic to be explored for ecological concern. Significant outcome may be achieved in biotechnological research and industries using these extremophiles. Unique genetic code is responsible to allow these types of microbes to sustain in harsh conditions. Extremophiles carry significant properties and they are explored and used for important product formation due to their high stability at high or low temperature, pH extremities, very high pressure and even in presence of deadly pollutants. Among various types of extremophiles, thermophiles, psychrophiles, acidophiles, alkalophiles, halophiles have great role in biorefineries.

2.2.1 Thermophiles/Psychrophiles

Thermophiles are the microorganisms that are able to grow between 60–80°C and hyperthermophiles are able to grow above the temperature of 80°C (Santos and Da Costa 2002). Thermophilic bacteria are generally found in hot springs where temperature is 70°C or more. Thermophiles generally have high metabolic activity, and can produce physically and chemically stable enzymes due to the compulsion to survive at extreme condition. They possess lower growth rate but results in similar end product like mesophilic species (Haki and Rakshit 2003). Thermophiles can thrive in volcanic sites with temperature 300°C or more, subterranean sites such as oil reservoirs, solar heated surface soils or in human made extreme environment such as compost piles, industrial discharges, water heaters, etc. (Oshima and Moriya 2008).

Psychrophiles are the most widely distributed microorganisms on Earth which survive in cold environment. Psychrophiles are able to survive constantly at temperature close to freezing point of water. Psychrophiles are distributed in wide range of microorganisms. Generally, in Antarctic and Arctic polar regions, extreme psychrophiles have been found. Moreover, psychrophilic microorganisms are largely found inside the oceans where the constant temperature is 4°C reservoir. Huge availability of psychrophiles offers an opportunity of biomining to utilize them in biotechnological research as well as industries (Leary 2008, Lohan and Johnston 2005).

2.2.2 Acidophiles/Alkalophiles

Acidophiles are the type of extremophiles which can sustain at very low pH (below pH 3) (Rothschild and Mancinelli 2001). Acidophiles are generally found in all domains of the world. Acidophiles exist in low pH sites like acid mine drainage, solfataric fields, acidothermal hot springs and fumaroles, coal spoils, and bioreactors (Seckbach and Libby 1970, Hallberg and Lindström 1994, Golyshina et al. 2000). The cellular activity of acidophiles at molecular level restricts proton entry by the cytoplasmic membrane and purging of protons and their effects by the cytoplasm. Additionally, acidophiles contain such enzyme activity which can balance the membrane condition at fluctuating acidic pH (Baker-Austin and Dopson 2007).

Alkaliphile are the microorganisms which grow and function very efficiently at basic environmental condition with pH 9 or above but their growth cease at neutral pH. Alkaliphiles generally include prokaryotes, eukaryotes, and archaea. Alkaliphiles are generally found in a normal environmental condition such as garden soil, high pH hotsprings, shallow hydrothermal system, and sewage. Unlike acidophiles, the cellular activity of alkaliphiles involves H+ influx to drive ATP synthesis. Alkaliphiles secrete alkaline cellulases and/or alkaline proteases which has applications in biological detergent industries.

2.2.3 Halophiles

Halophiles are generally distributed in all three domains of life like Archaea, Bacteria and Eucarya. They can exist in different salt concentration up to salt saturation. Due to their sustainability in extreme salt concentration, halophiles are used in salt making preparation. Halophiles are well-known for the traditional salt making process through seawater evaporation (Hillebrecht et al. 2004).

In salt making process, halophiles play great role in brine formation as well as crystallization. The capacity to survive in extreme salt concentration made the halophiles to be applied in several biorefineries.

3. Current Development

3.1 Extremophiles in Bioethanol Production

Fermentation is the main step in ethanol production process. Fermentation is carried out by a wide range of microorganisms such as fungi, bacteria, and yeasts. *S. cerevisiae* is one of the potential yeast which has been used in industry and household purposes from several decades. *S. cerevisiae* generally ferments sugar syrup and produce ethanol. *S. cerevisiae* is known as commercial yeast due to its superiority compared to bacteria, other yeasts, and filamentous fungi in various aspects such as it can sustain in wide range of acidic pH during fermentation which leads to less chance of contamination of bacterial species (Lin et al. 2012, Ortiz-Mu˜niz et al. 2010).

In spite of its wide range application, *S. cerevisiae* leading fermentation faces several challenges. Firstly, high concentration of ethanol inhibits the growth of *S. cerevisiae* and thus ceases ethanol production. Secondly and most importantly, *S. cerevisiae* has the ability to ferment only hexose sugars while lignocellulosic hydrolysate consists of hexose as well pentose sugars. Thus, maximum sugar content of lignocellulosic hydrolysate is wasted which leads to less ethanol production than expected. It is reported that at high concentration of hexose and pentose sugars, *S. cerevisiae* utilizes only hexoses thus reducing the overall bioethanol production (Guti´errez-Rivera et al. 2012).

To solve this problem, researchers have taken initiative by co-culturing *S. cerevisiae* ITV-01 and *Pichia stipitis* NRRLY-7124 to improve productivity of ethanol production by utilizing pentose and hexose, both sugars (Guti´errez-Rivera et al. 2012). However, co-culturing faces several challenges because growth condition for different microorganisms are different.

Furthermore, traditional biomass processing schemes involve enzymatic or microbial hydrolysis which includes four steps of conversions at mild condition: the microbial production of extracellular enzymes (cellulases and hemicellulases); saccharification of pretreated feedstock; the fermentation of hexose sugars (glucose, mannose and galactose); and the fermentation of pentose sugars (xylose and arabinose). These four transformations can be summed up in a single step which is called consolidated bioprocessing (CBP). CBP is gaining increasing recognition as a potential breakthrough for low cost biomass processing.

In this context, extremophiles play a great role. Several thermophiles are found in nature which are able to catabolize hexose as well as pentose of the lignocellulosic hydrolysate and cellulosic substrate and, as a consequence, bioethanol yield is improved.

Thermophilic clostridia are capable of ethanol fermentation and can grow at temperature range between 60°C and 65°C (Lamed and Zeikus 1980). Additionally, this strain is able to hydrolyse lignocellulosic substrate by secreting different types of cellulases and hemicellulases such as endo-β-glucanases, exoglucanases, and β-glucosidases (Demain 2005). It is reported that *Clostridium thermocellum* is capable of converting cellulose to cellobiose and cellodextrins and hemicellulose to xylose, xylobiose and other pentose sugars and simultaneously to ferment them into ethanol, acetate, lactate, H_2 and CO_2 as well (Demain 2005). However, *C. thermocellum* is not absolutely efficient to ferment hemicelluloses derived pentose sugars to ethanol (Taylor 2005). Mixed culture therefore can solve the problem of fermenting all the sugars derived from lignocellulosic feedstock. Several themophillic bacteria such as *Thermoanaerobacterium thermosaccharolyticum* (Mori 1990), *Thermoanaerobacter thermohydrosulphuricum* (Mori 1990), *Thermoanaerobacter ethanolicus* (Demain 2005), and *Geobacillus stearothermophilus* (Sharma 1991) are available which can be cultured with *C. thermocellum* to ferment pentose sugars for bioethanol production. *Thermoanaerobacterium* has also been reported as hemicellulolytic thermophilic anaerobe which can convert xylose to ethanol in an efficient manner. It is capable of utilizing pentose sugars such as xylose to produce ethanol (Lacis and Lawford 1991). *Thermoanaerobacter ethanolicus* is originated from *Clostridium* species which can ferment both D-glucose and D-xylose to ethanol, and it has low ethanol tolerance capacity (Burdette et al. 2002, Fong et al. 2006).

Among various thermophillic microbes, Geobacillus has gained huge attention due to its high ethanol tolerance capacity, high yield and co-fermentation ability. Geobacillus is able to produce alcohol at 70°C. A strain of *Geobacillus thermoglucosidasius* has been reported as most efficient thermophillic anaerobe with high ethanol yield (Fangrui and Hanna 1999).

3.2 Extremophiles for Biodiesel Production

Transesterification is the traditional and commercial process for biodiesel production, though it possesses lots of bottlenecks such as difficulties in downstream processing, soap formation and necessity of pretreatment step, etc. Transesterification process involves nucleophilic attack on triglycerides which is degraded into mono glycerides, di glycerides and glycerol (by-product). After product formation, biodiesel is generally separated by decanting the mixture overnight (Ertan et al. 2014). Free fatty acid (FAA) content is very high (> 1%) in commercially used feedstocks which leads to soap formation. Hence, pretreatment step is required to reduce the amount of FFA content (Behzad and Farid 2007). Furthermore, current feedstocks are allowed to go through several processing before transesterification process (Chisti 2007).

In this context, microalgae are a concerned area for biodiesel production. Microalgae has high lipid content and several microalgae grow in extreme condition as well.

Microalgae are eukaryotic which form algal biomass by utilizing sunlight, water and carbon dioxide through photosynthesis. Excessively high oil content of microalgae can be used to synthesize biodiesel using existing technology (Banerjee 2002, Sa´nchez Miro´n 1999). Several microalgae possess high lipid content exceeding 80% of the dry weight of algae biomass (Banerjee 2002, Sa´nchez Miro´n 1999). Approximately 1.535 kg m^3/d microalgal biomass is produced annually in a tropical region (Molina 1999). The rate of biomass productivity is so high that if the biomass contains lipid at least 30% of dry weight, it will result in oil yield approximately 123 m^3 in a cultivation field, which is very high. The particular amount of oil from microalgae is able to produce approximately 98.4 m^3 biodiesel per hectare. Hence, microalgae have been proved to be a potential source for biodiesel production compared to other sources. Additionally, growth rate of microalgae is very high with doubling time of only 24 h. It is reported that doubling time of microalgae in log phase reduced to 3.5 h which is very significant property for biodiesel production (Pulz and Gross 2004). Fresh and brackish water from lakes, rivers and aquifers, even waste water, can be used for microalgae cultivation. Growing of microalgae is considerably inexpensive (Pulz and Gross 2004). After cultivation, the algal biomass is treated with a water-immiscible solvent to extract algal oil and oil is used to produce biodiesel in a traditional manner (Pulz and Gross 2004). Microalgal biodiesel production in a large scale is economically feasible, though economic feasibility can be improved by using thermophilic microalgal strains and advanced technology such as photobioreactors.

Extremophilic microalgae play a major role to improve the feasibility of microalgal biodiesel production. *Cyanidium caldarium* and *Galdieriasulphuraria* are extremophiles which can thrive in both elevated temperatures and acidic pH, they grow fast at 50°C and pH 1 (De Luca et al. 1991, Hu et al. 2008). Additionally, extremophilic microalgae lower the chance of contamination within the bioreactors. It has been reported that *Ochromonasdanica* and *Nannocloropsissalina* have the capability to increase lipid content with increasing temperature (Chisti 2008). However, high lipid content of extremophile microalgae lowers the growth rate which is a major drawback for microalgal biodiesel production, which can be solved by application of genetic engineering (Robles-Medina et al. 2009, Jiang et al. 2009).

3.3 Extremophiles for Biobutanol Production

Due to the significant combustion features and excellent performance, biobutanol has become the most attractive biofuel compared to the existing biofuels. A drastic reduction in greenhouse emission has been seen after biobutanol combustion. Biobutanol has become a good choice to be used in internal combustion engines. However, large scale biobutanol production faces more challenges as most of the wild type bacterial strains are less tolerant to the biobutanol toxicity. *Closridium* sp., a well-known bacterial strain for biobutanol production, also has several bottlenecks as it cannot sustain at high concentration of butanol as well as other by products. Toxic effect of butanol to the bacterial cell membrane is the major drawback for commercial biobutanol production. Therefore, butanol tolerant microbial strain has to be explored. All the challenges of biobutanol production can be nullified except biobutanol toxicity to cell membranes.

Several synergistic approaches have been initiated to explore the non-native bacterial strains which can grow in high solvent toxicity. In this context, *Clostridium beijerinckii* derived mutant strain has been reported as the potential mutant strain with high butanol tolerance capacity. This mutant strain has shown high yield compared to wild type and has lowered product inhibition as well. The robust strain has been reported as the good invention for biobutanol production from low-cost cellulosic feedstocks. Development of other potential mutant strain from clostridium species is still in lab scale (Lee et al. 2009, Das 2009).

3.4 Extremophiles for Biohydrogen Production

Biologically, hydrogen production has the potential to fulfill the future energy need. Hydrogen is an efficient, economical and ecofriendly source of renewable energy. Fossil fuel can be substituted with biohydrogen as it possesses high calorific value and emits no greenhouse gas as well.

It is well-known as the safest energy source for transportation and for environmental concern as well. Demand for this safest energy source is growing at an exponential phase through the world (> 45 million tons per annum). To meet the current energy demand, commercial sectors are focusing on a wide range of process including physical, chemical, physiochemical, and biological processes for H_2 production (Rittmann and Herwig 2012). Biohydrogen is generally produced through water electrolysis, steam reformation, catalytic steam gasification of biomass, biomass pyrolysis, supercritical water gasification, photolysis of water, and microbial fermentation.

Most commonly used biological process for biohydrogen production is dark fermentation, though it has few drawbacks like high chance of bacterial contamination and less efficiency.

Furthermore, dark fermentation requires simple reactor, easy operation, and wide range of raw materials. Dark fermentation results in high yield compared to other biological methods (Kumar et al. 2015, Saripan and Reungsang 2013).

In this context, exploring extremophilic anaerobes may be a good choice because they produce biohydrogen anaerobically in non-aseptic systems under extreme environmental condition which blocks the growth of H_2-consuming microorganisms such as acetogenic bacteria and methanogens. Additionally, extremophilic microorganisms are capable of utilizing a wide range of organic waste such as hot, highly saline, acidic or alkaline for biohydrogen production. Recently, it was reported that thermophiles are very efficient for biohydrogen production at elevated temperature range.

It has been reported that the efficiency of thermophilic microorganism mediated biohydrogen production is more than the mesophilic one because high temperature evolves good thermodynamic conditions, high mass transfer rate, substrate solubilization and inhibition of by-product formation as well as bacterial contamination, though biohydrogen production at elevated temperature is not commercially feasible because of low yield (de Vrije and Claassen 2003, Hallenbeck 2005, Jones 2008). Reported thermophilic bacteria for hydrogen production are *Thermotoga neapolitana* and *Caldicellulosiruptor saccharolyticus* which can utilize wide range of substrates. Some of the thermophiles are able to generate considerable amount of biohydrogen (4.0 mol-H2/mol-glucose), i.e., very near to theoretical yield (Munro et al. 2009, d'Ippolito et al. 2010). Due to the presence of membrane bound NADPH-dependent hydrogenase, enzyme thermophiles or hyperthermophiles are able to work in a thermodynamically feasible condition which results in higher biohydrogen yield from complex feedstock. Complex biomass is utilized by several thermophiles such as

C. saccharolyticus and *T. Neapolitan* for hydrogen production. De Vrije et al. studied biohydrogen production by thermophiles from a complex feedstock where 2.9–3.4 mole of H_2/mole of hexose was obtained (de Vrije et al. 2009). Among various extreme thermophiles, *C. saccharolyticus* DSM 8903 is hyperthermophile which can result in higher yield of biohydrogen from untreated and dried biomass like sweet sorghum, sugarcane bagasse, maize leaves, wheat straw, silphium, and pine wood. With the addition of higher efficiency, several thermophiles are explored which are able to utilize hexose as well as pentose sugars. *C. saccharolyticus* and *Thermoanaerobacterium, Thermosaccharolyticum* can produce biohydrogen from C_5 and C_6 sugars (Ren et al. 2008).

3.5 Extremophiles for Biogas Production

Utilization of various types of waste for biogas production is a traditional process. Waste product includes animal's droppings in farmhouse, biodegradable waste from slaughter house, waste cooking oil, biodegradable household waste or municipal solid waste, etc. (Demain 2009, Holm-Nielsen et al. 2009).

Hydrolysis, acidogenesis, acetogenesis, and methanogenesis are four consequent steps for biogas production. Hydrolysis step leads to conversion of complex feedstocks to their simple forms. Complex substrate may be carbohydrate, protein and fats. In acidogenesis, the monomers and long-chain fatty acids are degraded into short chain volatile acids such as valeric acid, propionic acid, and butyric acid. The short-chain acids produce acetic acid, hydrogen in acetogenesis phase. Finally, during methanogenesis phase, acetic acid, hydrogen, and carbon dioxide lead to produce methane and carbon dioxide. Rajendran et al. reported that acetic acid plays a major role for biogas production (Rajendran et al. 2013). Psychrophiles, mesophiles, thermophiles are capable of biogas production. Biogas production in household generally occurs at psychrophilic range which is quite slow where as mesophillic and thermophillc digestion is carried out at industrial scale.

Various types of thermophilic microorganisms are found for methane production such as *Methanobacterium* sp., *Methanosarcina thermophila* and *Methanothermococcus okinawensis* (Einspanier et al. 2004, Ariesyady et al. 2007).

4. Future Prospects

Biofuel production is the emerging sector throughout the world. Alternative fuel must be explored to fulfill the gradual rising of energy demand. For economically and ecologically feasible biofuel production, microbial route is the best way to opt. Traditional microorganisms, i.e., mesophiles generally face challenges as they are not capable of tolerating extreme conditions. Extremophiles possess special features which allow them to sustain under absolute odd conditions. Extremophiles have been explored for cost effective biofuel production. Unique characteristics of extremophiles can nullify the drawbacks of large scale biofuel production.

5. Conclusion

With the growing demand of crude oil, it can be understood that limited source of fossil fuel will run out very soon. Alternative sources need to be explored to fulfill the energy demand in future. Though a number of researches are on the way for commercially feasible biofuel production, it has not been reached up to the mark yet. Process technologies for large scale production are to be modified too. Traditional biofuel production is mostly dependent on raw materials with food value which is the main drawback for economically viable biofuel production. 1st generation biofuel production cannot be the substitute of fossil fuel sources. Therefore, 2nd generation of biofuel production is adapted with improved and feasible process technologies. Microorganism plays a great role in fermentation process because economical biofuel production is beyond imagination without microorganisms. Mesophilic organisms are generally used in research field as well as commercial sector because of their well-established structure and metabolic pathways. However, mesophiles are unable to work under extreme condition which is a normal phenomenon in industries. Hence, newly explored thermophiles have created craze in research area for their capacity to thrive under extreme conditions. Thermophiles are able to work at such process conditions which are economically and environmentally favorable. However, more information has to be gathered, and more researches are to be carried out about their metabolic pathway to improve their robustness and efficiency. Genetic engineering with these thermophiles can generate a new era in the field of biofuel. More investigations are to be carried out, and modern technologies are to be applied on thermophiles to produce cost effective biofuel in near future.

References

Ali, M.Y.M., M.M. Hanafiah, Y.H. Wen, M. Idris, N.I.H.A. Aziz, A.A. Halim and K.E. Lee. 2015. Biogas production from different substrates under anaerobic conditions, 3rd International Conference on Agricultural and Medical Sciences. CAMS-2015. 54–56.

Al-Zuhair, S. 2007. Production of biodiesel: Possibilities and challenges. Biofuels, Bioprod. Biorefin. 1: 57–66.

Antoni, D., V. Zverlov and W. Schwarz. 2007. Biofuels from microbes. Appl. Microbiol. Biotechnol. 77: 23–35.

Ariesyady, H.D., T. Ito and S. Okabe. 2007. Functional bacterial and archaeal community structures of major trophic groups in a full-scale anaerobic sludge digester. Water Res. 41: 1554–1568.

Babcock, R.E., E.C. Clausen, M. Popp and B.W. Schulte. 2014. Yield Characteristics of Biodiesel produced from chicken fat-tall oil blended feedstocks. ⟨https://www.researchgate.net/publication/253638502⟩.

Baker-Austin, C. and M. Dopson. 2007. Life in acid: pH homeostasis in acidophiles. Trends Microbiol. 15: 165–171.

Ballesteros, I., M.J. Negro, J.M. Oliva, A. Cabanas, P. Manzanares and M. Ballesteros. 2006. Ethanol production from steam-explosion pretreated wheat straw. Appl. Biochem. Biotechnol. 130: 496–508.

Banerjee, A. 2002. *Botryococcus braunii*: A renewable source of hydrocarbons and other chemicals. Crit. Rev. Biotechnol. 22: 245–279.

Behzad, S. and M.M. Farid. 2007. Review: Examining the use of different feedstock for the production of biodiesel. Asia-Pac. J. Chem. Eng. 2: 480–486.

Branduardi, P., F. de Ferra, V. Longo and D. Porro. 2014. Microbial n-butanol production from Clostridia to non-Clostridial hosts. Eng. Life Sci. 14: 16–26. https://doi.org/10.1002/elsc.201200146.

Burdette, D.S., S.H. Jung, G.J. Shen, R.I. Hollingsworth and J.G. Zeikus. 2002. Physiological function of alcohol dehydrogenases and long-chain (C-30) fatty acids in alcohol tolerance of *Thermoanaerobacter ethanolicu.* Appl. Environ. Microbiol. 68: 1914–1918.

Campbell, C.H. and J.H. Laherrere. 1998. The end of cheap oil. Scientific American. 78–83.

Cavendish, H. 1776. XII. An account of some attempts to imitate the effects of the torpedo by electricity. Philos. Trans. R. Soc. Lond. 66: 196–225.

Chisti, Y. 2007. Biodiesel from microalgae. Biotechnol. Adv. 25: 294–306.

Chisti, Y. 2008. Biodiesel from microalgae beats bioethanol. Trends Biotechnol. 26: 126–131.

Claassen, P.A.M., T. de Vrije and M.A.W. Budde. 2004. Biological hydrogen production from sweet sorghum by thermophilic bacteria, 2nd World Conference on Biomass for Energy, ETA-Florence and WIP-Munich, Rome.

Das, D. 2009. Advances in biohydrogen production processes: An approach towards commercialization. Int. J. Hydrog. Energy 34: 7349–7357.

Demirbas, A. 2005. Bioethanol from cellulosic materials: A renewable motor fuel from biomass. Energy Sources 27: 327–33.

De Luca, P., A. Masacchio and R. Taddei. 1981. Acidophilic algae from the fumaroles of Mount Lawu (Java) locus classius of *Cyanidium caldarium* Geitler. Giornale Botan. Ital. 115: 1–9.

Demain, A.L., M. Newcomb and J.H.D. Wu. 2005. Cellulase, clostridia, and ethanol. Microbiol. Mol. Biol. Rev. 69: 124–154.

Demain, A.L., M. Newcomb and JH.D. Wu. 2005. Cellulase, Clostridia, and Ethanol. Microbiol. Mol. Biol. Rev. 69(1): 124–154.

Demain, A.L. 2009. Biosolutions to the energy problem. J. Ind. Microbiol. Biotechnol. 36: 319–332.

Demirbas, A. 2003. Biodiesel fuels from vegetable oils via catalytic and non-catalytic supercritical alcohol transesterifications and other methods: A survey. Energy Convers. Manag. 44(13): 2093–2109.

de Vrije, T., G.G. de Haas, G.B. Tan, E.R.P. Keijsers and P.A.M. Claassen. 2002. Pretreatment of Miscanthus for hydrogen production by *Thermotoga elfii*. Int. J. Hydrogen Energy 27: 1381–1390.

de Vrije, T. and P.A.M. Claassen. 2003. Dark hydrogen fermentations. pp. 103–123. *In*: Reith, J.H., R.H. Wijffels and H. Barten (eds.). Biomethane & Bio-hydrogen: Status and Perspectives of Biological Methane and Hydrogen Production. Dutch Biological Hydrogen Foundation, The Hague.

de Vrije, T., R.R. Bakker and M.A.W. Budde. 2009. Efficient hydrogen production from lignocellulosic energy crop Miscanthus by the extreme thermophilic bacteria *Caldicellulosiruptor saccharolyticus* and *Thermotoga neapolitana*. Biotechnol. Biofuels 2: 12.

d'Ippolito, G., L. Dipasquale and F.M. Vella. 2010. Hydrogen metabolism in the extreme thermophile *Thermotoga neapolitana*. Int. J. Hydrog. Energy 35: 2290–2305.

Einspanier, R., B. Lutz, S. Rief, O. Berezina, V. Zverlov, W.H. Schwarz and J. Mayer. 2004. Tracing residual recombinant feed molecules during digestion and rumenbacterial diverstiy in cattle fed transgene maize. Eur. Food Res. Technol. 218: 269–273.

Ertan, A., C. Mustafa and S. Huseyin. 2014. Biodiesel production from vegetable oil and waste animal fats in a pilot plant. Waste Manag. 34: 2146–54.

Fangrui, M. and M.A. Hanna. 1999. Biodiesel production: A review. Bioresour. Technol. 70: 1–15.

Fong, J.C., C.J. Svenson, K. Nakasugi, C.T. Leong, J.P. Bowman, B. Chen, D.R. Glenn, B.A. Neilan and P.L. Rogers. 2006. Isolation and characterization of two novel ethanol-tolerant facultative-anaerobic thermophilic bacteria strains from waste compost. Extremophiles 10: 363–372.

Gaffron, H. 1939. Reduction of carbon dioxide with molecular hydrogen in green algae. Nature 143: 204.

Gaffron, H. and J. Rubin. 1942. Fermentative and photochemical production of hydrogen in algae. J. Gen. Physiol. 26: 219–240.

Golyshina, O.V., T.A. Pivovarova, G.I. Karavaiko, T.F. Kondrat'eva, E.R.B. Moore, W.R. Abraham, H. Lunsdorf, K.N. Timmis, M.M. Yakimov and P.N. Golyshin. 2000. Ferroplasma acidiphilum gen. nov., sp. nov., an acidophilic, autotrophic, ferrous-iron-oxidizing, cell-wall-lacking, mesophilic member of the Ferroplasmaceae fam. nov., comprising a distinct lineage of the Archaea. Int. J. Syst. Evol. Microbiol. 50: 997–1006.

Gonz alez-Ramos, D., M. van den Broek, A.J. van Maris, J.T. Pronk and J.M. Daran. 2013. Genome-scale analyses of butanol tolerance in *Saccharomyces cerevisiae* reveal an essential role of protein degradation. Biotechnol. Biofuels. 6: 48. https://doi.org/10.1186/1754-6834-6-48.

Greenwell, H.C., M. Loyd-Evans and C. Wenner. 2012. Biofuels, science and society. Interface Focus 3: 1–4.

Guti´errez-Rivera, B., K. Waliszewski-Kubiak, O. Carvajal-Zarrabal and M.G. Aguilar-Uscanga. 2012. Conversion efficiency of glucose/xylose mixtures for ethanol production using *Saccharomyces cerevisiae ITV01* and *Pichia stipitis* NRRL Y-7124. J. Chem. Technol. Biotechnol. 87(2): 263–270.

Haki, G.D. and S.K. Rakshit. 2003. Developments in industrially important thermostable enzymes. Bioresour Technol. 89: 17–34.

Hallenbeck, P.C. 2005. Fundamentals of the fermentative production of hydrogen. Water Sci. Technol. 52: 21–29.

Hallberg, K.B. and E.B. Lindström. 1994. Characterization of *Thiobacillus caldus*, sp. nov., a moderately thermophilic acidophile. Microbiol. 140: 3451–3456.

Hillebrecht, J.R., K.J. Wise, J.F. Koscielecki and R.R. Birge. 2004. Directed evolution of bacteriorhodopsins for device applications. Meth. Enzymol. 388: 333–347.

Holm-Nielsen, J.B., T. Al Seadi and P. Oleskowicz-Popiel. 2009. The future of anaerobic digestion and biogas utilization. Bioresour. Technol. 100: 5478–5484.

Hu, Q., M. Sommerfeld, E. Jarvis, M. Ghirardi, M. Posewitz, M. Seibert and A. Darzins. 2008. Microalgal triacylglycerols as feedstocks for biofuel production: Perspectives and advances. Plant J. 54: 621–639.

International Energy Agency (IEA). 2007. World Energy Outlook World Energy Outlook. International Energy Agency. Paris, France.

IEA Statistics. 2016. CO2 Emissions from Fuel Combustion-Highlights, IEA, Paris, 2011 http://www.iea.org/co2highlights/co2highlights.pdf, Accessed date: 16 October.

IEA. 2017. World Energy Outlook 2017. OECD/IEA 2017, Paris. www.iea.org.

Janaun, J. and N. Ellis. 2010. Perspectives on biodiesel as a sustainable fuel. Renew Sustain Energy Reviews 14: 1312–20. doi: 10.1016/j.rser.2009.12.011.

Jiang, Y., C. Xu, F. Dong, Y. Yang, W. Jiang and S. Yang. 2009. Disruption of the acetoacetate decarboxylase gene in solvent-producing *Clostridium acetobutylicum* increases the butanol ratio. Metab. Eng. 11: 284–291.

Jones, P. 2008. Improving fermentative biomass-derived H2-production by engineered microbial metabolism. Int. J. Hydrog. Energy 33: 5122–5130.

Kataoka, N., T. Tajima, J. Kato, W. Rachadech and A.S. Vangnai. 2011. Development of butanol-tolerant *Bacillus subtilis* strain GRSW2-B1 as a potential bioproduction host. AMB Express. 1: 10. https:// doi.org/10.1186/2191-0855-1-10.

Kim, S. and B.E. Dale. 2004. Global potential bioethanol production from wasted crops and crop residues. Biomass and Bioenerg. 26: 361–75.

Klass, D.L. 2004. Biomass for renewable energy and fuels. Encyclop. Energ. 1(1): 212–193.

Kumar, S., A. Bhalla and M. Bibra. 2015. Thermophilic biohydrogen production: Challenges at the industrial scale. *In*: Krishnaraj, N. and J.S. Yu (eds.). Bioenergy: Opportunities and challenges. Apple Academic Press, Oakville, 3–35.

Lacis, L.S. and H.G. Lawford. 1991. *Thermoanaerobacter ethanolicus* growth and product yield from elevated levels of xylose or glucose in continuous cultures. Appl. Environ. Microbiol. 57: 579–585.

Lamed, R. and J.G. Zeikus. 1980. Glucose fermentation pathway of *Thermoanaerobium brockii.* J. Bacteriol. 141: 1251–1257.

Leary, D. UNU-IAS Report: Bioprospecting in the Arctic, http://www.ias.unu.edu/(2008).

Lee, J.Y., Y.S. Jang, J. Lee, E.T. Papoutsakis and S.Y. Lee. 2009. Metabolic engineering of *Clostridium acetobutylicum* M5 for highly selective butanol production. Biotechnol. J. 4: 1432–1440.

Lee, R.A. and J.M. Lavoie. 2013. From first- to third-generation of biofuels: Challenges of producing a commodity from a biomass of increasing complexity. Animal Frontiers 3(2): 6–11.

Lin, Y., W. Zhang, C. Li, K. Sakakibara, S. Tanaka and H. Kong. 2012. Factors affecting ethanol fermentation using *Saccharomyces cerevisiae* BY4742. Biomass Bioenerg. 47: 395–401.

Liu, S. and N. Qureshi. 2009. How microbes tolerate ethanol and butanol. N. Biotechnol. 26: 117–121. https://doi.org/10.1016/j.nbt.2009.06.984.

Lohan, D. and S. Johnston. 2005. UNU-IAS Report: Bioprospecting in Antarctica. http://www.ias.unu.edu/.

Luque, R., L. Herrero-Davila, J.M. Campelo, J.H. Clark, J.M. Hidalgo, D. Luna, J.M. Marinas and A.A. Romero. 2008. Biofuels: A technological perspective. Energy Environ. Sci. 1: 542–564.

Lynd, L.R. and M.Q. Wang. 2003. A product e nonspecific framework for evaluating the potential of biomass-based products to displace fossil fuels. J. Ind. Ecol. 7: 17–32.

Mariano, A.P., D.F. de Angelis, F.M. Filho, D.I.P. Atala, M.R.W. Maciel and R.M. Filho. 2008. An alternative process for butanol production: Continuous flash fermentation. Chem. Prod. Process Model. 3: 1–14.

Marousek, J. 2013. Use of continuous pressure shock waves apparatus in rapeseed oil processing. Clean Technol. Environ. Policy 15: 721–25. doi: 10.1007/s10098-012-0549-3.

Marousek, J. 2014. Novel technique to enhance the disintegration effect of the pressure waves on oilseeds. Ind. Crops Prod. 53: 1–5. doi: 10.1016/j.indcrop.2013.11.048.

Mehrasbi, M., J. Mohammadi, M. Peyda and M. Mohammad. 2017. Covalent immobilization of *Candida antarctica* lipase on coreshell magnetic nanoparticles for production of biodiesel from waste cooking oil. Ren. Ener. 101: 593–602.

Molina, E. 1999. Photobioreactors: Light regime, mass transfer, and scaleup. J. Biotechnol. 70: 231–247.

Mona, S. and A. Kaushik. 2015. Screening metal-dye-tolerant photoautotrophic microbes from textile wastewaters for biohydrogen production. J. Appl. Phycol. 27: 1185–1194.

Mori, Y. 1990. Characterization of a symbiotic coculture of *Clostridium thermohydrosulfuricum* YM3 and *Clostridium thermocellum* YM4, Appl. Environ. Microbiol. 56: 37–42.

Munro, S.A., S.H. Zinder and L.P. Walker. 2009. The fermentation stoichiometry of *Thermotoga neapolitana* and influence of temperature, oxygen, and pH on hydrogen production. Biotechnol. Prog. 25: 1035–1042.

Nagy, V. and E. Szabó. 2011. Biogas from organic wastes, studia universitatis vasile goldis arad. Seria Stiintele Vietii 21(4): 891–887.

OECD. 2008. Biofuel Support Policies: An Economic Assessment. Paris: OECD Publishing.

Ortiz-Muʼniz, B., O. Carvajal-Zarrabal, B. Torrestiana-Sanchez and M.G. Aguilar-Uscanga. 2010. Kinetic study on ethanol production using *Saccharomyces cerevisiae* ITV-01 yeast isolated from sugar canemolasses. J. Chem. Technol. Biotechnol. 85(10): 1361–1367.

Oshima, T. and T. Moriya. 2008. A preliminary analysis of microbial and biochemical properties of high-temperature compost. Ann. N Y Acad. Sci. 1125: 338–344.

Pulz, O. and W. Gross. 2004. Valuable products from biotechnology of microalgae. Appl. Microbiol. Biotechnol. 65: 635–648.

Rajendran, K., S. Aslanzadeh, F. Johansson and M.J. Taherzadeh. 2013. Experimental and economical evaluation of a novel biogas digester. Energy Convers. Manag. 74: 183–191.

Ren, N., A. Wang and L. Gao. 2008. Bioaugmented hydrogen production from carboxymethyl cellulose and partially delignified corn stalks using isolated cultures. Int. J. Hydrog. Energy 33: 5250–5255.

Rittmann, S. and C. Herwig. 2012. A comprehensive and quantitative review of dark fermentative biohydrogen production. Microb. Cell Fact. 11: 115.

Robles-Medina, A., P.A. Gonzalez-Moreno, L. Esteban-Cerdan and E. Molina-Grima. 2009. Biocatalysis: Towards ever greener biodiesel production. Biotechnol. Adv. 27: 398–408.

Rosenbaum, M., U. Schröder and F. Scholz. 2005. Utilizing the green alga *Chlamydomonas reinhardtii* for microbial electricity generation: A living solar cell. Appl. Microbiol. Biotechnol. 68: 753–756.

Rothschild, L.J. and R.L. Mancinelli. 2001. Life in extreme environments. Nature 409: 1092–1101.

Santos, H. and M.S. Da Costa. 2002. Compatible solutes of organisms that live in hot saline environments. Environ. Microbiol. 4: 501–509.

Sa´nchez Miro´n, A. 1999. Comparative evaluation of compact photobioreactors for large-scale monoculture of microalgae. J. Biotechnol. 70: 249–270.

Saripan, A.F. and A. Reungsang. 2013. Biohydrogen production by *Thermoanaerobacterium thermosaccharolyticum* KKU-ED1: Culture conditions optimization using mixed xylose/arabinose as substrate. Electron. J. Biotechnol. 16. https://doi.org/10.2225/vol16-issue1-fulltext-1.

Sarkar, N., S.K. Ghosh, S. Bannerjee and K. Aikat. 2012. Bioethanol production from agricultural wastes: An overview. Renew. Energy 37: 19–27.

Seckbach, J. and W.F. Libby. 1970. Vegetative life on Venus? Or investigations with algae which grow under pure CO2 in hot acid media at elevated pressures. Space Life Sci. 2: 121–143.

Shaishav, S., R. Singh and T. Satyendra. 2013. Biohydrogen from algae: Fuel of the future. Int. Res. J. Environ. Sci. 2: 44–47.

Shapovalova, O.I. and L.A. Ashkinaib. 2008. Biobutanol: Biofuel of second generation. Russ. J. Appl. Chem. 81: 2232–2236.

Sharma, G. 1991. Prospects for ethanol production from cellulose with *Clostridium thermocellum–Bacillus stearothermophilus* co-cultures. Biotechnol. Lett. 13: 761–764.

Singh, L. and D.M. Mahapatra. 2019. Waste to Sustainable Energy: MFCs–Prospects Through Prognosis. CRC Press.

Stevens, D.J. 2001. Hot Gas Conditioning: Recent Progress With Larger-scale Biomass Gasification Systems;Update and Summary of Recent Progress. No. NREL/SR-510-29952. National Renewable Energy Lab, Golden,CO (US) https://doi.org/10.2172/786288.

Taylor, M.P., K.L. Eley, S. Martin, M.I. Tuffin, S.G. Burton and D.A. Cowan. 2009. *Thermophilic ethanologenesis:* Future prospects for second-generation bioethanol production. Trends Biotechnol. 27: 398–405.

Torquati, B., S. Venanzi, A. Ciani, F. Diotallevi and V. Tamburi. 2014. Environmental sustainability and economic benefits of dairy farm biogas energy production: A case study in umbria. Sustainability 6(10): 6713–6696.

Uihlein, A. and L. Schbek. 2009. Environmental impacts of a lignocellulosic feedstock biorefinery system: An assessment. Biomass Bioenerg. 33: 793–802.

Wisselink, H.W., M.J. Toirkens, Q. Wu, J.T. Pronk and A.J.A. van Maris. 2009. Novel evolutionary engineering approach for accelerated utilization of glucose, xylose, and arabinose mixtures by engineered *Saccharomyces cerevisiae* strains. Appl. Environ. Microbiol. 75: 907–914.

Zhang, S., X.T. Bi and R. Clift. 2015. Life cycle analysis of a biogas-centred integrated dairy farm-greenhouse system in British Columbia. Process Saf. Environ. Protect. 93: 30–18.

12

Biofuel Production from Biomass using Extremophilic Microorganisms

Sunny Kumar

1. Introduction

An extremophile is an organism with optimal growth in extreme environmental conditions that it is challenging for a carbon-based life form with water as a solvent, such as all life on Earth, to survive (Rothschild and Mancinelli 2001). Life in the form of extremophiles may have begun on Earth in hydrothermal vents far below the ocean's surface. The discovery of extremophiles points out the extraordinary adaptability of primitive life-forms and raises the prospect of finding at least microbial life elsewhere in the Solar System and beyond (Tucker et al. 2018). Extremophiles live in a varied range of extreme environments. The products obtainable from extremophiles such as proteins, enzymes (extremozymes) and compatible solutes are of great interest to biotechnology because of the potential applications in biotechnology. There are many types of extremophiles microorganisms in nature, each corresponding to the way its environment differs from mesophilic conditions. Many extremophiles fall under multiple categories and are classified as polyextremophiles. For example, organisms living inside hot rocks deep under Earth's surface are thermophilic and piezophilic such as *Thermococcus barophilus* (Marteinsson et al. 1999). A polyextremophile living at the summit of a mountain in the Atacama Desert might be a radioresistant *xerophile*, a *psychrophile*, and an *oligotroph*. Polyextremophiles are well known for their ability to tolerate both high and low pH levels (Yadav et al. 2015). Extremophiles include members of all three domains of life, such as bacteria, archaea, and eukarya (Rampelotto 2013). The extremophilic organism and enzymes can be develop and improve the biofuels production using biotechnology (Barnard et al. 2010). The combustion engines require the input of fuel to get the energy to run the vehicle. This production was mainly used to (i) extract the energy available from fuel, (ii) remove inorganic or organic impurities, (iii) extract energy from biomass or petroleum, or (iv) power small-scale off-grid devices and sensors. These applications have evolved into chemical or microbial catalyzed generation of products with a higher added value from the produced electrons, such as hydrogen, methane, hydrogen peroxide, and short- and medium-chain fatty acids.

Biofuel from biomass using microorganisms is created to be an ideal substitute to extend the use and longevity of diesel. The main advantages of biofuel are that it provides energy security and benefit the local community. Biofuel can be stored, burned, and pumped in the same manner

Department of Chemical Engineering, Institute of Chemical Technology, Marathwada Campus, Jalna, India.
Emails: chemicalsunnyraj@gmail.com, s.kumar@marj.ictmumbai.edu.in

as petroleum fuel. Energy security is the most constant supply that is available and affordable for consumers and industry. The use of biofuel will help to boost the growth of the economy. Biofuel will also reduce emission of carcinogenic compounds and simultaneously develop clean environment. Biofuel is less toxic than diesel as its attribute make it less likely to harm the environment and cost less damage. Due to its low volatility, biofuel can be safer to handle than petroleum fuel. Presently, many countries have started the production of biofuel, which improves the economies of relative countries (Kruger et al. 2018). In 2019, worldwide biofuel production reached 161 billion litres (43 billion gallons US), 6% up from 2018, and biofuel provided 3% of the world's fuel for road transport. To reduce their dependency on the petroleum, various agencies and countries want to produce more biofuel to meet world demand. From 2020 to 2030, global biofuel output has to increase by 10% each year to reach the International Energy Agency (IEA)'s goal. The IEA has a target of 3% growth annually for the next 5 years.

2. Various Types of Biofuels

There are various extremophiles microorganisms which can be used for biofuel production from biomass as shown in Fig. 1. The utilization of microorganisms for biofuel production reduces the energy cost for production and is environmental friendly. The research requires the discovery of extremophile to be able to treat the desired biomass and this will require sampling of diverse extreme environments likely to contain suitable microorganisms. The microorganisms will also vary with the concentration cellulose, hemicellulose, lignin and starch in the agriculture waste and fermentation and produce the biofuel such as bioethanol, biobutanol, biodiesel, and biohydrogen/biogas as shown in the Fig. 2. These biomass molecules are complex structures which are not easily fermented. Therefore, physical and chemical pretreatment steps are also required for production of biofuel.

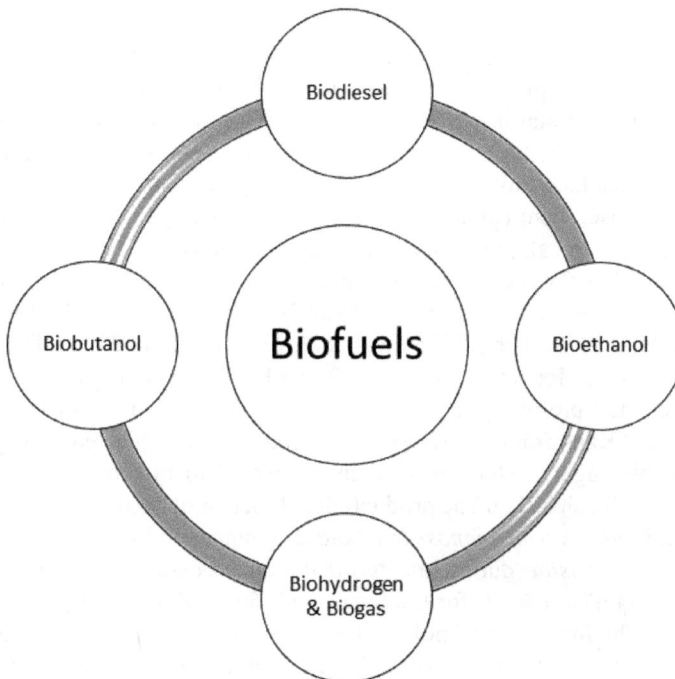

Fig. 1. Various types of biofuels.

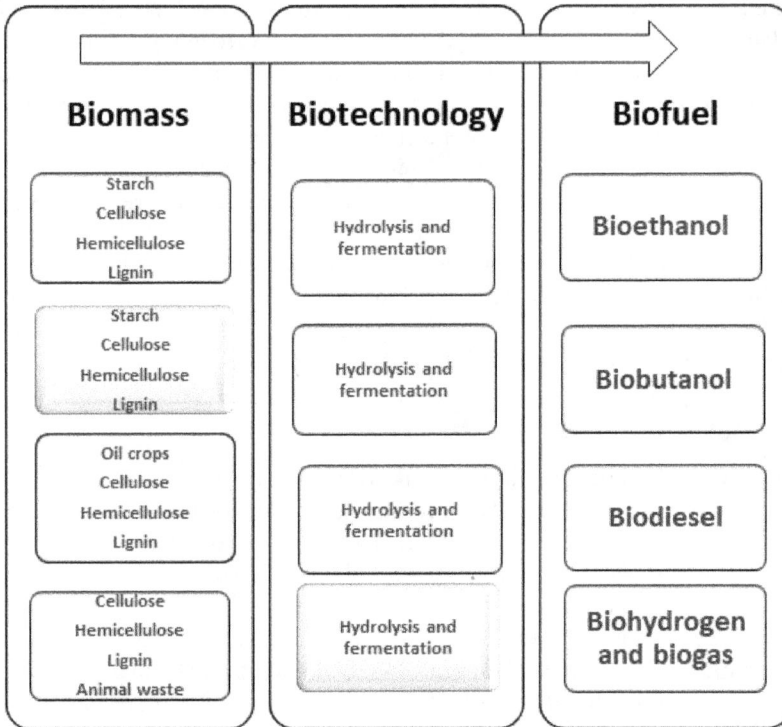

Fig. 2. Various ways of production of biofuels from biomass.

3.1 Bioethanol

Biomasses have complex structures that are resistant to chemical and biological breakdown. There are several factors, such as crystalline structure, degree of lignification, and structural heterogeneity, which resist the breakdown of the structure. Therefore, pretreatment is a significant step in the biorefinery process to produce biofuels. There are many pretreatment processes used for various compositions of cellulose, hemicellulose, and lignin in biomass. The common pretreatment processes are physical, chemical, physicochemical, and biological.

Enzymatic hydrolysis of waste hamburger could be used for bioethanol production as shown in Fig. 3. Enzyme loading volumes determine the hydrolysis efficiency. The highest ethanol yield of 0.271 g ethanol/g WH (0.34 g ethanol/g TS) was achieved (Han et al. 2020). The ethanol yield can also enhanced by delignification and synergistic effect of cellulase and xylanase.

The laccase mediated pretreatment resulted in 6.7% delignification of rice straw (Bhardwaj et al. 2020). Acid treated Kans grass (*Saccharum sponteneum*) was subjected to enzymatic hydrolysis to produce fermentable sugars, which were then fermented to bioethanol using *Saccharomyces cerevisiae* (Kataria and Ghosh 2011). The production of bioethanol from residual Ataulfo (*Mangifera indica* L.) mango pulp uses *vermicompost leachate* as a nutritional supplement for ethanologenic yeast *Saccharomyces cerevisiae* during the fermentation process. The bioethanol concentration was achieved at upto 44.67 ± 1.6 g/L for treatment with 50% of *vermicompost leachate* (Aleman-Ramirez et al. 2019). The low cost technology for bioethanol production is important for scale up the plant and demand of bioethanol (Carrillo-Nieves et al. 2019). The improvement in enzymatic hydrolysis treatment generally helps to increase production of bioethanol, which simultaneously decreases the enzyme amount and required time.

Ultrasound can be effectively used to improve the hydrolysis process by reducing the structural rigidity of lignocellulose and by eliminating the mass-transfer resistance, which can contribute to an increase in the product yield with reduced processing time and enzyme consumption (Subhedar and

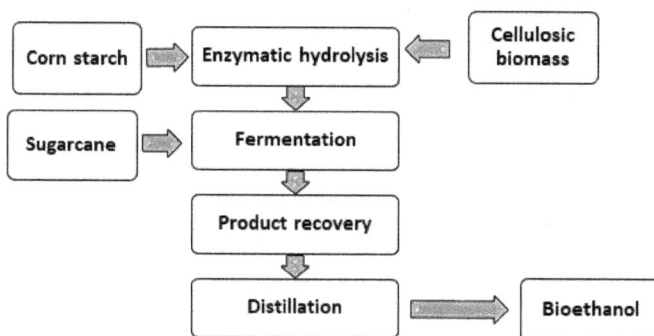

Fig. 3. Schematics represents the production of bioethanol from corn starch, sugarcane and cellulosic biomass.

Gogate 2013). Consolidated bioprocessing of cellulose mixed with fermentable sugar is considered as a promising alternative to the use of cellulose as a sole substrate for bioethanol production (Li et al. 2017, van Seters et al. 2011). The bioethanol producing system suppresses the inhibition of fermentation by thermally processed residual lignin in a separate hydrolysis and fermentation and one-pot simultaneous saccharification and co-fermentation (SSCF).

The lignin-derived adsorbent and genetically engineered *Zymomonas mobiliz* was used for bioethanol, it was produced at 54 g/L from the pretreated biomass mash by one-pot simultaneous saccharification and co-fermentation. Prehydrolysis is used in conjunction with SSCF processes (Yoshioka et al. 2018). Ethanol produces slightly less greenhouse emissions than fossil fuels where carbon dioxide is recycled from the atmosphere to produce biomass. It can replace harmful fuel additives such as methyl tertiary butyl ether and produce jobs for farmers and refinery workers.

But the industrial utilization of lignocelluloses for bioethanol production is hindered by the lack of ideal microorganisms which can efficiently ferment both pentose and hexose sugars. For a commercially viable ethanol production method, an ideal microorganism should have broad substrate utilization, high ethanol yield and productivity, should have the ability to withstand high concentrations of ethanol and high temperature, should be tolerant to inhibitors present in hydrolysate and have cellulolytic activity. Genetically modified or engineered microorganisms are thus used to achieve complete utilization of the sugars in the hydrolysate and better production benefits. The processes usually employed in the fermentation of lignocellulosic hydrolysate are simultaneous saccharification and fermentation and separate hydrolysis and fermentation.

A combination of *Candida shehatae* and *Saccharomyces cerevisiae* was reported as suitable for the SSCF process. Sequential fermentation with two different microorganisms in different time periods of the fermentation process for better utilization of sugar has also been employed using *S. cerevisiae* in the first phase for hexose utilization and *C. shehatae* in the second phase for pentose utilization but ethanol yields achieved are not high.

Some native or wild type microorganisms used in the fermentation are *S. cerevisiae*, *Escherichia coli*, *Zymomonas mobilis*, *Pachysolen tannophilus*, *C. shehatae*, *Pichia stipitis*, *Candida brassicae*, *Mucor indicus*, etc. Among all the best-known yeast and bacteria employed in ethanol production from hexoses, *S. cerevisiae* and *Z. mobilis*, respectively. But *S. cerevisiae* cannot utilize the main C-5 sugar, xylose of the hydrolysate. Different microorganisms have shown different yields of ethanol depending on their monomer utilization.

3.2 Biodiesel

The biodiesel can be produced by an enzymatic based reaction using waste oil as a raw material which is an efficient procedure for resolving the problems of energy demand and environmental pollution as shown in Fig. 4. The cost of enzymes is the main hindrance to the industrialization of this process. The enzymes such as lipases were used with a combination of amino coated magnetite nanoparticles.

Biodiesel production from waste cooking oils by immobilized enzymes increased about 20% more than free enzymes (Badoei-dalfard et al. 2019). The biodiesel production processes come from wet and lipid-rich *C. protothecoides*. In the presence of malic enzyme, accumulation happens in the lipases for a few microorganisms. As a results, genetically engineered *C. protothecoides* can help malic enzyme in improve lipid accumulation during biodiesel production from microalgal biomass.

A dehydration step using methanol pretreatment coupled with H_2SO_4 or lipase catalysed biodiesel production enabled 93–98% biodiesel yields directly from microalgal biomass without extraction of lipid under optimal conditions (Yan et al. 2019).

Black soldier fly larvae containing high fat content are a favourable nonedible feedstock for biodiesel production. The fat can be extracted with a high amount of solvent, a long extraction time, and thermal conditions (Su et al. 2019).

A biodiesel conversion of 80% was obtained by chemical catalysis (1 h at 100°C) using free fatty acid and methanol, but a biodiesel conversion of more than 90% (3 h at 30°C) in case magnetic cross-linked enzyme aggregates (Picó et al. 2018). Castor oil was converted by a liquid lipase catalyst into biodiesel. Higher amounts of methanol decrease the enzyme activity, decreasing the yield (Andrade et al. 2017).

Macauba (*Acrocomia aculeata*) is a plant with a high potential of oil supply for biodiesel production but has high acidity (35–43%). The lipase (*Lipozyme* 435) was used for the esterification of free fatty acids in the oil to reduce the acidity under temperature range (20–45°C) and water content (560–30000 ppm) ranges (Teixeira et al. 2017).

Generally, lipase biocatalyst with immobilization is used in biodiesel production where immobilization of enzymes improves activity and stability with adsorption-crosslinking methods. The highest unit activity (24.69 U/g resin) is achieved by immobilized lipase on anion-exchange macroporous resin with addition of chitosan on resin, which gives a 50.79% yield (Alamsyah et al. 2017).

Many microbial sources can produce the lipases which can be used as catalysts for the conversion of oil to biodiesel. Biodiesel and glycerol obtained by enzymatic conversion have shown a higher purity as compared to other processes. Low energy is required for enzymatic conversion of oil to biodiesel because less purification steps are involved in processing which reduces the operating cost (Chesterfield et al. 2012). A wide range of feedstock oils can be converted to biodiesel using a good catalytic conversion by lipase enzyme (Guldhe et al. 2015).

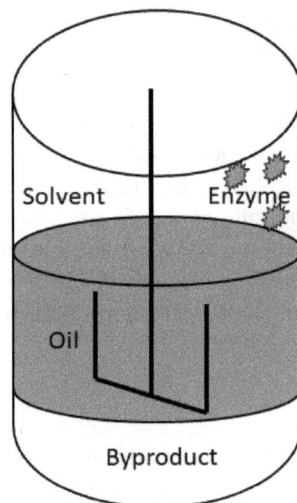

Fig. 4. Biodiesel production process.

The phoenix tree seed oil was also biomass for biodiesel preparation. Sun and Li 2020 found that free liquid lipozyme TL100L showed the best performance for the transesterification and gave a maximum (98.8 ± 1.1%) biodiesel yield. There are other transesterification methods such as homogeneous alkali-catalyzed, homogeneous acid-catalyzed, and non-catalytic reactions, but enzyme-catalyzed reactions have a low cost and less purification steps.

Adsorption, covalent and affinity binding, entrapment, and cell immobilisation are all examples of enzyme immobilisation that improve catalytic efficiency (Moazeni et al. 2019). The components for biodiesel production with enzymatic reactions are lipase, oil, and an acyl acceptor. Previously, some of the methods were studied to improve biodiesel production through enzymatic reaction with a combination of lipases, enzyme pretreatment, enzyme post treatment, methanol addition technique, use of solvents, and adding of silica gel. The genetic modification of a strain of lipase should have a high tolerance to solvent, good catalytic efficiency, and resistivity to harsh conditions.

Another acyl acceptor for biodiesel production is an ester such as methyl acetate and ethyl acetate. Methyl and ethyl acetate do not cause negative effect on lipase activity compared to methanol or ethanol (thus no need for stepwise addition) and will produce a higher value byproduct called triacetin or triacetyl glycerol, which do not have a negative effect on the reaction. Triacetin is a useful product that can be used in many fields, such as medicine, food, cosmetics, pesticides, and cigarettes. The comparative result of biodiesel production was shown in Table 1 for different reactions.

Table 1. Comparison of biodiesel production using different reactions.

Methods	Advantages	Disadvantages
Enzymatic reaction (immobilised lipase)	Medium yield, can convert FFA, low energy usage, high product, reusable catalyst	Inhibition by alcohol or byproduct, high cost of enzymes, slow reaction
Non-catalyzed reaction (supercritical alcohol)	Fast reaction, high yield, can convert FFA, no catalyst, easy production	High temperature and pressure, high cost of reactor, high alcohol to oil ratio
Chemical catalyzed reaction	High yield, can convert FFA, low cost	Wastewater, need product purification steps, difficult catalyst recovery saponification

3.3 Biobutanol

Biobutanol has emerged as an excellent biofuel and has comparatively low carbon emissions. Butanol has a wide range of applications, such as solvent in industries, chemical intermediate in various reactions, extractant and as a potential high energy density biofuel. Butanol is a colourless, flammable four carbon straight chain alcohol with better combustion properties than alcohol-based biofuels, including a high octane number, good calorific value, lower volatility, lower fire point, and high viscosity. Biobutanol is also an alternative biofuel which can be produced from biomass by the following steps: pretreatment, detoxification, fermentation, cell separation, and solvent recovery (Birgen et al. 2019), as shown in Fig. 5.

Commonly, immobilized *Clostridium acetobutylicum* bacterium is used for biobutanol production from lignocellulosic biomass (Tsai et al. 2020). There are various renewable feedstocks such as rice straw, sugarcane bagasse and microalgal hydrolysate used through acetone-butanol-ethanol fermentation through separate hydrolysis and fermentation (Huzir et al. 2018). A biobutanol titer and yield of 9.10 g/L and 0.42 mol/mol glucose (0.17 g biobutanol/g glucose), respectively, was obtained from rice straw, while sugarcane bagasse achieved a biobutanol titer and yield of 8.40 g/L and 0.40 mol/mol glucose (0.16 g biobutanol/g glucose), respectively.

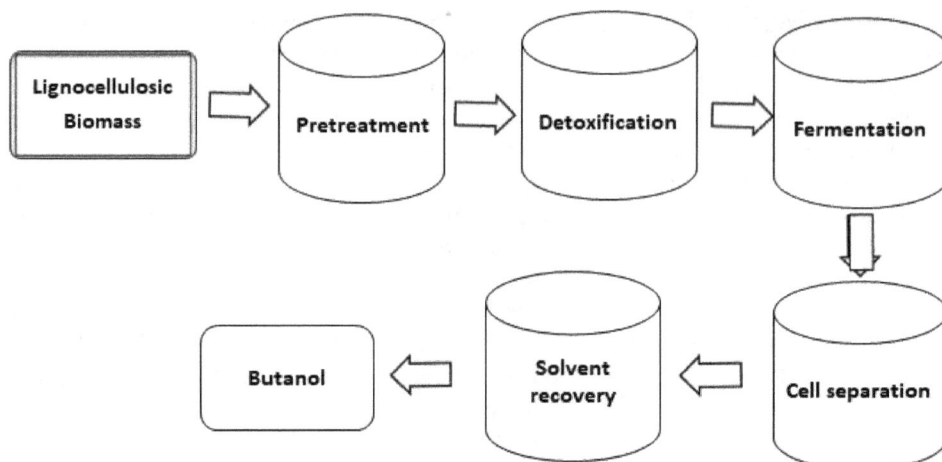

Fig. 5. Schematic of the production of biobutanol.

The potential of hydrolysates of cyanobacteria *Lyngbya limnetica* and *Oscillatoria obscura* for biobutanol production was evaluated under varying operating conditions using *Clostridium beijerinckii* ATCC 35702 culture as the fermenting microorganism (Kushwaha et al. 2020). The high biobutanol production (8.873 g/L) could be produced with glucose supplemented hydrolysate, while the highest yield (0.421 g/g sugar) was found with pure cyanobacterial hydrolysate.

The different types of anaerobic biobutanol production are fed-batch fermentation, continuous fermentation, and two-stage continuous fermentation. Commonly, the butanol recovery methods are adsorption, gas stripping, and pervaporation. *Clostridia* are the main fermentative organisms used to produce biobutanol (Pugazhendhi et al. 2019). Ornamental tree waste, Cassia fistula pods for biobutanol production using *Clostridium acetobutylicum* TISTR 2375 through various methodologies compositional analysis, extraction, saccharification, and fermentation (Khunchit et al. 2020).

A two-stage sequential bioprocess for the production of biobutanol uses low-cost substrates, corn steep liquor and industrial grade maize starch (Mukherjee et al. 2020). Acetone-butanol-ethanol fermentation has been performed for many years by solventogenic *Clostridium beijerinckii* and *C. acetobutylicum* (Jiang et al. 2019). For industrial scale biobutanol production was also reported from microalgae (Yeong et al. 2018).

High-efficiency bioconversion of kitchen garbage to biobutanol uses an enzymatic cocktail procedure (Ibrahim et al. 2018). The efficiency of *Clostridium acetobutylicum*-mediated biobutanol production was then measured using two modes, which are separate hydrolysis and fermentation and saccharification fermentation with specific pH (Chen et al. 2017).

Microalgae are considered a potential feedstock for biobutanol production due to their high growth rate and high carbohydrate content (Wang et al. 2017). Pea pod waste from the vegetable sector was used for biobutanol production using *C. acetobutylicum* B 527 through several steps (Nimbalkar et al. 2018).

For high biobutanol production, various parameters should be improved, such as choice of feedstock, low product yield, product toxicity to strain, multiple end-products and downstream processing of alcohol mixtures, plus metabolic engineering for improvement of the fermentation process and products (Ibrahim et al. 2017).

The rice straw can be a potential feedstock to produce biobutanol by fermentation. Maximum biobutanol production require the pretreatment process to access the cellulosic content for efficient enzymatic hydrolysis, the concept of consolidated bioprocessing, and genetic engineering tools. Table 2 shows the production yield of biobutanol from various organisms' treatments and processes.

Table 2. List of microorganisms and feedstocks for biobutanol production.

Process	Organism Involved	Butanol Yield	Substrate
Semi solid	*Pichia. pastoris*	9.5–10.5 g/L	Corn-starch
Continuous fermentation	*Clostridium acetobutylicum* YMI	5.89 g/L	De-oiled rice bran
Co-culture	*Bacillus subtilis* and *clostridium acetobutylicum*	8.28 g/L	Agave hydrolysates
Batch	*Clostridium beijerinckii*	7.3g/L	Brewer spent grains
Batch	*Clostridium carboxidivorans*	2.66 g/L	Carbon dioxide

3.4 Biohydrogen

Biohydrogen is a very important biofuel that is produced by two methods: biological and thermochemical, as shown in Fig. 6. The biological methods are enzymatic, photo fermentation, anaerobic fermentation, and microbial electrolysis. Photo fermentation is further classified as direct photolysis, indirect photolysis, light fermentation, and dark fermentation (Kuppam et al. 2017).

The dark fermentation of palm oil mill effluent can produce biohydrogen under mesophilic conditions. The enzyme was extracted from castor bean seeds (*Ricinus communis* L.) that can improve the amount of nutrients in effluent. The biohydrogen yield of 2.58 mmol H_2/g_{DQO} uses many approaches, such as impact of nanomaterials on cellulosic biohydrogen production, pretreatment technology, enzyme modification, and sugar production to enhance the biohydrogen yield (Garritano et al. 2017). There are various ways of producing biohydrogen from waste biomass such as dark fermentation, photo-fermentation, and hybrid-fermentation, from a variety of waste biomass (Srivastava et al. 2020).

The pretreated waste of the paper industry can produce biohydrogen from the saccharification and fermentation processes under different conditions. The solid wastes of the paper industry were pretreated with H_2SO_4 at 120°C. Moreno-Davila et al. 2019 perform the experiments for the biohydrogen production for different enzyme loadings (10 to 70 FPU/mL) at 5 pH.

Microalgae is one of the biofuel feedstocks where hydrogenases and nitrogenases produce biohydrogen in microalgae (Mona et al. 2020). Industrial wastewater is a cheap substrate for profitable biohydrogen production. Pretreatment of inoculum enriches the selective hydrogen producing microbes (Preethi et al. 2019). The process for utilizing residual organics with nanoparticle addition also shows improved biohydrogen production (Rajesh Banu et al. 2020).

Anaerobic sludge can also be used to produce biohydrogen from waste bread. Initially, the waste bread was first hydrolysed by the crude enzymes which were generated by *Aspergillus awamori* and *Aspergillus oryzae* through solid-state fermentation. One-gram of waste bread could produce

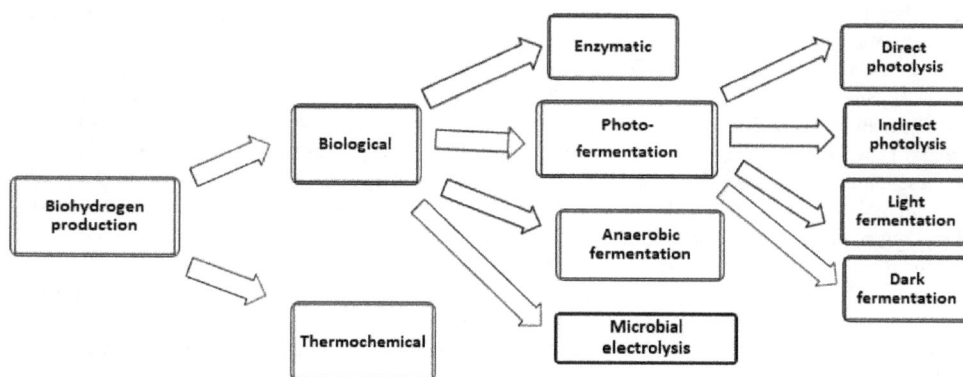

Fig. 6. Schematic of the production of biohydrogen.

109.5 mL of hydrogen (Han et al. 2016). Improving biohydrogen production includes microbial culture immobilization, bacterial strains, bioreactor modifications, the optimization of process conditions (temperature, pH, over loading rate, and hydraulic retention time), culture selection and enrichments, suitable substrate, and the metabolic engineering of biohydrogen specialists (Arimi et al. 2015, Azwar et al. 2014).

4. Conclusions

This chapter focuses on the significance of development of extremophile microorganisms for biofuel production, such as bioethanol, biodiesel, biobutanol, and biohydrogen from biomass. It will be very difficult to synthesize gasoline and diesel using enzyme-based technology. But the blending materials can be produced by the enzyme's technology, such as ethanol, butanol, acetone, and dimethyl ether. This chapter shows the historical background and development of enzymes, especially microbial catalysts. Biofuel may become useful as a power supply in areas where the infrastructure is not well developed, for example, a portable power supply generating electricity. Biofuel production from biomass will also involve farmer contributions and provide economic benefits.

References

Alamsyah, G., V.A. Albels, M. Sahlan and H. Hermansyah. 2017. Effect of chitosan's amino group in adsorption-crosslinking immobilization of lipase enzyme on resin to catalyze biodiesel synthesis. Energy Procedia. 136: 47–52.

Aleman-Ramirez, J.L., B.Y. Pérez-Sariñana, S. Torres-Arellano, S. Saldaña-Trinidad, A. Longoria and P.J. Sebastian. 2019. Bioethanol production from Ataulfo mango supplemented with vermicompost leachate. Catal. Today.

Andrade, T.A., M. Errico and K.V. Christensen. 2017. Influence of the reaction conditions on the enzyme catalyzed transesterification of castor oil: A possible step in biodiesel production. Bioresour. Technol. 243: 366–374.

Arimi, M.M., J. Knodel, A. Kiprop, S.S. Namango, Y. Zhang and S.-U. Geißen. 2015. Strategies for improvement of biohydrogen production from organic-rich wastewater: A review. Biomass Bioenergy 75: 101–118.

Azwar, M.Y., M.A. Hussain and A.K. Abdul-Wahab. 2014. Development of biohydrogen production by photobiological, fermentation and electrochemical processes: A review. Renew. Sustain. Energy Rev. 31: 158–173.

Badoei-dalfard, A., S. Malekabadi, Z. Karami and G. Sargazi. 2019. Magnetic cross-linked enzyme aggregates of Km12 lipase: A stable nanobiocatalyst for biodiesel synthesis from waste cooking oil. Renew. Energy 141: 874–882.

Barnard, D., A. Casanueva, M. Tuffin and D. Cowan. 2010. Extremophiles in biofuel synthesis. Environ. Technol. 31: 871–888.

Bhardwaj, N., B. Kumar, K. Agrawal and P. Verma. 2020. Bioconversion of rice straw by synergistic effect of in-house produced ligno-hemicellulolytic enzymes for enhanced bioethanol production. Bioresour. Technol. Rep. 10: 100352.

Birgen, C., P. Dürre, H.A. Preisig and A. Wentzel. 2019. Butanol production from lignocellulosic biomass: Revisiting fermentation performance indicators with exploratory data analysis. Biotechnol. Biofuels 12: 167.

Carrillo-Nieves, D., M.J. Rostro Alanís, R. de la Cruz Quiroz, H.A. Ruiz, H.M.N. Iqbal and R. Parra-Saldívar. 2019. Current status and future trends of bioethanol production from agro-industrial wastes in Mexico. Renew. Sustain. Energy Rev. 102: 63–74.

Chen, Hua, H. Shen, H. Su, HongZhen Chen, F. Tan and J. Lin. 2017. High-efficiency bioconversion of kitchen garbage to biobutanol using an enzymatic cocktail procedure. Bioresour. Technol. 245: 1110–1121.

Chesterfield, D.M., P.L. Rogers, E.O. Al-Zani and A.A. Adesina. 2012. A novel continuous extractive reactor for biodiesel production using lipolytic enzyme. Procedia Eng., International Energy Congress 2012 49: 373–383.

Garritano, A. do N., L.R.V. de Sá, É.C.G. Aguieiras, D.M.G. Freire and V.S. Ferreira-Leitão. 2017. Efficient biohydrogen production via dark fermentation from hydrolized palm oil mill effluent by non-commercial enzyme preparation. Int. J. Hydrog. Energy 42: 29166–29174.

Guldhe, A., B. Singh, T. Mutanda, K. Permaul and F. Bux. 2015. Advances in synthesis of biodiesel via enzyme catalysis: Novel and sustainable approaches. Renew. Sustain. Energy Rev. 41: 1447–1464.

Han, W., J. Huang, H. Zhao and Y. Li. 2016. Continuous biohydrogen production from waste bread by anaerobic sludge. Bioresour. Technol. 212: 1–5.

Han, W., Y. Liu, X. Xu, J. Huang, H. He, L. Chen, S. Qiu, J. Tang and P. Hou. 2020. Bioethanol production from waste hamburger by enzymatic hydrolysis and fermentation. J. Clean. Prod. 264: 121658.

Huzir, N.M., M.M.A. Aziz, S.B. Ismail, B. Abdullah, N.A.N. Mahmood, N.A. Umor and S.A.F. Syed Muhammad. 2018. Agro-industrial waste to biobutanol production: Eco-friendly biofuels for next generation. Renew. Sustain. Energy Rev. 94: 476–485.

Ibrahim, M.F., N. Ramli, E. Kamal Bahrin and S. Abd-Aziz. 2017. Cellulosic biobutanol by Clostridia: Challenges and improvements. Renew. Sustain. Energy Rev. 79: 1241–1254.

Ibrahim, M.F., S.W. Kim and S. Abd-Aziz. 2018. Advanced bioprocessing strategies for biobutanol production from biomass. Renew. Sustain. Energy Rev. 91: 1192–1204.

Jiang, Y., Y. Lv, R. Wu, Y. Sui, C. Chen, F. Xin, J. Zhou, W. Dong and M. Jiang. 2019. Current status and perspectives on biobutanol production using lignocellulosic feedstocks. Bioresour. Technol. Rep. 7: 100245.

Kataria, R. and S. Ghosh. 2011. Saccharification of Kans grass using enzyme mixture from *Trichoderma reesei* for bioethanol production. Bioresour. Technol. 102: 9970–9975.

Khunchit, K., S. Nitayavardhana, R. Ramaraj, V.K. Ponnusamy and Y. Unaprom. 2020. Liquid hot water extraction as a chemical-free pretreatment approach for biobutanol production from Cassia fistula pods. Fuel 279: 118393.

Krger A., C. Schäfers, C. Schroder and G. Antranikian. 2018. Toward a sustainable biobased industry—Highlighting the impact of extremophiles, 2018. New Biotechnol. 40: 144–153.

Kuppam, C., S. Pandit, A. Kadier, C. Dasagrandhi and J. Velpuri. 2017. Biohydrogen production: Integrated approaches to improve the process efficiency. pp. 189–210. *In*: Kalia, V.C. and P. Kumar (eds.). Microbial Applications Vol. 1: Bioremediation and Bioenergy. Springer International Publishing, Cham.

Kushwaha, D., N. Srivastava, D. Prasad, P.K. Mishra and S.N. Upadhyay. 2020. Biobutanol production from hydrolysates of cyanobacteria *Lyngbya limnetica* and *Oscillatoria obscura*. Fuel 271: 117583.

Li, Y.-J., Y.-Y. Lu, Z.-J. Zhang, S. Mei, T.-W. Tan and L.-H. Fan. 2017. Co-fermentation of cellulose and sucrose/ xylose by engineered yeasts for bioethanol production. Energy Fuels 31: 4061–4067.

Marteinsson, V.T., J.-L. Birrien, A.-L. Reysenbach, M. Vernet, D. Marie, A. Gambacorta, P. Messner, U.B. Sleytr and D. Prieur. 1999. *Thermococcus barophilus* sp. nov., a new barophilic and hyperthermophilic archaeon isolated under high hydrostatic pressure from a deep-sea hydrothermal vent. Int. J. Syst. Evol. Microbiol. 49: 351–359.

Moazeni, F., Y.-C. Chen and G. Zhang. 2019. Enzymatic transesterification for biodiesel production from used cooking oil, a review. J. Clean. Prod. 216: 117–128.

Moreno-Dávila, I.M.M., L.J. Ríos-González, J.A. Rodríguez-de la Garza, T.K. Morales-Martínez and Y. Garza-García. 2019. Biohydrogen production from paper industry wastes by SSF: A study of the influence of temperature/ enzyme loading. Int. J. Hydrog. Energy, XVII Mexican Hydrogen Society Congress Special Issue 44: 12333–12338.

Mukherjee, M., G. Goswami, P.K. Mondal and D. Das. 2020. Biobutanol as a potential alternative to petroleum fuel: Sustainable bioprocess and cost analysis. Fuel 278: 118403.

Nimbalkar, P.R., M.A. Khedkar, P.V. Chavan and S.B. Bankar. 2018. Biobutanol production using pea pod waste as substrate: Impact of drying on saccharification and fermentation. Renew. Energy 117: 520–529.

Picó, E.A., C. López, Á. Cruz-Izquierdo, M. Munarriz, F.J. Iruretagoyena, J.L. Serra and M.J. Llama. 2018. Easy reuse of magnetic cross-linked enzyme aggregates of lipase B from *Candida antarctica* to obtain biodiesel from Chlorella vulgaris lipids. J. Biosci. Bioeng. 126: 451–457.

Preethi, Usman, T.M.M., J. Rajesh Banu, M. Gunasekaran and G. Kumar. 2019. Biohydrogen production from industrial wastewater: An overview. Bioresour. Technol. Rep. 7: 100287.

Pugazhendhi, A., T. Mathimani, S. Varjani, E.R. Rene, G. Kumar, S.-H. Kim, V.K. Ponnusamy and J.-J. Yoon. 2019. Biobutanol as a promising liquid fuel for the future—recent updates and perspectives. Fuel 253: 637–646.

Rajesh Banu, J., S. Kavitha, R. Yukesh Kannah, R.R. Bhosale and G. Kumar. 2020. Industrial wastewater to biohydrogen: Possibilities towards successful biorefinery route. Bioresour. Technol. 298: 122378.

Rampelotto, P.H. 2013. Extremophiles and Extreme Environments. Life 3: 482–485.

Rothschild, L.J. and R.L. Mancinelli. 2001. Life in extreme environments. Nature 409: 1092–1101.

Srivastava, N., M. Srivastava, P.K. Mishra, M.A. Kausar, M. Saeed, V.K. Gupta, R. Singh and P.W. Ramteke. 2020. Advances in nanomaterials induced biohydrogen production using waste biomass. Bioresour. Technol. 307: 123094.

Su, C.-H., H.C. Nguyen, T.L. Bui and D.-L. Huang. 2019. Enzyme-assisted extraction of insect fat for biodiesel production. J. Clean. Prod. 223: 436–444.

Subhedar, P.B. and P.R. Gogate. 2013. Intensification of enzymatic hydrolysis of lignocellulose using ultrasound for efficient bioethanol production: A review. Ind. Eng. Chem. Res. 52: 11816–11828.

Sun, S. and K. Li. 2020. Biodiesel production from phoenix tree seed oil catalyzed by liquid lipozyme TL100L. Renew. Energy 151: 152–160.

Teixeira, D.A., C.R. da Motta, C.M.S. Ribeiro and A.M. de Castro. 2017. A rapid enzyme-catalyzed pretreatment of the acidic oil of macauba (*Acrocomia aculeata*) for chemoenzymatic biodiesel production. Process Biochem. 53: 188–193.

Tsai, T.-Y., Y.-C. Lo, C.-D. Dong, D. Nagarajan, J.-S. Chang and D.-J. Lee. 2020. Biobutanol production from lignocellulosic biomass using immobilized *Clostridium acetobutylicum*. Appl. Energy 277: 115531.

Tucker, Y.T. and T. Mroz. 2018. Microbes in Marcellus Shale: Extremophiles living more than two kilometers inside the Earth? 2018. Fuel 234: 1205–1211.

van Seters, J.R., J.P.J. Sijbers, M. Denis and J. Tramper. 2011. Build your own second-generation bioethanol plant in the classroom! J. Chem. Educ. 88: 195–196.

Wang, Y., S.-H. Ho, H.-W. Yen, D. Nagarajan, N.-Q. Ren, S. Li, Z. Hu, D.-J. Lee, A. Kondo and J.-S. Chang. 2017. Current advances on fermentative biobutanol production using third generation feedstock. Biotechnol. Adv., Metabolic Engineering Frontiers Emerging with Advanced Tools and Methodologies 35: 1049–1059.

Yadav, A.N., P. Verma, M. Kumar, K.K. Pal, R. Dey, A. Gupta, J.C. Padaria, G.T. Gujar, S. Kumar, A. Suman, R. Prasanna and A.K. Saxena. 2015. Diversity and phylogenetic profiling of niche-specific Bacilli from extreme environments of India. Ann. Microbiol. 65: 611–629.

Yan, J., Y. Kuang, X. Gui, X. Han and Y. Yan. 2019. Engineering a malic enzyme to enhance lipid accumulation in Chlorella protothecoides and direct production of biodiesel from the microalgal biomass. Biomass Bioenergy 122: 298–304.

Yeong, T.K., K. Jiao, X. Zeng, L. Lin, S. Pan and M.K. Danquah. 2018. Microalgae for biobutanol production—Technology evaluation and value proposition. Algal Res. 31: 367–376.

Yoshioka, K., M. Daidai, Y. Matsumoto, R. Mizuno, Y. Katsura, T. Hakogi, H. Yanase and T. Watanabe. 2018. Self-Sufficient Bioethanol Production System Using a Lignin-Derived Adsorbent of Fermentation Inhibitors. ACS Sustain. Chem. Eng. 6: 3070–3078.

13

Bioprocessing of Agricultural Residues for Biofuel Production
Current Scenario

Richa Arora,[1] *Sonali Thakur,*[2] *Sweety Kaur*[3] and *Sachin Kumar*[4,]*

1. Introduction

The ample upswing in energy demand, visualized lift in energy utilization, diminution of fossil fuels reserves, and greenhouse gas emissions have served as the lashing powers to quest for new and renewable sources of energy (Ramadoss and Muthukumar 2015, Sharma et al. 2017a). With the ascent in total public and financial development, the energy utilization is anticipated to increase by 1.1% for each annum in 2030 (Nanda et al. 2015). Numerous lawmaking strategies have been premeditated to endorse the deployment of renewable energy resources, considering their impact on the environment (Behera et al. 2016). According to Guo et al. (2015), various biofuels including biogas, biohydrogen, biobutanol, bioethanol and biodiesel will contribute to nearly 30% of the requirement of global energy requirement by 2050.

The spurt for utilization of green technologies globally resulted in the development of new-fangled model, biorefining of biomass, which represents an integrated facility for conversion of biomass to yield biofuel and other value-added chemicals in lignocellulosic biorefineries (LCBR) (Arora et al. 2017, Duval and Lawoko 2014). Contemporary raw material for biofuels mainly centers on the utilization of lignocellulosic biomass (LCB) due to its renewability, cornucopia and overpowering of the competition between the crops for food and fuel. Various lignocellulosic feedstocks, e.g., agricultural wastes, forestry wastes and municipal wastes may be exploited for the production of biofuels. The recalcitrance of LCB could be reduced by employing various physicochemical pretreatment strategies which loosen up the structure and aids in enzymatic digestion (Behera et al. 2014a,b). Several researchers are attempting to integrate the key steps (pretreatment, enzymatic hydrolysis and (co)-fermentation) for biochemical conversion of LCB to biofuels (Arora et al. 2018, Sharma et al. 2016a,b). Although some progress has been made for the conversion of cellulosic fraction, the hemicellulosic, lignin and silica fraction still needs to be exploited and valorized to value-added products (Clauser et al. 2015).

---

[1] Department of Microbiology, Punjab Agricultural University, Ludhiana-141004, India.
[2] School of Bioengineering and Biosciences, Lovely Professional University, Phagwara, Punjab, India.
[3] Nestle India Limited, Rajarhat, Kolkata-700156, India.
[4] Biochemical Conversion Division, SardarSwaran Singh National Institute of Bio-Energy, Kapurthala, Punjab, India.
* Corresponding author: sachin.biotech@gmail.com, sachin.kumar20@gov.in

Apart from the raw material, another important modification is the exploitation of extremophilic microorganisms for both enzymatic hydrolysis and (co)-fermentations (Haki and Rakshit 2003). The pluses in the use of extreme microorganisms are stability of enzymes under extreme conditions with very high specific activity and high mass transfer rates (Elleuche et al. 2014). Moreover, the overall cost of the process w.r.t. downstream processing is greatly reduced at high temperature, which further aids in reducing the toxic levels of the solvents produced after fermentation (Arora et al. 2015a,b,c). Also, due to rapid diffusion of the nutrients at elevated temperatures, the growth rate and productivity of fermenting cells is also enhanced (Sharma et al. 2017b). Many researchers have reported the utilization of thermophilic/thermotolerant microorganisms responsible for co-fermentation of six carbon (C6) and five carbon (C5) sugars, which leads to economical production of biofuels (Arora et al. 2015a). Besides this, to enhance the efficiency of a bioprocess in LCBR, the by-products must be valorized to value-added compounds (Cotana et al. 2014).

The present chapter deals with the latest advancements and current scenario mainly focusing on the major challenges for biochemical conversion of agricultural waste to biofuels. Besides this, it also deals with the commercial facets and valorization of by-products of hemicellulose, lignin and silica. Moreover, stress is laid upon the use of extremophilic microorganisms as the source of enzymes for saccharification and (co)-fermentations for the development of economical bioprocess.

2. Agricultural Residues and Biofuel Production: An Overview

The fundamental paces for biochemical conversion of agricultural wastes to biofuel comprises: (i) pretreatment to remove the recalcitrance from the matrix; (2) enzymatic hydrolysis for the extraction of fermentable sugars; and (3) (co)-fermentation of C6 and C5 fractions (Fig. 1). A variety of agricultural wastes have been reported in literature for the production of biofuels, viz. bioethanol, biobutanol and biogas as shown in Table 1.

Pretreatment aims to enhance the porosity of the biomass, solubilize lignin, and increase the enzymatic digestibility of cellulosic and hemicellulosic fraction of agricultural wastes (Behera et al. 2014a). Yang and Wyman (2008) reported the promising potential of chemical pretreatment w.r.t. separation of solid-liquid fraction and porosity of the biomass. On the other hand, the foremost challenge lies in its unsafe disposal in the environment and chemical recycling (Brodeur et al. 2011, Harmsen et al. 2010). Apart from size decrease, a few strategies result in different impacts depending upon the pretreatment like the impact of maceration is more because of shearing than cutting of the fibers, and the fundamental impact of ultrasonic pretreatment is particle mass decrease at low frequency sound waves. High-recurrence sound waves likewise cause the development of radicals, for example, OH^*, HO_2^*, H^*, which brings about oxidation of dense substances (Ariunbaatar et al. 2014).

Following the pretreatment step, next phase is saccharification of agricultural wastes to monomeric forms of sugars by using synergistic enzymatic cocktails (Behera et al. 2013). Even though extensive research is being carried out to hydrolyse the agricultural waste, the technology suffers from several limitations including affluent inducers, monitoring systems, culture conditions, and removal of undesired metabolites, etc., which make the overall saccharification very pricey at commercial level. In addition to this, another lacuna is with the use of conventional mesophilic microorganisms which results in lesser diffusion at lower temperatures and a constraint of higher enzyme loading and lengthy incubation times (Lambertz et al. 2014). In order to overcome these constraints, recent research is more focused on the engagement of high temperature tolerating enzymes to enhance the rate of reaction and better process economics (Arora et al. 2014a,b). Apart from this, acidophilic/alkalophilic thermozymes may be exploited so that the process of pretreatment and saccharification may be clubbed together in a single reactor.

Lastly, the fermentable sugars obtained after saccharification of agricultural waste are subjected to fermentation/digestion depending upon the biofuel desired to be produced. In this phase also, the conformist microorganisms are mesophilic by nature with very narrow range of optimum

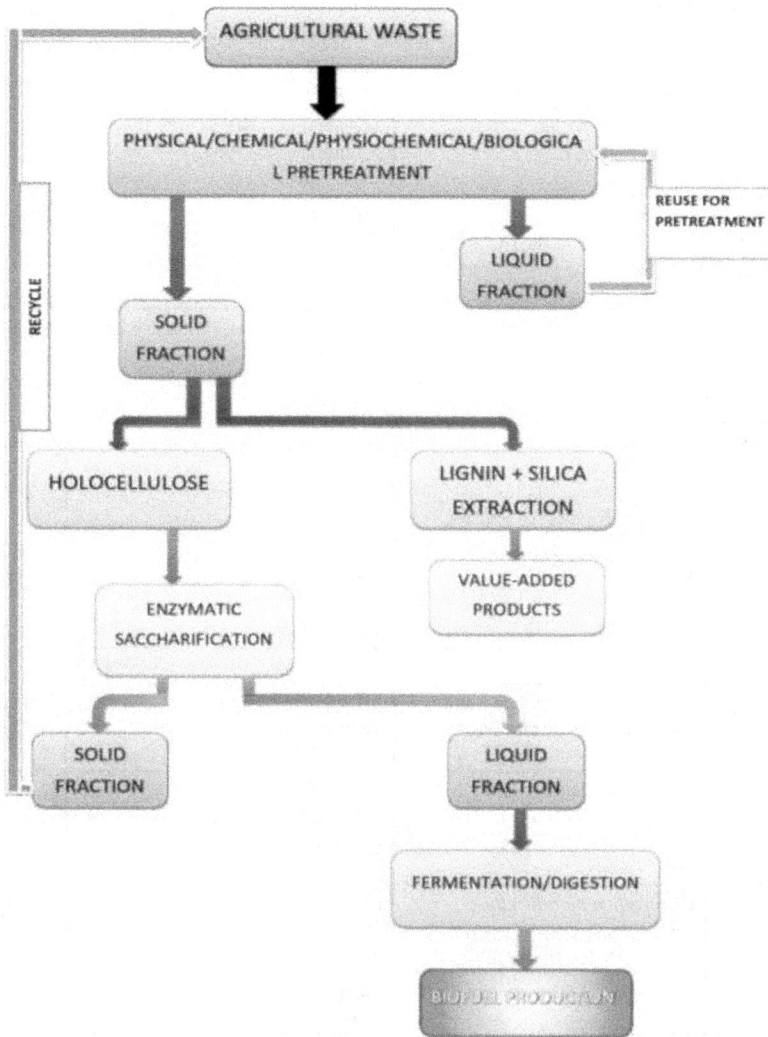

Fig. 1. Key steps for biofuel production from agricultural waste.

temperature with lesser growth rate. Besides this, they cannot ferment both C6 and C5 fractions of lignocellulosic biomass and are sensitive to higher concentrations of substrate/product. These challenges can be overcome by exploiting thermophilic or thermotolerant microorganisms with broad substrate range, higher yield and productivity with lesser cost during recovery (Arora et al. 2015a).

3. Biochemical Conversion of Agricultural Residues

3.1 Advances in Pretreatment

Bioprocessing of agricultural wastes can be made cost-effective and effectual by exploiting several pretreatment strategies (alone or in combination) due to direct correlation between separations of carbohydrate fraction from lignin moiety, which is generally solubilized (Narayanaswamy et al. 2013). Various factors which affect the type of pretreatment to be followed for a particular agricultural waste comprise of content of hemicellulosic and lignin fractions, crystallinity of waste agricultural biomass, accessible surface area, degree of polymerization (DP), etc.

Table 1. Various agricultural wastes used for the production of liquid biofuels.

Bioethanol			
Agricultural waste	**Microorganism involved**	**Fermentation conditions**	**References**
Taro (*Colocasia esculenta*) waste	*Kluyveromyces marxianus* K21 (BCRC 21363)	Temperature – 40°C Agitation – 150 rpm Time – 24 h–72 h	Wu et al. 2016
Banana lignocellulosic waste (rachis and pseudostem)	*Saccharomyces cerevisiae*	pH – 4.8 Temperature – 50°C Agitation – 150 rpm Time – 72 h	Guerrero et al. 2018
Lignocellulosic biomass – rice straw and wheat straw	*Kluyveromyces* sp.	pH – 4.8 Temperature – 42°C Agitation – 120 rpm Time – 72 h	Narra et al. 2015
Cashew apple bagasse	*Kluyveromyces marxianus* (ATCC 36907)	pH – 4.5-5.0 Temperature – 40°C Agitation – 150 rpm Time – 72 h	de Barros et al. 2017
Rice straw cellulignin	*Kluyveromyces marxianus* NRRL Y-6860	pH – 4.8 Temperature – 45°C Agitation – 100 rpm Time – 72 h	Castro and Roberto 2014
Biobutanol			
Birchwood	*Kluyvera* sp. Strain OM3 and *Clostridium* sp. Strain BOH3	pH – 8 Temperature – 60°C Time – 72 h	Xin and He 2013
Napiegrass	*Clostridium thermocellum* BCRC 14411	pH – 6.5 Temperature – 60°C Agitation –150 rpm Time – 240 h	Lin et al. 2011
Cellulose	*Clostridium thermocellum*	pH – 7.5 Temperature – 50°C Time – 24 h	Lin et al. 2015
Cellobiose	*Geobacillus thermoglucosidasius*	Temperature – 50°C Time – 48 h	Lin et al. 2014
Corn stover	*Thermoanaerobacterium thermosaccharolyticum* W16	pH – 7 Temperature – 60°C	Ren et al. 2010

Physical pretreatment, e.g., milling, grinding, chipping, irradiating, lessens the crystallinity of cellulose and enhances the porosity of agricultural waste. Although size reduction is an exorbitant and energy consuming step, it reduces the overall cost due to high mass and heat transfer rate (Karunanithy and Muthukumarappan 2010, Zhu et al. 2010).

Besides physical pretreatment, chemical pretreatment using acids, alkalis, salts, etc., is employed for delignification and separation of hemicellulose and cellulose. However, care should be taken to avoid the use of concentrated acids owing to the problem of corrosion of reactor and difficult downstream processing (Tabil et al. 2011). Apart from this, many researchers have started exploiting combination of several physical and chemical pretreatment techniques, e.g., wet oxidation, organosolv process, carbon-dioxide explosion, steam explosion, ammonia recycle percolation (ARP) and ammonia fiber explosion (AFEX), to achieve good results (Brodeur et al. 2011, Sathitsuksanoh et al. 2012). Moreover, the use of ionic liquids has gained much popularity in the recent years for pretreatment (Zhang et al. 2015).

Another important class of pretreatment technique, biological pretreatment, mainly involving the use of ligninolytic enzymes has resulted in higher specificity, higher product yield, and lesser energy requirement (Narayanaswamy et al. 2013). However, major techno-economical challenge with biological pretreatment lies in the requirement of long time period and maintenance of aseptic conditions.

Since pretreatment is the most cost and energy consuming step, the selection of a particular pretreatment strategy is dependent upon the location and composition of agricultural waste in use bearing in mind the cost of pretreatment reactor, chemicals/solvents used with ensuing neutralization method, power consumption and retention time (Littlewood et al. 2013, Mussatto et al. 2010). Table 2 shows the mechanism of action of several pretreatment techniques with equipments/chemicals required.

3.2 Advances in Enzymatic Hydrolysis

Since holocellulosic part of agricultural waste is composed of cellulose and hemicellulose, a synergistic cocktail of both cellulases and hemicellulases is required for efficient hydrolysis. On the basis of oxygen requirement by enzyme producing microorganisms, aerobic (*Actinomycetales*) and anaerobic (*Clostridiales*) cellulase producing microorganisms have different mechanism of action (Fig. 2). In case of aerobic cellulase producing microorganisms, extracellular enzymes are released which initiate the hydrolysis of cellulose fiber to form cellodextrin moieties and the latter are further oxidized to water and carbon-dioxide (Horn et al. 2012). On the other hand, anaerobic cellulase producing microorganisms make a physical contact with the cellulose fiber via giant enzyme-containing structure, protruding outside the cell, i.e., cellulosome which results in saccharification of cellulose to cellodextrins which undergo fermentation (Sadhu and Maiti 2013).

The futuristic approach for cellulose degradation is the use of natural/synthetic cellulosomes (Hyeon et al. 2013). They play a major role in more substrate uptake, specific and synergistic interaction of cellulosomal enzymes with the substrate (Chanal et al. 2011). Many researchers have investigated several cellulosomal complexes in *Clostridium josui, C. acetobutylicum, C. papyrosolvens, C. cellulovorans, C. clariflavum, C. cellulolyticum, Ruminococcusalbus, R. flavefaciens, Bacteroides cellulosolvens*, and *Acetivibrio cellulolyticus* (Cho et al. 2010, Hemme et al. 2010, Karpol et al. 2013, Smith and Bayer 2013, Ichikawa et al. 2014). Table 3 shows the construction of synthetic cellulosomes for hydrolysis of agricultural wastes.

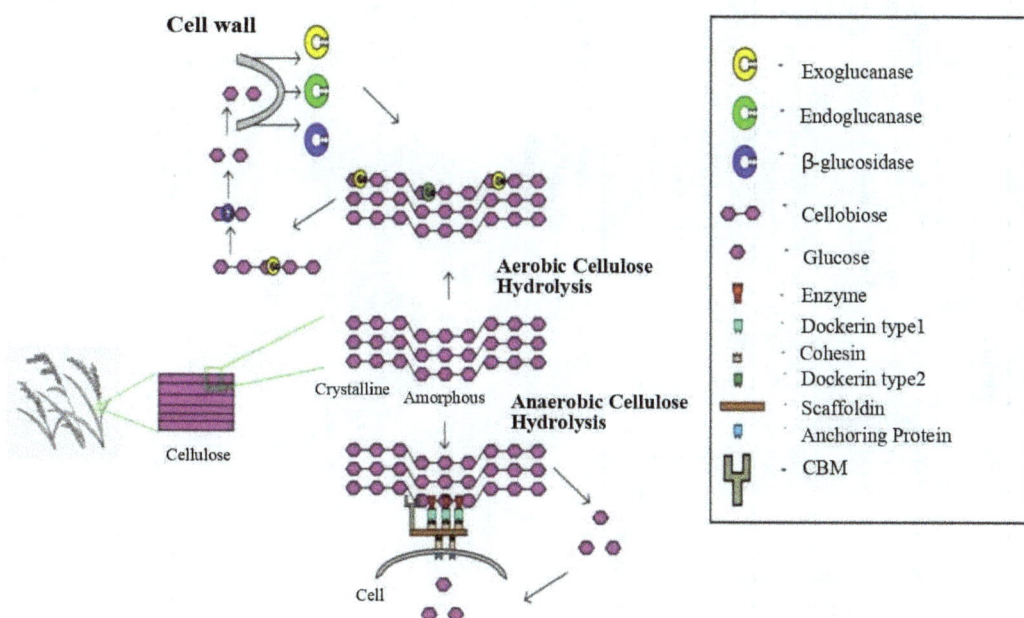

Fig. 2. Mechanism of action of aerobic and anaerobic cellulase producing microorganisms.

Table 2. Mechanism of action and equipments and/or chemicals required for various pretreatment technologies.

Pretreatment type	Mechanism of action	Equipment/chemicals	References
Physical			
Mechanical	Milling processes (like two roll milling, hammer milling, colloid milling and vibratory milling), grinding and chipping are used to increase surface area by reducing particle size and crystallinity of lignocellulosic biomass	Ball mill, centrifugal mill	(Maurya et al. 2015, Motte et al. 2015)
Chemical			
Acidic	Dilute acid pretreatment in hydrolysis of hemicellulose to its monomeric units, rendering the cellulose more available	Diluted sulfuric acid, diluted hydrochloric acid, diluted nitric acid	(Agbor et al. 2011, Loow et al. 2016)
Alkaline	When a lignocellulosic material is soaked in an alkaline solution, e.g., sodium, potassium or ammonium hydroxide, a swelling action of the pores occurs at elevated temperatures. The alkaline treatment breaks the bonds between lignin and the carbohydrate polymers and it can also partly solubilize lignin	Sodium hydroxide	(Galbe and Zacchi 2012, Taherdanak and Zilouei 2014)
Physico-chemical			
Steam explosion	Biomass is quickly heated by steam under high pressure for a specific time period and then pressure is rapidly released which causes the steam to swell inside lignocellulosic material, this results in breakdown of cell wall structure	Pilot-plant system is used, comprising of a reactor vessel and a flash tank with detachable bucket to gather pretreated portion	(Singh et al. 2015, Baêta et al. 2017)
Ammonia fiber explosion	At moderate temperature of 60–100°C and at high pressure, liquid ammonia is added to biomass for 30–60 min. This, followed by sudden pressure release, causes break in biomass fibers and cellulose decrystallization	Bench-top high pressure Parr reactor, liquid ammonia	(Maurya et al. 2015, Abdul et al. 2016)
Wet oxidation	Air and water are added to lignocellulosic biomass at temperature higher than 120° for 30 min, which causes partial degradation of cellulose, lignin is cleaved and oxidized and hemicellulose is converted to low molecular weight sugars	Hermetically closed stainless-steal reactor equipped with a mixer	(Maurya et al. 2015, Biswas et al. 2015)
Biological			
Enzymatic treatment	Fungi produces enzymes—cellulase, hemicellulase and ligninase which degrades the main constituents of agricultural biomass, i.e., cellulose, hemicellulose and lignin, respectively	White Rot fungus inoculum	(Saini et al. 2015, Baral and Shah 2017)

Table 3. Construction of synthetic cellulosomes for enhanced saccharification.

New component/ Cellulosome	Scaffoldin	Enzymes/(source)	Mode of construction	Substrate	Saccharification rate	References
Chimeric CBM3-CgIT	*C. thermocellum* scaffoldin	β-glucosidase (CgIT)/ (*T. brockii*)	Fusion of CBM3 from scaffolding protein Cip A into N-terminal region of CgIT	Rice straw	70%	Waeonukul et al. 2013
Nanocluster	CdSe-ZnS core-shell quantum dots QDs	Cel A/(*C. thermocellum*); Cel E/(*C. cellulolyticum*)	Metal affinity between core-shell QDs and polyhistidine tag	-	-	Tsai et al. 2013
B or DM-casein-EG (Cel5A) conjugate	Casein based scaffold	Microbial Transglutaminase (*Thermobifidafusca*)	Genetically engineered cellulase, EG (Cel5A) and (Cel6A) were conjugated on casein	Crystalline cellulose (Avicel)	80%, 100%	Budinova et al. 2017
Chimeric LPMOs	CohA – bearing scaffoldin	Lytic polysaccharide monooxygenases (LPMOs)	Cohesin - dockerin interaction	Avicel	1.7 fold increased	Arfi et al. 2014
Chimeric Cellulosome	Engineered Chimeric scaffoldin	*Clostridium cellulolyticum*, CelA, CelF	Cellulosomes of two clostridia, *Clostridium thermocellum* and *Clostridium cellulolyticum*, wherein cellulolytic enzymes are incorporated into the complexes by means of high-affinity species-specific cohesin-dockerin interactions	Crystalline cellulose	2-3 folds increase	Fierobe et al. 2001
Protein A-CBM-enzyme conjugates	*Staphylococcal* protein A	Enhanced Green Fluorescent Protein (EGFP) and Fusariumsolanipisi lipase cutinase	IgG-affibody interaction	-	-	Eklund et al. 2004
Rosettazyme	Group II Chaperonin (Rosettazomes)	*Clostridium thermocellum*, Cel9B, Cel9K, Cel9R, Cel48S	Scaffold proteins have cohesin modules that bind conserved dockerin modules on the enzymes	Avicel	2.3-2.4 fold increase	Mitsuzawa et al. 2009
Nanocluster	Streptavidin-immobilized inorganic nanoparticles	EglA (*Aspergillus niger*),	Catalytic (EglA) and CBD modules were clustered on streptavidin and on CdSe nanoparticles via biotin–avidin interactions	Microcrystalline cellulose	7.2 fold increase	Kim et al. 2011
Nanocluster	Polystyrene nanosphere	Cellulase (*Trichodermaviride*)	EDC and Sulfo-NHS coupling chemistry	Carboxymethyl cellulose (CMC), microcrystalline cellulose	50%	Blanchette et al. 2012
Cellulase-ankyrin fusions	Ankyrin scaffold	CelA (*Clostridium thermocellum*), Cel12A (*Thermotogamaritima*)	Protein fusion	CMC	6.2 fold increase	Cunha et al. 2013
DNA-cellulase covalent conjugates	Double strand DNA	Cel5A (*Thermobifidafusca*)	MTG mediated cross-linking	Crystalline cellulose (Avicel)	5.7 fold increase	Mori et al. 2013
DNA-cellulase noncovalent conjugates	Double strand DNA	CelA(*Clostridium thermocellum*)	Zinc-finger protein guided assembly	Cellulose	3 fold increase	Sun et al. 2014

The other important carbohydrate fraction of agricultural waste is hemicellulose which can be hydrolysed by hemicellulase producing microorganisms (*Thermoanaerobacteriales*). Although many studies have been reported for the isolation and characterization of hemicellulases, hemicellulolytic systems are lesser exploited and understood. Therefore, more studies need to be done to investigate the regulation mechanism of hemicellulases.

3.3 Advances in (co)-fermentation

To subsume the technical hurdles, current research is more focused on the use of extremophilic microorganisms. Several thermotolerant and thermophilic microorganisms have been reported in the literature including *Thermoanaerobacter ethanolicus, Aeromonas hydrophila, Thermoanaerobacterium saccharolyticum, Clostridium acetobutylicum, C. thermosaccharolyticum,* and *Kluyveromyces marxianus* (Chandel et al. 2011, Shaw et al. 2008). Table 4 shows the advantages and limitations of industrially important fermenting microorganisms at high temperature, reported in literature.

Apart from the use of thermotolerant or thermophilic microorganisms, current research is more focused on integrating different steps for the development of cost-effective bioprocess (Dien et al. 2006). Various researchers have exploited ethanol fermentation in different configurations including separate hydrolysis and fermentation (SHF) and simultaneous saccharification and (co)-fermentation (SS(C)F) and also consolidated bioprocessing (CBP) to some extent (Arora et al. 2015a, Paulova et al. 2015).

SHF is performed in separate reactors, thereby allowing independent optimal conditions for saccharification and fermentation but, as the saccharification progresses, it suffers from feedback inhibition (Tengborg et al. 2001). On the other hand, SSF configuration allows hydrolysis and fermentation in the same reactor, compromising between the optimum conditions of the enzymatic reaction and the fermenting organism, thereby negating the effect of feedback inhibition of the enzymes (Wingren et al. 2003). However, a major drawback of the SSF is that the fermenting organism cannot be recirculated since it is mixed with the biomass (Taherzadeh and Karimi 2007, Olofsson et al. 2008, Tomás- Pejó et al. 2008). SSCF is an excellent processing technology to produce bioethanol from lignocellulose since it could improve bioethanol production by alleviating the end-product inhibition (Saha et al. 2011) as well as ameliorating the consumption of pentose (Gubicza et al. 2016).

The semi-simultaneous saccharification and fermentation (SSSF) has been recently studied which involves pure enzymatic pre-hydrolysis and SSF (Cotana et al. 2015, Goncalves et al. 2014)

Table 4. Advantages and limitations of industrially important fermenting microorganisms at high temperature.

Microorganism	Temperature	Limitations	Advantages	References
Kluyveromyces marxianus	40–42°C	Intolerance to temperature fluctuation; ethanol intolerance; cannot hydrolyze inulin-type oligosaccharides with high DP	High temperature range for fermentation activity	Murata et al. 2015, Charoensopharat et al. 2015
Blastobotrys adeninivorans	50°C	Difficulty in recycling	Ferment sugars in a single reactor	Antil et al. 2015
Saccharomyces cerevisiae	40°C	Unable to utilize xylose; intolerance to high concentration of organic material; produces small amounts of by-products; fermentation temperature lower than the optimum growth temperature (42°C)	Can utilize various substrates	Huang et al. 2015
Pichia kudriavzevii	45°C	Cannot metabolize complex sugars	Ethanol production from any lignocellulosic biomass	Oberoi et al. 2012

for the release of sugars from the biomasses and providing these sugars for yeasts in the beginning of SSF. This strategy also aims to reduce viscosity of the solid–liquid mixture prior to the addition of the microorganism (Ohgren et al. 2007).

Another process is simultaneous saccharification, filtration and fermentation (SSFF) which overcomes the problem of enzyme inhibition. In addition, both the enzyme cocktail and the fermenting organism in SSFF can be used at their different optimal conditions. Furthermore, it would be possible to reuse the fermenting organism several times (Ishola et al. 2013).

CBP is the finest configuration which integrates all the steps of bioethanol production, involving an engineered microorganism capable of saccharification and co-fermentation with high ethanol yield (Hasunuma and Kondo 2012). Nevertheless, this configuration is not completely developed at commercial level owing to long fermentation period, poor ethanol yield, low tolerance to ethanol and formation of fermentation inhibitors (Zaldivar et al. 2001).

However, despite the process design employed, one of the main drawbacks of these technologies is the low ethanol concentration produced. Many technical problems including high viscosity, low amount of free water due to its absorption into the biomass, high content of inhibitors, problems with mixing, nutrient levels, heat, mass transport, type of enzymes used, etc., contribute to this (Modenbach and Nokes 2012). Therefore, to find an optimum combination of all variables for a particular feedstock is challenging.

4. Major Stumbling Blocks

The agricultural waste based biofuel production suffers from various challenges. A well-organized system is required for the collection and storage of biomass for successfully running a commercial plant. Besides this, a sturdy governmental policy is required for the segregation of different agricultural wastes for biofuel production. Moreover, the technological barriers need to be filled up for the development of efficient bioprocess in a biorefinery. The price gaps amongst bio and traditional jet fuels, supportability and money related issues in commercialization are real worries in the improvement of substitute jet fuels (Nair and Paulose 2014). Monetary and ecological issues including land-water use, ozone harming substance and particulate emissions, and fuel-food rivalry are the main barriers come across. Maintainability is a major obstruction to beat which depends primarily on the accessibility of feed stock and the fuel production method that bring out social, economic and ecological effects (Hari et al. 2015).

4.1 Construction of Efficient Cellulosome Producing Microorganisms

Cellulosome generating microorganisms arise from various anaerobic habitats which can be thermophilic or mesophilic and belong to several genera and species. The ideal cellulosome constructing bacterium is the thermophilic *Clostridium thermocellum*, which has the most-considered, best-portrayed cellulosome system. *C. thermocellum* has 8 scaffoldin genes and its very organized cellulosome complex comprises up to 63 enzymes. The bacterium *Acetivibrio cellulolyticus* (mesophilic) contains a considerably more entangled cellulosome framework that has 16 scaffoldins (Artzi et al. 2014). From the rumen of cow, *Ruminococcus flavefaciens* was isolated and produces a cellulosomal framework that has a huge number of parts, including 222 dockerin-containing proteins that are partitioned into six distinctive subtypes (Rincon et al. 2010). As of late, the *Ruminococcus champanellensis* (cellulolytic bacterium), which was found in the human gut, was additionally seen to produce a developed cellulosome framework (Cann et al. 2016, Moraïs et al. 2016). The hydrolysis of agricultural waste using native cellulosome-producing microorganisms is not adequate to support an industrial set-up. The proficient hydrolysis is dependent upon the specific set of enzymes for a particular substrate (Stern et al. 2014). Moreover, the anaerobic nature of cellulosome producing microorganisms further increases the cost of the overall process (Lambertz et al. 2014). Therefore, researchers are making efforts to construct the cellulosomes artificially with better control over the regulatory mechanisms of cellulosomal genes

(Nataf et al. 2010). Hence, several studies are being carried out to improve this technology and subsequently use in the biorefineries for biofuel production (Fontes and Gilbert 2010).

4.2 Genetic Engineering

Numerous strategies are being followed to alter the molecular set-up of the microorganisms to improve the yield and productivity of biofuel production. Attempts have been made to modify the genes for pyruvate assimilation (pyruvate decarboxylase (PD), pyruvate dehydrogenase (PDH), pyruvate ferridoxin oxidoreductase (PFO)) and improve the pentose metabolism by inserting C5 assimilation genes (Oloson et al. 2015, Demeke et al. 2013). But the major challenge with the microorganisms is instability of genetically-engineered microorganisms after few generations. Product variety, yield, concentration, and output may be developed in genetically-engineered microorganisms for improved fuel production. Moreover, the cost of fuel generation might be lessened by engineering microorganisms to utilize modest substrates (Majidian et al. 2017). A recombinant strain TMY-HFPX of *Zymomonas mobilis* with numerous enhanced qualities was settled and appeared to accomplish a proficient ethanol concentration. *E. coli* KO11 carried genes from *Z. mobilis* for coding pyruvate decarboxylase and alcohol dehydrogenase (Wang et al. 2016). *Clostridium acetobutylicum* M5 was engineered to attain improved butanol production (Liao et al. 2015). Mesophilic bacteria, for example *Clostridium butyricum*, *Enterobacter aerogenes*, and *E. coli*, have attracted specific consideration for biohydrogen production using genetic engineering (Khanna et al. 2011).

4.3 Co-fermentation Hurdles

Another important challenge in the development of efficient bioprocess is the co-fermentation of C6 and C5 fractions at industrial level (Kazi et al. 2010). Among different strategies being followed by several researchers, one of the efficient approach is evolutionary engineering in which the pentose fermenting microorganism is adapted to grow in increasing concentrations of xylose (Sharma et al. 2016a,b). Apart from this, microorganisms may be subjected to mutation and genetic engineering before adaptation for increasing the growth rate (Liu and Hu 2010). Even though momentous work has been done for biochemical conversion of pentose sugars to bioethanol, commercialization is still a challenge. Numerous endeavors have been dedicated to the introduction and advancement of heterologous metabolic pathways like the oxidoreductase-based pathway with xylose reductase (XR) and xylitol dehydrogenase (XD) and the isomerase-based pathway with xylose isomerase (XI) used in *S. cerevisiae* (Ko et al. 2016), though xylose fermentation to ethanol by engineered *S. cerevisiae* is slow and usually results in low ethanol yields than glucose fermentation. Advancements have been made by overexpression of endogenous xylulokinase (XK) and by regulating the cofactor specificity of XR towards NADH (Nielsen et al. 2016).

4.4 Low Tolerance to Products and Pretreatment Inhibitors

Additional challenge in the industrial set up of biofuel production is the sensitivity of the microorganisms to higher concentrations of the fermentation products and pretreatment inhibitors produced. Several efforts are being made to adapt the microorganisms to furfurals and hydroxymethyl furfurals (HMFs) produced in the bioprocess (Sudhakar et al. 2016). Sugar degrading products, for example the common aliphatic carboxylic acids, formic acid, and levulinic acid, and the furan aldehydes furfural and HMF, show relatively low toxicity, but can exist in high concentrations reliant on the pretreatment conditions and the feedstock (Jonsson and Martin 2016). The catalytic activity of cellulolytic enzymes can be hindered by non-profitable binding to components of the solid portion, like lignins and remaining hemicelluloses (Pareek et al. 2013). Accumulations of usual varieties of feedstock that are of enthusiasm for bioconversion utilizing a sugar platform idea, for example, *Populus trichocarpa* trees, can be selected to recognize assortments that display lower recalcitrance (Studer et al. 2011). Purification or conditioning of lignocellulosic hydrolysates (LH) and slurries are a standout amongst the most ground-breaking approaches to balance inhibitory issues (Jönsson et al. 2013). Screening of collections of microorganisms assembled from normal or modern conditions can be utilized to distinguish strains with high protection from inhibitors.

5. Feasibility at Commercial Level

The establishment of a bioprocess at industrial scale is dependent upon two important parameters, i.e., eco-friendly process and cost-effectiveness (Margeot et al. 2009, Behera et al. 2016). However, biofuel production from agricultural waste is quite challenging owing to the costly raw material and several technological obstructions discussed above (Balat 2011). Besides this, the economic analysis of the bioprocess developed is very cumbrous as several pretreatment techniques and conversion processes are involved. The overall bioprocess economics can be evaluated by total annual expenditure by the amount of biofuel produced (Hamelinck et al. 2005). Goh et al. (2010) reported that pretreatment is the most expensive step in terms of operational and capital costs. The choice of a pretreatment strategy is entirely hooked upon the site and accessibility (Mussatto et al. 2010). Moreover, by integrating several steps of the process, exploitation of high yielding enzymes and fermenting microorganisms, the overall process can be made economical (Dien et al. 2006). With the usage of thermophilic or thermotolerant microorganisms, metabolic heat is released, both during growth and fermentation, which may be exploited to maintain the reactor temperature if it is completely insulated, thereby aiding in overall process economics (Kumar et al. 2013).

6. Concluding Remarks

Agricultural wastes extend massive potential for biofuel production, particularly the liquid biofuels through biochemical conversion processes. Nevertheless, several technological gaps must be jam-packed via various approaches at the molecular level. Enhanced expression systems for enzymatic hydrolysis and improved strains to undergo co-fermentation are in progress. Moreover, metabolic engineering techniques also help in the development of stalwart microbial strains. Further, the approaches of flux balance analysis help in improved understanding of the physiological systems for enhancing the biofuel yields and productivities from various agricultural wastes.

Acknowledgement

One of the authors (Sonali) is very thankful to Lovely Professional University, Phagwara for providing Masters Registration (Reg. No. 11400774) and financial support to carry out the research activities.

Abbreviations

C5: five carbon; C6: six carbon; DP: degree of polymerization; CBP: consolidated bioprocessing; LCB: lignocellulosic biomass; LCBR: lignocellulosic refineries; SHF: separate hydrolysis and fermentation; SSCF: Simultaneous saccharification and co-fermentation; SSF: simultaneous saccharification and fermentation; SSFF: simultaneous saccharification filtration and fermentation; SSSF: semi-simultaneous saccharification and fermentation; XD: xylitol dehydrogenase; XK: xylulokinase; XR: xylose reductase; XI: Xylose isomerase; AFEX: ammonia fiber explosion; ARP: ammonia recycle percolation; HMF: hydroxymethyl furfural; PD: pyruvate decarboxylase; PDH: pyruvate dehydrogenase; PFO: pyruvate ferredoxin oxidoreductase; LH: lignocellulosic hydrolysate; NADH: Nicotinamide adenine dinucleotide hydrogen

References

Abdul, P.M., J.M. Jahim, S. Harun, M. Markom, N.A. Lutpi, O. Hassan, V. Balan, B.E. Dale and M.T.M. Nor. 2016. Effects of changes in chemical and structural characteristic of ammonia fibre expansion (AFEX) pretreated oil palm empty fruit bunch fibre on enzymatic saccharification and fermentability for biohydrogen. Bioresour. Technol. 211: 200–208.

Agbor, V.B., N. Cicek, R. Sparling, A. Berlin and D.B. Levin. 2011. Biomass pretreatment: Fundamentals toward application. Biotechnol. Adv. 29(6): 675–685.

Antil, P.S., R. Gupta and R.C. Kuhad. 2015. Simultaneous saccharification and fermentation of pretreated sugarcane bagasse to ethanol using a new thermotolerant yeast. Ann. Microbiol. 65(1): 423–429.

Arfi, Y., M. Shamshoum, I. Rogachev, Y. Peleg and E.A. Bayer. 2014. Integration of bacterial lytic polysaccharide monooxygenases into designer cellulosomes promotes enhanced cellulose degradation. Proc. Natl. Acad. Sci. USA 111(25): 9109–9114.

Ariunbaatar, J., A. Panico, G. Esposito, F. Pirozzi and P.N. Lens. 2014. Pretreatment methods to enhance anaerobic digestion of organic solid waste. Appl. Energy 123: 143–156.

Arora, R., S. Behera and S. Kumar. 2014a. Comparative study of fermentation efficiency for bio-ethanol production by isolates. pp. 149–155. *In*: Kumar, S., A.K. Sarma, S.K. Tyagi and Y.K. Yadav (eds.). Recent Advances in Bioenergy Research, SSS-NIRE, Kapurthala, November 22–24, 2013.

Arora, R., S. Behera, N.K. Sharma, R. Singh, Y.K. Yadav and S. Kumar. 2014b. Biochemical conversion of rice straw (*Oryza sativa* L.) to bioethanol using thermotolerant isolate *K. marxianus* NIRE-K3. Proceedings of Exploring & Basic sciences for Next Generation Frontiers, Elsevier held on November 14-15, 2014 at LPU, Punjab, pp. 143–146.

Arora, R., S. Behera and S. Kumar. 2015a. Bioprospecting thermophilic/thermotolerant microbes for production of lignocellulosic ethanol: A future perspective. Ren. Sustain. Energy Rev. 51: 699–717.

Arora, R., S. Behera, N.K. Sharma and S. Kumar. 2017. Augmentation of ethanol production through statistically designed growth and fermentation medium using novel thermotolerant yeast isolates. Renew. Energy 109: 406–421.

Arora, R., S. Behera and S. Kumar. 2015b. Bioprospecting thermostable cellulosomes for efficient biofuel production from lignocellulosic biomass. Bioresour. Bioprocess. 2: 38. doi: 10.1186/s40643-015-0066-4.

Arora, R., S. Behera, N.K. Sharma and S. Kumar. 2015c. A new search for thermotolerant yeasts, its characterization and optimization using response surface methodology for ethanol production. Front. Microbiol. 6: 889. doi: 10.3389/fmicb.2015.00889.

Arora, R., N.K. Sharma and S. Kumar. 2018. Valorization of by-products following the biorefinery concept: Commercial aspects of by-products of lignocellulosic biomass. pp. 163–178. *In*: Chandel, A.K. and M.H.L. Silveira (eds.). Advances in Sugarcane Biorefinery Technologies, Commercialization, Policy Issues and Paradigm Shift for Bioethanol and By-Products. Elsevier Sci.

Artzi, L., B. Dassa, I. Borovok, M. Shamshoum, R. Lamed and E.A. Bayer. 2014. Cellulosomics of the cellulolytic thermophile *Clostridium clariflavum*. Biotechnol. Biofuels 7(1): 100.

Baêta, B.E.L., P.H. de Miranda Cordeiro, F. Passos, L.V.A. Gurgel, S.F. de Aquino and F. Fdz-Polanco. 2017. Steam explosion pretreatment improved the biomethanization of coffee husks. Bioresour. Technol. 245: 66–72.

Balat, M. 2011. Production of bioethanol from lignocellulosic materials via the biochemical pathway: A review. Energy Convers. Manage. 52: 858–875.

Baral, N.R. and A. Shah. 2017. Comparative techno-economic analysis of steam explosion, dilute sulfuric acid, ammonia fiber explosion and biological pretreatments of corn stover. Bioresour. Technol. 232: 331–343.

Behera, S., R. Arora and S. Kumar. 2013. Bioprospecting the cellulases and *xylanases thermozymes* for the production of biofuels. 2013 AICHE annual meeting, November 3–8, Hilton Sanfrancisco Union Square, San Francisco, CA. (http://www3.aiche.org/Proceedings/content/Annual-013/extendedabstracts/332768.pdf).

Behera, S., R. Arora, N. Nandhagopal and S. Kumar. 2014a. Importance of chemical pretreatment for bioconversion of lignocellulosic biomass. Ren. Sustain. Energy Rev. 36: 91–106.

Behera, S., R. Arora, N.K. Sharma and S. Kumar. 2014b. Fermentation of glucose and xylose sugar for the production of ethanol and xylitol by the newly isolated NIRE-GX1 yeast. pp. 175–182. *In*: Kumar, S., A.K. Sarma, S.K. Tyagi, Y.K. Yadav (eds.). Proceedings of the 3rd National conference on Recent Advances in Bio-energy Research. Recent Advances in Bioenergy Research, SSS-NIRE, Kapurthala, November 22–24, 2013.

Behera, S., R. Singh, R. Arora, N.K. Sharma, M. Shukla and S. Kumar. 2015. Scope of algae as third generation biofuels. Front. Bioeng. Biotechnol. 2: 90. doi: 10.3389/fbioe.2014.00090.

Behera, S., N.K. Sharma, R. Arora and S. Kumar. 2016. Effect of evolutionary adaption on xylosidase activity in thermotolerant yeast isolates *Kluyveromyces marxianus* NIRE-K1 and NIRE-K3. Appl. Biochem. Biotechnol. 179: 1143–1154.

Biswas, R., H. Uellendahl and B.K. Ahring. 2015. Wet explosion: A universal and efficient pretreatment process for lignocellulosic biorefineries. Bioenergy Res. 8(3): 1101–1116.

Blanchette, C., C.I. Lacayo, N.O. Fischer, M. Hwang and M.P. Thelen. 2012. Enhanced cellulose degradation using cellulase-nanosphere complexes. PloS one 7(8): e42116.

Brodeur, G., E. Yau, K. Badal, J. Collier, K.B. Ramachandran and R. Ramakrishnan. 2011. Chemical and physiocochemical pretreatment of lignocellulosic biomass: A review. Enz. Res. 11: 1–17.

Budinova, G.A., Y. Mori, T. Tanaka and N. Kamiya. 2017. Casein-based scaffold for artificial cellulosome design. Process Biochem.

Cann, I., R.C. Bernardi and R.I. Mackie. 2016. Cellulose degradation in the human gut: *Ruminococcus champanellensis* expands the cellulosome paradigm. Environ. Microbiol. 18(2): 307–310.

Castro, R.C.A. and I.C. Roberto. 2014. Selection of a thermotolerant *Kluyveromyces marxianus* strain with potential application for cellulosic ethanol production by simultaneous saccharification and fermentation. Appl. Biochem. Biotechnol. 172(3): 1553–1564.

Chanal, A., F. Mingardon, M. Bauzan, C. Tardif and H.P. Fierobe. 2011. Scaffoldin modules can serve as "cargo" domains to promote the secretion of heterologous cellulosomal cellulases by *Clostridium acetobutylicum*. Appl. Environ. Microbiol. 77: 6277–6280

Chandel, A.K., G. Chandrasekhar, K. Radhika, R. Ravinder and P. Ravindra. 2011. Bioconversion of pentose sugars into ethanol: A review and future directions. Biotechnol. Mol. Biol. Rev. 6: 8–20.

Charoensopharat, K., P. Thanonkeo, S. Thanonkeo and M. Yamada. 2015. Ethanol production from Jerusalem artichoke tubers at high temperature by newly isolated thermotolerant inulin-utilizing yeast *Kluyveromyces marxianus* using consolidated bioprocessing. Antonie Leeuwenhoek 108(1): 173–190.

Cho, W., S.D. Jeon, H.J. Shim, R.H. Doi and S.O. Han. 2010. Cellulosomic profiling produced by *Clostridium cellulovorans* during growth on different carbon sources explored by the cohesin marker. J. Biotechnol. 145: 233–239.

Clauser, N.M., S. Gutierrez, M.C. Area, F.E. Felissia and M.E. Vallejos. 2015. Small-sized biorefineries as strategy to add value to sugarcane bagasse. Chem. Eng. Res. Des. 107: 137–146.

Cotana, F., G. Cavalaglio, A. Nicolini, M. Gelosia, V. Coccia, A. Perrozzi and L. Brinchi. 2014. Lignin as co-product of second generation bioethanol production from ligno-cellulosic biomass. Energy Procedia 45: 52–60.

Cotana, F., G. Cavalaglio, M. Gelosia, V. Coccia, A. Petrozzi, D. Ingles and E. Pompili. 2015. A comparison between SHF and SSSF processes from cardoon for ethanolproduction. Ind. Crop. Prod. 69: 424–432.

Cunha, E.S., C.L. Hatem and D. Barrick. 2013. Insertion of endocellulase catalytic domains into thermostable consensus ankyrin scaffolds: Effects on stability and cellulolytic activity. Appl. Environ. Microbiol. 79(21): 6684–6696.

de Barros, E.M., V.M. Carvalho, T.H.S. Rodrigues, M.V.P. Rocha and L.R.B. Gonçalves. 2017. Comparison of strategies for the simultaneous saccharification and fermentation of cashew apple bagasse using a thermotolerant *Kluyveromyces marxianus* to enhance cellulosic ethanol production. Chem. Eng. 307: 939–947.

Demeke, M.M., H. Dietz, Y. Li, M.R. Foulquie-Moreno, S. Mutturi, S. Deprez, T.D. Abt, B.M. Bonini, G. Liden, F. Dumortier, A. Verplaetse, E. Boles and J.M. Thevelein. 2013. Development of a D-xylose fermenting and inhibitor tolerant industrial *Saccharomyces cerevisiae* strain with high performance in lignocellulose hydrolysates using metabolic and evolutionary engineering. Biotechnol. Biofuels 6: 89.

Dien, B.S., H.J.G. Jung, K.P. Vogel, M.D. Casler, J.A.F.S. Lamb, L. Iten, R.B. Mitchell and G. Sarath. 2006. Chemical composition and response to diluteacid pretreatment and enzymatic saccharification of alfalfa, reed canarygrass and switchgrass. Biomass Bioenergy 30: 880–891.

Duval, A. and M. Lawoko. 2014. A review on lignin-based polymeric, micro- and nano-structured materials. React. Funct. Polym. 85: 78–96.

Eklund, M., K. Sandström, T.T. Teeri and P.Å. Nygren. 2004. Site-specific and reversible anchoring of active proteins onto cellulose using a cellulosome-like complex. J. Biotechnol. 109(3): 277–286.

Elleuche, S., C. Schröder, K. Sahm and G. Antranikian. 2014. Extremozymes—biocatalysts with unique properties from extremophilic microorganisms. Current Opinion in Biotechnology 29: 116–123.

Fierobe, H.P., A. Mechaly, C. Tardif, A. Belaich, R. Lamed, Y. Shoham, J.P. Belaich and E.A. Bayer. 2001. Design and production of active cellulosomechimeras Selective incorporation of dockerin-containing enzymes into defined functional complexes. Int. J. Biol. Chem. 276(24): 21257–21261.

Fontes, C. and H.J. Gilbert. 2010. Cellulosomes: Highly efficient nanomachines designed to deconstruct plant cell wall complex carbohydrates. Annu. Rev. Biochem. 79: 655–681.

Galbe, M. and G. Zacchi. 2012. Pretreatment: The key to efficient utilization of lignocellulosic materials. Biomass Bioenergy. 46: 70–78.

Goh, C.S., K.T. Tan, K.T. Lee and S. Bhatia. 2010. Bio-ethanol from lignocellulose: Status, perspectives and challenges in Malaysia. Bioresour. Technol. 101: 4834–4841.

Goncalves, F.A., H.A. Ruiz, C.C. Nogueira, E.S. Santos, J.A. Teixeira and G.R. Macedo. 2014. Comparison of delignified coconuts waste and cactus for fuel-ethanolproduction by the simultaneous and semi-simultaneous saccharification and fermentation strategies. Fuel 131: 66–76.

Gubicza, K., I.U. Nieves, W.J. Sagues, Z. Barta, K.T. Shanmugam and L.O. Ingram. 2016. Techno-economic analysis of ethanol production from sugarcane bagasse using a liquefaction plus simultaneous saccharification and co-fermentation process. Bioresour. Technol. 208: 42–48.

Guerrero, A.B., I. Ballesteros and M. Ballesteros. 2018. The potential of agricultural banana waste for bioethanol production. Fuel 213: 176–185.

Guo, M., W. Song and J. Buhain. 2015. Bioenergy and biofuels: History, status and perspective. Renew. Sust. Energy Rev. 42: 712–725.

Haki, G.D. and S.K. Rakshit. 2003. Developments in industrially important thermostable enzymes: A review. Bioresour. Technol. 89(1): 17–34.

Hamelinck, C.N., G. Van Hooijdonk and A.P.C. Faaij. 2005. Ethanol from lignocellulosic biomass: Techno-economic performance in short-, middle- and long-term. Biomass Bioenergy 28: 384–410.

Hari, T.K., Z. Yaakob and N.N. Binitha. 2015. Aviation biofuel from renewable resources: Routes, opportunities and challenges. Renew. Sust. Energ. Rev. 42: 1234–1244.

Harmsen, P., W. Huijgen, L. Bermudez-Lopez and R. Bakker. 2010. Literature review of physical and chemical pretreatment processes for lignocellulosic biomass. 1st ed. Wageningen: Food and Biobased Research, Netherlands, 857–570.

Hasunuma, T. and A. Kondo. 2012. Consolidated bioprocessing and simultaneous saccharification and fermentation of lignocellulose to ethanol with thermotolerant yeast strains. Process Biochem. 47: 1287–1294.

Hemme, C.L., H. Mouttaki, Y.J. Lee, G. Zhang, L. Goodwin, S. Lucas, A. Copeland, A. Lapidus, T. Glavina del Rio, H. Tice, E. Saunders, T. Brettin, J.C. Detter, C.S. Han, S. Pitlick, M.L. Land, L.J. Hauser, N. Kyrpides, N. Mikhailova, Z. He, L. Wu, J.D. Van Nostrand, B. Henrissat, Q. He, P.A. Lawson, R.S. Tanner, L.R. Lynd, J. Wiegel, M.W. Fields, A.P. Arkin, C.W. Schadt, B.S. Stevenson, M.J. Mclnerney, Y. Yang, H. Dong, D. Xing, N. Ren, A. Wang, R.L. Huhnke, J.R. Mielenz, S.Y. Ding, M.E. Himmel, S. Taghavi, D. van der Lelie, E.M. Rubin and J. Zhou. 2010. Sequencing of multiple clostridial genomes related to biomass conversion and biofuel production. J. Bacteriol. 192: 6494–6496.

Horn, J.H., G. Vaaje-Kolstad, B. Westereng and V.G.H. Eijsink. 2012. Novel enzymes for the degradation of cellulose. Biotechnol Biofuels 5: 1–12.

Huang, Y., X. Qin, X.M. Luo, Q. Nong, Q. Yang, Z. Zhang, Y. Gao, F. Lv, Y. Chen, Z. Yu and J.L. Liu. 2015. Efficient enzymatic hydrolysis and simultaneous saccharification and fermentation of sugarcane bagasse pulp for ethanol production by cellulase from *Penicilliumoxalicum* EU2106 and thermotolerant *Saccharomyces cerevisiae* ZM1-5. Biomass Bioenergy 77: 53–63.

Hyeon, J.E., S.D. Jeon and S.O. Han. 2013. Cellulosome-based, Clostridium-derived multi-functional enzyme complexes for advanced biotechnology tool development: Advances and applications. Biotechnol. Adv. 31: 936–944.

Ichikawa, S., S. Karita, M. Kondo and M. Goto. 2014. Cellulosomal carbohydrate-binding module from *Clostridium josuibinds* to crystalline and non-crystalline cellulose, and soluble polysaccharides. FEBS Lett. 588: 3886–3890.

Ishola, M.M., A. Jahandideh, B. Haidarian, T. Brandberg and M.J. Taherzadeh. 2013. Simultaneous saccharification, filtration and fermentation (SSFF): A novel method for bioethanol production from lignocellulosic biomass. Bioresour. Technol. 133: 68–73.

Jönsson, L.J., B. Alriksson and N.O. Nilvebrant. 2013. Bioconversion of lignocellulose: Inhibitors and detoxification. Biotechnol. Biofuels 6(1): 16.

Jönsson, L.J. and C. Martín. 2016. Pretreatment of lignocellulose: Formation of inhibitory by-products and strategies for minimizing their effects. Bioresour. Technol. 199: 103–112.

Karpol, A., M.K. Jobby, M. Slutzki, I. Noach, S. Chitayat, S.P. Smith and E.A. Bayer. 2013. Structural and functional characterization of a novel type-III dockerin from *Ruminococcus flavefaciens*. FEBS Lett. 587: 30–36.

Karunanithy, C. and K. Muthukumarappan. 2010. Influence of extruder temperature and screw speed on pretreatment of corn stover while varying enzymes and their ratios. Appl. Biochem. Biotechnol. 162: 264–279.

Kazi, F.K., J.A. Fortman, R.P. Anex, D.D. Hsu, A. Aden, A. Dutta and G. Kothandaraman. 2010. Techno-economic comparison of process technologies for biochemical ethanol production from corn stover. Fuel 89: S20–S28.

Khanna, N., S.M. Kotay, J.J. Gilbert and D. Das. 2011. Improvement of biohydrogen production by *Enterobacter cloacae* IIT-BT 08 under regulated pH. J. Biotechnol. 152(1-2): 9–15.

Kim, D.M., M. Umetsu, K. Takai, T. Matsuyama, N. Ishida, H. Takahashi, R. Asano and I. Kumagai. 2011. Enhancement of cellulolytic enzyme activity by clustering cellulose binding domains on nanoscaffolds. Small 7(5): 656–664.

Ko, J.K., Y. Um, H.M. Woo, K.H. Kim and S.M. Lee. 2016. Ethanol production from lignocellulosic hydrolysates using engineered *Saccharomyces cerevisiae* harboring xylose isomerase-based pathway. Bioresour. Technol. 209: 290–296.

Kumar, S., P. Dheeran, S.P. Singh, I.M. Mishra and D.K. Adhikari. 2013. Cooling system economy in ethanol production using thermotolerant yeast *Kluyveromyces* sp. IIPE453. Am. J. Microbiological Res. 1: 39–44.

Lambertz, C., M. Garvey, J. Klinger, D. Hessel, H. Klose, R. Fischer and U. Commandeur. 2014. Challenges and advances in the heterologous expression of cellulolytic enzymes: A review. Biotechnol. Biofuels 7: 135.

Liao, C., S.O. Seo, V. Celik, H. Liu, W. Kong, Y. Wang, H. Blaschek, Y.S. Jin and T. Lu. 2015. Integrated, systems metabolic picture of acetone-butanol-ethanol fermentation by *Clostridium acetobutylicum*. Proc. Natl. Acad. Sci. USA 112(27): 8505–8510.

Lin, C.W., C.H. Wu, D.T. Tran, M.C. Shih, W.H. Li and C.F. Wu. 2011. Mixed culture fermentation from lignocellulosic materials using thermophilic lignocellulose-degrading anaerobes. Process Biochem. 46(2): 489–493.

Lin, P.P., K.S. Rabe, J.L. Takasumi, M. Kadisch, F.H. Arnold and J.C. Liao. 2014. Isobutanol production at elevated temperatures in thermophilic *Geobacillus thermoglucosidasius*. Metab. Eng. 24: 1–8.

Lin, P.P., L. Mi, A.H. Morioka, K.M. Yoshino, S. Konishi, S.C. Xu, B.A. Papanek, L.A. Riley, A.M. Guss and J.C. Liao. 2015. Consolidated bioprocessing of cellulose to isobutanol using *Clostridium thermocellum*. Metab. Eng. 31: 44–52.

Littlewood, J., R.J. Murphy and L. Wang. 2013. Importance of policy support and feedstock prices on economic feasibility of bioethanol production from wheat straw in the UK. Renew. Sust. Energy Rev. 17: 291–300.

Liu, E. and Y. Hu. 2010. Construction of a xylose-fermenting *Saccharomyces cerevisiae* strain by combined approaches of genetic engineering, chemical mutagenesis and evolutionary adaptation. Biochem. Eng. J. 48: 204–210.

Loow, Y.L., T.Y. Wu, J.M. Jahim, A.W. Mohammad and W.H. Teoh. 2016. Typical conversion of lignocellulosic biomass into reducing sugars using dilute acid hydrolysis and alkaline pretreatment. Cellulose 23(3): 1491–1520.

Majidian, P., M. Tabatabaei, M. Zeinolabedini, M.P. Naghshbandi and Y. Chisti. 2017. Metabolic engineering of microorganisms for biofuel production. Renew. Sust. Energ. Rev.

Margeot, A., B. Hahn-Hagerdal, M. Edlund, R. Slade and F. Monot. 2009. New improvements for lignocellulosic ethanol. Curr. Opin. Biotechnol. 20: 372–380.

Maurya, D.P., A. Singla and S. Negi. 2015. An overview of key pretreatment processes for biological conversion of lignocellulosic biomass to bioethanol. 3 Biotech 5(5): 597–609.

Mitsuzawa, S., H. Kagawa, Y. Li, S.L. Chan, C.D. Paavola and J.D. Trent. 2009. The rosettazyme: A synthetic cellulosome. J. Biotechnol. 143(2): 139–144.

Modenbach, A.A. and S.E. Nokes. 2012. The use of high-solids loadings in biomass pretreatment—a review. Biotechnol. Bioeng. 109(6): 1430–1442.

Moraïs, S., Y.B. David, L. Bensoussan, S.H. Duncan, N.M. Koropatkin, E.C. Martens, H.J. Flint and E.A. Bayer. 2016. Enzymatic profiling of cellulosomal enzymes from the human gut bacterium, *Ruminococcus champanellensis*, reveals a fine-tuned system for cohesin-dockerin recognition. Environ. Microbiol. 18(2): 542–556.

Mori, Y., S. Ozasa, M. Kitaoka, S. Noda, T. Tanaka, H. Ichinose and N. Kamiya. 2013. Aligning an endoglucanase Cel5A from Thermobifidafusca on a DNA scaffold: Potent design of an artificial cellulosome. Chem. Comm. 49(62): 6971–6973.

Motte, J.C., C. Sambusiti, C. Dumas and A. Barakat. 2015. Combination of dry dark fermentation and mechanical pretreatment for lignocellulosic deconstruction: An innovative strategy for biofuels and volatile fatty acids recovery. Appl. Energy 147: 67–73.

Murata, Y., H. Danjarean, K. Fujimoto, A. Kosugi, T. Arai, W.A. Ibrahim, O. Suliman, R. Hashim and Y. Mori. 2015. Ethanol fermentation by the thermotolerant yeast, *Kluyveromyces marxianus* TISTR5925, of extracted sap from old oil palm trunk. AIMS Energy 3(2): 201–213.

Mussatto, S.I., G. Dragone, P.M.R. Guimarães, J.P.A. Silva, L.M. Carneiro, I.C. Roberto, T. Domingues and J.A. Teixeira. 2010. Technological trends, global market, and challenges of bio-ethanol production. Biotechnol. Adv. 28: 817–830.

Nair, S. and H. Paulose. 2014. Emergence of green business models: The case of algae biofuel for aviation. Energy Policy 65: 175–184.

Nanda, S., R. Azargohar, A.K. Dalai and J.A. Kozinski. 2015. An assessment on the sustainability of lignocellulosic biomass for biorefining. Renew. Sust. Energ. Rev. 50: 925–941.

Narayanaswamy, N., P. Dheeran, S. Verma and S. Kumar. 2013. Biological pretreatment of lignocellulosic biomass for enzymatic saccharification. pp. 3–34. *In*: Fang, Z. (ed.). Pretreatment Techniques for Biofuels and Biorefineries, Berlin, Heidelberg, Springer Verlag.

Narra, M., J.P. James and V. Balasubramanian. 2015. Simultaneous saccharification and fermentation of delignified lignocellulosic biomass at high solid loadings by a newly isolated thermotolerant *Kluyveromyces* sp. for ethanol production. Bioresour. Technol. 179: 331–338.

Nataf, Y., L. Bahari, H. Kahel-Raifer, I. Borovok, R. Lamed, E.A. Bayer, A.L. Sonenshein and Y. Shoham. 2010. *Clostridium thermocellum cellulosomal* genes are regulated by extracytoplasmic polysaccharides via alternative sigma factors. Proc. Natl. Acad. Sci. USA 107: 18646–18651.

Nielsen, F., G. Zacchi, M. Galbe and O. Wallberg. 2016. Prefermentation improves ethanol yield in separate hydrolysis and cofermentation of steam-pretreated wheat straw. Sustain. Chem. Process. 4(1): 10.

Oberoi, H.S., N. Babbar, S.K. Sandhu, S.S. Dhaliwal, U. Kaur, B.S. Chadha and V.K. Bhargav. 2012. Ethanol production from alkali-treated rice straw via simultaneous saccharification and fermentation using newly isolated thermotolerant *Pichia kudriavzevii* HOP-1. J. Ind. Microbiol. Biotechnol. 39(4): 557–566.

Ohgren, K., J. Vehmaanpera, M. Siika-Aho, M. Galbe, L. Viikari and G. Zacchi. 2007. High temperature enzymatic prehydrolysis prior to simultaneous saccharification and fermentation of steam pretreated corn stover for ethanol production. Enzyme MicrobTechnol. 40: 607–13.

Olofsson, K., M. Bertilsson and G. Liden. 2008. A short review on SSF—an interesting process option for ethanol production from lignocellulosic feedstocks. Biotechnol. Biofuels 1(1): 7.

Olson, D.G., R. Sparling and L.R. Lynd. 2015. Ethanol production by engineered thermophiles. Curr. Opin. Biotechnol. 33: 130–141.

Pareek, N., T. Gillgren and L.J. Jönsson. 2013. Adsorption of proteins involved in hydrolysis of lignocellulose on lignins and hemicelluloses. Bioresour. Technol. 148: 70–77.

Paulova, L., P. Patakova, B. Branska, M. Rychtera and K. Melzoch. 2015. Lignocellulosic ethanol: Technology design and its impact on process efficiency. Biotechnol. Adv. 33: 1091–1107.

Ramadoss, G. and K. Mmuthukumar. 2015. Influence of dual salt on the pretreatment of sugarcane bagasse with hydrogen peroxide for bioethanol production. Chem. Eng. J. 260: 178–187.

Ren, N.Q., G.L. Cao, W.Q. Guo, A.J. Wang, Y.H. Zhu, B.F. Liu and J.F. Xu. 2010. Biological hydrogen production from corn stover by moderately thermophile *Thermoanaero bacteriumthermo saccharolyticum* W16. Int. J. Hydrog. Energy 35(7): 2708–2712.

Rincon, M.T., B. Dassa, H.J. Flint, A.J. Travis, S. Jindou, I. Borovok, R. Lamed, E.A. Bayer, B. Henrissat, P.M. Coutinho and D.A. Antonopoulos. 2010. Abundance and diversity of dockerin-containing proteins in the fiber-degrading rumen bacterium, *Ruminococcus flavefaciens* FD-1. PLoS One 5(8): e12476.

Sadhu, S. and T.K. Maiti. 2013. Cellulase production by bacteria: A review. British Microbiol. Res. J. 3: 235–258.

Saha, B.C., N.N. Nichols and M.A. Cotta. 2011. Ethanol production from wheat straw by recombinant *Escherichia coli* strain FBR5 at high solid loading. Bioresour. Technol. 102: 10892–10897.

Saini, J.K., R. Saini and L. Tewari. 2015. Lignocellulosic agriculture wastes as biomass feedstocks for second-generation bioethanol production: Concepts and recent developments. 3 Biotech 5(4): 337–353.

Sathitsuksanoh, N., Z. Zhu and Y.H. Zhang. 2012. Cellulose solvent and organic solvent-based lignocelluloses fractionation enabled efficient sugar release from a variety of lignocellulosic feedstocks. Bioresour. Technol. 117: 228–233.

Sharma, N.K., S. Behera, R. Arora and S. Kumar. 2016a. Enhancement in xylose utilization using *Kluyveromyces marxianus* NIRE-K1 through evolutionary adaptation approach. Bioprocess Biosyst. Eng. 39: 835–843.

Sharma, N.K., S. Behera, R. Arora and S. Kumar. 2016b. Potential role of xylose transporters in industrial yeast for bioethanol production: A perspective review. pp. 81–93. *In*: Kumar, S., S.K. Khanal and Y.K. Yadav. (eds.). Proceedings of the First International Conference on Recent Advances in Bioenergy Research. Proceedings in Energy, Springer India.

Sharma, N.K., S. Behera, R. Arora and S. Kumar. 2017a. evolutionary adaptation of *Kluyveromyces marxianus* nire-K3 for enhanced Xylose Utilization. Front. Energy Res. 5: 32.

Sharma, N.K., S. Behera, R. Arora, S. Kumar and R.K. Sani. 2017b. Xylose transport in yeast for lignocellulosic ethanol production: Current status. J. Biosci. Bioeng.

Shaw, A.J., F.E. Jenney, M.W.W. Adams and L.R. Lynd. 2008. End-product pathways in the xylose fermenting bacterium, *Thermoanaero bacterium saccharolyticum*. Enzyme Microb. Technol. 42: 453–458.

Singh, J., M. Suhag and A. Dhaka. 2015. Augmented digestion of lignocellulose by steam explosion, acid and alkaline pretreatment methods: A review. Carbohydr. Polym. 117: 624–631.

Smith, S.P. and E.A. Bayer. 2013. Insights into cellulosome assembly and dynamics: From dissection to reconstruction of the supramolecular enzyme complex. Curr. Opin. Struct. Biol. 23: 686–694.

Stern, J., M. Anbar, S. Morais, R. Lamed and E.A. Bayer. 2014. Insights into enhanced thermostability of a cellulosomal enzyme. Carbohyd Res. 389: 78–84.

Studer, M.H., J.D. DeMartini, M.F. Davis, R.W. Sykes, B. Davison, M. Keller, G.A. Tuskan and C.E. Wyman. 2011. Lignin content in natural Populus variants affects sugar release. Proc. Natl. Acad. Sci. USA 108(15): 6300–6305.

Sudhakar, M.P., R. Merlyn, K. Arunkumar and K. Perumal. 2016. Characterization, pretreatment and saccharification of spent seaweed biomass for bioethanol production using baker's yeast. Biomass Bioenergy 90: 148–154.

Sun, Q., B. Madan, S.L. Tsai, M.P. DeLisa and W. Chen. 2014. Creation of artificial cellulosomes on DNA scaffolds by zinc finger protein-guided assembly for efficient cellulose hydrolysis. Chem. Comm. 50(12): 1423–1425.

Tabil, L., P. Adapa and M. Kashaninejad. 2011. Biomass feedstock pre-processing-Part-1: Pretreatment. pp. 411–438. *In*: Bernaedes, M.A.D.S. (ed.). Biofuel's Engineering Process Technology, InTech.

Taherdanak, M. and H. Zilouei. 2014. Improving biogas production from wheat plant using alkaline pretreatment. Fuel 115: 714–719.

Tengborg, C., M. Galbe and G. Zacchi. 2001. Reduced inhibition of enzymatic hydrolysis of steam-pretreated softwood. Enzyme Microb. Technol. 28(9-10): 835–844.

Tomás-Pejó, E., J.M. Oliva, M. Ballesteros and L. Olsson. 2008. Comparison of SHF and SSF processes from steam-exploded wheat straw for ethanol production by xylose-fermenting and robust glucose-fermenting *Saccharomyces cerevisiae* strains. Biotechnol. Bioeng. 100(6): 1122–1131.

Tsai, S.L., M. Park and W. Chen. 2013. Size-modulated synergy of cellulase clustering for enhanced cellulose hydrolysis. Biotechnol. J. 8(2): 257–261.

Waeonukul, R., A. Kosugi, P. Prawitwong, L. Deng, C. Tachaapaikoon, P. Pason, K. Ratanakhanokchai, M. Saito and Y. Mori. 2013. Novel cellulase recycling method using a combination of *Clostridium thermocellum cellulosomes* and *Thermoanaerobacter brockii* β-glucosidase. Bioresour. Technol. 130: 424–430.

Wang, H., S. Cao, W.T. Wang, K.T. Wang and X. Jia. 2016. Very high gravity ethanol and fatty acid production of *Zymomonasmobilis* without amino acid and vitamin. J. Ind. Microbiol. Biotechnol. 43(6): 861–871.

Wingren, A., M. Galbe and G. Zacchi. 2003. Techno-economic evaluation of producing ethanol from softwood: comparison of SSF and SHF and identification of bottlenecks. Biotechnol. Progr. 19(4): 1109–1117.

Wu, W.H., W.C. Hung, K.Y. Lo, Y.H. Chen, H.P. Wan and K.C. Cheng. 2016. Bioethanol production from taro waste using thermo-tolerant yeast *Kluyveromyces marxianus* K21. Bioresour. Technol. 201: 27–32.

Xin, F. and J. He. 2013. Characterization of a thermostable xylanase from a newly isolated Kluyvera species and its application for biobutanol production. Bioresour. Technol. 135: 309–315.

Yang, B. and C.E. Wyman. 2008. Pretreatment: The key to unlocking low-cost cellulosic ethanol. Biofuels Biopr. Bioref. 2: 26–40.

Zaldivar, J., J. Nielsen and L. Olsson. 2001. Fuel ethanol production from lignocelluloses: A challenge for metabolic engineering and process integration. Appl. Microbiol. Biotechnol. 56: 17–34.

Zhang, P., S.J. Dong, H.H. Ma, B.X. Zhang, Y.F. Wang and X.M. Hu. 2015. Fractionation of corn stover into cellulose, hemicellulose and lignin using a series of ionic liquids. Ind. Crops Prod. 76: 688–696.

Zhu, W., J.Y. Zhu, R. Gleisner and X.J. Pan. 2010. On energy consumption for size-reduction and yields from subsequent enzymatic saccharification of pretreated lodge-pole pine. Bioresour. Technol. 101: 2782–2792.

Index

For Product Safety Concerns and Information please contact our EU
representative GPSR@taylorandfrancis.com
Taylor & Francis Verlag GmbH, Kaufingerstraße 24, 80331 München, Germany

www.ingramcontent.com/pod-product-compliance
Lightning Source LLC
Chambersburg PA
CBHW061357210326
41598CB00035B/6013